AF866411

MB&F THE FIRST FIFTEEN YEARS

2005–2020

MB&F THE FIRST FIFTEEN YEARS

A Catalogue Raisonné of MB&F Watches
by Suzanne Wong and William Massena

Éditions
de La Martinière

ABRAMS

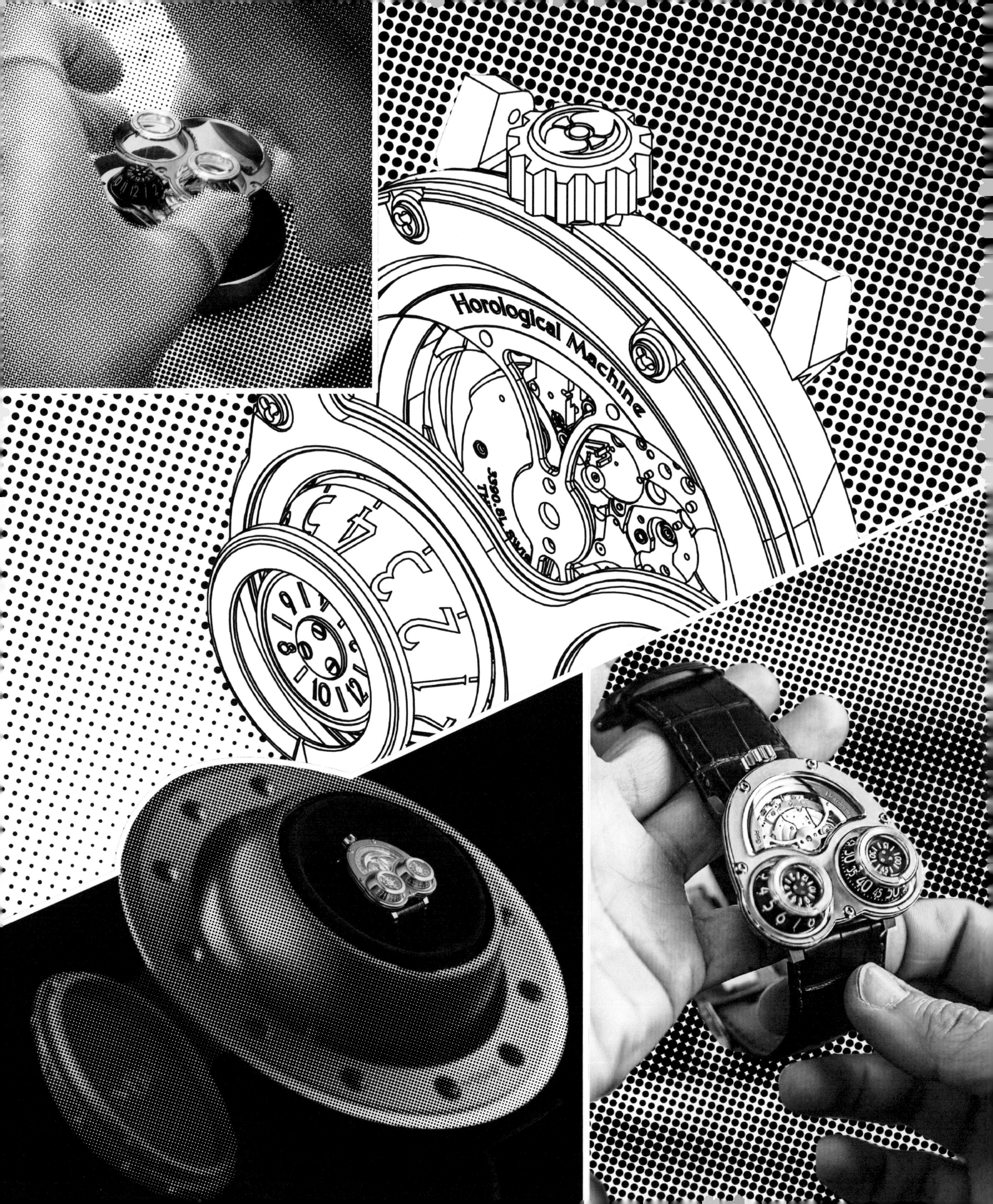
Horological Machine

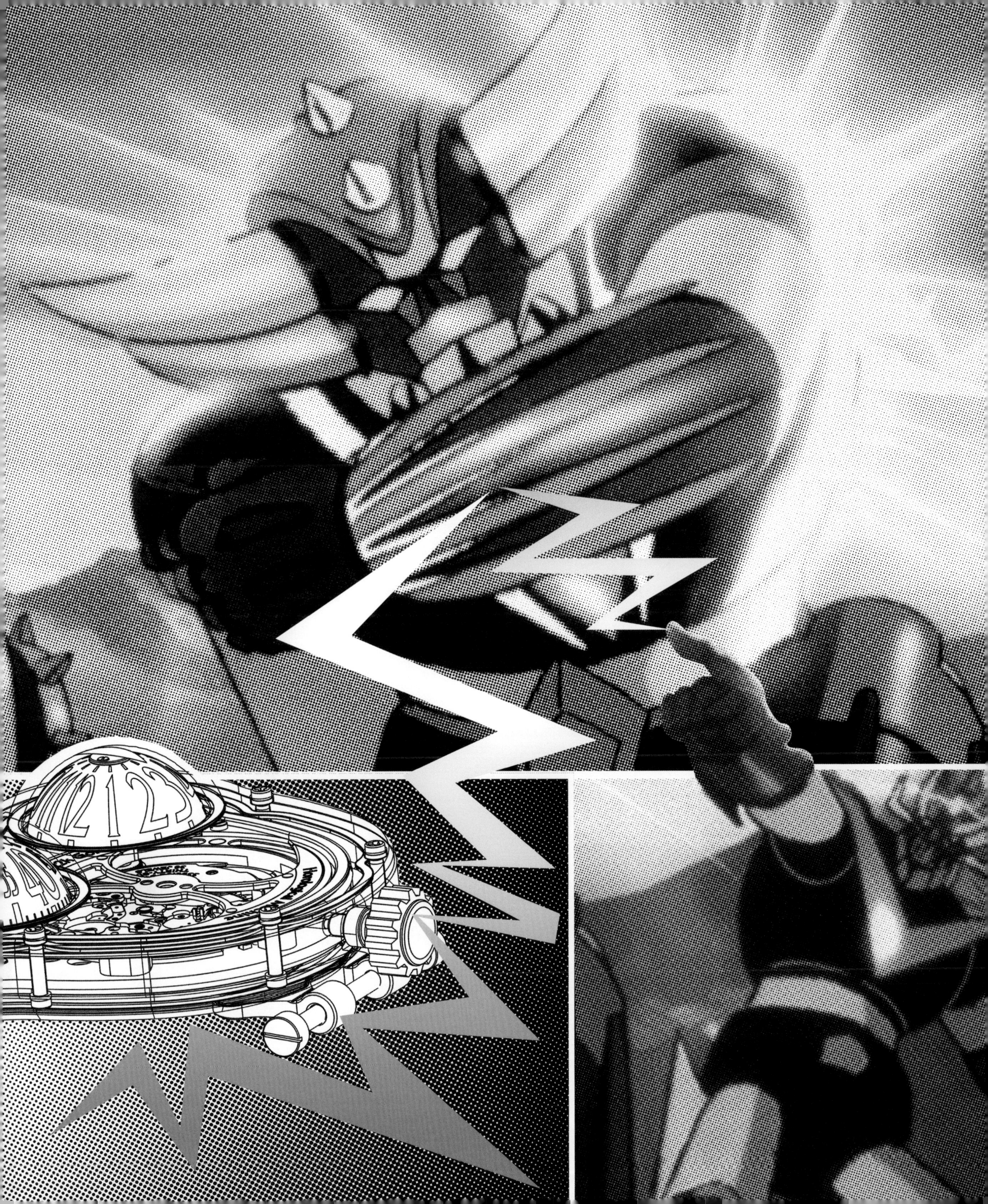

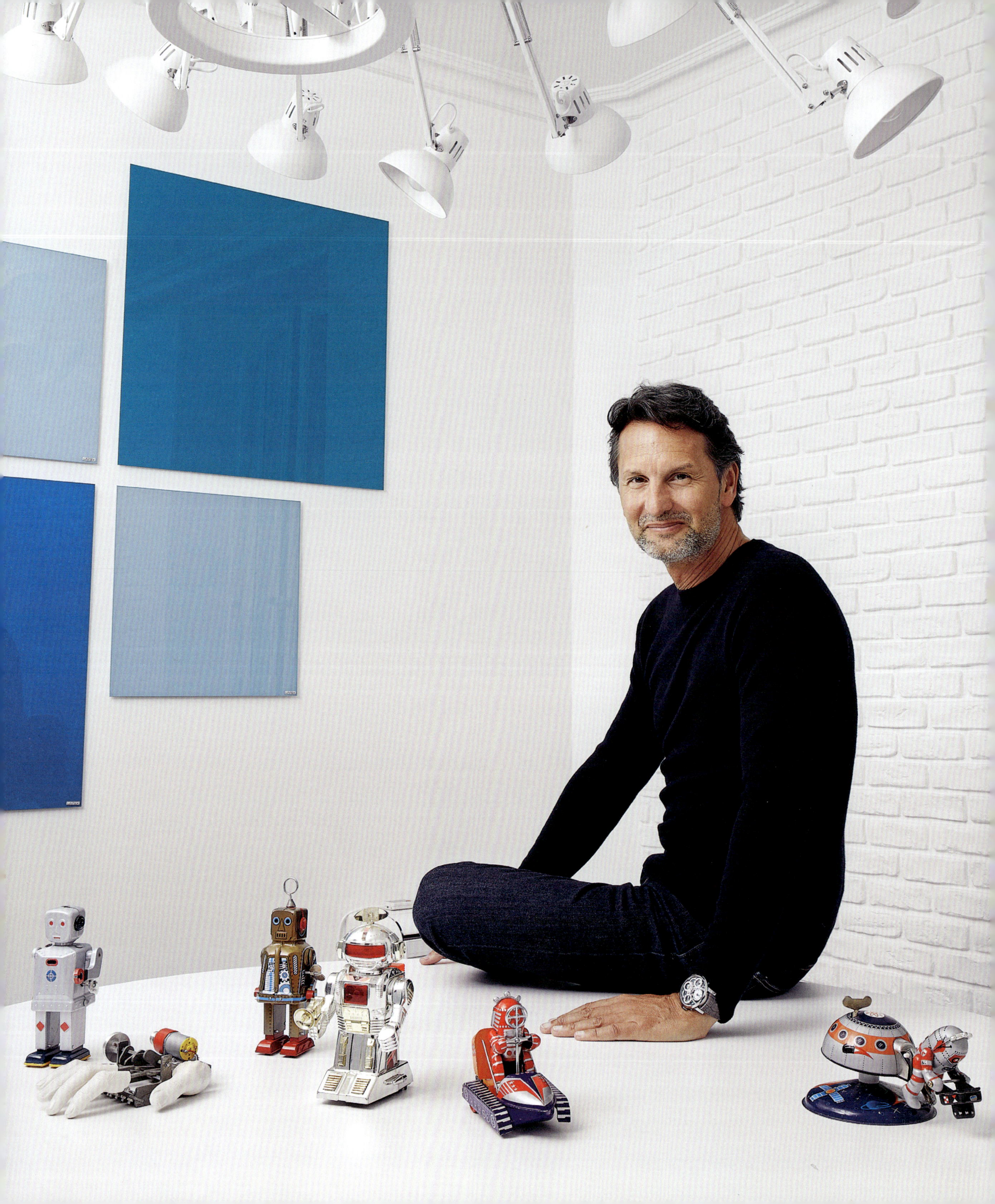

-"You should write a book!"

How many times have I heard this whilst sharing a few of the terrifying or exhilarating anecdotes on the MB&F journey? The story is incredible and deserves to be told, but an unnerving feeling that biographies are only written at the end of a journey had prevented me from considering it up until now.

What about a Catalogue Raisonné then?

It is true, the body of work created by MB&F in such a short time is pretty astounding. Eighteen calibres for over 160 product references is surely something that deserves to be documented, right? Not to mention the hundreds of artisans, engineers and watchmakers that have been along for the ride these last 15 years. This is the story I love to tell.

I was once asked what three words or expressions would best define the MB&F journey so far. It was an incredibly tough question and after some reflection, I realized that it all boils down to: "a personal journey", "challenge and try" and "encounters with incredible human beings." I hope the 312 pages of this Catalogue Raisonné will contribute to sharing that journey, illustrating the challenges and the amazing people I have met along the way — and serve as a reference for the decades to come.

Maximilian Büsser
Owner & Creative Director, MB&F

CONTENTS

Portrait of Maximilian Büsser
as a kid, around 1970

Space capsule dwarfed by the
disc of the moon from Jules
Vernes "Autour de la Lune,"
Paris, 1865

Goldorak UFO Robot
(Gurendaizâ), 1976

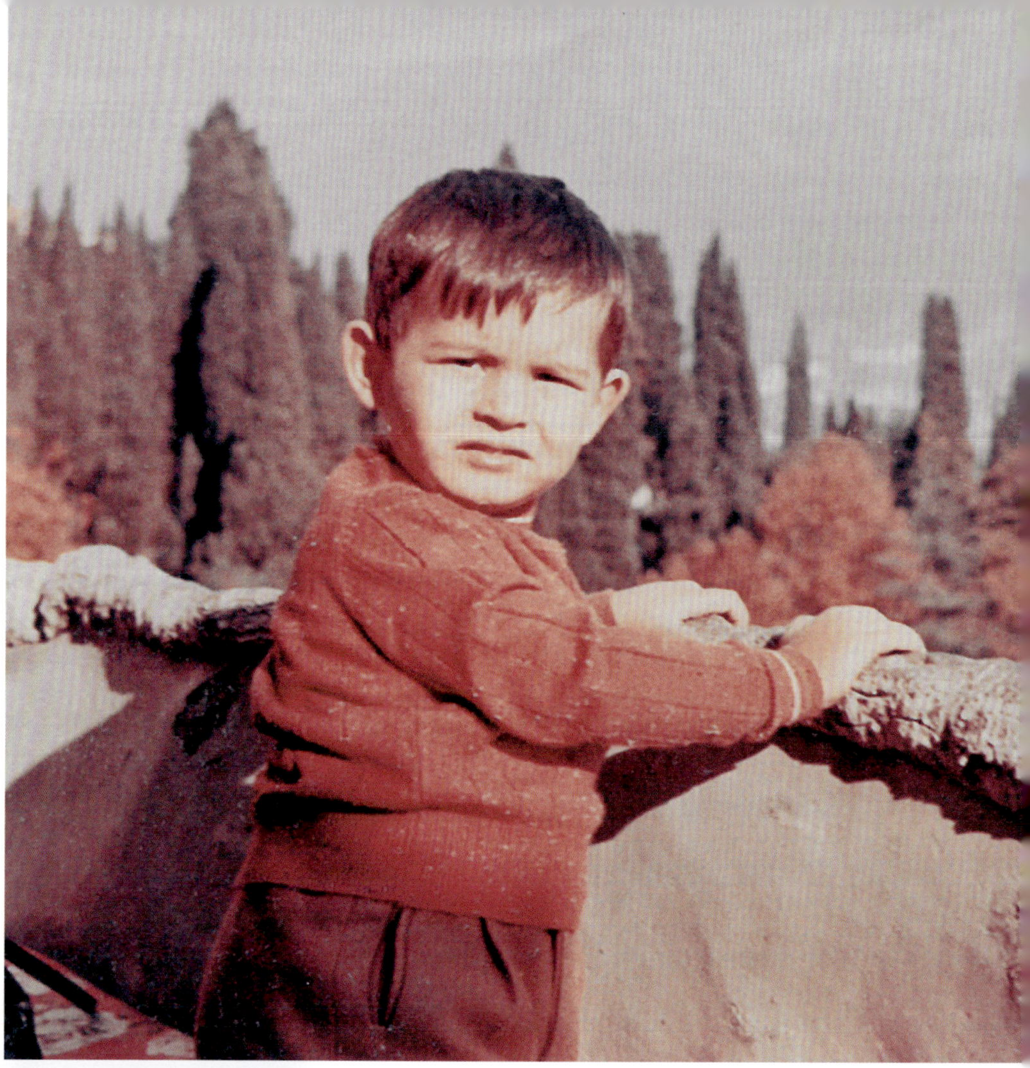

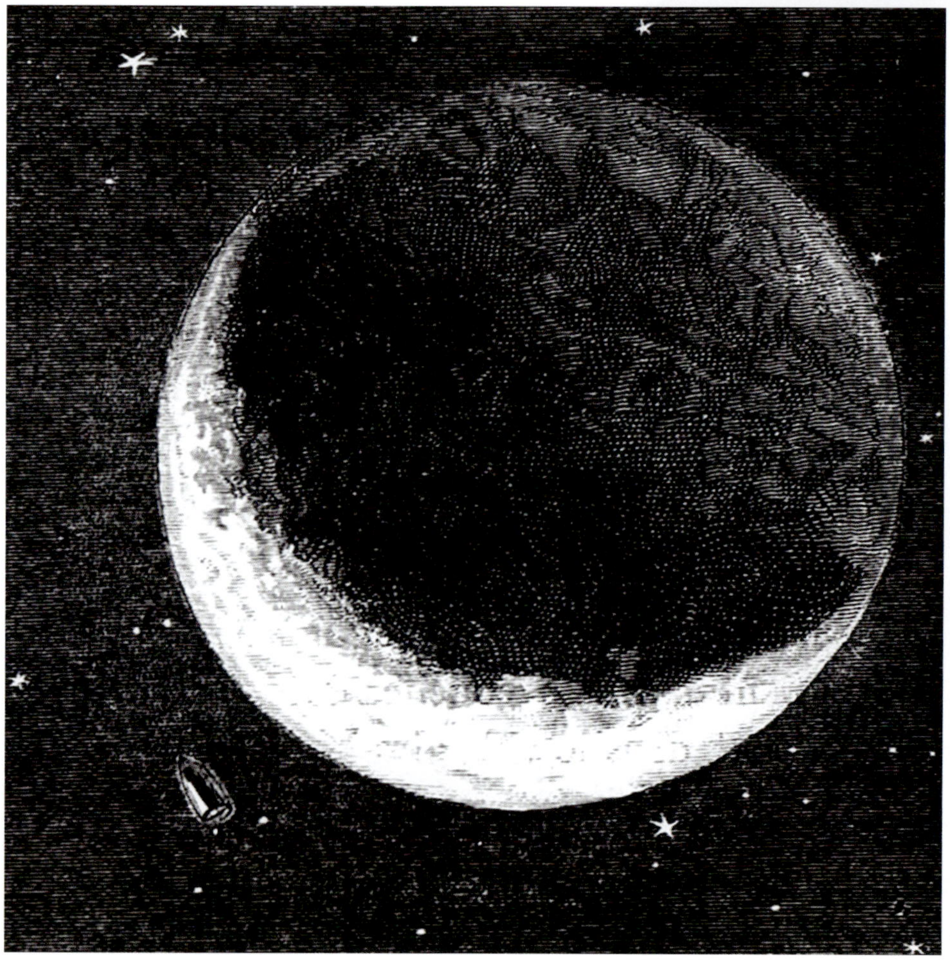

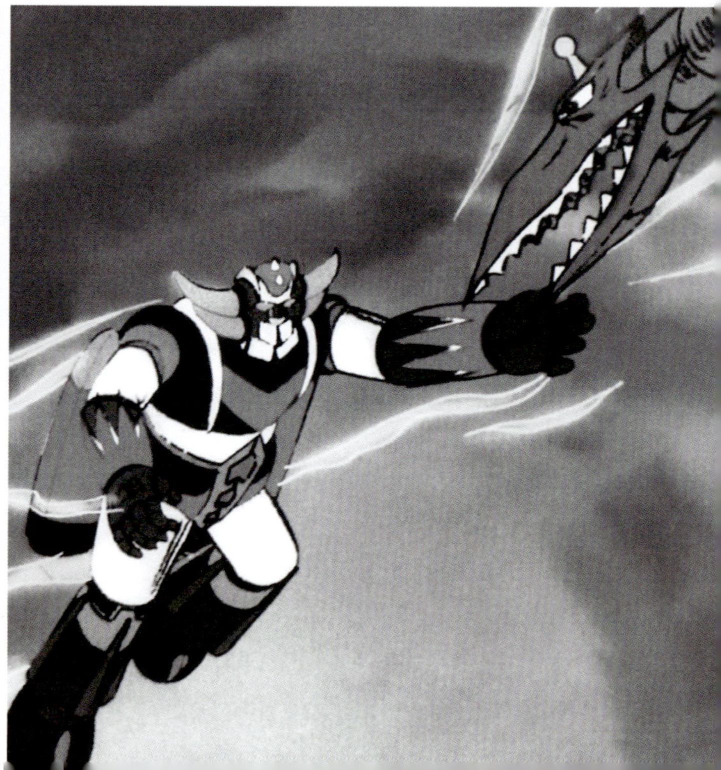

BEFORE MB&F

There is only one true way to tell the story of MB&F, and that is to tell the story of Maximilian Büsser. Before the Legacy Machines, before the Horological Machines, before the advent of a company that would come to define the face of 21st century independent watchmaking, there was a young boy living in the Vaudoise countryside, in the southwest of Switzerland.

He was born an only child, on 21 January 1967 in Milan, Italy, to Mario, a Swiss father, and Dinoo, an Indian mother. His name was Maximilian Büsser, but it almost wasn't. If family tradition had been followed, he would have been named Melchior, after his grandfather — as every eldest Büsser son in alternating generations had been named for five centuries. But the last Melchior Büsser so detested the name that he decided to break with half a millennium of patronymic precedent. He insisted on being addressed as Max, and named his son Mario (instead of Balthazar, as mandated by the same family tradition).

Maximilian Büsser was next in line, the first among all firstborn Büsser sons in five hundred years to bear this name, a Swiss-Indian child with two cultural heritages, set on a path of rule-breaking from the day he entered the world.

1970–1985

Max's earliest memory is one of discovery — Dinoo feeding the three-year-old Max a plate of scrambled eggs, and making a game of it. Hidden under the formless mass of yellow curds, anathema to any toddler of taste, the plate was illustrated with a scene of fish swimming in a pond, and the game was to reveal the scene by finishing all the eggs. Max remembers the eggs, standing (quivering) between him and his objective, with extreme disfavour. "I dislike scrambled eggs to this day," is his assessment of that early, embryonic, existential challenge.

The Büssers returned to Switzerland in 1970, when young Max was three and a half. The first creative seeds of MB&F were planted during these fertile, formative years.

Five things that imprinted themselves on the childhood of Max Büsser

1. Animated science-fiction series (UFO Robot Grendizer, Captain Future, Star Trek)
2. Close companions both animate and inanimate (the family golden retriever, a plush toy panda bear)
3. Afternoons spent daydreaming about being a superhero and saving the world
4. The science-fiction adventure novels of Jules Verne
5. The designs of the great Italian automotive carrosseries (Gruppo Bertone, Touring Superleggera)

There was no great spirit of entrepreneurship in the Büsser family then, no trace of the world of high luxury and six-figure mechanical dreams that Max would come to inhabit. He found the door to this world by accident, through an education that left him intellectually and emotionally unfulfilled and an encounter with a classmate that left him perplexed. He found his fate, as many of us do, through a detour.

1985–1991

Childhood and early adolescence were characterized by a passion for automobiles, sketching page after page of car designs from the age of eight to eighteen, and dreaming of one day attending the ArtCenter College of Design in Pasadena, California. But the modest resources at Max's disposal obliged him to pursue his further education locally. He enrolled in the École Polytechnique Fédérale de Lausanne (EPFL), a prominent Swiss research institute and university that is further distinguished today as one of the premier startup incubators and accelerators in Switzerland. With zero intention of becoming an engineer, Max nevertheless plunged straight into a technical course, eventually earning a Master's degree in micro engineering.

He started university in 1985, the year he turned eighteen. A milestone age that his parents celebrated in time-hallowed fashion by buying their child a watch. Max was given a budget of some 700 francs and told to pick something out for himself.

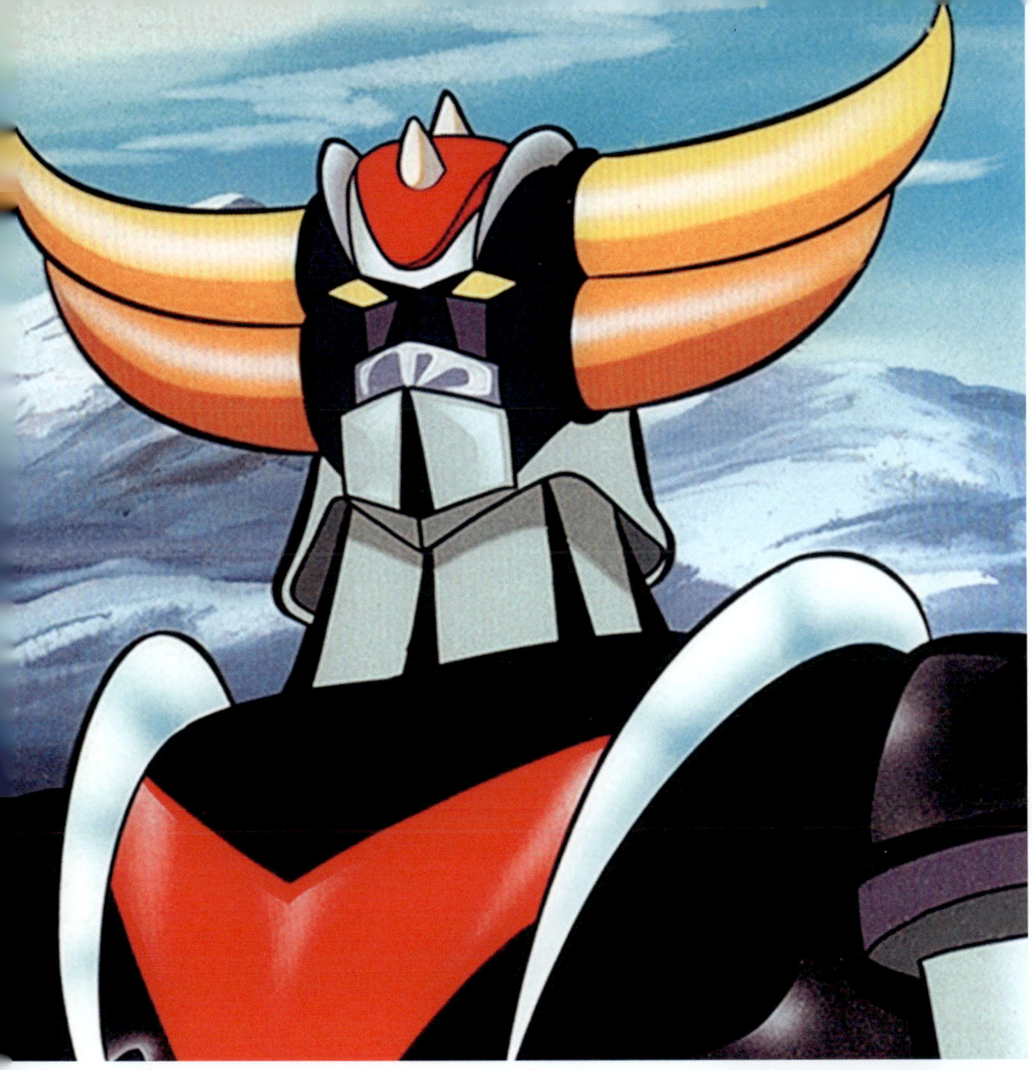

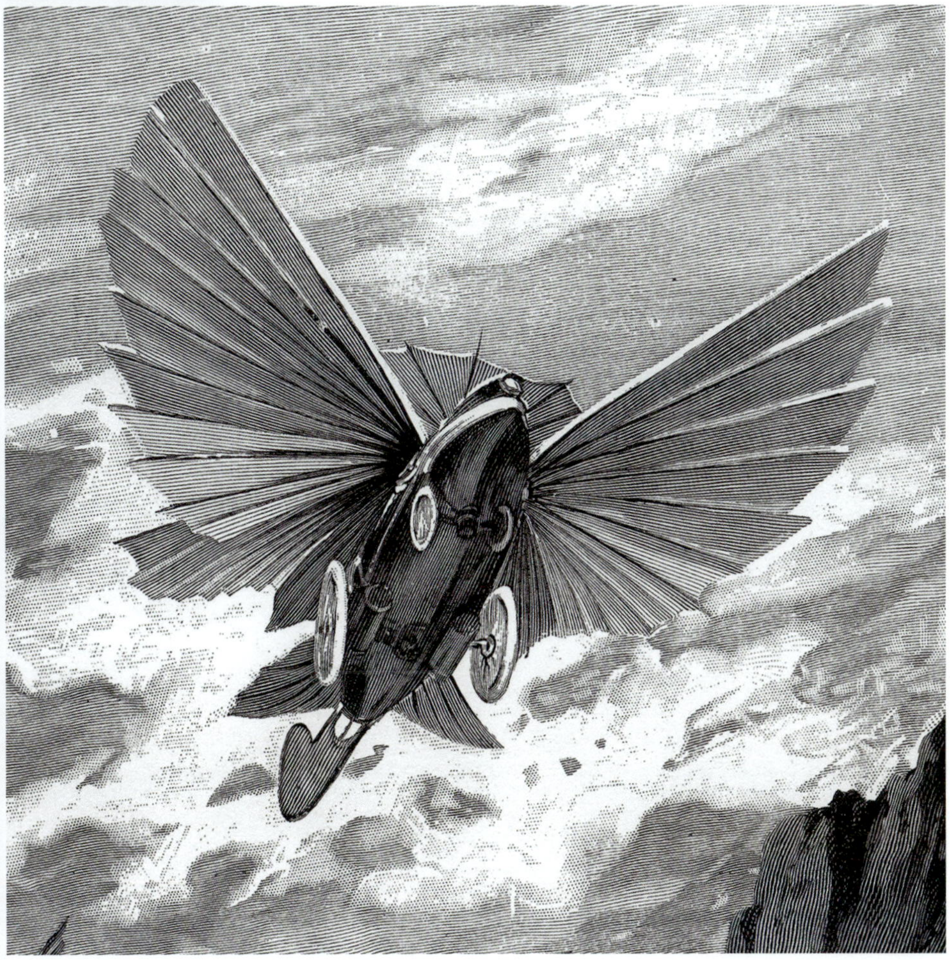

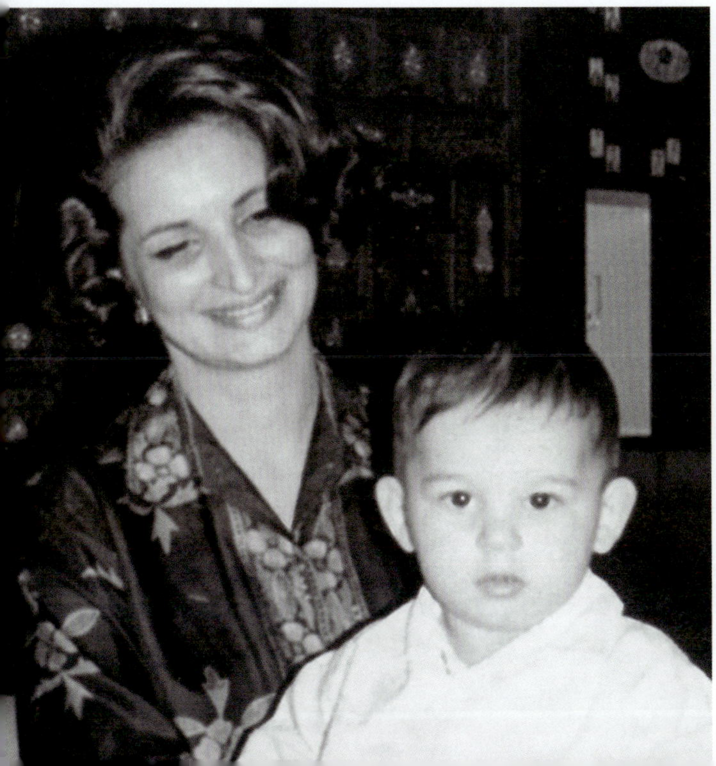

Goldorak Robot, 1976

Flying car from Jules Verne "Maître du monde", Paris, 1904

Portrait of Maximilian Büsser and his mother Dinoo, Milano, Italy, 1968

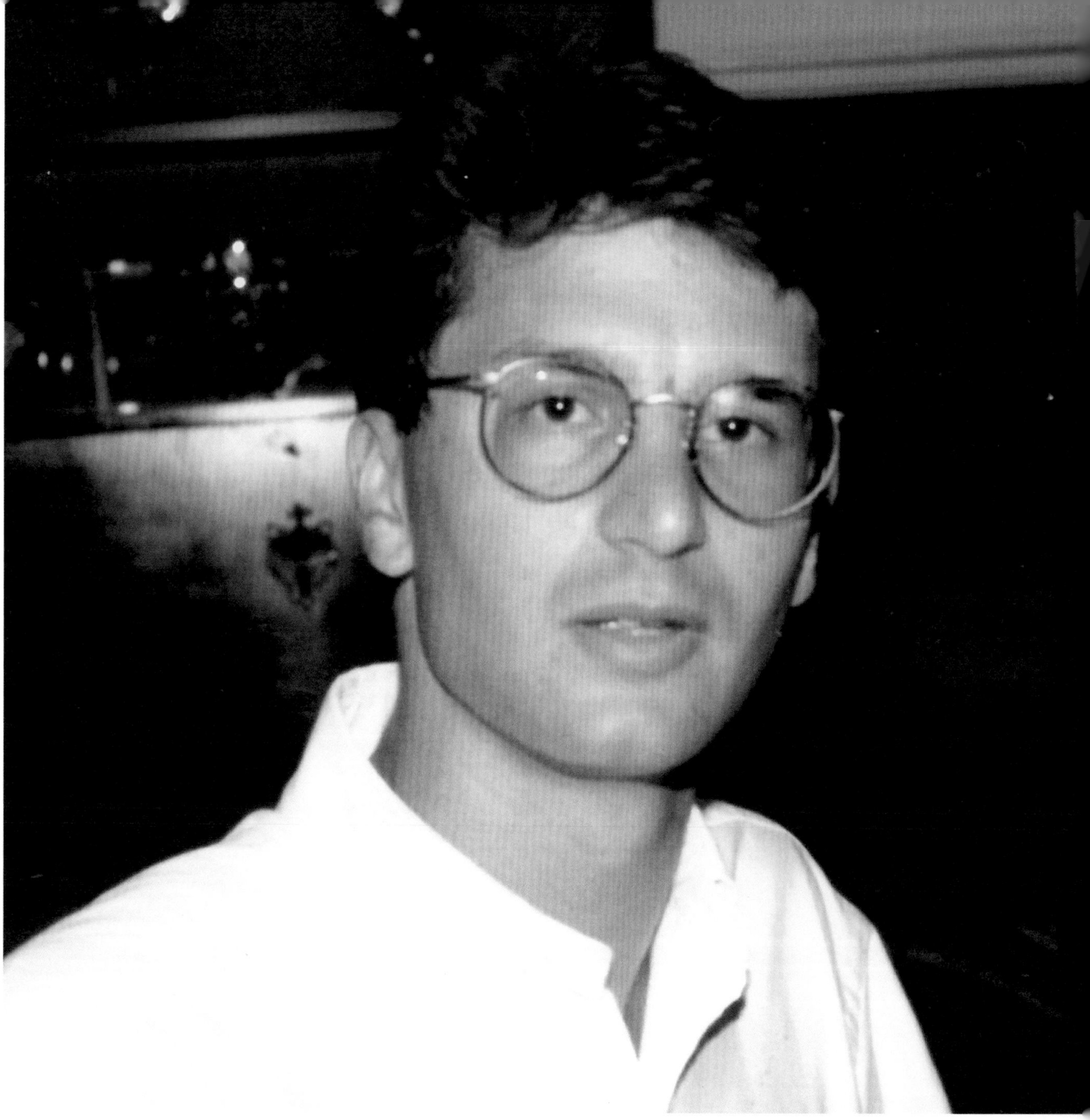

Portrait of Maximilian Büsser
as a student, 1987

Up to this point, despite growing up in Switzerland, the land of watchmaking, Max had had very little contact with watches and knew almost nothing about them. He began paying attention to what people around him wore on their wrists, and one day as he sat down in class, he turned to look at what the student beside him had on.

I said to this guy, "What are you wearing?"

He said, "It's a Rolex."

I thought to myself, "What's a Rolex?" At the time I had no idea what he was talking about, but I went ahead and asked him how much it cost anyway. He told me — and I'll always remember this — "4,700 francs." I was astounded. What did I just hear? "Did you say 4,700 francs?"

I was eighteen, I had all these little jobs on the side, as a cinema usher, as a private tutor. I sold hi-fi systems on weekends. And even with all that, 4,700 francs is probably what I would have made in an entire year. I thought, this guy has one year's worth of wages on his wrist and it's not even a quartz watch? What kind of idiot spends that amount of money on something that isn't even the latest technology?

"Did you say 4,700 francs?"

He said, "But it's a two-tone GMT Master." More words that meant nothing to me. Credit to him, he continued to try and explain to me what his watch was about. He even brought me to the Bucherer boutique, and he advised me to try and stretch my budget to 1,100 francs so that I could get an Air-King in steel.

Of course, still not really understanding, I said, "Why would I spend more on an automatic watch that doesn't have a date indication and looks unappealing — all steel — when I could get a gold-plated quartz Tissot with all the latest bells and whistles?"

Guess what? I got the Tissot. And when it was time to come up with my first big project at university, which according to the assignment rubric had to be linked to psychology and engineering, I chose to focus on luxury watchmaking as a sociological phenomenon.

A major part of Max's university project was interviewing the top executives at watch brands. In the mid-1980s, the Swiss mechanical watchmaking industry was just starting to claw its way back to a position of relevance and existence. Watch brands were small companies with localized mindsets and a long way to go before they would become the sleek multinational corporations we know today. Their leaders responded positively to the interest expressed in this young student's letters.

The interviews and interactions that came out of those letters changed Max Büsser's life in two ways. First of all, they ignited his passion for watchmaking. Secondly, they gave him his first professional opportunity in the watch industry — but we'll get to that a little later.

At the end of his fourth year at university, before completing his thesis, Max went on sabbatical to fulfill his obligation as a Swiss citizen by enlisting for compulsory military service. In the course of his training, he met with a near-fatal accident. An army jeep he was driving overturned. He was thrown from the vehicle, which then landed on top of him. Two and a half tons of metal with 50km/h of momentum slammed into the ground, with Max in between.

He survived, still conscious as he was pulled from the wreckage. And when he was back on his feet, after six weeks in hospital and months of physical therapy, he jumped on a bus to Lausanne and bought himself his first mechanical timepiece — a 2,700-franc Ebel chronograph powered by the legendary Zenith El Primero movement. His right shoulder had exploded in the impact of the accident and had been reconstructed; he was still in bandages and plaster, but he was able to put the watch on his left wrist, and that was all that mattered.

He observed with amusement that he was now on the flip side of that situation all those years ago when his classmate had told him the cost of his Rolex GMT Master. Not many people around him understood why anyone would spend that kind of money on a mechanical watch. But Max saw things differently now.

One wintery day in 1991, Max was on the slopes, skiing with friends around Les Ruinettes in Verbier. While he was there, he ran into Henry-John Belmont, who at the time was the managing director (in the days before the CEO title became popular) of Jaeger-LeCoultre, the historic watchmaking company founded in 1833. Max had met with Belmont a few times before, in the course of researching his university thesis, and they took this chance encounter as an opportunity to have a coffee and chat.

When asked what his career plans were, Max told Belmont that he was considering applying for a position in the fast-moving consumer goods (FMCG) sector. Procter & Gamble (P&G), perhaps, or Nestlé. Imagine, if you can, just for five seconds, the idea of Max Büsser at Nestlé.

But we talk about all sorts of things. And, at the end of the day, I say, almost as a joke, "You know what? If P&G doesn't give me a job, you can give me a job, Mr Belmont."

He laughs. I laugh. It's a joke, right? A week later, I get a call asking me if I want to come to Le Sentier, because Mr Belmont would like to meet. I get in my beaten-up Opel Corsa and drive up to the Vallée de Joux, and I have the interview of my life. It lasts three hours, and Henry-John Belmont doesn't ask me a single question. You have to remember what it was like back then; times were tough, he was trying to save that company.

He takes me around the factory, telling me all the changes they planned to make, all the new watches they were going to create. We're walking around, on these uneven wood floors that make the machines move about as they're being operated. I heard the future of Jaeger-LeCoultre laid out before me that day — their plans to create the Géographique at some point they wanted to make a tourbillon, a minute repeater...

At the end of those three hours, he says to me, "Okay, you've got the job." And I say, "Sorry, what?"

"Okay, you've got the job." And I say, "Sorry, what?"

He tells me, "I need a young guy like you, an engineer, someone who's passionate about watchmaking. You can be my product manager."

I say to him in the most respectful way possible that I'd have to think about it. It had never crossed my mind to enter the watch industry. Because of my mixed heritage, I always had the idea of working for a multinational company, of living in Asia. The one thing I was absolutely sure of was that I wanted to leave Switzerland. Joining a struggling company in the Vallée de Joux wasn't exactly on my radar.

And then he tells me something that has stayed with me my entire life. "Do you want to be one out of 200,000 in a big corporation? Or do you want to be one of the five of us who are going to save Jaeger-LeCoultre?"

Henry-John Belmont gave me my first professional sense of purpose. I called him the next day. I said, "Okay."

What followed were seven exhilarating years of intense learning and passion-driven work. He wasn't in it for the money, and neither was anyone else around him — there was no money to be had back then, anyway.

He saw that watchmaking was where engineering and emotion came together. The watchmakers around him admitted that what they made was not practical — but it was beautiful. To someone like Max, who grew up watching science-fiction cartoons and worshipping the unrestrained exuberance of Italian car design, this genre of technology, nourished by imagination, felt like home. Their truths resonated with how he believed the world should be. He had found his people.

Günter Blümlein was the industrialist patriarch of Les Manufactures Horlogères (LMH) that consisted of Jaeger-LeCoultre, IWC and A. Lange & Söhne. At team meetings, Max would speak his mind robustly, in contrast to the towering deference and acquiescence that most others adopted around the famously imposing Blümlein. In the middle of one particular extended debate about sports watches, Blümlein paused and looked hard at Max. Above the palpable fog of "R.I.P. Max Büsser" vibes growing steadily in the room, Blümlein intoned the eight words that would sear themselves into Max's watchmaking ethos from that very day onwards.

"Mr Büsser, creativity is not a democratic process." End of story.

In 1998, Egon Zehnder, the big guys in the field of management and recruitment consultancy, came calling. Headhunters had approached Max before, and he had sent them packing, because they never offered him anything truly interesting. This time was different. Egon Zehnder was known to recruit only for top positions. What could they want with the marketing and sales manager of Jaeger-LeCoultre? Curiosity piqued, Max went to meet them. The interview was an hour and a half long.

They said to him, "You're too young. You're ten years too young. But we think you can do this job."

Max was 31 years old when he was appointed managing director of Harry Winston Rare Timepieces.

When Max came to Harry Winston, he was forcibly reminded of the day that Henry-John Belmont took him around the Jaeger-LeCoultre factory. He thought, this is what it's like to be tasked with turning a company around with only a tiny handful of employees. He knew he would receive little or no support from company headquarters in New York. Watches made up just 3–4 percent of Harry Winston's annual turnover. Sink or swim, they told him. 1998–2005

Max swam for his life, and for the lives of his minuscule team in Geneva. In seven years, Harry Winston Rare Timepieces increased tenfold in headcount and in revenue — eight people became eighty, eight million became eighty million. In 2001, Max pioneered the Opus Project, an initiative that ushered in a new era for contemporary, avant-garde watchmaking, with unprecedented transparency in communicating about the external watchmakers who worked on the project.

The concept was simple, yet groundbreaking. One watchmaker per Opus, one Opus per year. Independent watchmakers, with exceptional talent and vision but unaffiliated with any of the big watch companies, were invited to showcase the absolute pinnacle of their skills in watches of Harry Winston design.

Under Max's leadership, the Opus Project featured the following watchmakers, each of them now an established horological icon: François-Paul Journe, Antoine Preziuso, Vianney Halter, Christophe Claret and Felix Baumgartner. It was also at this time that Max formed one of the closest professional and personal relationships of his life with designer extraordinaire Eric Giroud.

From the perspective of others, Max Büsser was on top of the world. At the Baselworld trade fair, Harry Winston had become one of the top must-see brands. Max was featured in glowing newspaper interviews and glossy features in magazines. In the small but unimaginably wealthy world of high-end mechanical watchmaking, Max was a star.

From Max's perspective, things looked a little different. Mario, his father, had passed away in late 2001, never really getting to see his son's glittering professional success. In processing the complicated aftermath of losing a parent, Max came to realize that he could only be truly happy creating for himself, and not for others. He realized that he was surrounded by people who told him how proud they were of him and of his work, but that he himself took no share in that pride. And he realized that despite being brought up by people who were deeply honest and respectful, the corporate system was forcing him to work with people who did not share the same values.

Once, Max ventured to remark to his mother Dinoo, "I never heard you or Dad make a snide or ironic comment about anyone."

She looked at him with a look that registered itself indelibly in Max's memory, and responded, "Why would we ever say something like that about anybody?"

Five sentences that stayed with Max Büsser all his life

1 "4,700 francs." Max's university friend, 1985
2 "Do you want to be one out of 200,000 in a big corporation? Or do you want to be one of the five of us who are going to save Jaeger-LeCoultre?" Henry-John Belmont, 1991
3 "Mr Büsser, creativity is not a democratic process." Günter Blümlein, 1994
4 "Who gives a damn about your sitting room?" Eric Giroud, 2004
5 "Why would we ever say something like that about anybody?" Dinoo Büsser, 2006

Max began nurturing the idea of creating a brand that would be just for him, the purest and truest expression of Max Büsser. While on a long-haul flight back from Singapore in 2003, he made the first sketches of the watch he dreamed of creating. At around the same time, Eric Giroud was destabilizing every idea that Max had about art.

Along with his professional success, Max had picked up a taste for contemporary art. One fateful day in 2004, he asked Giroud, a gifted designer with razor-sharp wit, to come around with him to some art galleries and give his opinion on a few things Max wanted to buy. They visited gallery after gallery, Max pointing out the pieces that he liked and asking Giroud what he thought. Gallery after gallery, Giroud responded with a flat "I don't think anything of it." Eventually, after one final "je n'en pense rien," Giroud followed up with a brief elucidation — "You just don't get it. Come with me."

Max followed Giroud to the Andata Ritorno Contemporary Art Laboratory (which to Max looked more like an abandoned building than anything else, a far cry from the slick modern galleries they had visited that day), where the Thai artist Siripoj Chamroenvidhya was exhibiting. The art was typical of his oeuvre, primarily dense charcoal sketches of uninhabited landscapes. Max felt depressed by the dark and, to his thinking, creepy drawings. He felt as if he had wandered into Sleepy Hollow and was expecting to see the Headless Horseman galloping out of some dim corner. Instead, he received a metaphorical lobotomy.

"You know what, Eric? I'm never going to put any of this in my sitting room," was Max's sincere cri de coeur.

Giroud pulled out his (metaphorical) scalpel. "Who gives a damn about your sitting room?" Bone saw. "There it is. You don't get it. What you've been showing me is not art. It's marketing, it's interior decoration." Plasma knife. "The guys who create that stuff, they know that bourgeois guys like you will buy it because they want their interiors to look like they've come out of Architectural Digest." Surgical stapler. "This artist here doesn't give a damn about your interiors. He doesn't give a damn about you." Anesthetic reverser. "He doesn't even care if you like what he does. His art comes from him needing to express something."

In early 2005, Harry Winston Rare Timepieces enjoyed their most successful Baselworld fair yet. The Opus 5 and the Project Z2 were huge hits. The company had taken eighteen months' worth of production orders. Max received a brand-new shiny contract, courtesy of new management at the New York headquarters who were eager to lock down the chief engineer of their fastest-growing brand segment. It was a dream contract, for the first 24 pages out of 25 — stock options, bountiful bonuses, generous redundancy pay, the whole package. On the last page, however, was the pill that 24 pages of jam could not get Max to swallow. Non-compete clauses, in case of departure. Non-solicitation clauses, for clients, for employees, for suppliers. If Max signed, that would signal the end of the dream of independent creation that had secretly sustained him.

Early May 2005, Max flew to New York and submitted his official resignation to the board of Harry Winston. His last day with the company was 15 July 2005; ten days later he registers the company that will become MB&F, and the rest is — it is not history, it is too vigorous and exciting for that word, and sixteen years is not such a long time. The rest is now.

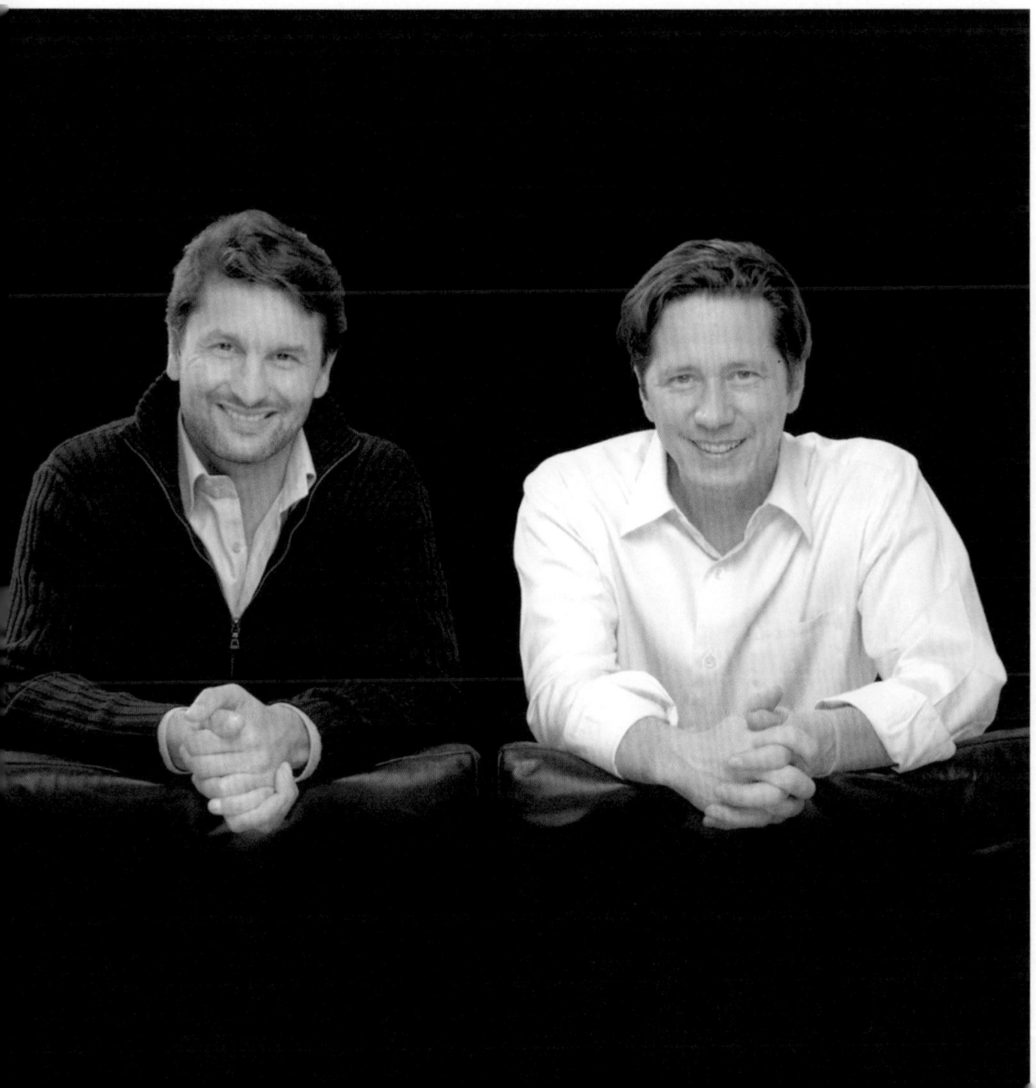

Portrait of Maximilian Büsser and Eric Giroud, 2007

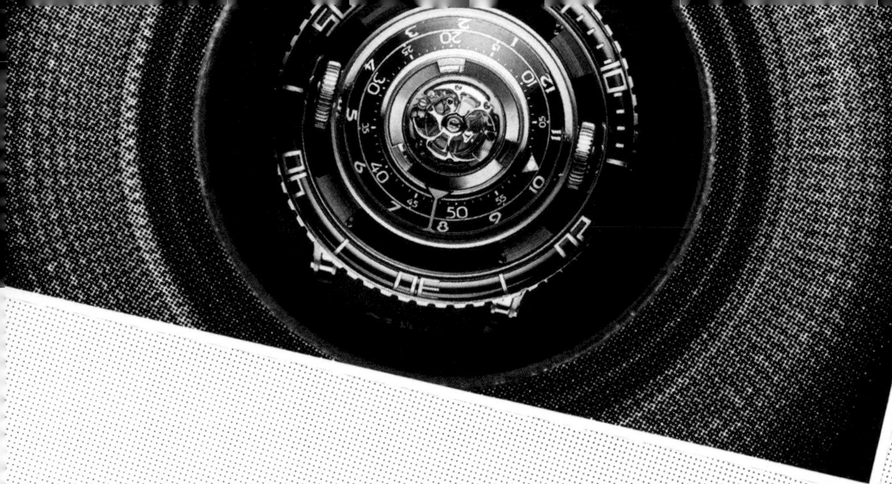
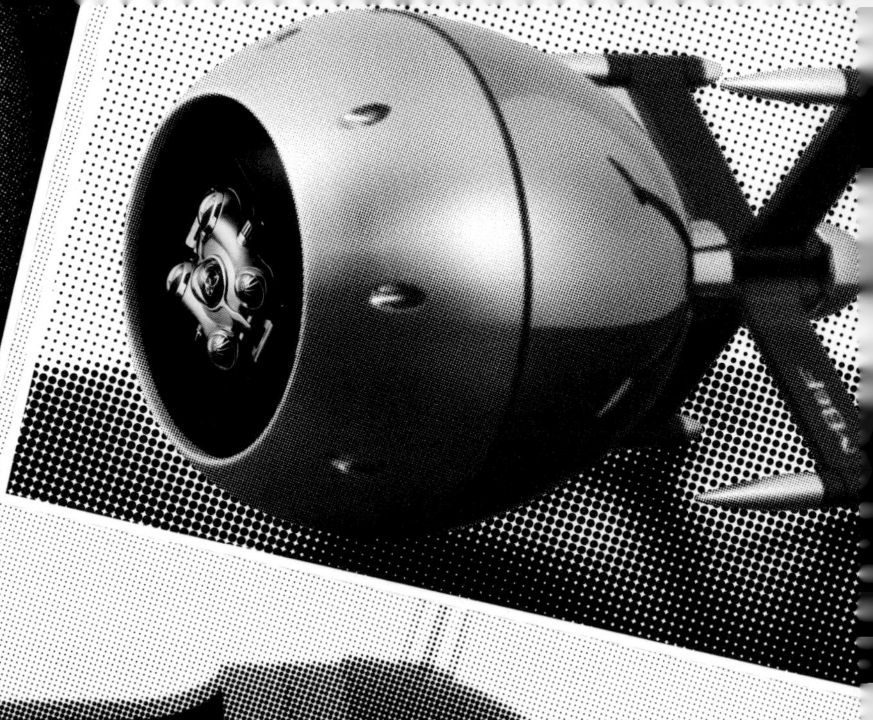
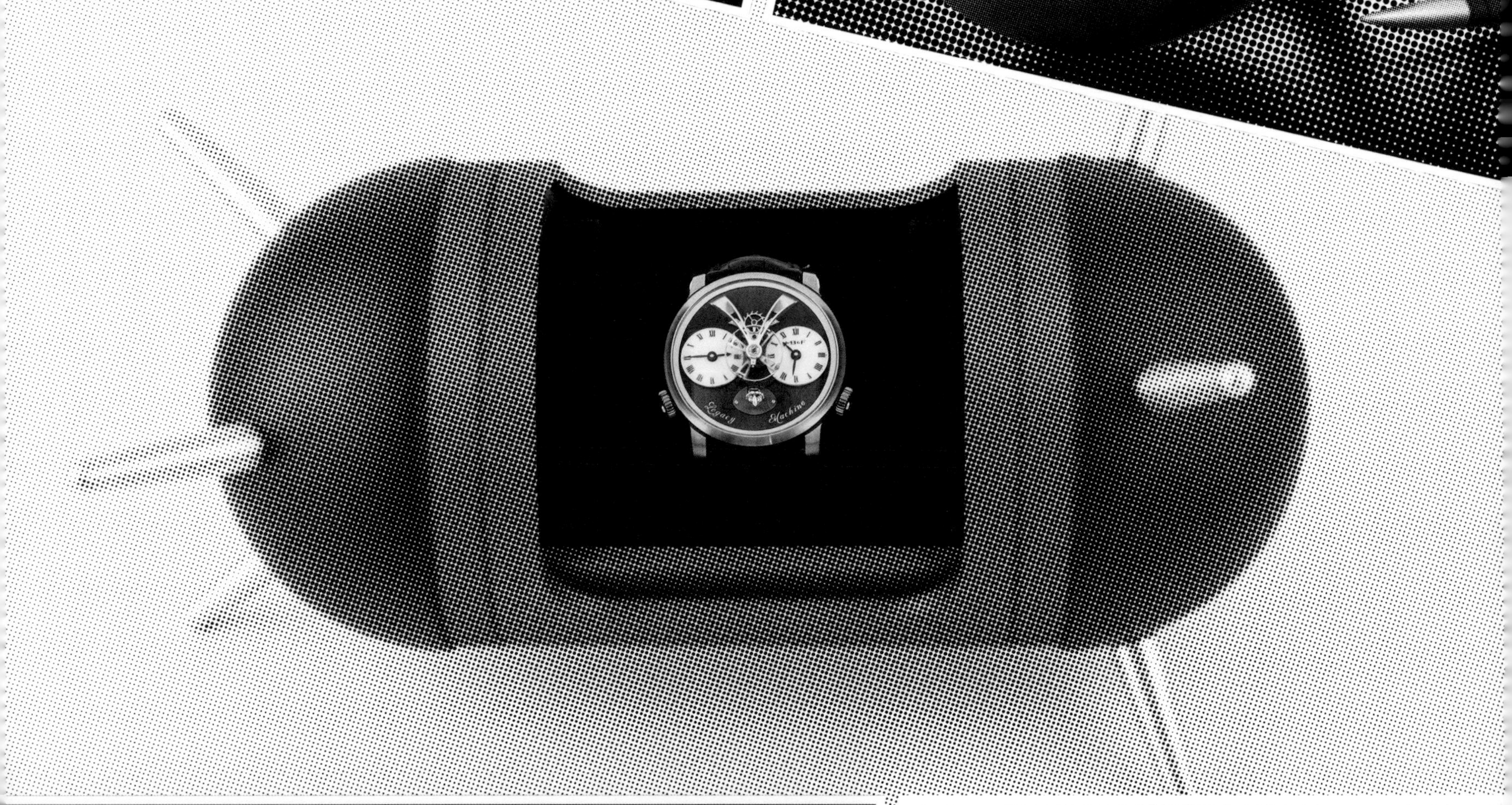
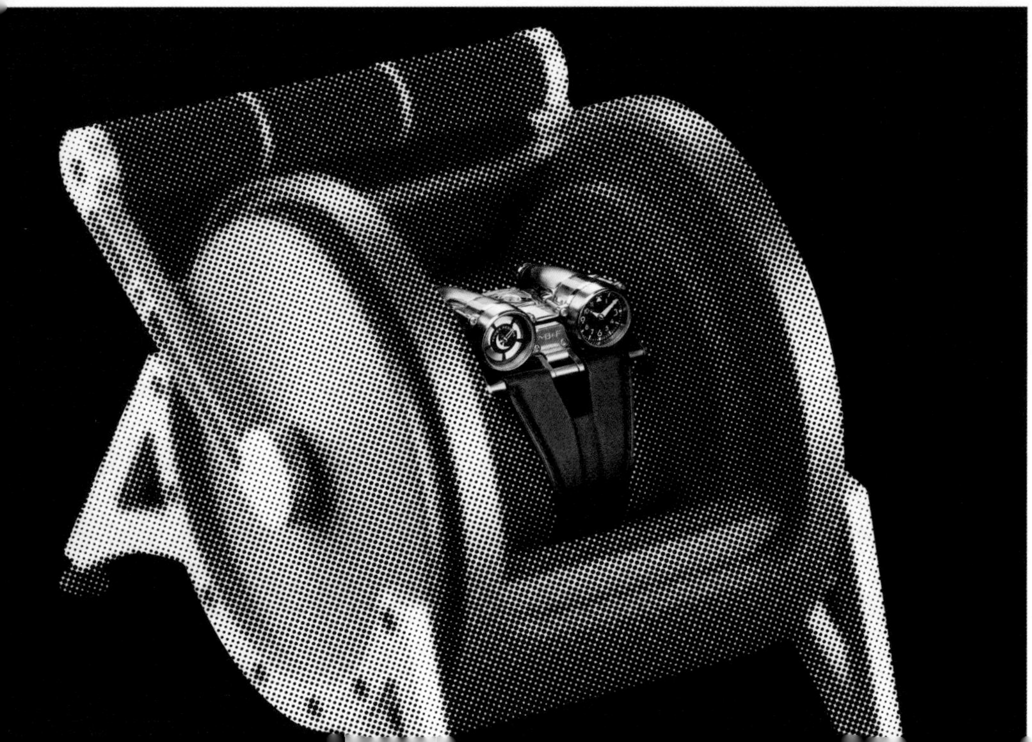
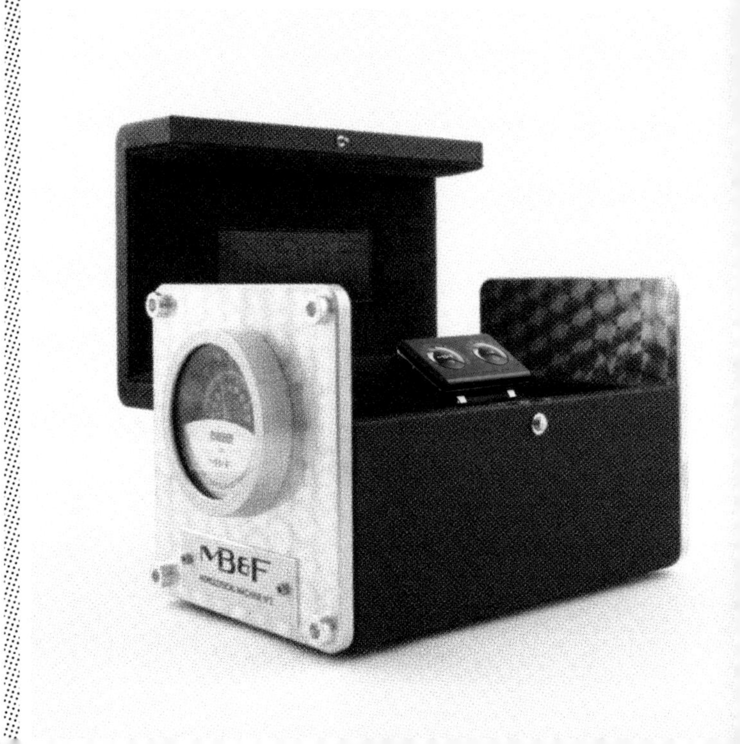
MB&F

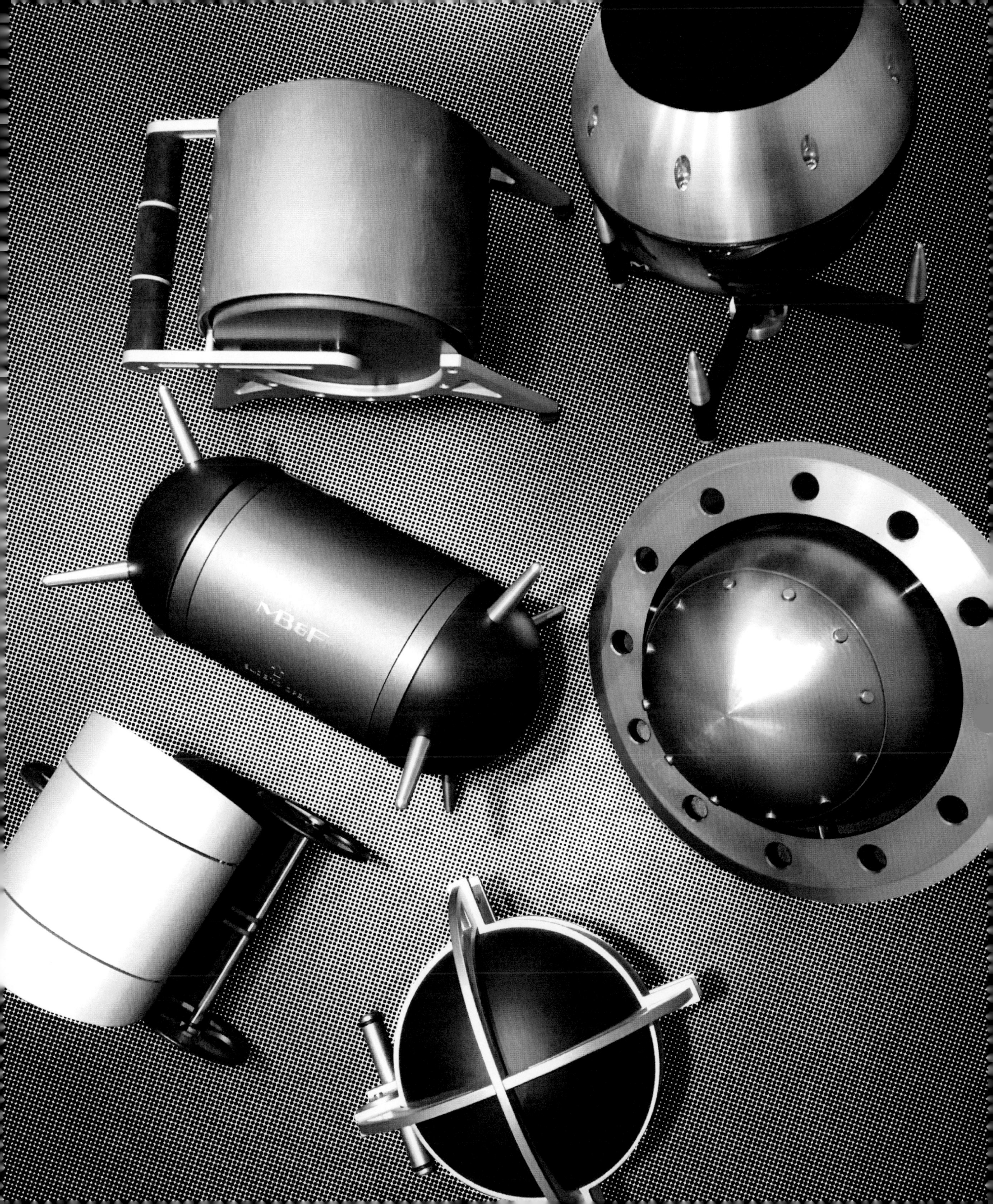
MB&F

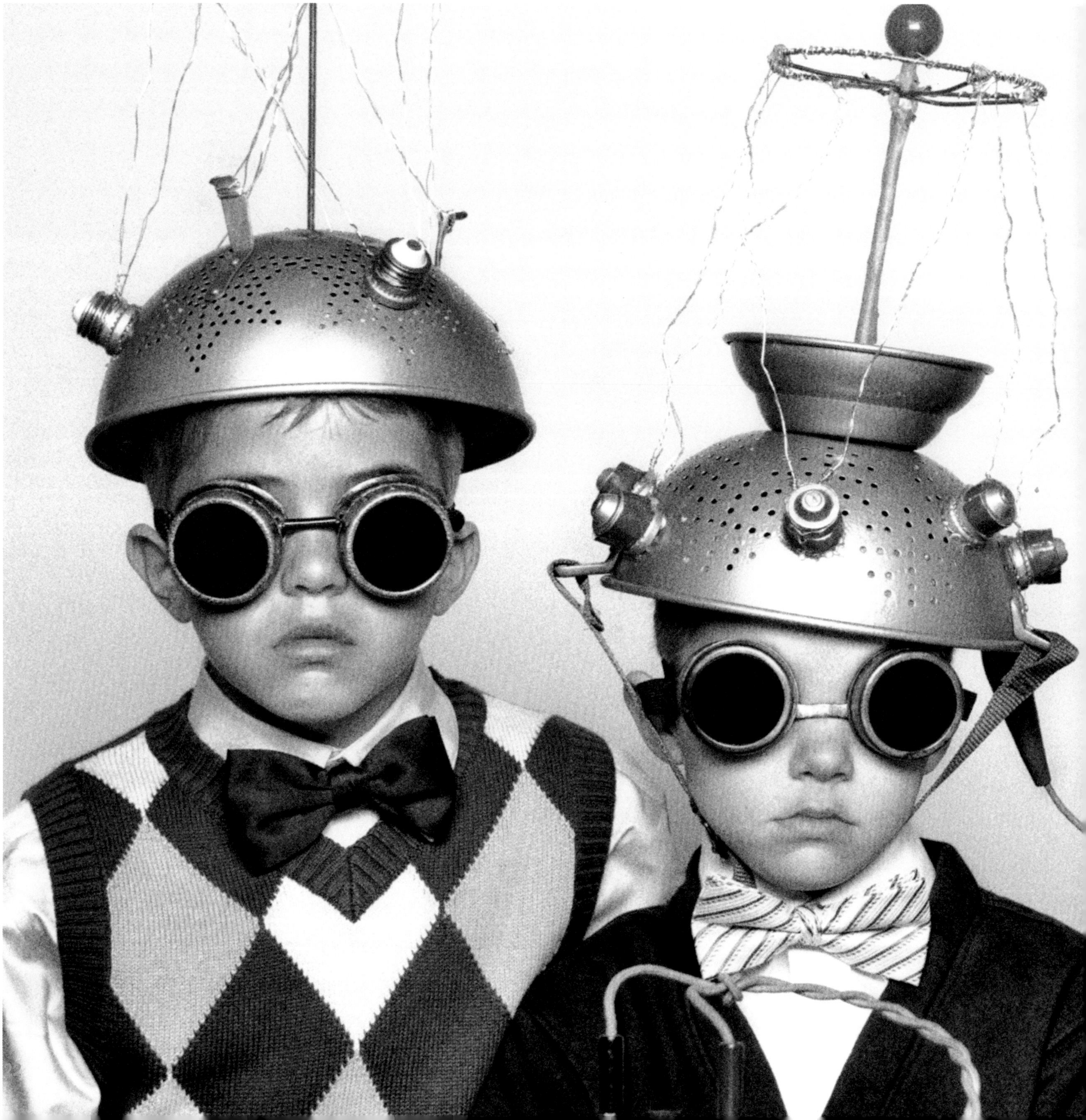

MAXIMILIAN BUSSER AND FRIENDS

What does it take to create a brand that will change the face of modern mechanical watchmaking? In early 2005, Maximilian Büsser stood on the precipice of decision: whether to continue his professional ascent at Harry Winston, on an upwards trajectory that was virtually guaranteed, or whether to risk everything to follow a dream that had smoldered inside him over the past three years.

Trying to fortify his decision-making with some form of underlying rationale, he surrounded himself with spreadsheets, tabulations of projected profits and losses, expected cash flow, attempting to reassure himself that his dream could somehow thrive, out in the real world. But the more he wrestled with the figures, the more it seemed that there would never be enough. However much capital he had at the start, he still needed more. In the end, when faced with the hard and unyielding facts of the matter, Max decided to believe in himself anyway.

Sometimes, with enough faith, you can change the world. Sometimes, you can even create new ones.

Max sketched out his first MB&F creation in 2003, a design that symbolized the coming together of two worlds, joined by exceptional watchmaking. As fate would have it, he was on a plane between Singapore and Geneva at the time, literally in transit between two very different places that nevertheless held watchmaking very dear to their hearts. Max called his company MB&F, Maximilian Büsser and Friends, enshrining this concept of connected, collaborative creation in the very identity of the brand itself. Little did he know how quickly, and how deeply, these ideals would be put to the test. MB&F had just been born, but would anyone ever see it take its first steps?

From the very beginning, Max knew that he had to put everything he had into MB&F. There was no Plan B, and that was wholly deliberate. His entire life savings went into the company, a total of 700,000 francs, which even then he knew wasn't going to be enough. And so, in November 2005, he went around the world, visiting retailers that he had built relationships with during his time at Jaeger-LeCoultre and Harry Winston. These were foresighted and adventurous people, whom Max trusted to see the potential in his company and inject much-needed capital into it. He went armed with three things: an idea the likes of which the watchmaking world had never seen before, the knowledge that his only option was to succeed, and a small 3D plastic model of the watch.

It was, and still is, hard to say no to Max when the full force of his creativity and conviction is applied to getting you to say yes. Some said yes, but then balked when they realised that Max was effectively asking them to pay 35 percent upfront for an unseen watch from an unknown brand. After three weeks, he secured what he needed from five retailers: The Hour Glass (Singapore), Westime (Los Angeles), Ahmed Seddiqi & Sons (Dubai), Chronopassion (Paris) and Ghadah Jewellery (Kuwait). These were people who had known Max before; but who went above and beyond the limits of the average professional relationship, trusting in his vision.

With an additional 700,000 francs in his pocket, and the accompanying guarantee of 25 pre-sold pieces, Max launched production of Horological Machine N°1 (HM1).

Time moves differently when you're fuelled by passion and adrenaline. In April 2006, less than a year after he tendered his resignation to the board of Harry Winston, Max had a working prototype of HM1 that he could show to his retailers at the Baselworld trade fair. Nothing short of a disaster could now stop his dream from breaking through into reality.

Disaster struck in May 2006. The key supplier who was producing the HM1 movement (or engine, in MB&F parlance) was acquired by another watch brand, whose owner made it clear he had no interest in producing movements for other companies. Without an engine, HM1 was stalled, and the waters were rising — the first deliveries had been promised for November 2006.

Attempts were made at negotiation, but in the beginning of January 2007, barely a week after New Year's Day, when most companies were reopening their doors with fresh and post-festive optimism, Max was handed five incomplete movement kits for HM1. No assembly plans were included. It was his last significant interaction with the supplier and their new overlords. *Débrouillez-vous, Monsieur Büsser. Bonne chance (et bonne année).*

The watch world was at the height of its haute horlogerie frenzy in 2007; every company you'd ever heard of (and some that you hadn't) was offering tourbillon watches, minute repeaters, high-concept mechanical experiments that defied common sense and average incomes. Every single watchmaker in Max's phone book who could have helped him finish the movements was already busy. Things looked bleak for Max — no movements, no one to call for help, no more money. He nearly missed the solution that was right in front of him.

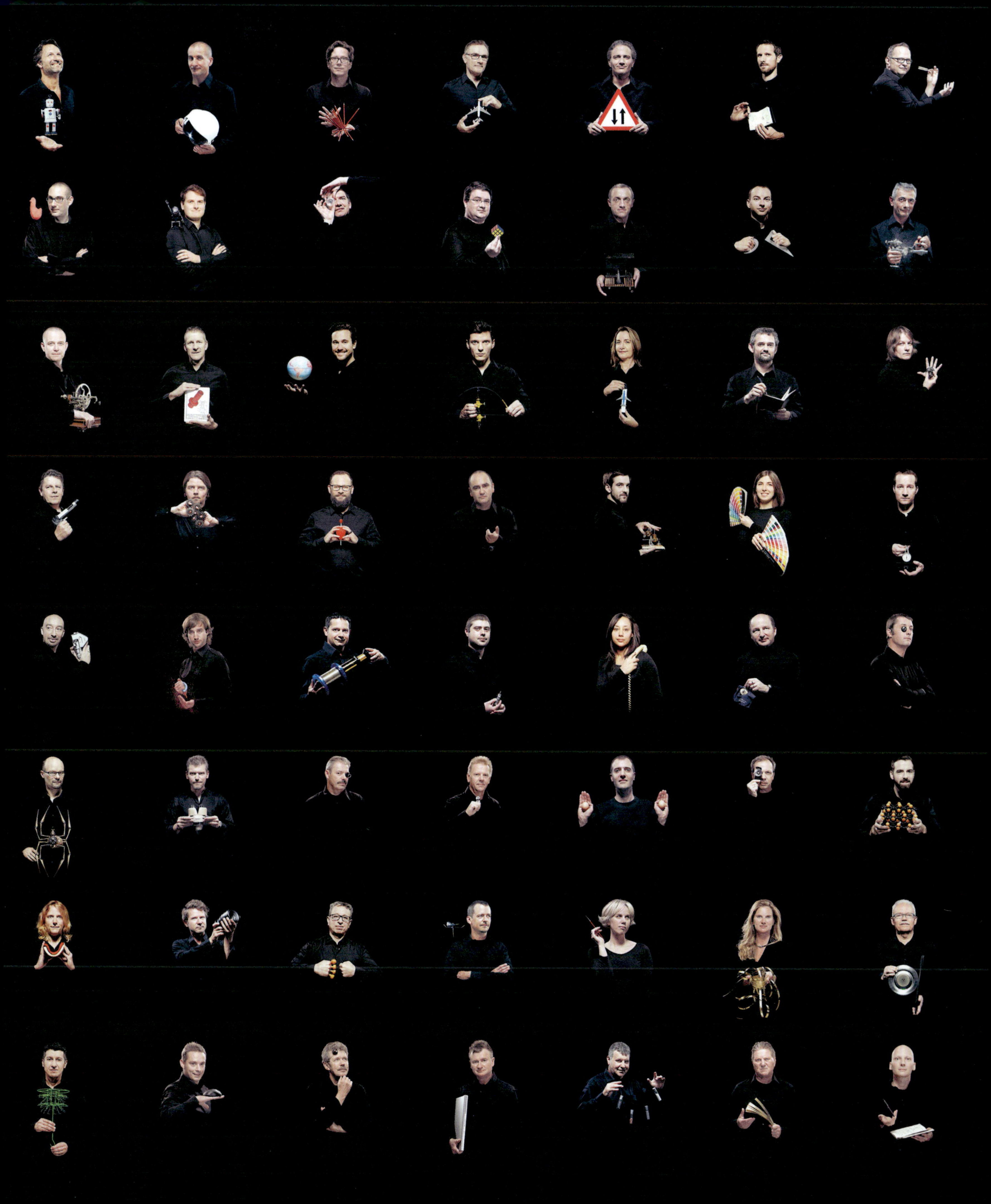

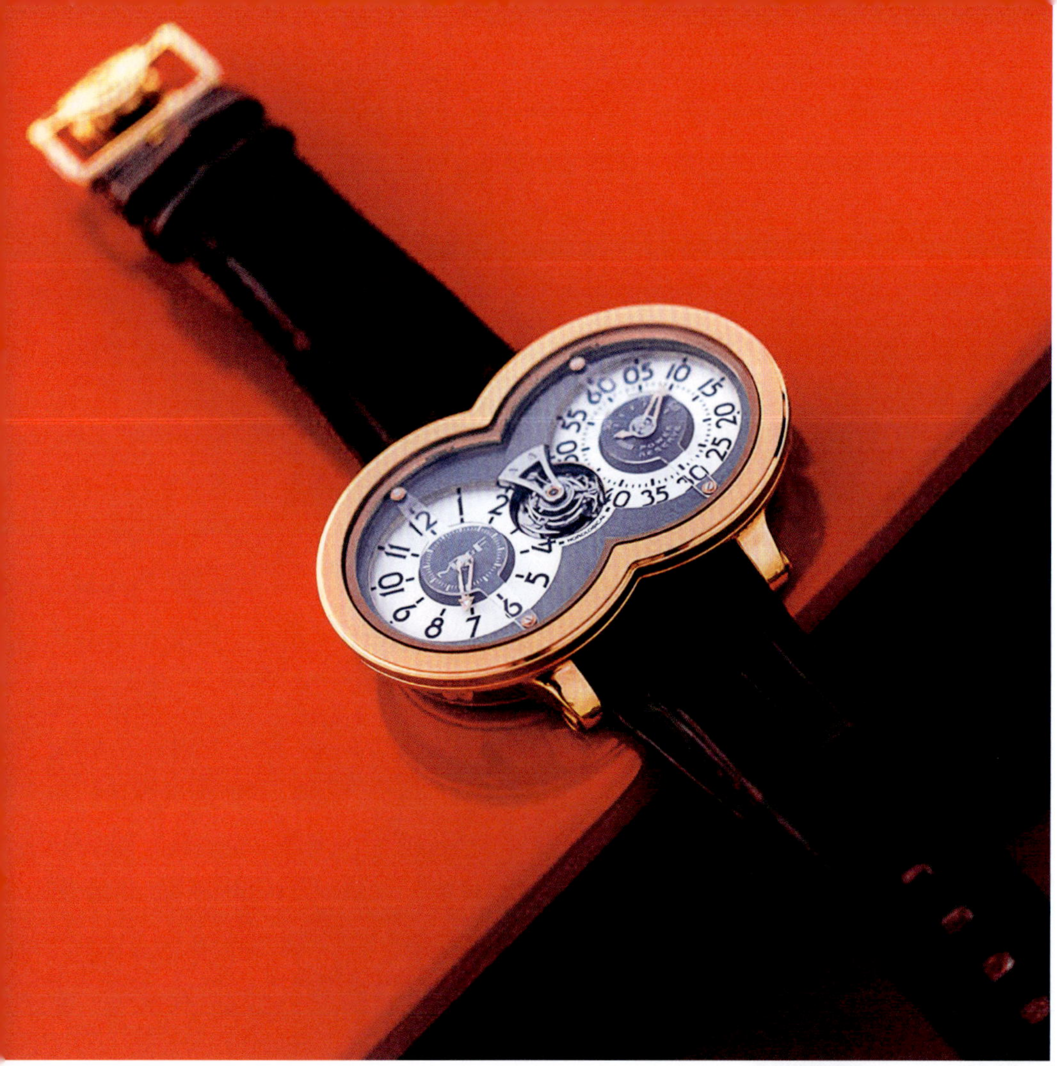

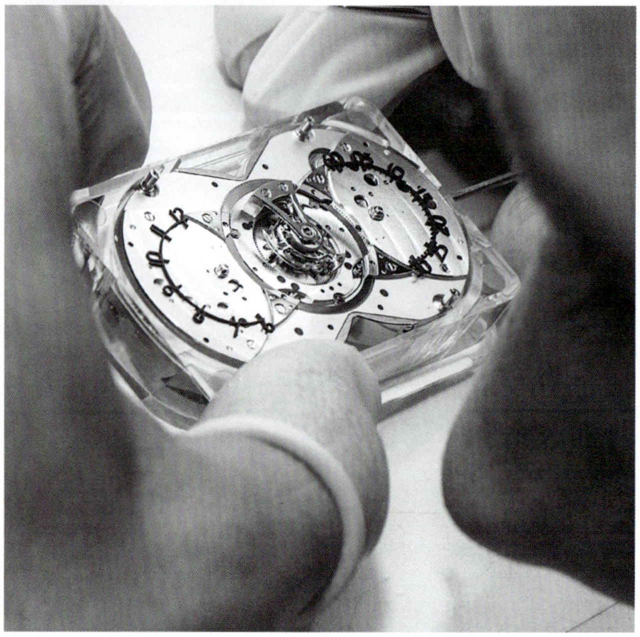

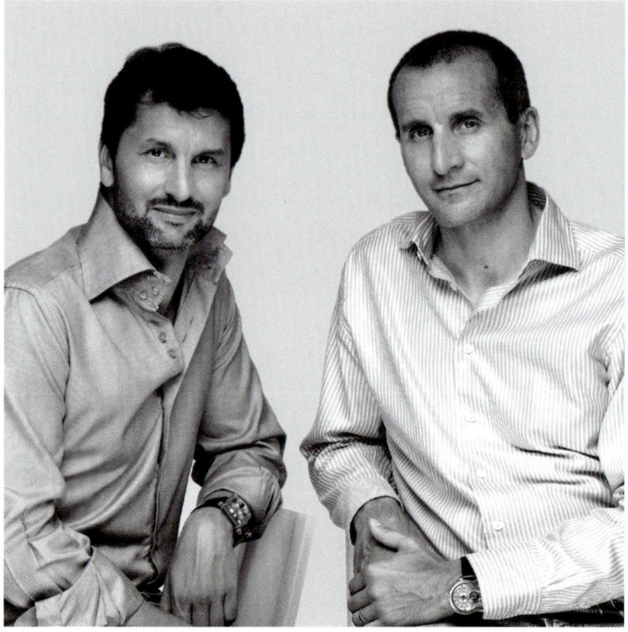

Lifestyle photography of an Horological Machine N°1

Watchmaker assembling an Horological Machine N°2 movement

Portrait of Maximilian Büsser and Serge Kriknoff, partner and chief-technology officer, 2009

With Max at that final meeting with the supplier was his friend, English watchmaker Peter Speake (at the time known as Peter Speake-Marin), whom Max had met during his Harry Winston days. "We'll fix this," Speake assured Max; Max numbly felt that these were little more than empty, if comforting, words. On the drive back from the meeting, Speake called up a small cadre of independent watchmakers, close friends of his, including a talented young Irish watchmaker called Stephen McDonnell who was head instructor at the WOSTEP school of watchmaking in Neuchâtel.

In ten days, the "Save MB&F" crack team was assembled, five watchmakers in total, including Laurent Besse, the original HM1 movement engineer. Everyone got an incomplete, unassembled kit and went away to fit the parts together as best they could, meeting every two weeks to share their progress, rewrite the assembly plans and puzzle through any intractable problems. And in six months, these men had delivered a miracle. The first two pieces of the MB&F Horological Machine N°1 were completed in June 2007.

Over the years, the Friends of MB&F have, together with Max, proven the truth and the strength of the brand's founding concept. In stark rejection of the corporate insularity and opacity that has defined almost every other watch company, MB&F continues to exist as the quod erat demonstrandum of its communal ethos. Eric Giroud, who memorably redefined Max's perspectives on art, design and the creative process, has been an equally formative influence on the aesthetic aspect of every single MB&F creation since Day One. The list of independent watchmakers who left their unmistakeable mark on MB&F is long and illustrious, including Laurent Besse, Jean-Marc Wiederrecht, Stepan Sarpaneva, Kari Voutilainen, Eric Coudray, Stephen McDonnell and Jean-François Mojon.

Open and transparent collaboration is not simply something that MB&F does. It is the definition of what MB&F is.

When Horological Machine N°1 made its long-awaited debut in 2007, the market went into raptures. Endless paeans were written about HM1 in watch magazines, and online collector forums documented celebratory parties when members received their purchases. Not only was the piece unlike anything else ever seen in the watch world, the MB&F story of transparency and collaboration resonated with a community that had grown skeptical of brands touting their own in-house supremacy. Coincidentally, this was the time that saw the launch of the first high-quality mass-market cameraphones (including the popular Nokia N95 with 5-megapixel digital camera resolution). The rise of MB&F was thus an intrinsic part of the online watch community's development, as signalled by the emerging phenomenon of the smartphone wristshot. After all, the wrist shot only makes sense if you have something on your wrist worth a shot, and there's no doubt that in 2007 and 2008, a Horological Machine ranked near or at the top of that category.

By November 2007, Max was giving his first previews of Horological Machine N°2. People were now starting to understand what MB&F was all about. While HM1 laid out the founding precepts of the brand — the converging worlds of Max Büsser and his Friends, radical disruption of the watchmaking status quo, futuristic interpretations of traditional haute horlogerie — HM2 delved into the actual creative process, a multitude of backstories, all inextricably tied to Max's childhood. It all began to spill out, like a starship overflowing with Tribbles. The winding rotors, shaped like the Double Harken twin-bladed halberd of UFO Robot Grendizer; the sapphire crystal pods inspired by science-fiction moon settlements; the bolted portholes and articulated sliding crown guards lifted directly from a young Max's Meccano set.

For the first time, Max was creating for himself. For the first time, he was being entirely selfish in what he put out there. And people loved it.

While the rest of the watch industry floundered in the roiling aftershock of the 2008 financial crisis, MB&F powered through disappointing sales orders in January 2009 to finish the year ahead of all predictions, thanks to the all-time collectors' favorite Horological Machine N°3. In the watch industry's worst year of the decade, MB&F delivered 100 new pieces, 85 of which spent barely any time in boutique windows before finding their way to a customer's wrist.

Despite scoring hit after hit, the prospect of launching Horological Machine N°4 in 2010 still terrified Max. It was so different from anything that anyone had done before. Then when it became the success that we now know it to be, it opened the floodgates to the full deluge of MB&F's aesthetic vision. Without HM4, there would have been no Horological Machine N°6 "Space Pirate" or Horological Machine N°9 "Flow". On a prosaic level, there might not have been any more Horological Machines at all. So much was riding on HM4 that the company might conceivably have gone bankrupt if the piece had flopped. And Max knew deep down that if HM4 hadn't been received well, he wouldn't have had the courage to attempt anything like it ever again.

At the 2010 edition of the Grand Prix d'Horlogerie de Genève (GPHG), Horological Machine N°4 took the prize in the category of Concept And Design. With commercial success on one side and critical acclaim on the other, the Horological Machines had shown they were here to stay.

"Did you see the crazy thing they just came out with"

The World of Legacy Machines

Rumors swirled around MB&F in 2011. They were way overdue for another launch, since their last major release was 2010's Horological Machine N°4. Speculation abounded online, with ardent MB&F followers posting theory upon theory about what was surely going to be the new HM5.

Even when influential watch magazine Revolution inadvertently revealed that the Next Big Thing from MB&F was going to be a round watch with white dials and Roman numerals, few paid attention. Absurd to think that an MB&F creation would be so conventional! Then October 2011 arrived, and the world got to meet Legacy Machine N°1: a round watch with white dials and Roman numerals.

As the watch community reacted with shock (mostly delight, with pockets of scattered confusion), one thing was abundantly clear. MB&F—the brand known for being different, the brand known for divisive designs, the brand known for provoking more than its fair share of headlines and comments of the sort "Did you see the crazy thing they just came out with"—had done it again. And, by some uniquely Büsserian alchemy, they had managed to be different, divisive and provocative with a classically styled timepiece.

Few realized that Legacy Machine N°1 had caused as much disruption behind the scenes at MB&F as it was currently doing in the wider world of watches. When Max first brought the Legacy Machine concept to the table, Eric Giroud stormed out of the room, saying he didn't see any point to it. Serge Kriknoff, Max's partner at MB&F who had been by his side since 2008 overseeing technical production, told Max that he hadn't joined MB&F to make a round watch. But Max held firm. Creativity is not a democratic process, after all. And now, with the Legacy Machine collection, there was a new equilibrium at MB&F.

While the Horological Machines were inspired by hypothetical versions of the future, the Legacy Machines represented the other side of that "what if" coin, plunging into alternative versions of the past. What would Max Büsser have created if he had been born in 1867 instead of 1967? His first watches would have debuted in 1907 and 1908 instead of 2007 and 2008. In place of science-fiction cartoons, Max turned to the adventure novels of Jules Verne, the extraterrestrial imaginings of H.G. Wells, the steampunk splendor of the 1900 World's Fair.

As the Legacy Machine collection evolved, with Legacy Machine N°2 debuting in 2013 and the Legacy Machine Perpetual in 2015, a clearer demarcation line began to emerge between the Horological Machines, which continued to emphasise design and aesthetic expression, and the Legacy Machines, which exhibited a heavier focus on mechanical innovation. This played out in Legacy Machine Perpetual (whose architect Stephen McDonnell had played a central role in assembling the HM1 engines in 2007), Legacy Machine Thunderdome, and the anniversary piece LMX.

There is a scene in the British indie film Human Traffic, a coming-of-age vignette that depicts social rebellion and youth culture at the turn of the millennium. In it, a main character pontificates about what separates the Sith from the Jedi, observing that the Emperor's desire to conquer outer space versus Yoda's quest to explore inner space illustrates the fundamental difference between the Light and Dark sides of the Force. Both came from the same source of cosmic energy. It was simply a different kind of journey.

The Legacy Machine collection also allowed Max to externalize one of the most intimate creations of his life, Legacy Machine FlyingT, a mechanical ode to the women in his life — mother, wife and daughters. For MB&F, Legacy Machine opened up another dimension of creativity. Worlds within worlds, as complex and beautiful as any multiverse conjured up by a young Max Büsser, daydreaming in his childhood bedroom.

The World of exceptional pieces

Even in the context of a brand whose entire existence is predicated on breaking the mold, there can be outliers, pieces that hover like satellites on the fringes of a galaxy far, far away. These satellites might not mean much, taken individually, in isolation. Their significance only becomes clear when you lay them out against the map of MB&F history. Placed beyond the main unswerving continuities of the Horological Machines and Legacy Machines, MB&F's exceptional pieces — Co-Creations and Performance Art — are lighter, more agile markers of the brand's rhythms.

The same principle applies to the field of musicology, when assessing the career and repertoire of an artist. While the greatest hits show the trajectory of success, it's the B-sides that really demonstrate the depth and richness of the story.

In 2009, the first MB&F Performance Art piece was announced, a unique edition of Horological Machine N°2 for Only Watch, the biennial charity watch auction. HM2 Only Watch was made in conjunction with Sage Vaughn, a Los Angeles-based artist, whose work explores the intersection of childlike innocence and the harsh realities of this world. As was often the case when the path of MB&F crossed into the realm of contemporary art, Eric Giroud made the link between Vaughn and Max, having discovered the former's work during his 2007 exhibition in Geneva. Reinterpreted by Vaughn, HM2 bore an intense and raw energy, featuring twists of blackened-gold barbed wire wrapping around an openworked butterfly in blued gold, trapped under a sapphire glass pane. It was a powerful allegory for the beneficiaries of the Only Watch auction, children living with Duchenne muscular dystrophy, an incurable degenerative disease.

While HM2 Only Watch was designed to resonate with the central premise of the charity auction, it also came to symbolize 2009 in general for MB&F. At the end of 2008, the watch industry was still enjoying the last highs of a turbo-charged consumer economy, with very little awareness of the imminent crash. Neither watch companies nor retailers had made any attempt to throttle back production and inventory, so when the time came to hit the brakes, the impact was hard, sudden and brutal. Based on projections made in the previous year, MB&F had planned to produce 175 pieces in 2009. When Max saw his retailers in January 2009, he took a total of 17 orders, less than 10 percent of what he had expected to make.

The company was barely three years in existence and its delicate wings were already in danger of being shredded by the merciless barbs of an economy in crisis. That year, Max embarked on an epic journey around the world, trying to disentangle MB&F from the downward spiral of a plummeting luxury market. In 2009 he travelled a grueling 260 out of 365 days, meeting press and customers, putting himself at the disposal of his retailers to keep MB&F from becoming another business casualty of that panic-soaked year.

It worked. The tireless travelling, the growing community of MB&F followers, the blockbusting Horological Machine N°3 — all of it together — it worked. The summer of 2010 found Max launching his most ambitious creation ever, Horological Machine N°4 "Thunderbolt". In autumn, came something equally ambitious, though on the other side of the spectrum: JwlryMachine, a high-jewelry reimagining of HM3 by Boucheron, a watch collaboration before watch collaborations were cool. HM3 had already taken a leap of faith, trusting in the strength of the brand identity by putting the MB&F logo on the caseback, a bold move given that a mere three years had passed since the debut of HM1. JwlryMachine went even further, obscuring the last overt indication of the brand, the engraved "Horological Machine N°3" on the dial-side rotor. No visible brand logo. No visible watch name. Fully set with gems, something that Max had adamantly insisted he would never do.

JwlryMachine was the first sign that there are no "nevers" in the MB&F lexicon. It was a sharp and unexpected departure from existing MB&F iconography, one full year before Legacy Machine N°1 came along. It was a high-jewelry feminine watch, nine years before Legacy Machine FlyingT debuted. It was the result of a collaboration with another luxury watch brand, ten years before LM101 MB&F × H. Moser & Cie. People may say that there's no predicting what Max will do next with MB&F, but that's not necessarily true. You just need to know how to read the signs.

The World of M.A.D.Galleries

The life progression of a profit-maximizing watch brand, if we may be allowed to generalize a little, is fairly straightforward. Start small, and achieve international distribution through retail partners. As the brand grows, strengthen identity, consolidate a loyal audience and increase production accordingly. Past a critical point of international recognition and sales volume, establish local offices in various key markets, open local boutiques with exclusive products to transfer sales and customer loyalty from retail partners to the brand. Keep on expanding, pushing the balance of market presence even further in favour of the brand, until the brand is more or less a completely autonomous multinational with full control over its entire product cycle, from conception and manufacture to distribution and sales. This is not the life that Max chose for MB&F.

When Max founded MB&F, his goal was to bring something completely different and fulfilling to the watch industry, but he also had certain figures in mind to benchmark company progress. He wanted MB&F to have 15 full-time employees, to produce 300 pieces per year, and to bring in 15 million Swiss francs in revenue. Having reached that endpoint, Max decided MB&F would cease to expand its activities. But there still remained the intermediate milestones of the brand to

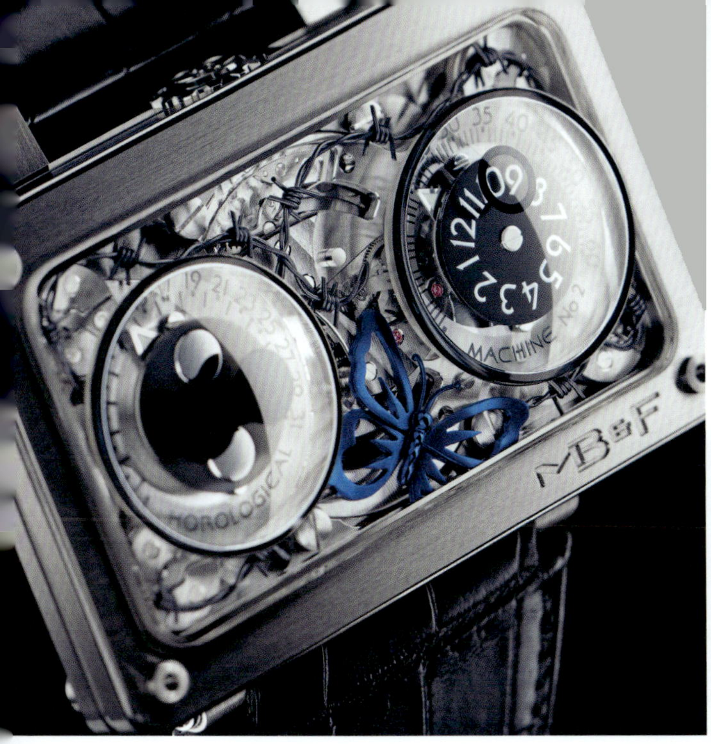

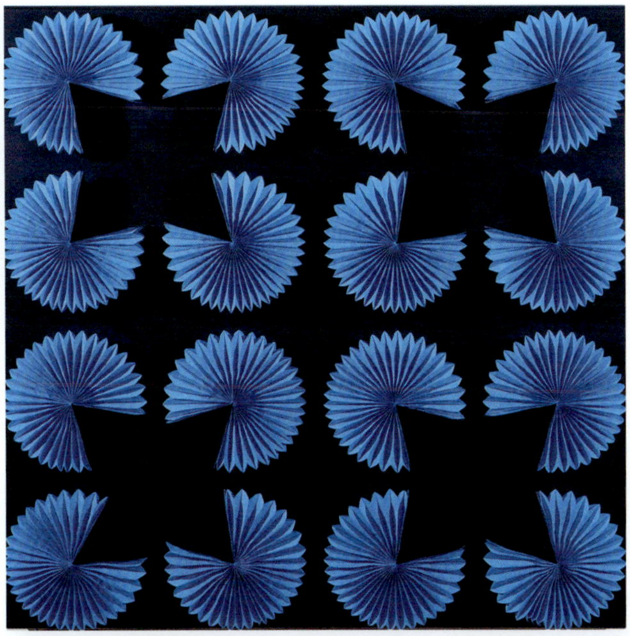

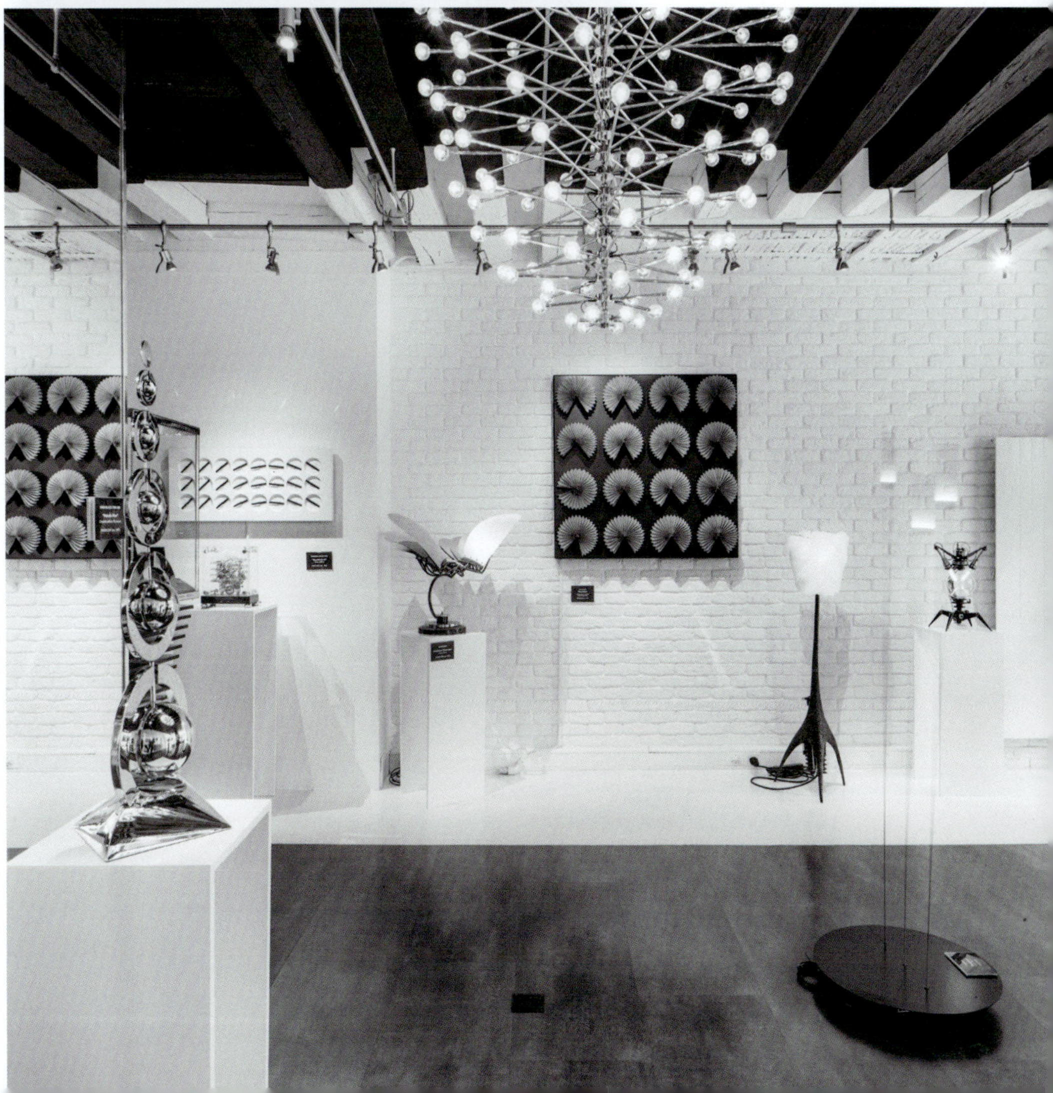

Close-up of the Horological Machine N°2 Only Watch, MB&F × Sage Vaughn

Royal Blue from the exhibition "Fuchsia, Orange & Royal Blue" by Nils Völker

MB&F M.A.D.Gallery in Geneva

MB&F M.A.D.Gallery in Dubai, operated by Ahmed Seddiqi & Sons

"Mechanical Entomology," exhibition at the M.A.D Gallery

Commodor from the exhibition "Viva la Robolución!" by +Brauer

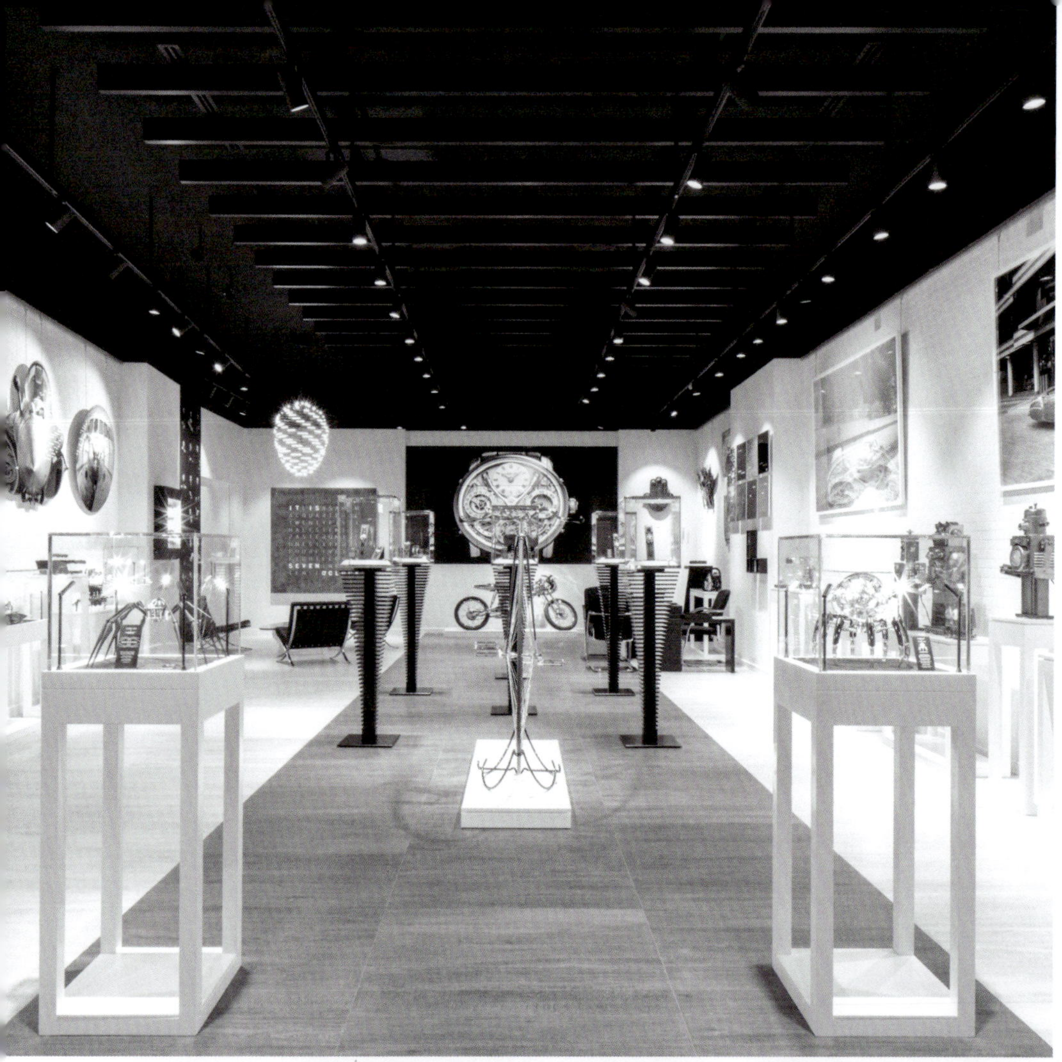

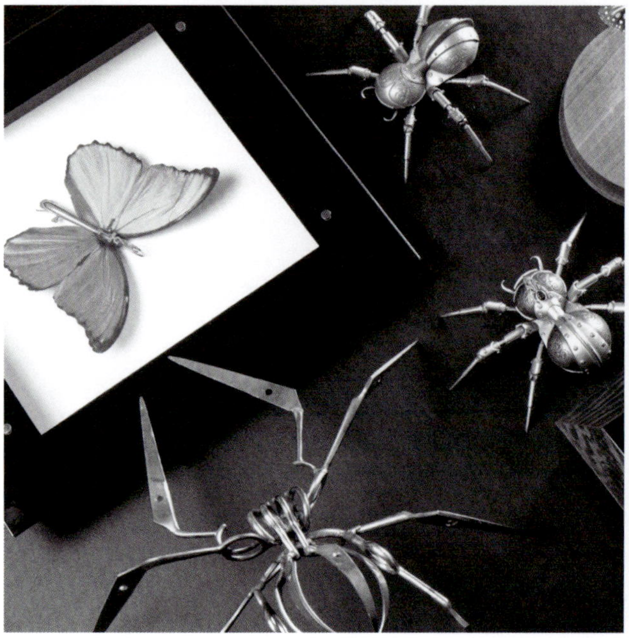

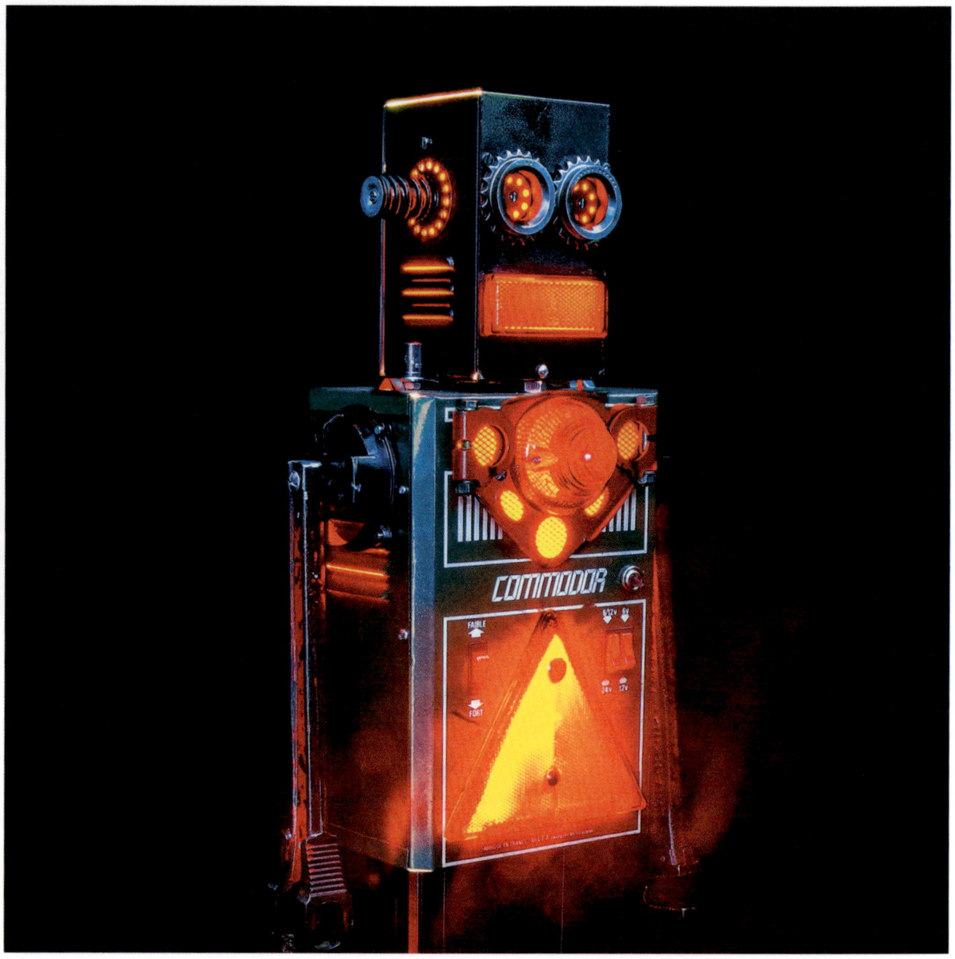

be defined. For instance, what about going into retail? The first MB&F M.A.D.Gallery opened its doors in Geneva in 2011, a foray into retail unlike any other retail experiment in the watch industry. It was emphatically not a boutique. You could buy an MB&F Machine there — there were certainly Machines on display, available for purchase — but it was primarily a showcase and retail space for artists. In what has come to be somewhat of a recurring theme in the MB&F story, neither Max nor anyone else in his team had the first idea about running an art gallery or a retail space. On opening day, the Geneva M.A.D.Gallery didn't even have a credit card machine. The hope was that this experiment wouldn't lose too much money.

Within seven years, one gallery had become four. In 2014, Taipei became the second M.A.D.Gallery location, after Geneva. Two years later, Dubai claimed its own M.A.D.Gallery. Hong Kong made the fourth, in 2018. The acronym M.A.D. originally stood for Mechanical Art Devices, but the widening scope of the galleries meant that they quickly outgrew this taxonomy. Close to 30 visitors came to the Geneva gallery every day, a veritable deluge compared to the anticipated trickle of two or three visitors per day.

While a host of creatives have seen their work pass through the various M.A.D.Galleries — from futurist wood sculptor James McNabb and steampunk lamp maker Frank Buchwald to bio-mechanical entomologist Gaby Wormann and dissected-reality photographer Fabian Oefner — the galleries also generated a portal of creativity for MB&F. Through the M.A.D.Gallery concept, Max was able to create more than watches. MB&F started working with companies such as L'Épée 1839, renowned Swiss clockmakers, and Reuge, the music-box experts from Sainte-Croix.

By 2013, MB&F boasted those dream numbers that Max first set out to achieve: employee numbers, production numbers and sales numbers. In theory, MB&F would have stopped growing at this stage. But there are different types of growth, and not all of them are equal. Evolution is also a type of growth, one that MB&F has never stopped embracing. Bear in mind that the M.A.D.Galleries are not "points of sale", the retail jargon used to describe the places you go to buy the things produced by a brand. They are mission outposts, each one an ambassador for the MB&F ethos. The world of M.A.D.Galleries is no less a part of MB&F than its Machines, its Friends, its Maker.

What does it take to create a brand that will change the face of modern mechanical watchmaking? Many things, as it turns out. It can take an entire world — sometimes two. Sometimes more. For MB&F, it's always more.

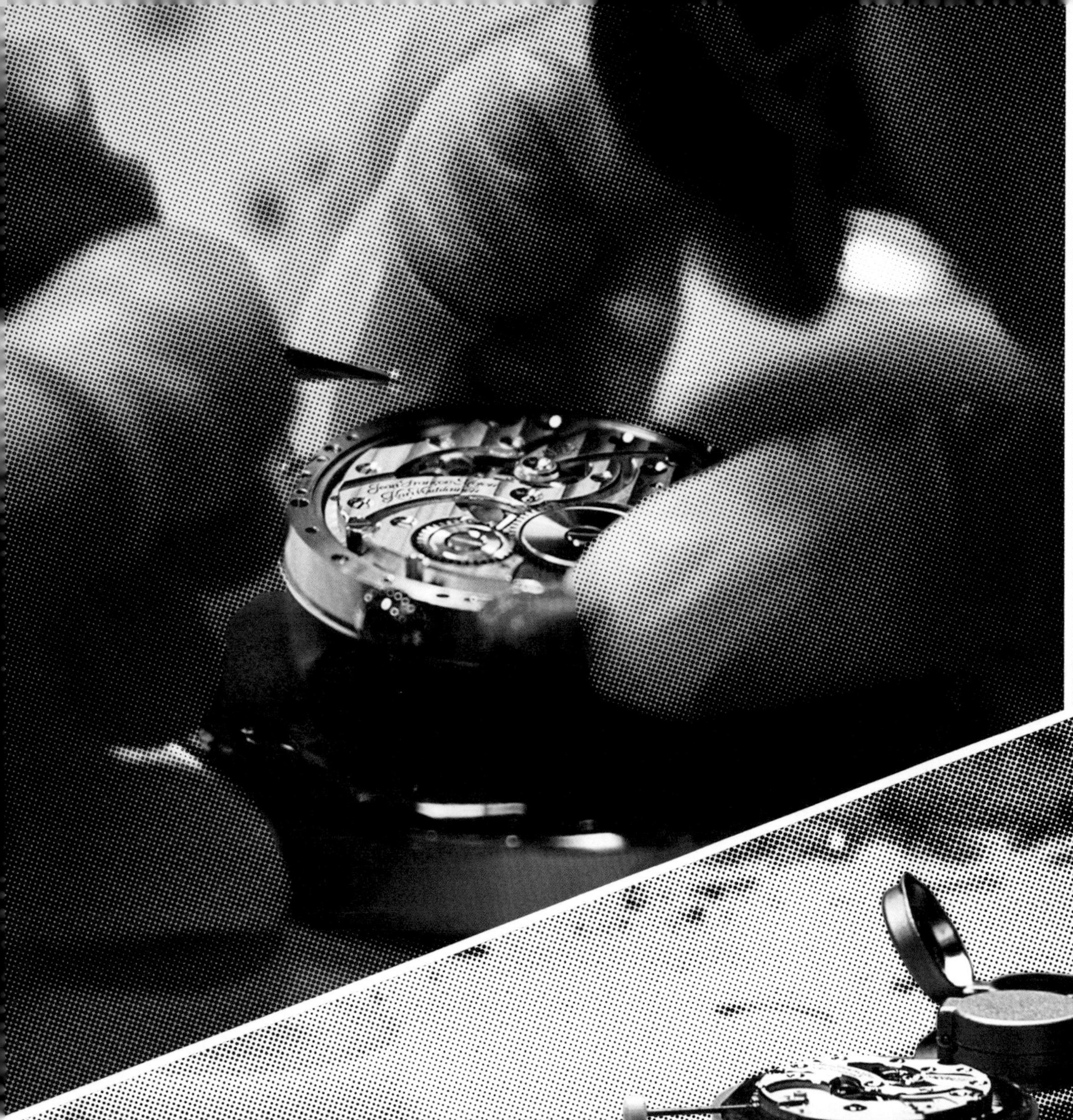
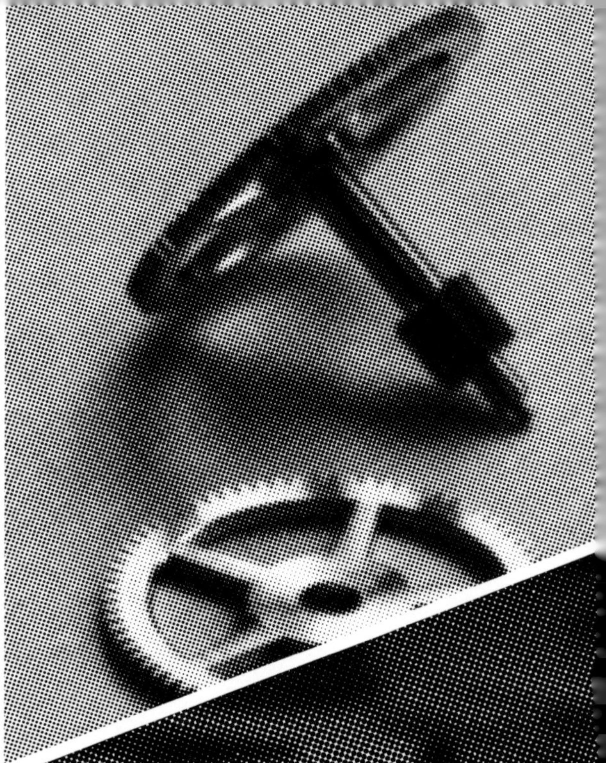
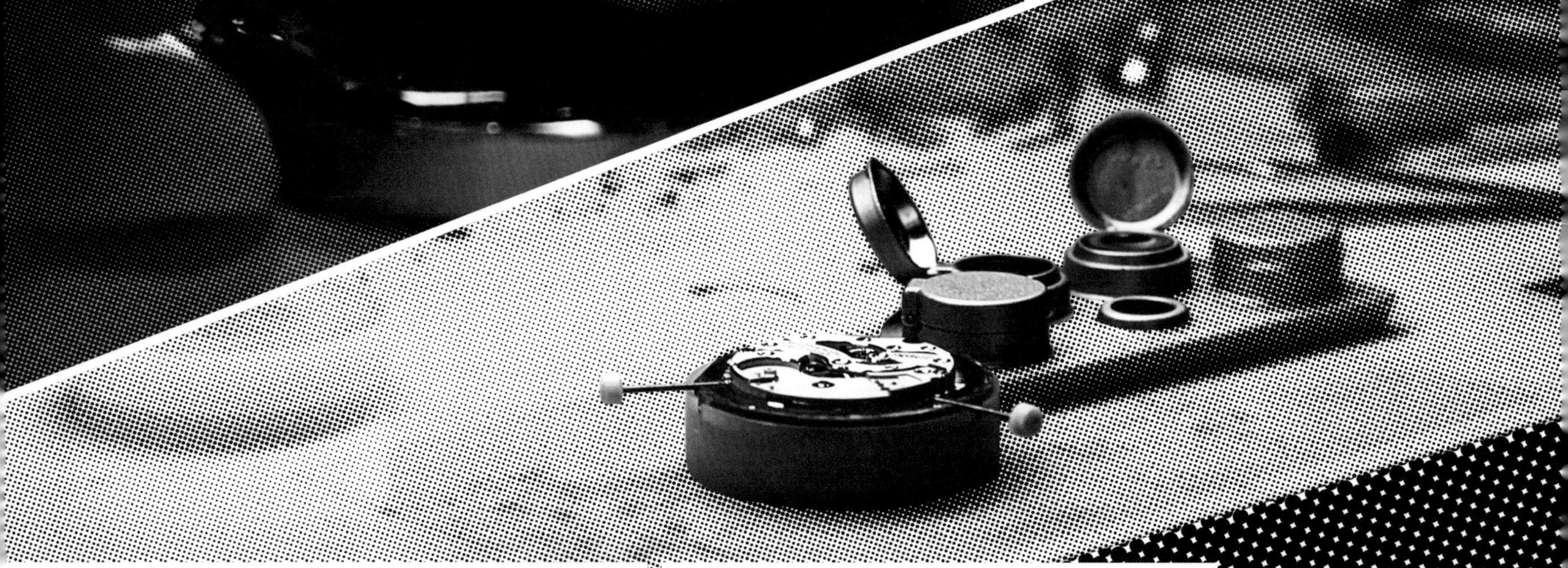
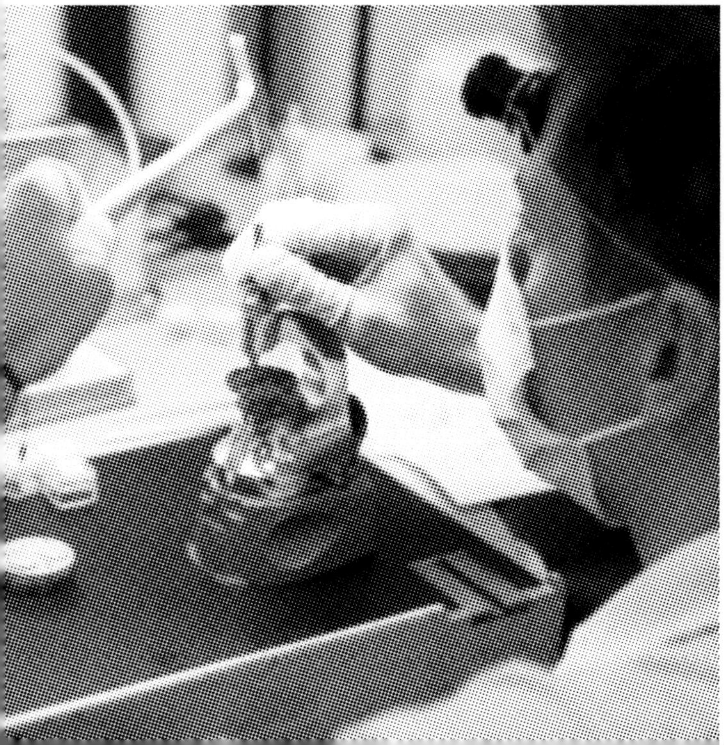
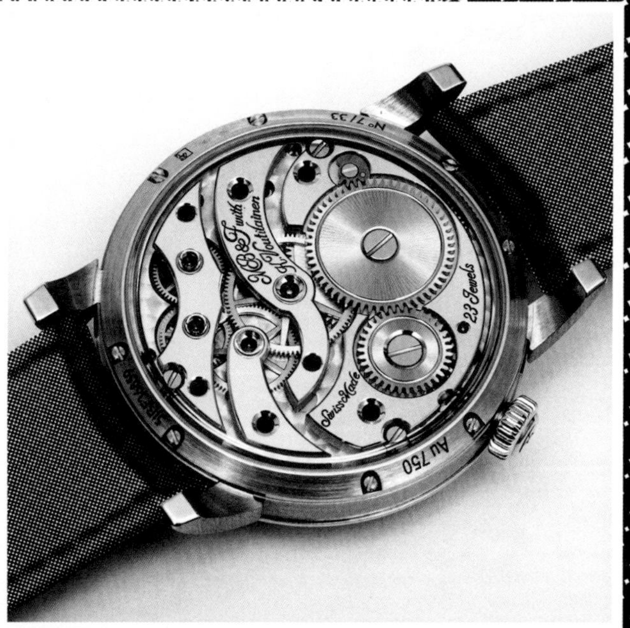
23 Jewels
Swiss Made
Au 750

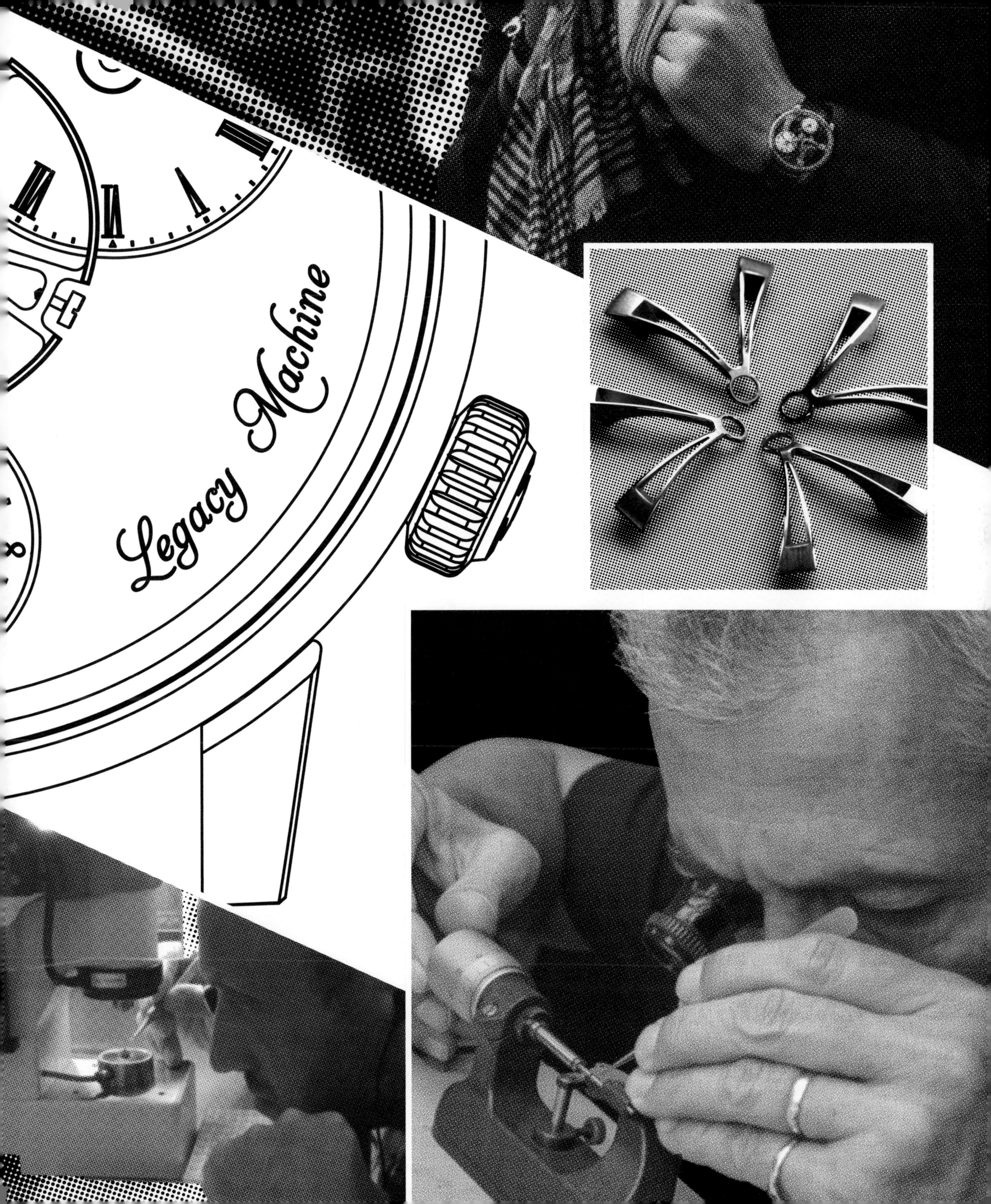
Legacy Machine

Drawing of an Horological Machine N°10 Bulldog playing with a Tri-Axe

Illustration of an Horological Machine N°4 Thunderbolt

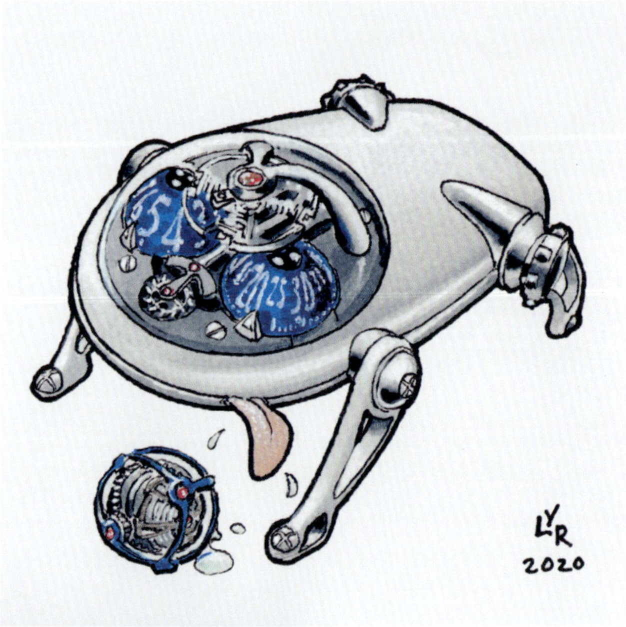

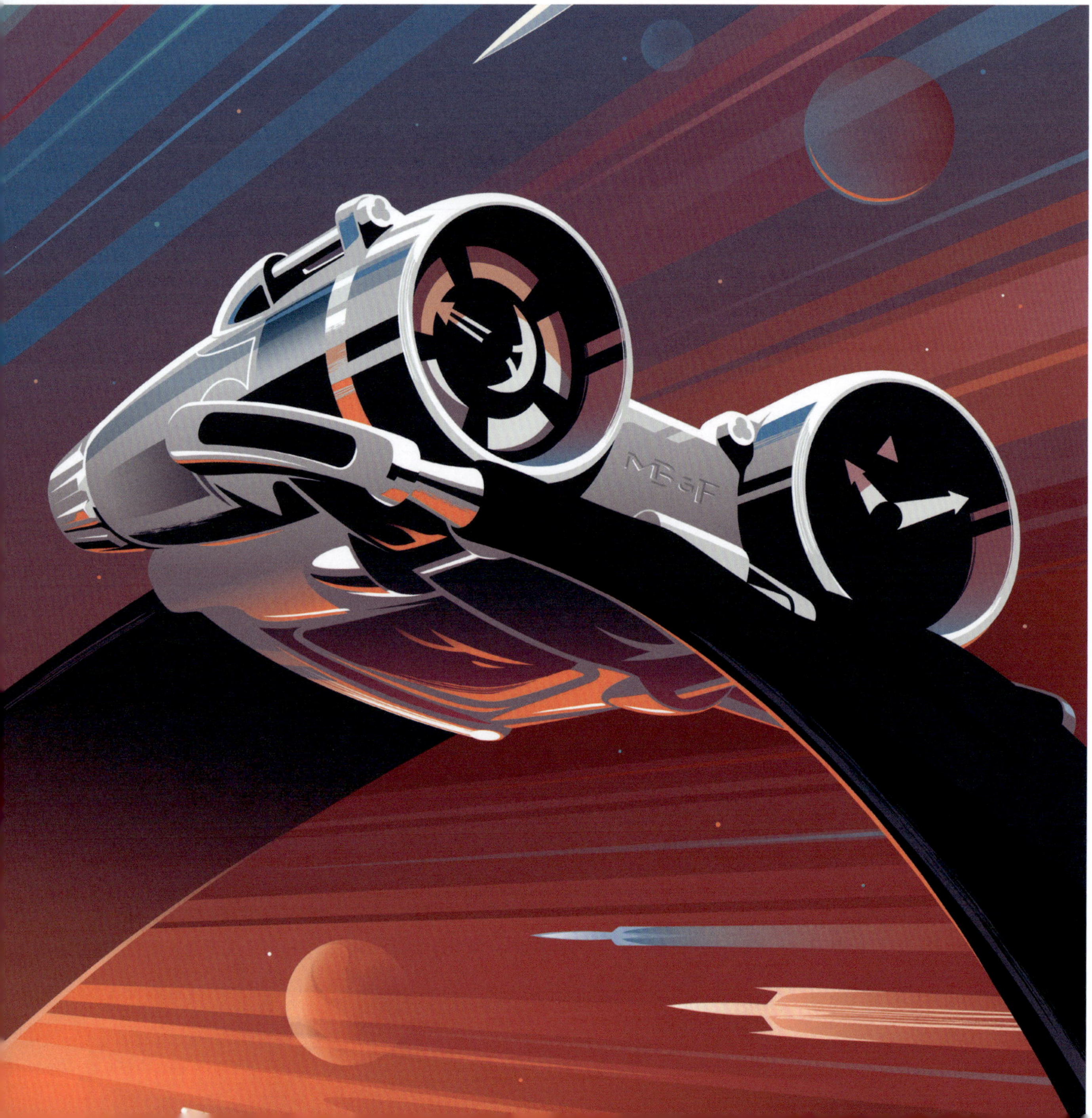

THE MB&F EFFECT

There are some key motifs that occur and recur with near-sentient regularity in the MB&F story. The concept of a playful, childlike approach. The idea of deep, disruptive change. The simultaneous focus on and deconstruction of the self. The pursuit of independence, of autonomy. The spirit of survival. Without fail, these themes have marked the history of the company. They mark it now, as it is being written. They will continue to mark its future.

The search for meaning is a journey that few choose to undertake, simply because it is infinite in scope and metaphysical exploration rarely results in happy endings. There are no cheerful philosophers. If you ask what MB&F means, your greatest chance of satisfaction is to learn that it stands for Maximilian Büsser And Friends — and stop right there. On a purely literal level, that really is all you need to know.

But let's say we persist in asking. What does Maximilian Büsser And Friends mean? It is the union of the individual and the collective, which is odd in an industry that has always emphasized one over the other, romanticizing its artisanal side while downplaying its industrial aspects. It is a strange paradigm, when you think about it. What kind of industry is it that avoids acknowledging its industrial nature? MB&F, a terse and cryptic initialism to the ears of anyone who hears it for the first time, is actually one of the most honest representations of watchmaking in all its centuries of history.

Max is nothing if not indefatigably, disarmingly honest. He left the corporate world to become an entrepreneur because he wished to reflect the values that his parents, Mario and Dinoo, had exemplified in his life. He told me, "What I wanted was to be free." And liberating people, even if it is just oneself, is already a worthwhile aim. But Max freed a lot more than himself when he decided to create MB&F. He freed an idea, and that is the most potent thing of all.

Years ago, I asked Jean-Marc Wiederrecht, founder of movement think-tank Agenhor and the man behind the HM2 and HM3 engines, who were the figures who had created the most inspiring body of work in watchmaking? He gave me two names. One was Abraham-Louis Breguet, the 18th-century watchmaker widely considered to be the father of modern horology. The second was Max Büsser, who Jean-Marc was quick to describe as a visionary and a leader.

> *"He shows us that it's not just that we can — but, in fact, we must — do otherwise than make only what others before us have already made. I don't know how he does it, but he simply knows what to do, and everyone follows."*

It's true that when I first met Max in 2010, the world of contemporary watchmaking did suddenly start to make more sense. Classical watchmaking had not altered its essential expression in two centuries, and there was a consciously cerebral side to it that was intriguing, but also intimidating. You always felt at risk of sounding stupid or fraudulent if you said the wrong thing about a tourbillon or a perpetual calendar. When I held Horological Machine N°4 in my hand, I felt something I had never before experienced in the context of a mechanical watch. I felt I was having fun.

In a decade and a half of existence, MB&F has done many things that were hailed as different, game-changing, audacious, unprecedented — and it's been said in as many languages as you can name. For a brand ostensibly labelled "indie", niche, or even esoteric, MB&F has undeniable reach, and it's because of what MB&F means.

Meaning is about what something is, but it is also about what something is not. Defining what you stand for also includes defining what you stand against. An MB&F Machine is a talisman against the apathy that most people today feel toward mechanical watchmaking, against the cynicism that many in this industry have about the business, and against the egotism that permeates much of the decision-making at the very highest levels. In place of apathy, it induces passion, as anyone who has ever encountered the brand will testify. In place of cynicism, it provokes curiosity, as demonstrated time and time again by the effervescent energy that greets each new MB&F release. In place of egotism, it communicates generosity — as you see in the collaborative spirit that is enshrined in the company name, the frankness with which Max shares aspects of his life in the telling of each Machine's story, in the ever-growing fraternity of the Friends and Tribe.

When you say someone changed the world, it is a statement of admiration. When you say someone changed your world, it is an expression of love. MB&F opened my eyes fully to mechanical watchmaking, an event that precipitated my choice of career and brought about a permanent change in my country of residence; MB&F changed my world. It's not something you hear every day about a watch company, but I know I'm not the only one who feels this way.

This is what MB&F means. It is to be different, but not alone.

When Max talks about the future of MB&F, it's usually in terms of the brand's upcoming launches. More than a few times, he's said to me, "I know exactly what we're doing for the next five years, but after that, I have no idea what's coming." Projections of what he actually intends for the company that bears his name are not so frequent.

Trying to predict what is ten years down the road is a high-stakes game, and Max is not the gambler he was at 38 when he took his life savings and plunged into entrepreneurship (spreadsheets be damned). One thing we do know is that MB&F is the kind of company that survives. It was founded by a survivor—of big army truck crashes and corporate crises—and if you consider the company's unofficial credo, that a creative adult is a child who survived, it employs and works with a vast community of survivors.

By Max's own count, MB&F has been through four near-bankruptcies, all of which he took as tough learning opportunities. In 2012, when a perfect storm of micro-problems converged to wipe almost eighty percent of their production orders off the books, Max and his team used that experience to rebound with a vengeance. One of those micro-problems happened to be a supply-chain failure, a delay in the delivery of a twenty-dollar wheel that prevented MB&F from completing and selling more than a million dollars worth of product: something entirely preventable, from Max's perspective.

In April 2014, the first machines that would allow MB&F to do their own prototyping and emergency parts manufacturing arrived on company premises. Live, learn, survive. You are never entirely at ease with the continual need to snap into high-alert mode, but Max is almost tranquil now when he speaks about it. "It's always the same with us." One year everything's going great, and then the next year, the hammer falls. Remember that year right before SIHH (Salon International de la Haute Horlogerie) when the Swiss National Bank decoupled the franc from the euro? Everyone else in the watch industry was going crazy. But we just gave each other that look — here we go again, get your battle dress on — and got to work. We knew we would deal with it one way or another."

"Deal with it" is another way to say adapt. "Deal with it" is what Max was told when he was handed five unassembled Horological Machine N°1 engines on the eighth of January, 2007. Dealing with it is what MB&F has done every year after that one, and there are very few indications that it will not continue dealing with it in the decades to come.

Will the company endure after Max? The long-term success of any idealistic endeavor is mostly (and, I think, correctly) judged by how well they survive their pioneer generation. Maximilian Büsser founded MB&F. And it isn't inconceivable that in another world, another reality, MB&F did not disrupt the watch industry or enjoy the success it has now, and maybe in that reality it wouldn't last long without Max. But when you change things so profoundly and connect so deeply with the imaginations of an entire community, you are more than a brand that sells watches. You are a story that lives and grows and inspires. And stories belong to everyone who believes in them.

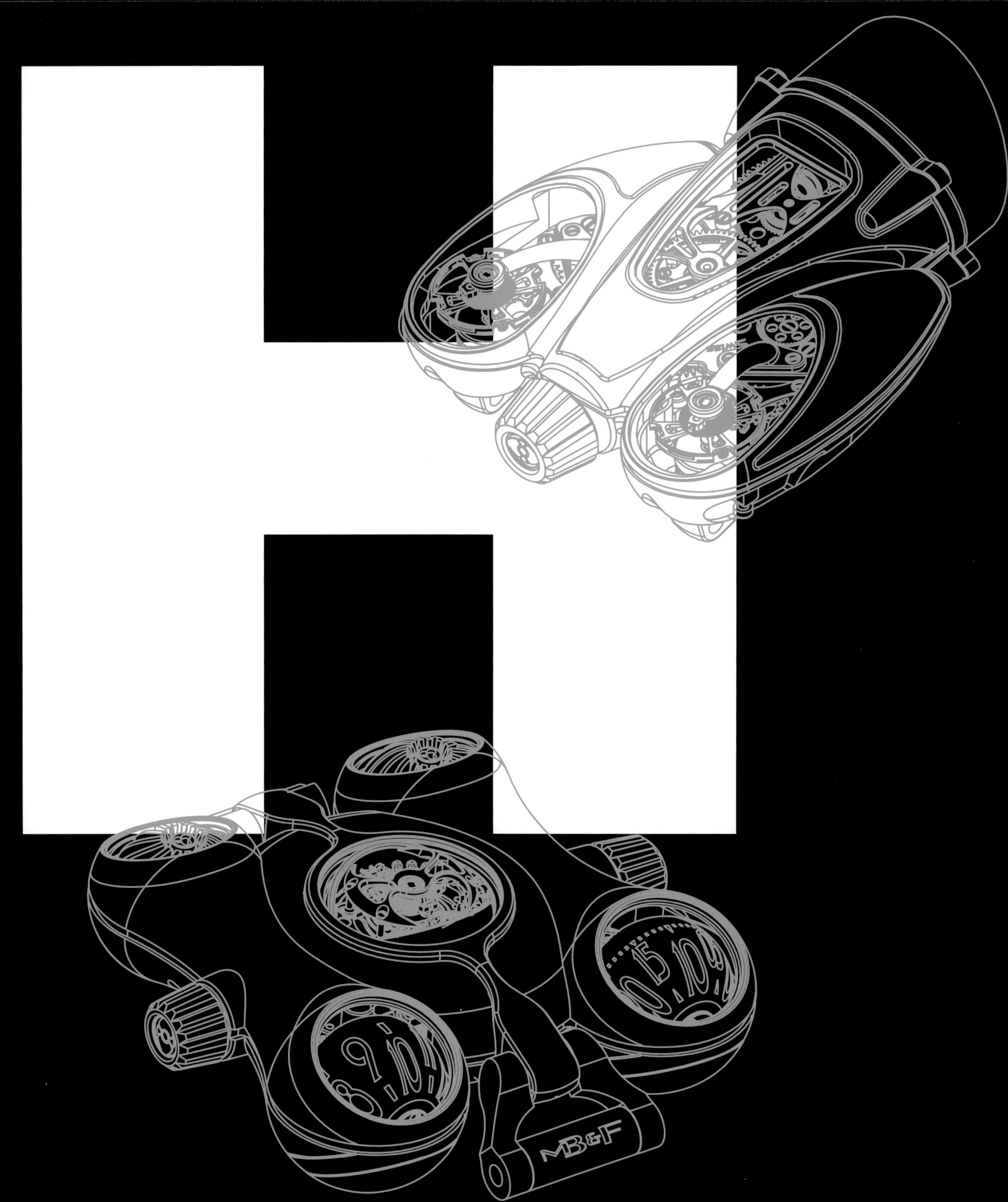
MB&F

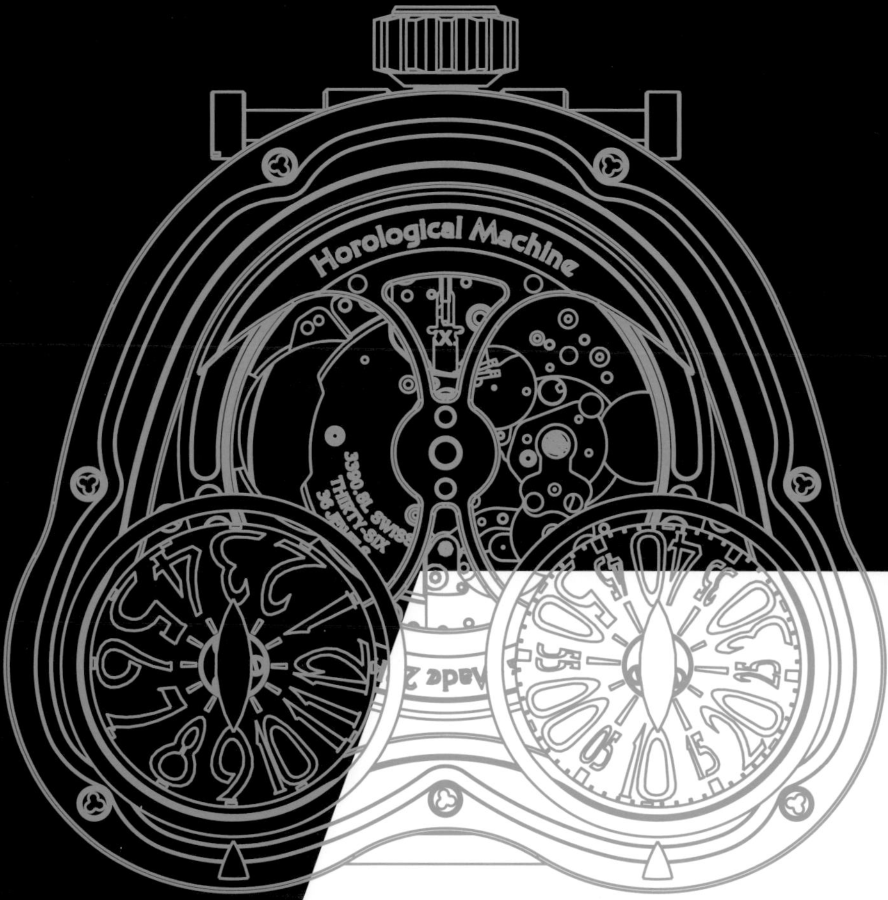

M

HIM

HOROLOGICAL MACHINES

The Horological Machine Collection from MB&F was officially launched in June 2007, two years after Max Büsser established the brand, with the unveiling of Horological Machine N°1. More than a timepiece, Horological Machine N°1 was a statement of intent. The intent to express the core ethos of MB&F, voice its values, starting with the desire for an unprecedented openness and transparency in communicating about MB&F's partners — the Friends that are the essential other half of Maximilian Büsser & Friends. Second was the blend of modern engineering with traditional haute horologerie, a blend that brought together the talents of movement constructor Laurent Besse (of modular fame) and watchmaker Peter Speake (then known as Peter Speake-Marin). It also referenced Max's own story as a micro-technology engineer with a passion for watches.

Of particular note in the Horological Machine Collection is Horological Machine N°3 (2009, Jean-Marc Wiederrecht), the Machine most beloved by the MB&F community and still the Machine most closely associated with MB&F in the popular mind. A reputation that persists even though HM3 officially ended production in 2015 with the HM3 MegaWind Final Edition. It made a surprise comeback in 2020 with the HM3 FrogX anniversary limited edition and stands as the most reinterpreted Machine in MB&F history (original Starcruiser and Sidewinder, various Frog versions, ReBel, MoonMachine, JwlryMachine, MegaWind).

Since the launch of the Legacy Machine collection (2011), the Horological Machine collection has tightened its focus on expressive design, with the Legacy Machines veering more toward pure watchmaking innovation. The aesthetic codes of the Horological Machine collection include the battle-axe motif that appears on winding crowns and rotors, the emblematic MB&F style of Arabic numerals, and—intangible but no less material—a sense of fun that reflects the unofficial mantra of MB&F: "A creative adult is a child who survived."

Sub-divisions within the Horological Machine Collection nod to phases of thematic development within MB&F: Automotive (HM5, HMX, HM8), Biomorphic (HM6, HM9) and Zoomorphic (HM3 Frog, HM7, HM9).

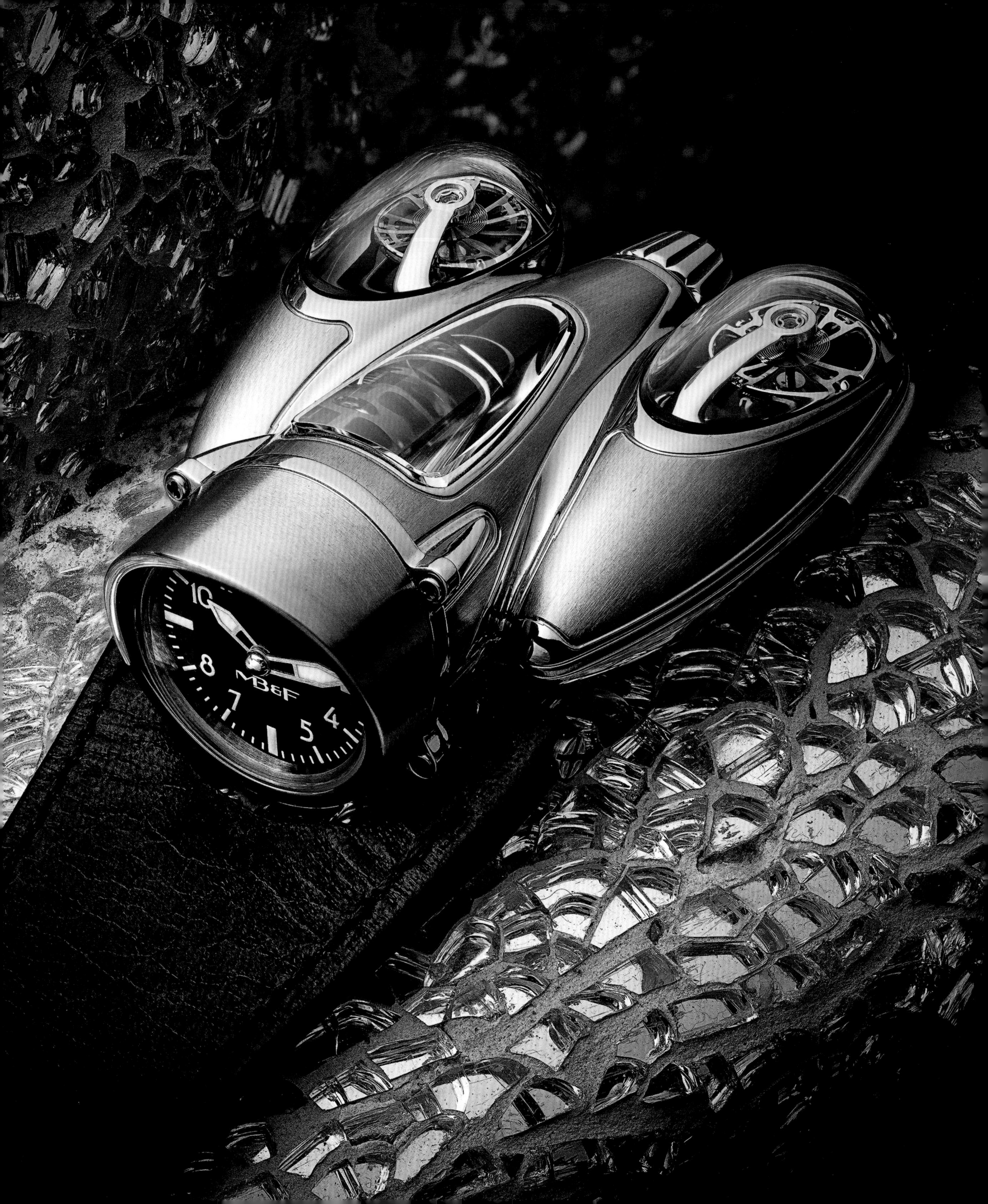
MB&F

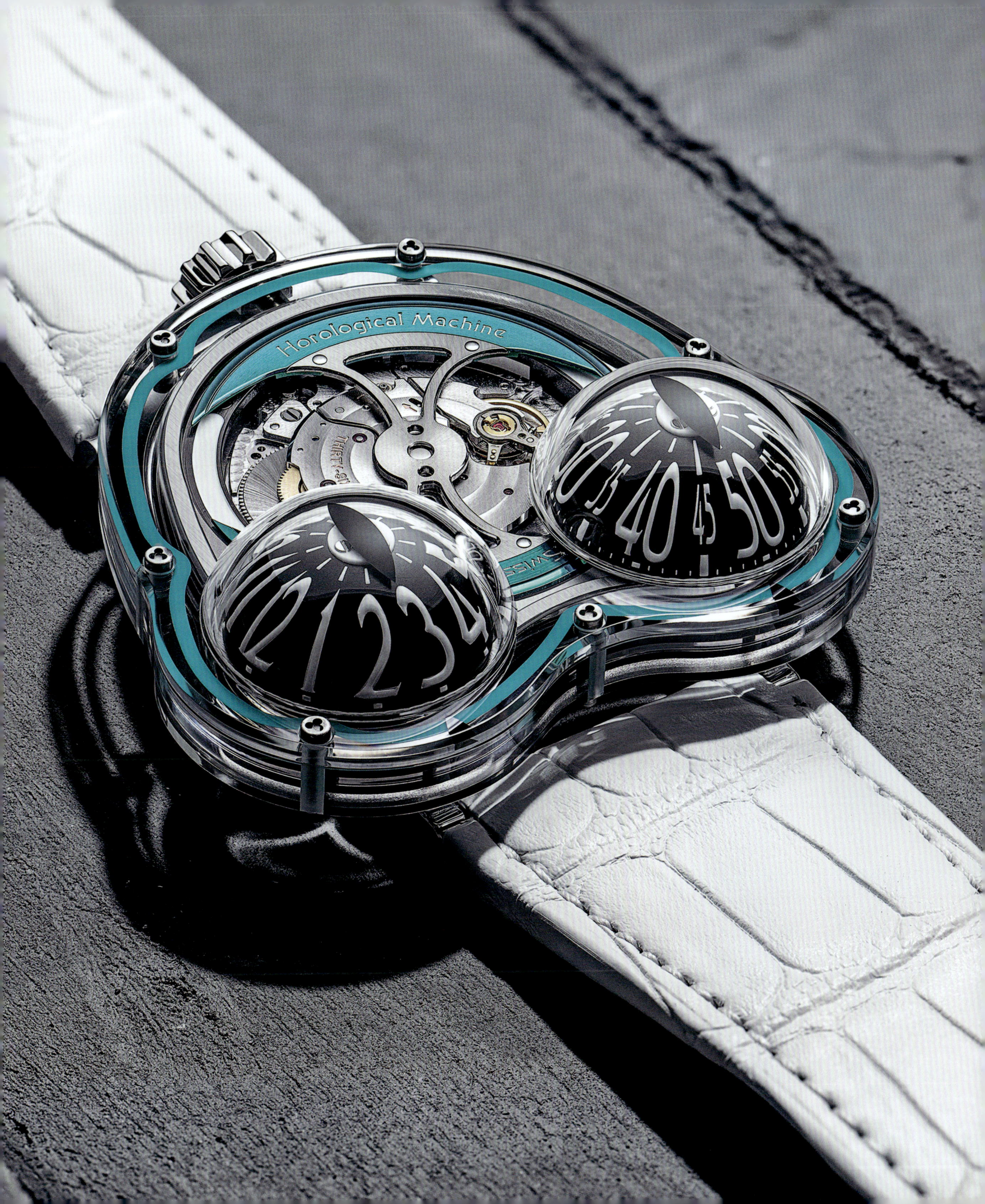

Horological Machine

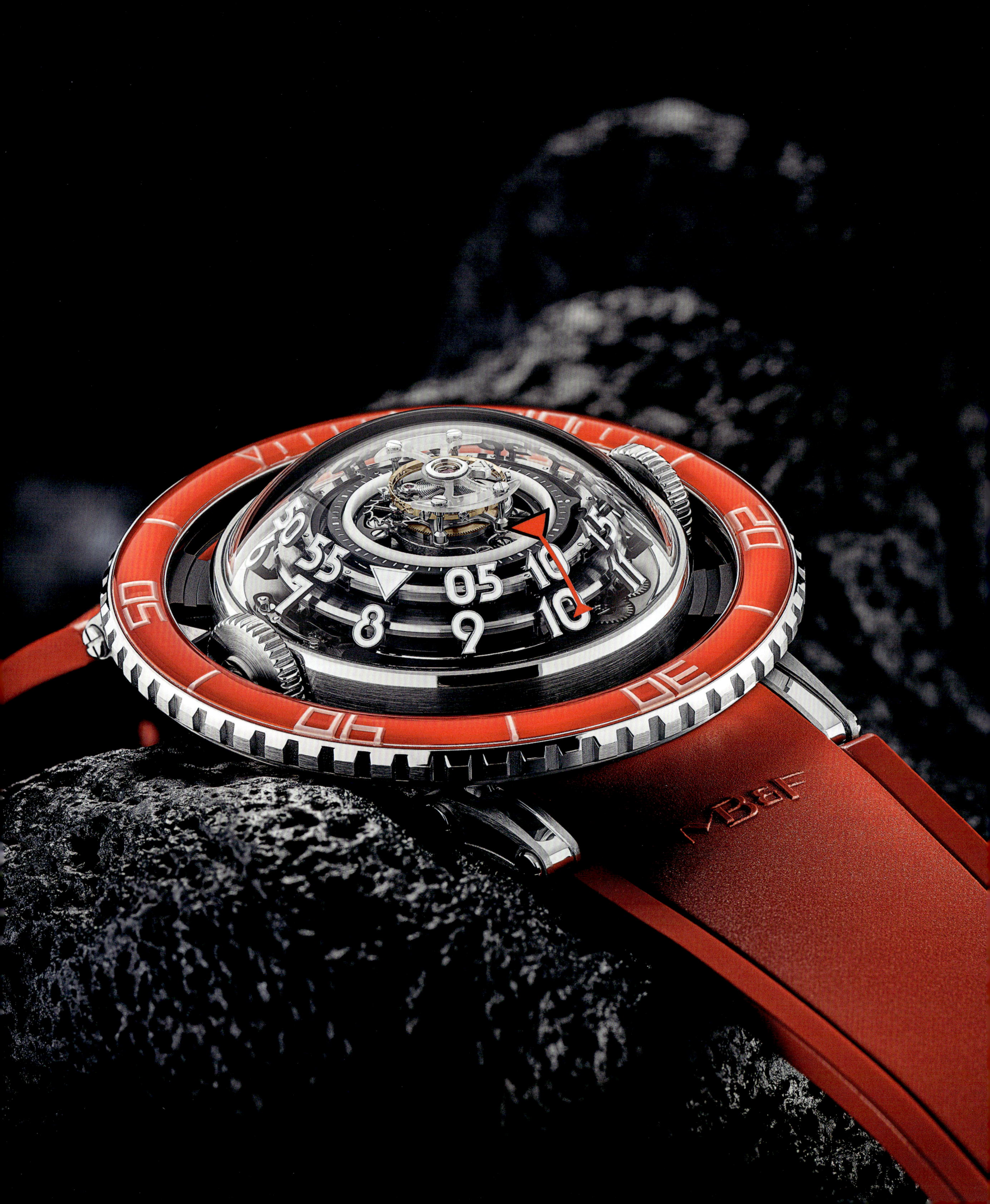
MB&F

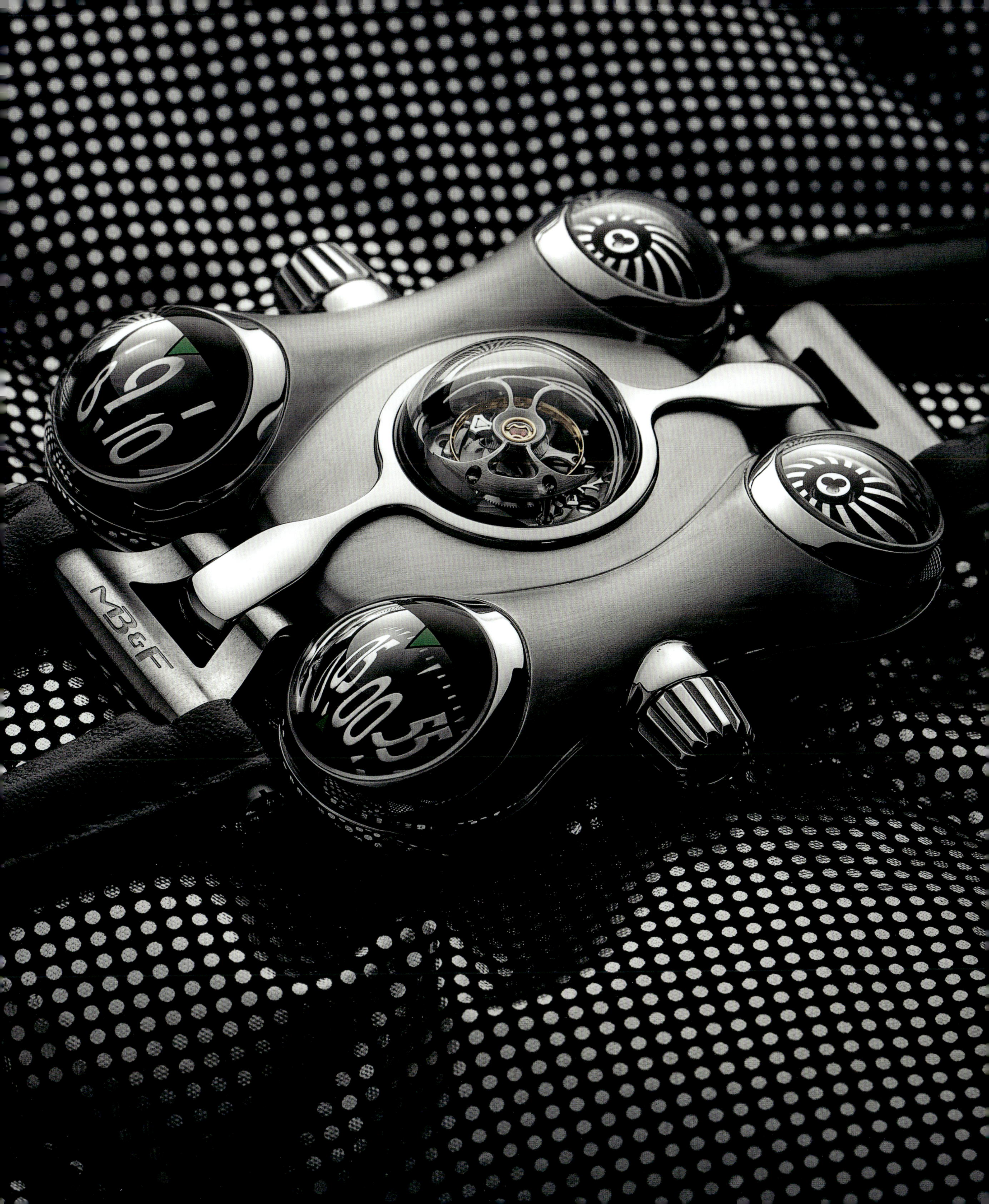

HM1

Art and engineering played equal roles in the creation of Horological Machine N°1, first launched in 2007 – an innovative, sophisticated, three-dimensional timepiece, liberated from the constraints of traditional horology. The HM1 was a complete original, unlike anything seen before in watchmaking.

Displaying hours and minutes on separate dials (hours to the left, and minutes to the right), the HM1 was as impressive technically as it was visually exciting. Its unusual form and volume capture the senses and draw the eye closer for further examination.

The figure-of-eight shape of the HM1 dictates the design of the movement. Each of its twin dials is powered by a pair of massive mainspring barrels, separated by a central, raised one-minute tourbillon, looking almost like the machine's third eye at the centre of the dial.

HOROLOGICAL MACHINE Nº1

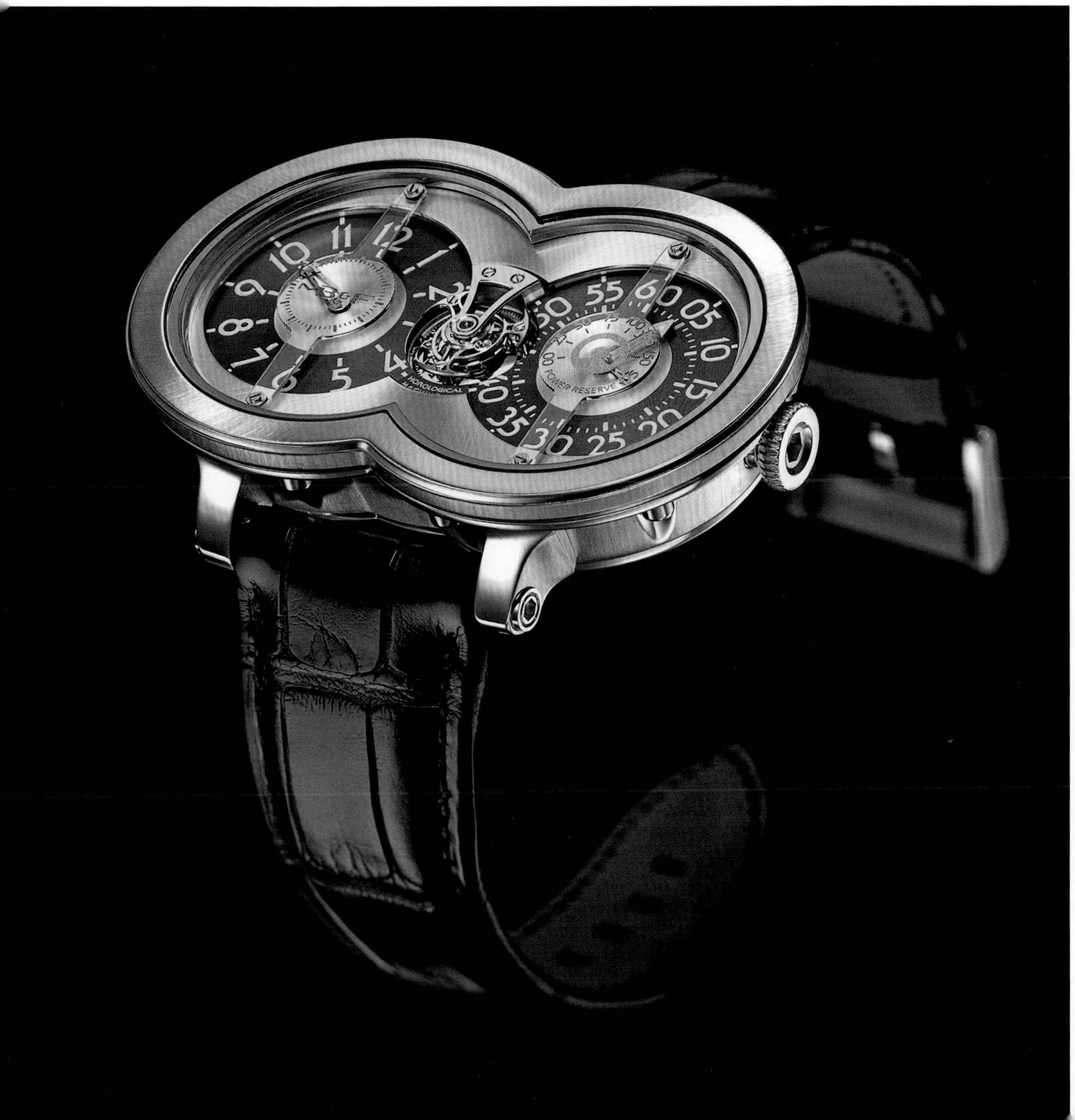

HM1 CALIBRE

Maximilian Büsser's concept sketches were painstakingly worked into physical reality by designer Eric Giroud working in partnership with movement engineer Laurent Besse and independent watchmaker Peter Speake-Marin. The HM1 draws on Besse's engineering skills and Speake-Marin's expertise in Haute Horlogerie to deliver a result that appears to be born of classical horology and science-fiction alike — somewhere between Jules Verne's 20,000 Leagues Under The Sea and Stanley Kubrick's 2001: A Space Odyssey. With 376 parts and 81 jewels, the movement of the HM1 was the first to transmit energy to its regulating system from two sources simultaneously. The four barrels, connected in both parallel and series (another first), reduce the torque applied to each mainspring while improving isochronism and increasing longevity.
A central one-minute tourbillon is already a complicated mechanism when traditionally positioned (in line with the dial inside the movement). But the three-dimensional character of the HM1 demanded something more. Hence its elevated tourbillon – which now takes centre stage.

Collection
Horological Machines
Series
Horological Machine N°1
Common name
HM1
Movement
Raised central one-minute tourbillon developed for MB&F by Peter Speake-Marin and Laurent Besse
Winding
Automatic

Power reserve
7 days
Components
376/81 jewels
Functions
Hours and minutes on separate dials
Add. features
22K gold "battle-axe" winding rotor, four mainspring barrels
Case
18K white gold, 18K red gold or black PVD, 18K white gold and titanium

Dimensions
41 × 64 × 14 mm
Case components
48
Water resistance
30 m/3 ATM
Strap
Alligator
Buckle
Folding buckle, 18K white gold, 18K red gold or titanium matching the case
Introduction
2007

Years in production
2007-2011
Quantity produced
95 + 3 prototypes

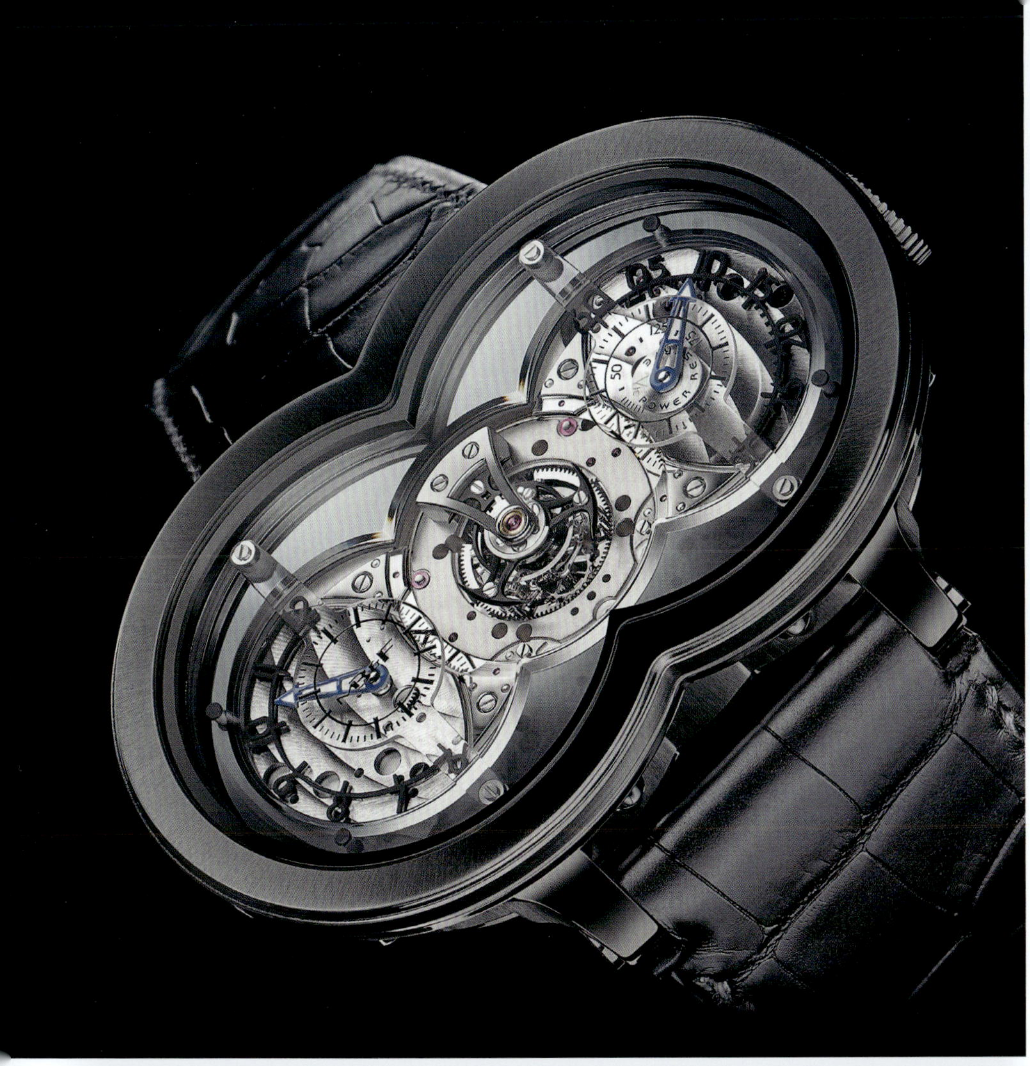

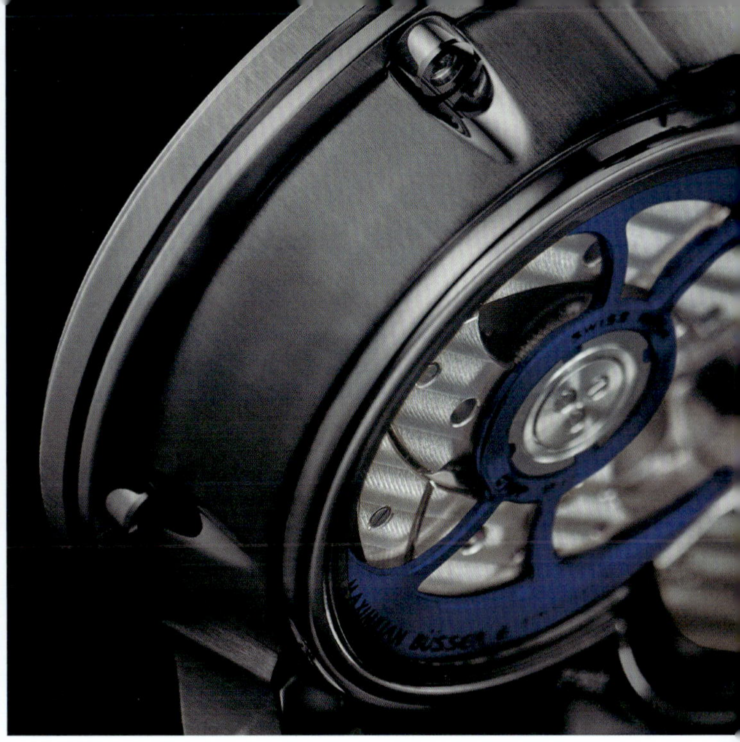

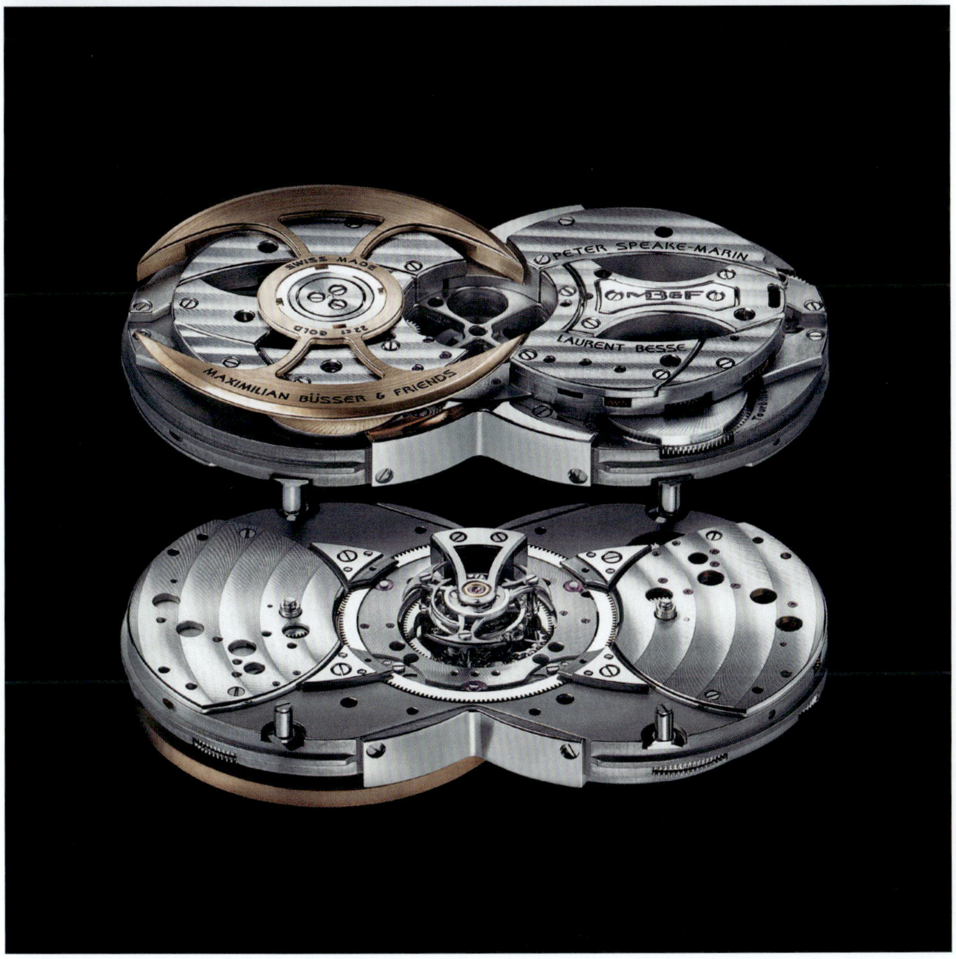
SWISS MADE
22K GOLD
MAXIMILIAN BÜSSER & FRIENDS
PETER SPEAKE-MARIN
MB&F
LAURENT BESSE

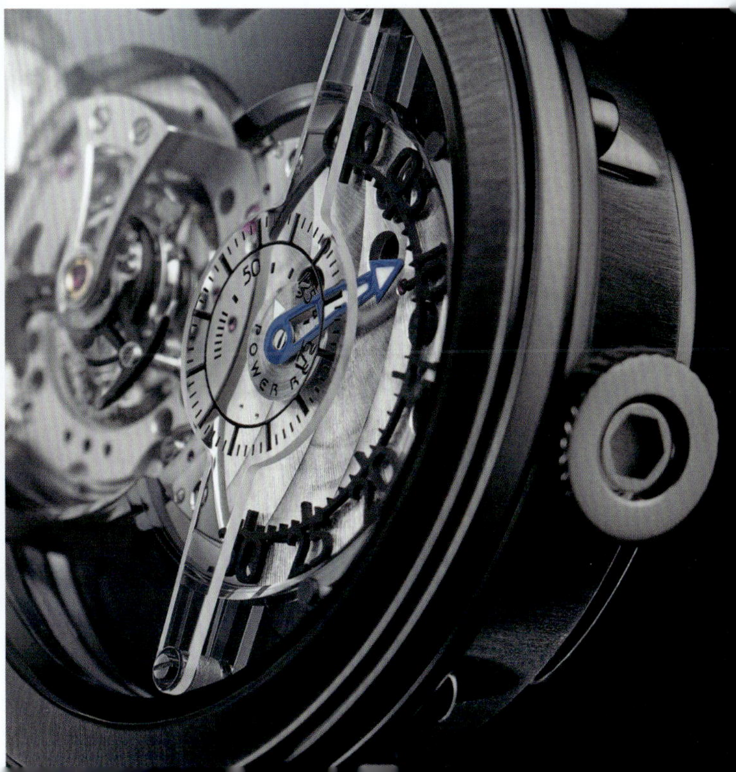
POWER RESERVE

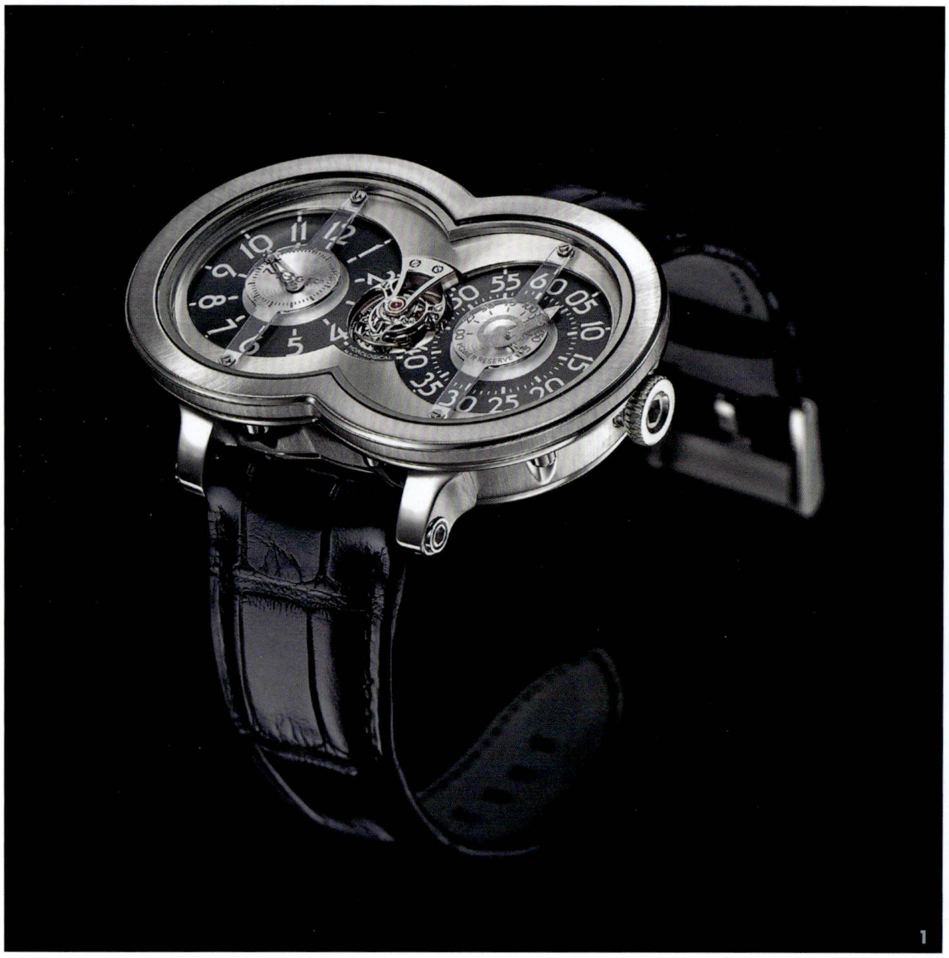

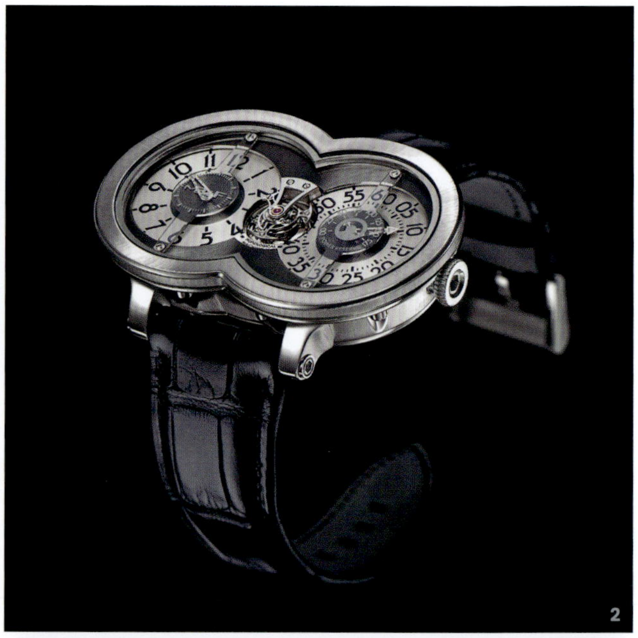

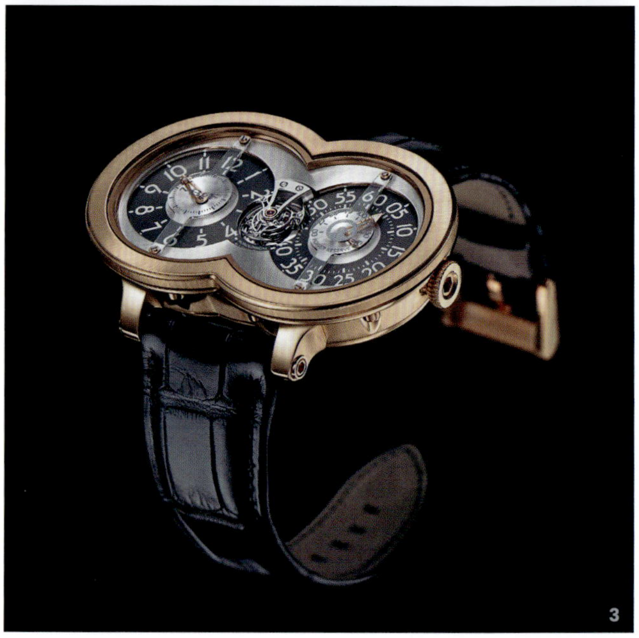

1 Horological Machine N°1
WG, silver dial
Reference: 10.T41WL.S
Material: 18K white gold with silver and ruthenium dials
Limitation: –
Introduction: 2007
Year in production: 2007
Quantity produced: 7

2 Horological Machine N°1
WG, ruthenium dial
Reference: 10.T41WL.R
Material: 18K white gold with silver and ruthenium dials
Limitation: –
Introduction: 2007
Years in production: 2007–2009
Quantity produced: 20

3 Horological Machine N°1
RG, silver dial
Reference: 10.T41RL.S
Material: 18K red gold with silver and ruthenium dials
Limitation: –
Introduction: 2007
Years in production: 2007–2009
Quantity produced: 25

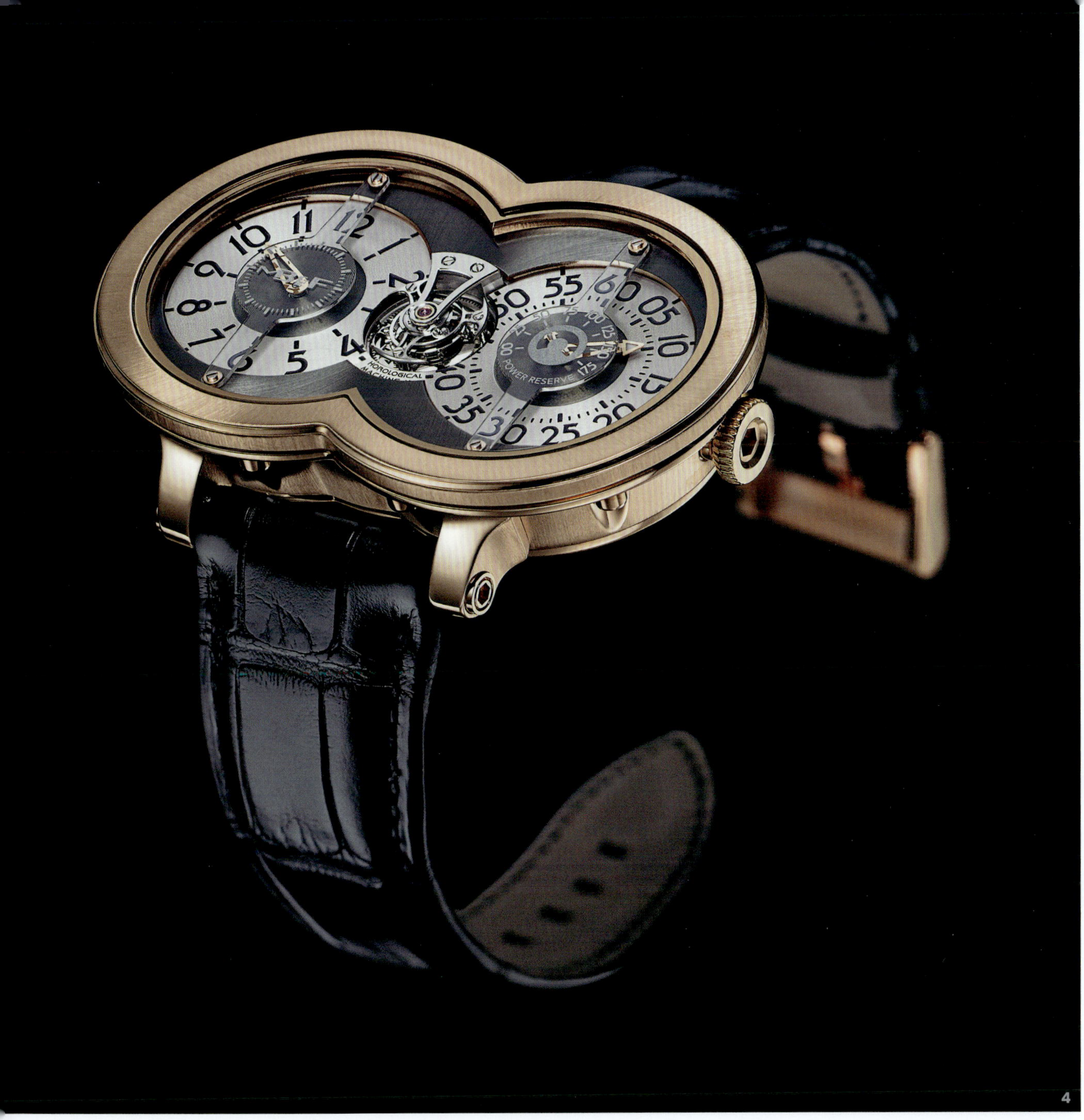

4

4 **Horological Machine N°1**
RG, ruthenium dial
Reference: 10.T41RL.R
Material: 18K red gold with silver and ruthenium dials
Limitation: –
Introduction: 2007
Years in production: 2007–2009
Quantity produced: 8

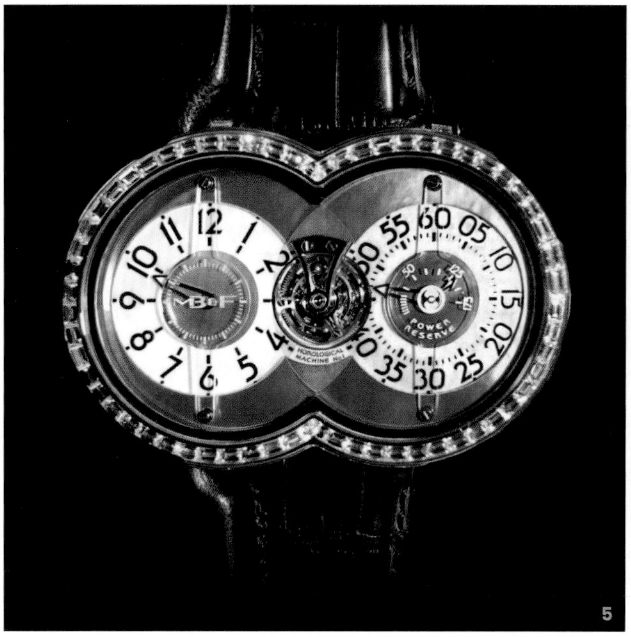

5

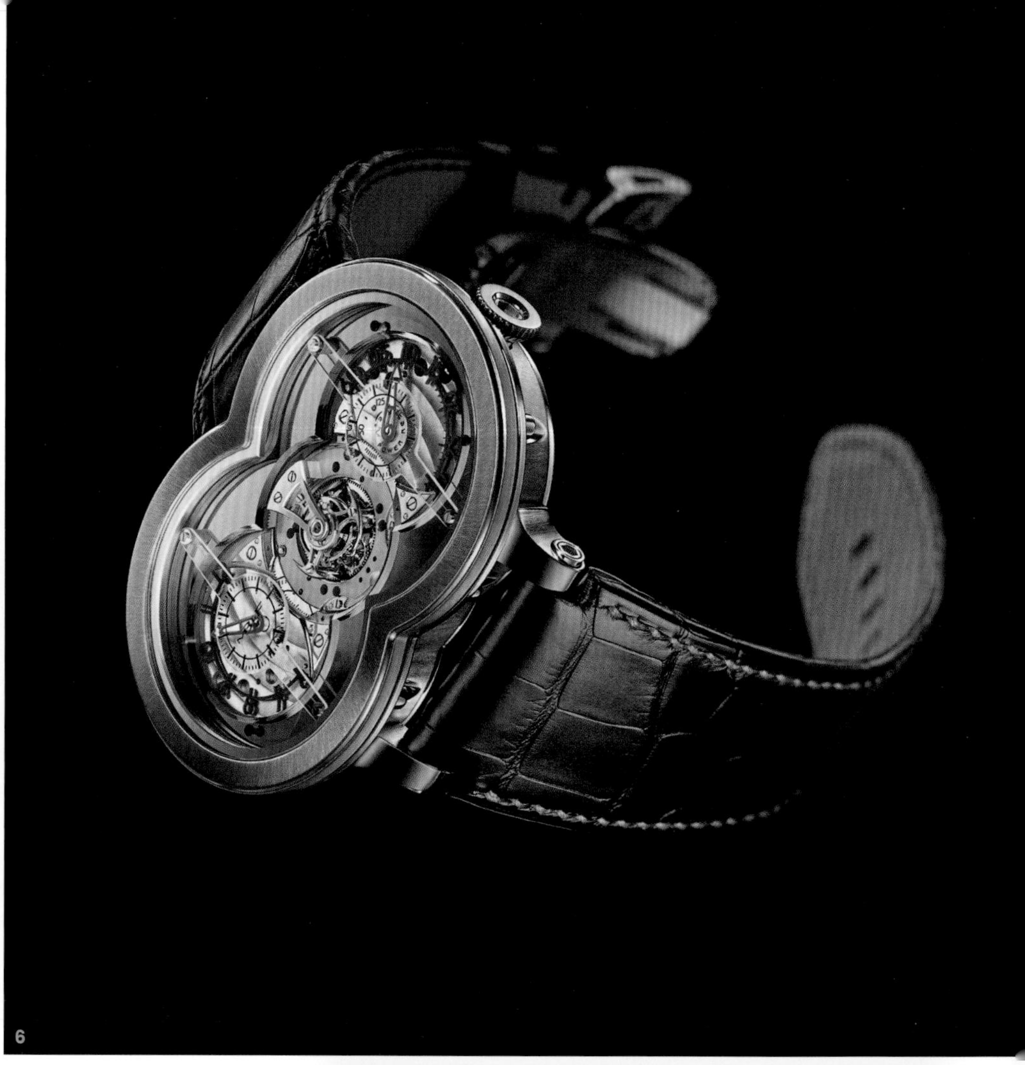

6

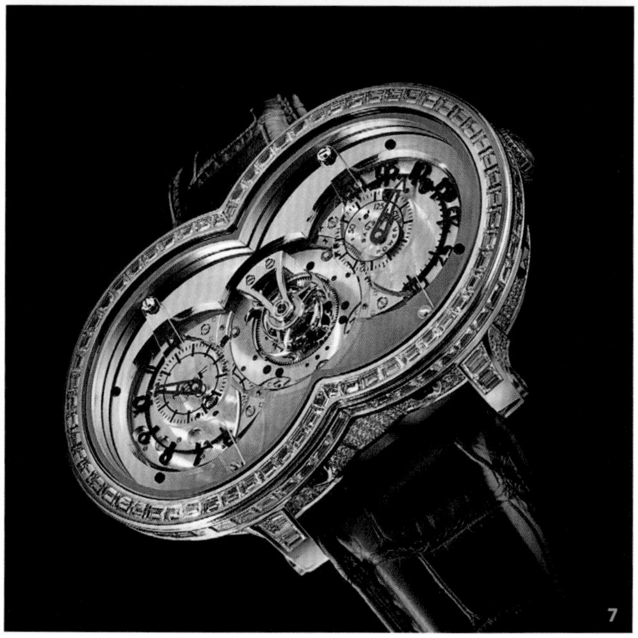

7

5 Horological Machine N°1
WG full diamond set,
Mother-of-pearl dial
Reference: 10.T41WL.M/BD
Material: 18K white gold case
fully set with diamonds,
mother-of-pearl dial
Limitation: 2
Introduction: 2008
Years in production: 2008-2009
Quantity produced: 2

6 Horological Machine N°1 Ti
Reference: 10.T41T
Material: grade 5 titanium
with sapphire crystal dial
Limitation: 10
Introduction: 2008
Years in production:
2008-2009
Quantity produced:
10 + 1 prototype

7 Horological Machine N°1
WG full diamond set, sapphire
crystal dial
Reference: 10.T41WL.O/BD
Material: 18K white gold case
fully set with diamonds,
sapphire crystal dial
Limitation: unique piece
Introduction: 2008
Year in production: 2008
Quantity produced: 1

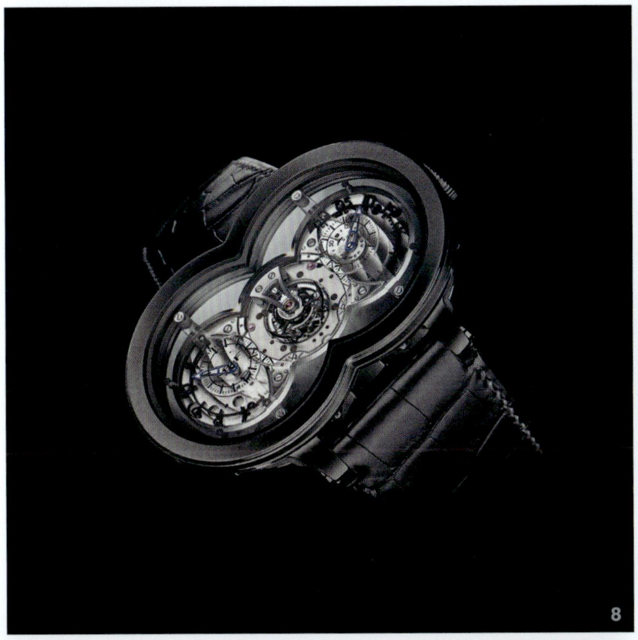

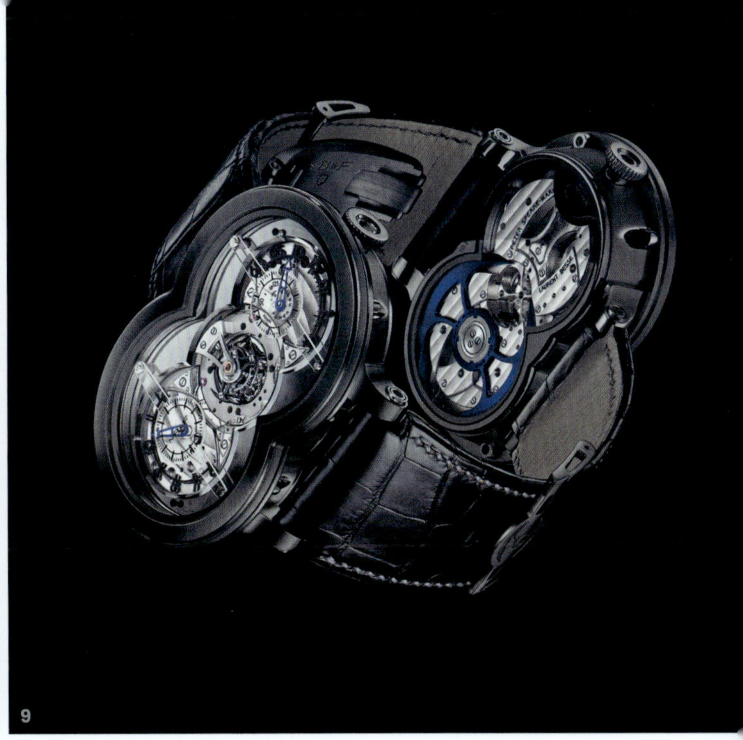

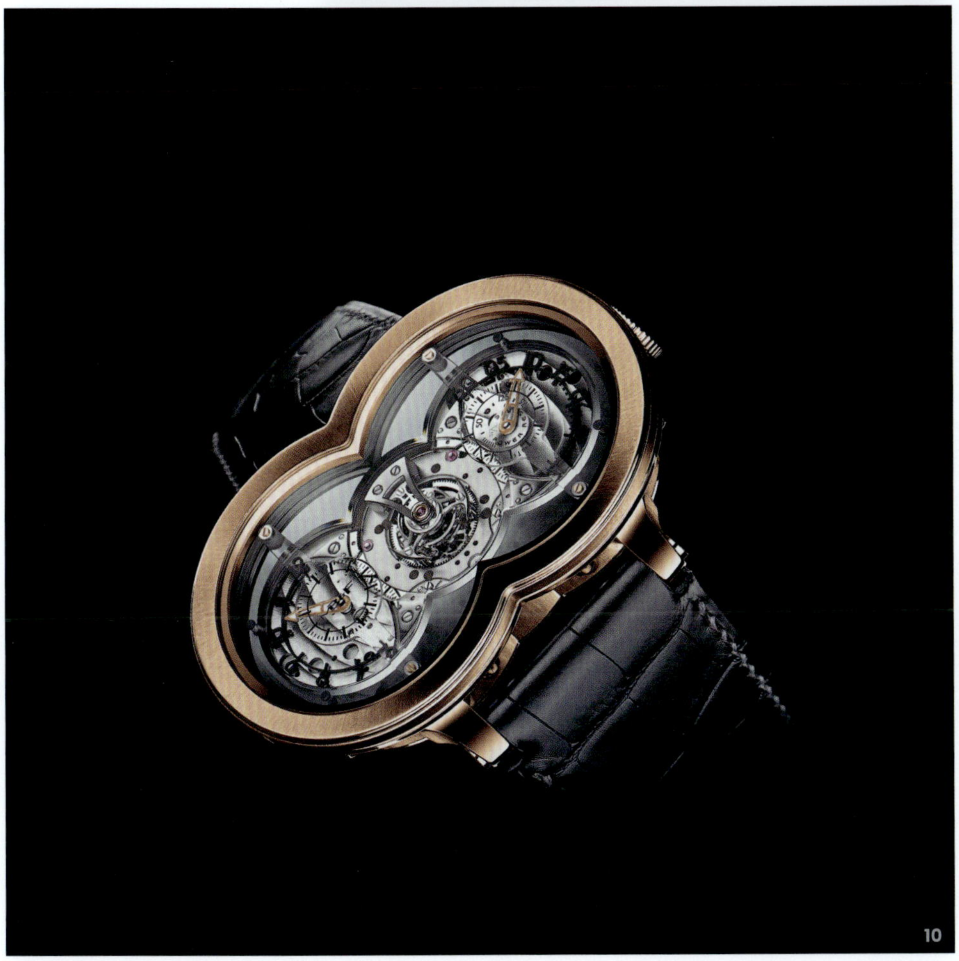

8 Horological Machine N°1
Ti Black
Reference: 10.T41TBL.O
Material: black PVD coated grade 5 titanium with sapphire crystal dial
Limitation: unique piece
Introduction: 2009
Year in production: 2009
Quantity produced: 1

9 Horological Machine N°1
Black
Reference: 10.T41WBL.O
Material: black PVD coated 18K white gold with sapphire crystal dial
Limitation: 10
Introduction: 2009
Years in production: 2009–2011
Quantity produced: 10 + 1 prototype

10 Horological Machine N°1 RS
Reference: 10.T41RL.O
Material: 18K red gold with sapphire crystal dial
Limitation: 10
Introduction: 2009
Years in production: 2009–2011
Quantity produced: 9 + 1 prototype

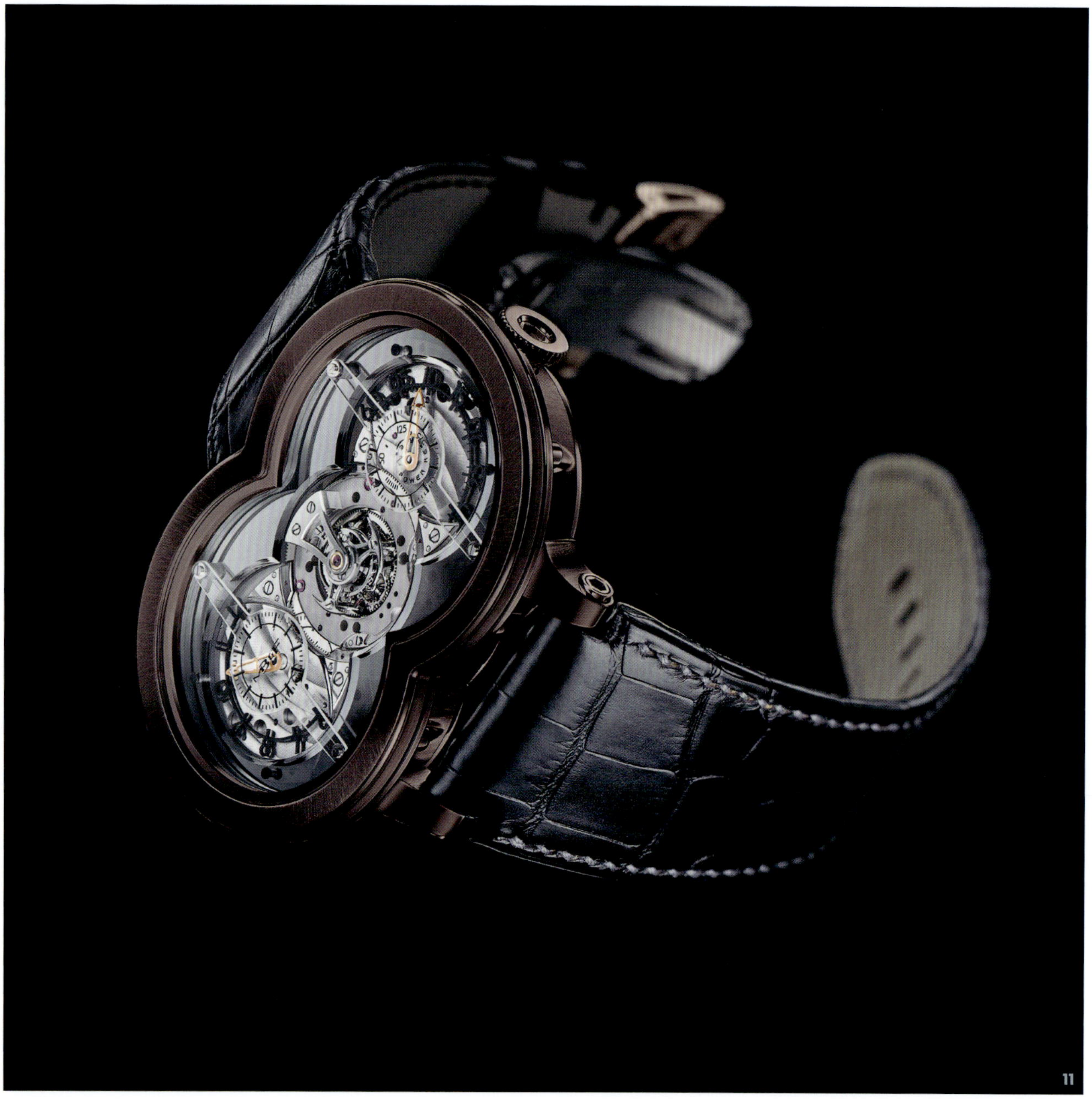

11 Horological Machine N°1
Chocolate
Reference: 10.T41TCL.O
Material: chocolate PVD coated
grade 5 titanium with
sapphire crystal dial
Limitation: unique piece
Introduction: 2010
Year in production: 2010
Quantity produced: 1

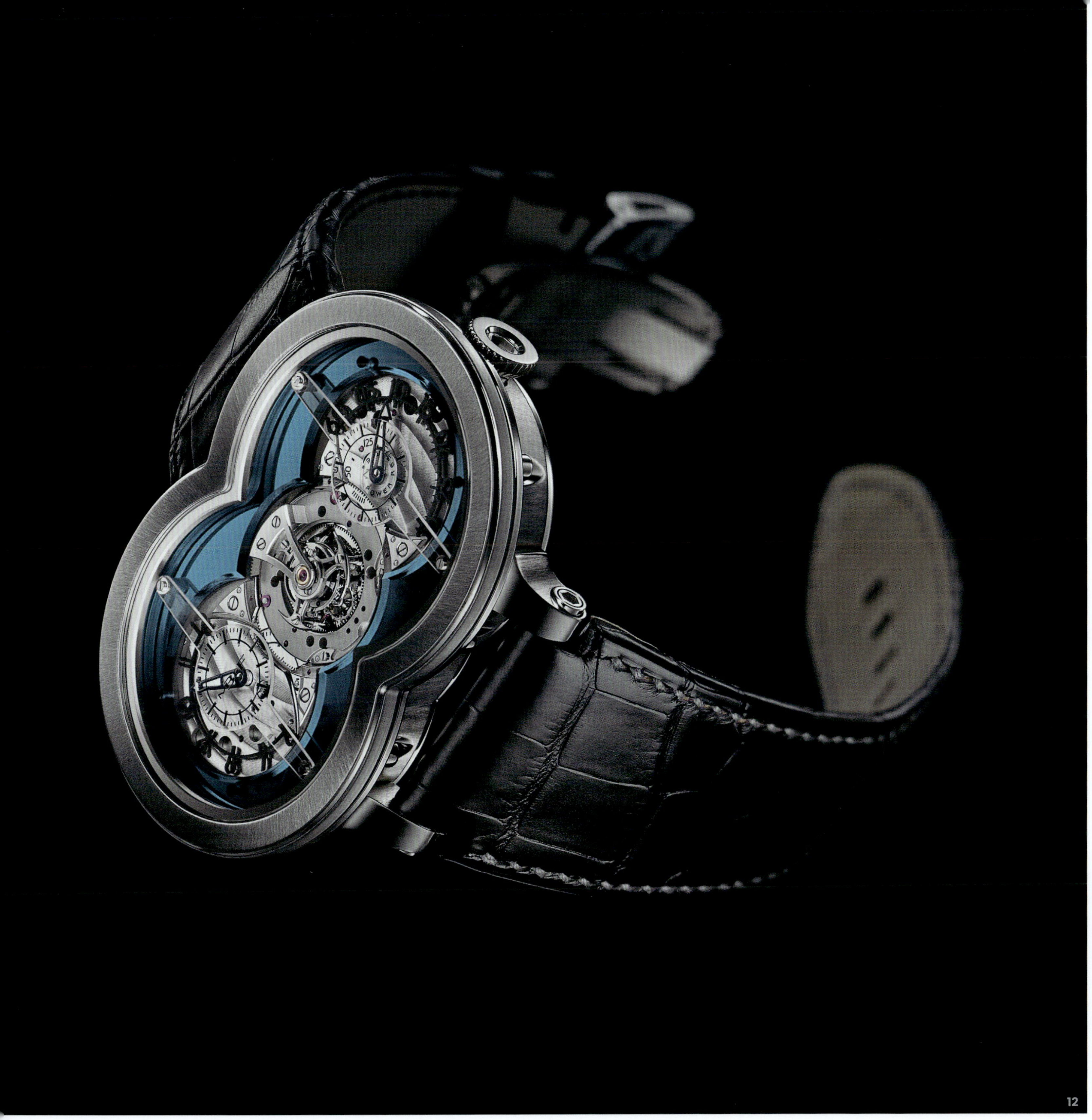

12

12 Horological Machine N°1 WG
White and Blue
Reference: 10.T41WL.OB
Material: 18K white gold
with blue sapphire
crystal dial
Limitation: unique piece
Introduction: 2010
Year in production: 2010
Quantity produced: 1

HM 2

Rectangular shape, flying buttresses, twin dials, modular construction... When first launched in 2008, Horological Machine N°2 was considered a revolution in the world of Haute Horlogerie. Inspired by the science-fiction comics of his youth, Maximilian Büsser came up with a machine that resembled the space stations he had dreamt about as a child. Whether on the moon or on Mars, these were places where space colonists lived under giant glass domes attached to the planetary surface. It is these same domes that would inspire Büsser to design the domed dials of the HM2. The twin dials show bi-hemisphere moon phase and date on the left dial with hours and minutes on the right dial, and are covered by clear sapphire crystal. The case employs more than 100 components, and is bolted together in a modular manner inspired by the Meccano sets of Büsser's childhood. This simplifies future refurbishment as each piece may be individually removed, serviced, and then refitted, increasing the life of the machine. The first cases were white gold and titanium, and red gold and titanium, followed later by ceramic with red gold, ceramic with titanium, clear sapphire with black PVD titanium, and clear sapphire with red gold.

HOROLOGICAL MACHINE Nº2

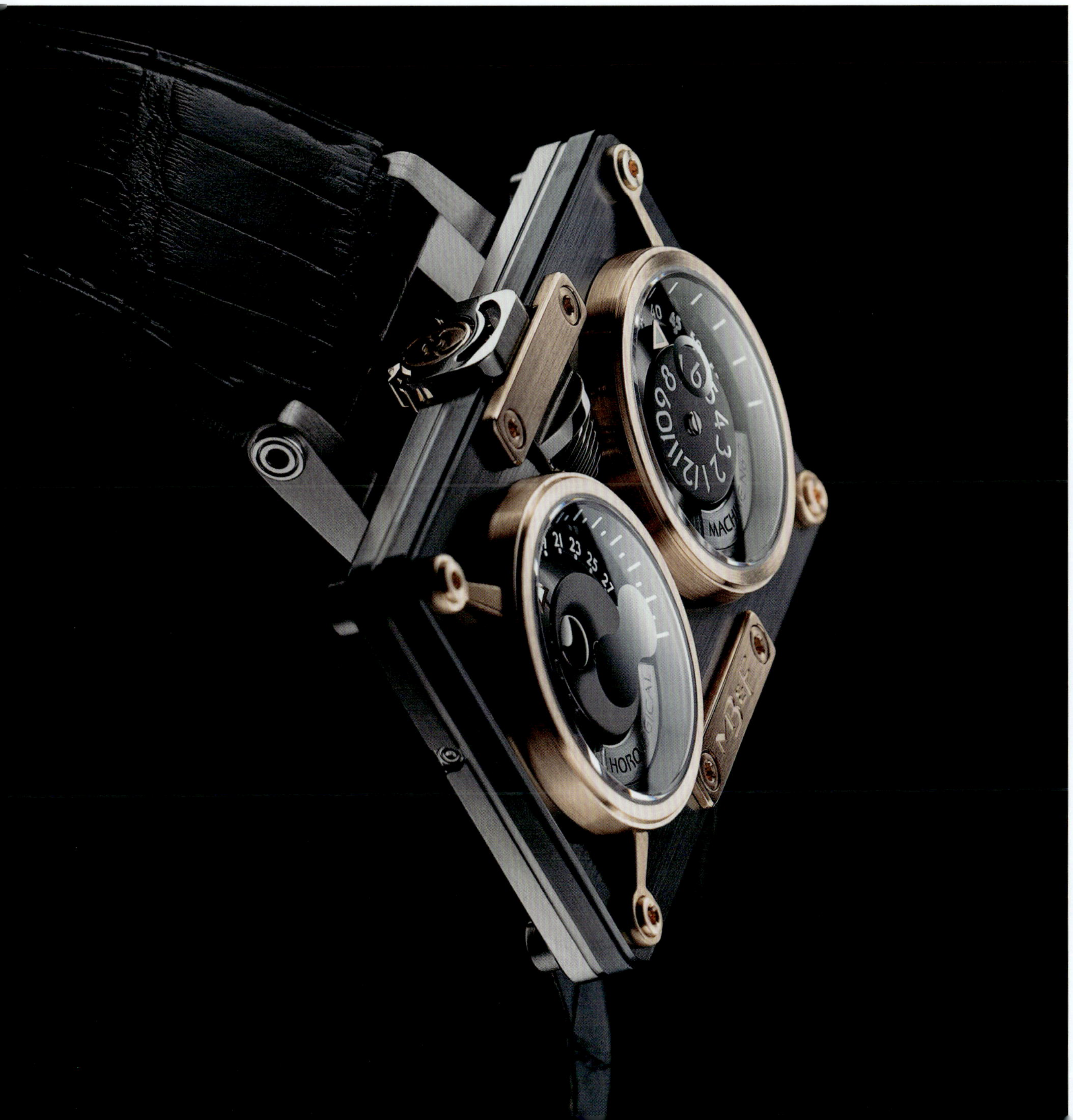

HM2 CALIBRE

The engine of the HM2, developed by master watchmaker Jean-Marc Wiederrecht, features the world's first mechanical movement to include an instantaneous jump hour, concentric retro grade minutes, retrograde date, a bi-hemisphere moon phase and automatic winding.
The rotor is the iconic MB&F battle-axe, optimized for the HM2 with one of its 22-karat gold blades machined to a remarkable, razor-sharp thickness of just 0.2mm. At first glance, the HM2 has the look of a 21st century machine, but the visible quality and careful hand execution of the movement plainly belong to the traditions of Haute Horlogerie.

Collection
Horological Machines
Series
Horological Machine N°2
Common name
HM2
Movement
Developed for MB&F by Jean-Marc Wiederrecht (Agenhor), powered by a Girard-Perregaux base

Winding
Automatic
Power reserve
45 hours
Components
349/44 jewels
Functions
Left dial: retrograde date and bi-hemisphere moon phase; right dial: jumping hours and concentric retrograde minutes

Add. features
22K gold "battle-axe" winding rotor
Case
18K white gold, 18K red gold, grade 5 titanium, ceramic or sapphire crystal
Dimensions
59 × 38 × 13 mm
Components
120
Water resistance
30 m/3 ATM

Strap
Alligator
Buckle
Folding and pin buckle, titanium, 18K white or red gold matching the case
Introduction
2008
Years in production
2008–2020
Quantity produced
217+ 2 prototypes

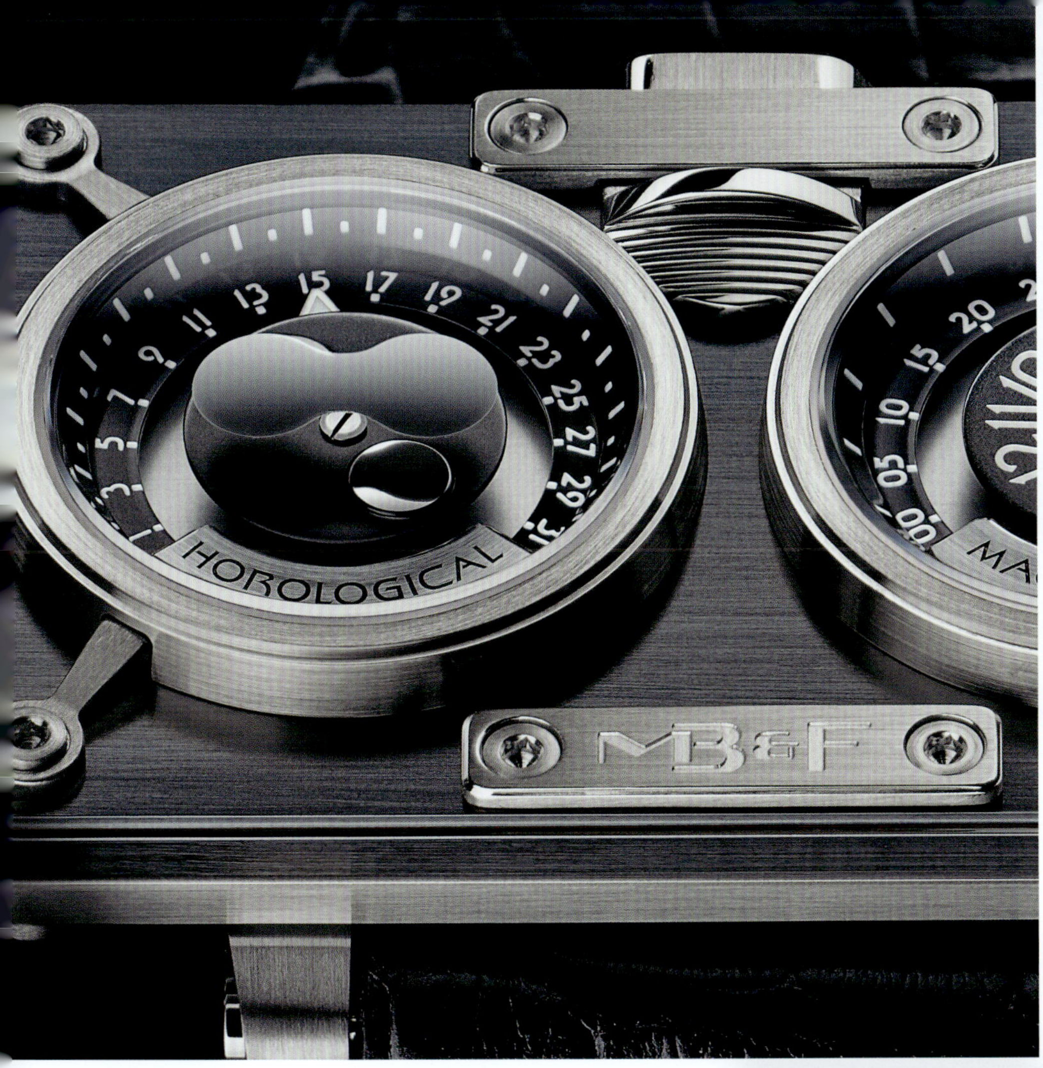
HOROLOGICAL
MB&F

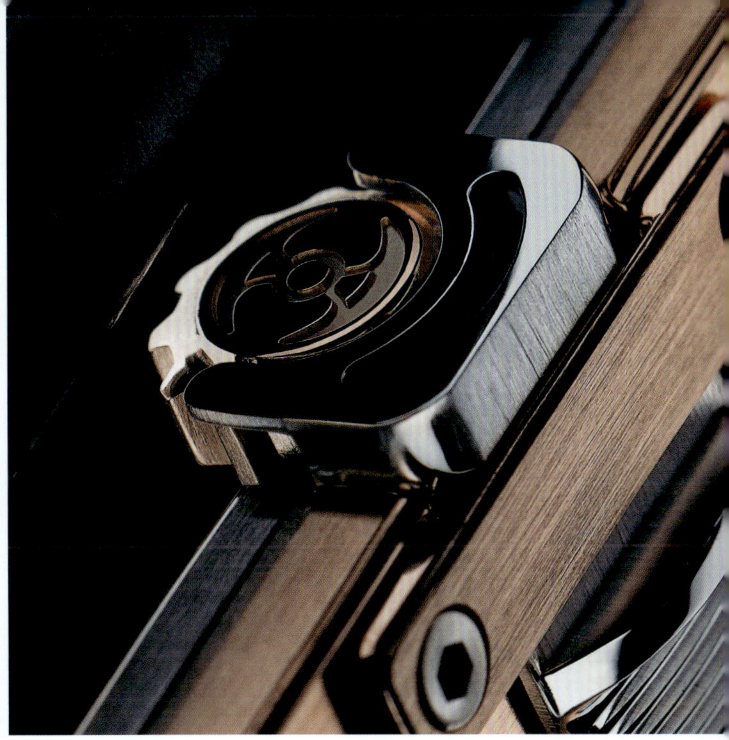

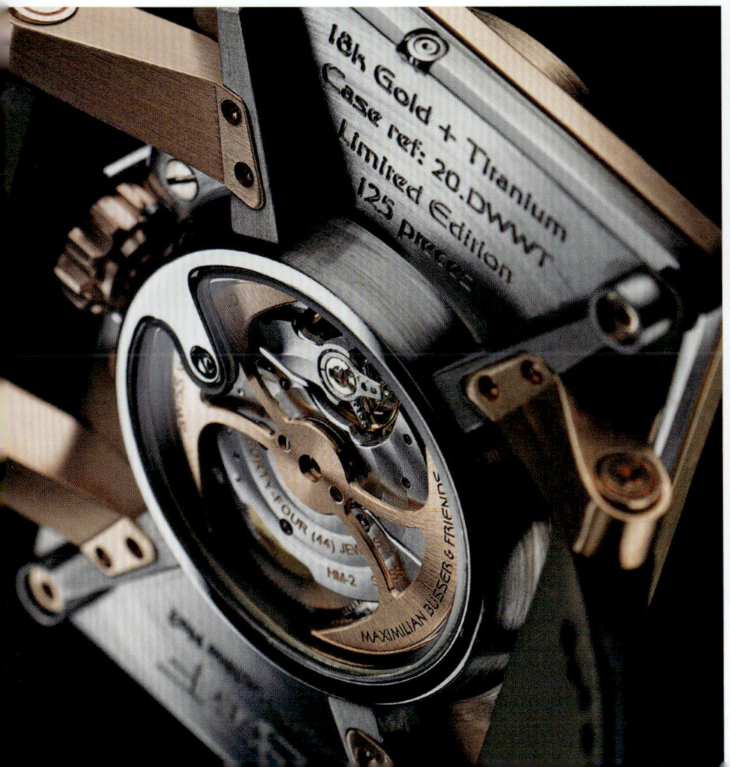
18k Gold + Titanium
Case ref: 20.DWWT
Limited Edition
125 pieces
MAXIMILIAN BUSSER & FRIENDS

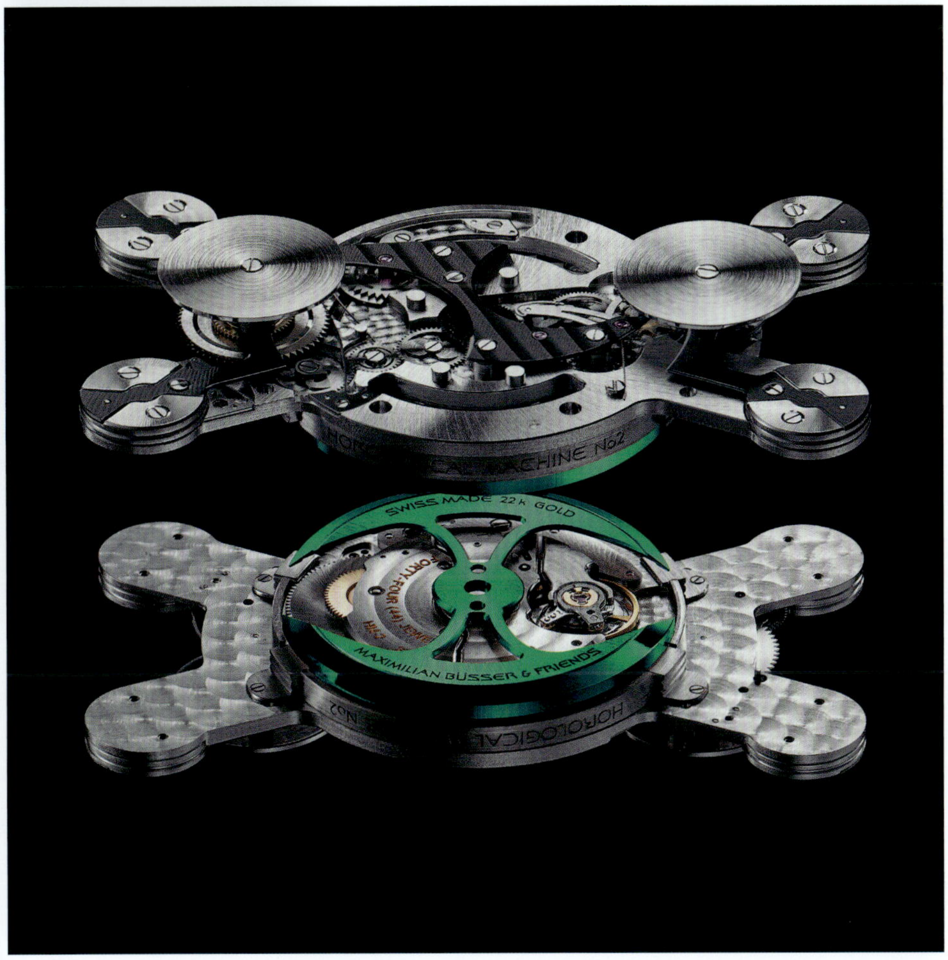
SWISS MADE 22k GOLD
MAXIMILIAN BUSSER & FRIENDS

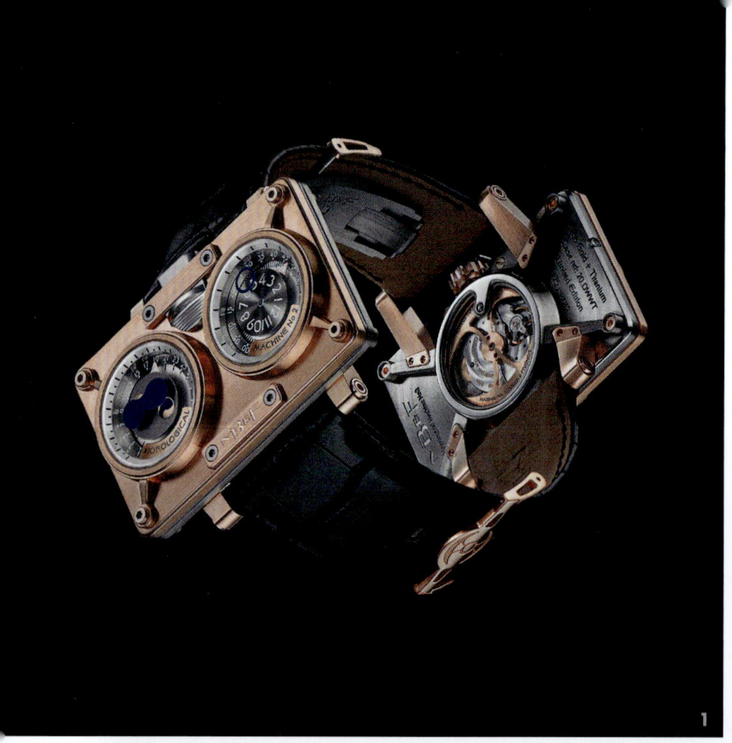

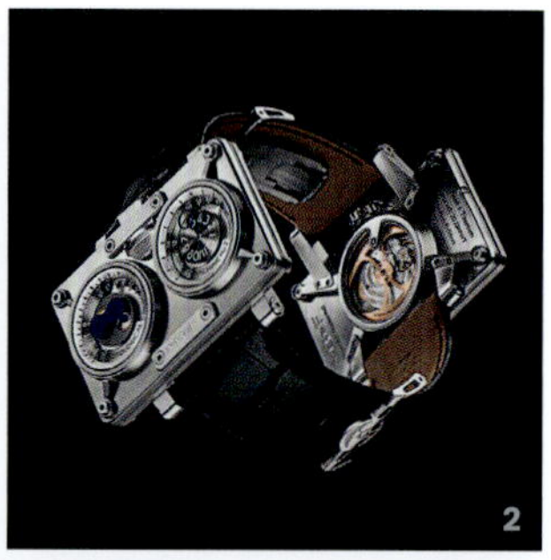

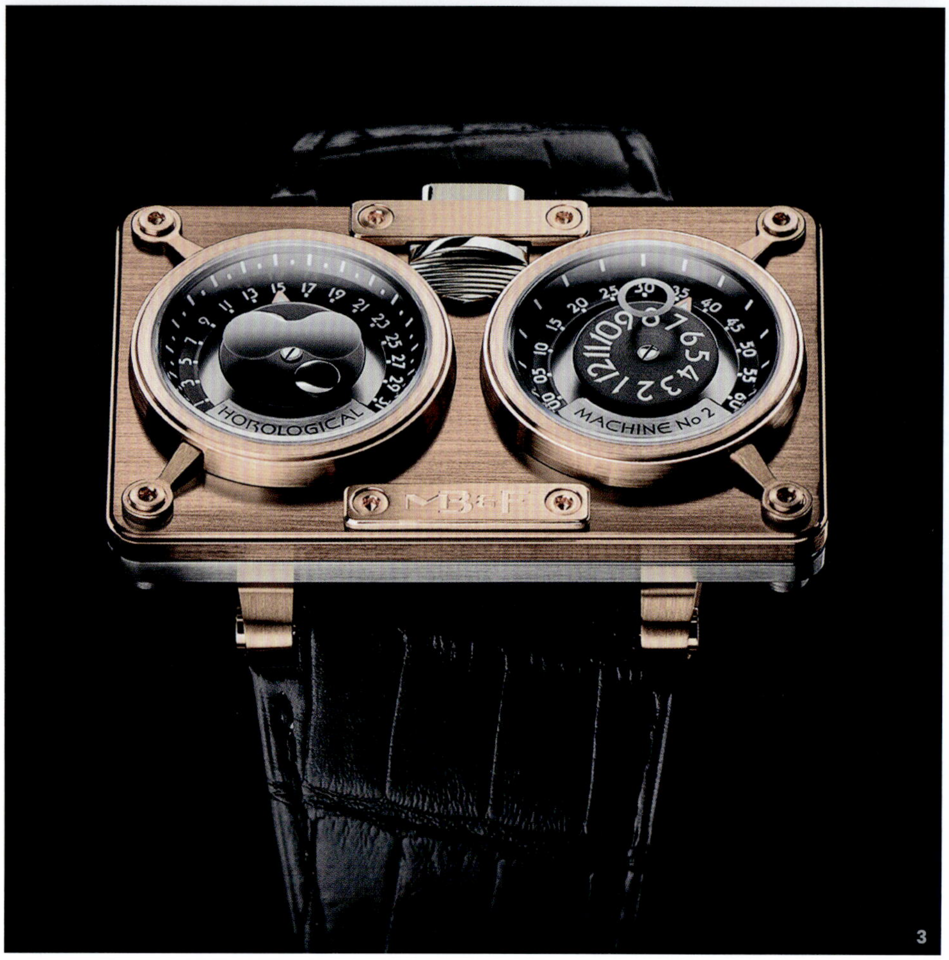

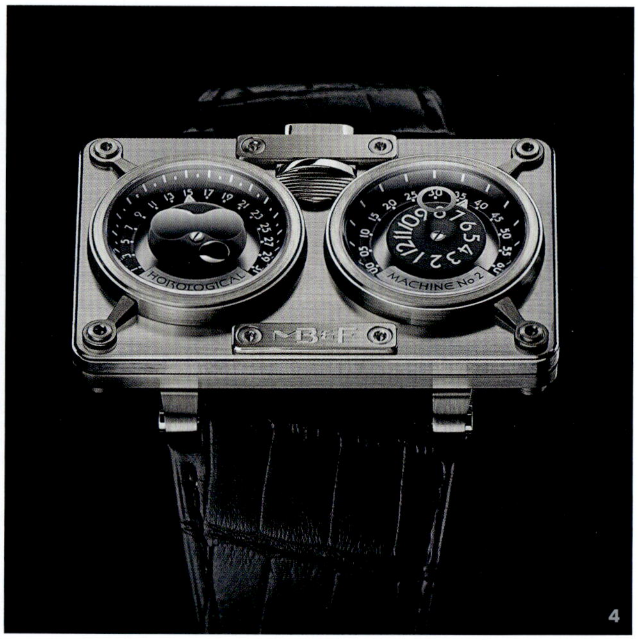

1 Horological Machine N°2 RT
Reference: 20.DRRTL.R
Material: 18K red gold and grade 5 titanium
Limitation: 125 (in total included both 20.DRRTL.R and 20.DRRTL.B)
Introduction: 2008
Year in production: 2008
Quantity produced: 29

2 Horological Machine N°2 WT
Reference: 20.DWWTL.R
Material: 18K white gold and grade 5 titanium
Limitation: 125 (in total included both 20.DWWTL.R and 20. DWWTL.B)
Introduction: 2008
Year in production: 2008
Quantity produced: 26

3 Horological Machine N°2 RT, black dials
Reference: 20.DRRTL.B
Material: 18K red gold and grade 5 titanium
Limitation: 125 (in total included both 20.DRRTL.R and 20.DRRTL.B)
Introduction: 2008
Years in production: 2008–2009
Quantity produced: 17

4 Horological Machine N°2 WT, black dials
Reference: 20.DWWTL.B
Material: 18K white gold and grade 5 titanium
Limitation: 125 (in total included both 20.DWWTL.R and 20. DWWTL.B)
Introduction: 2008
Years in production: 2008–2009
Quantity produced: 14

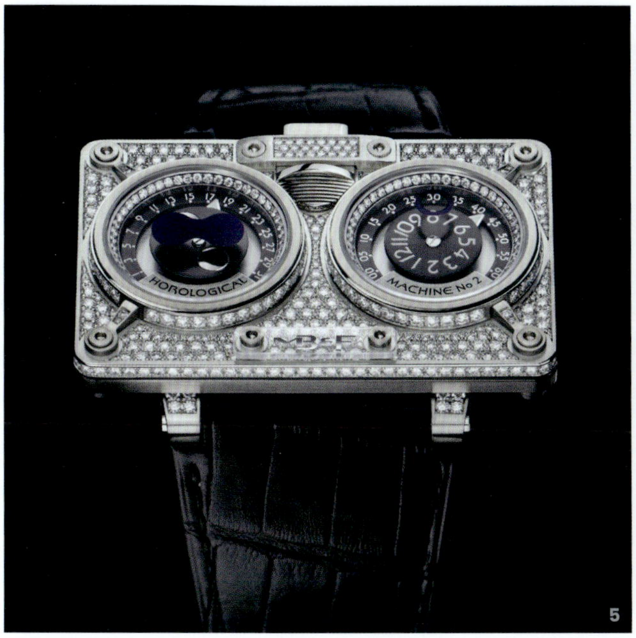

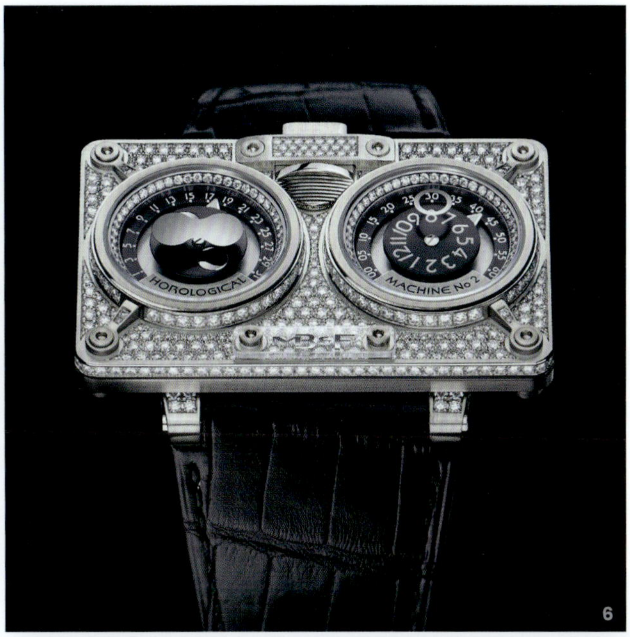

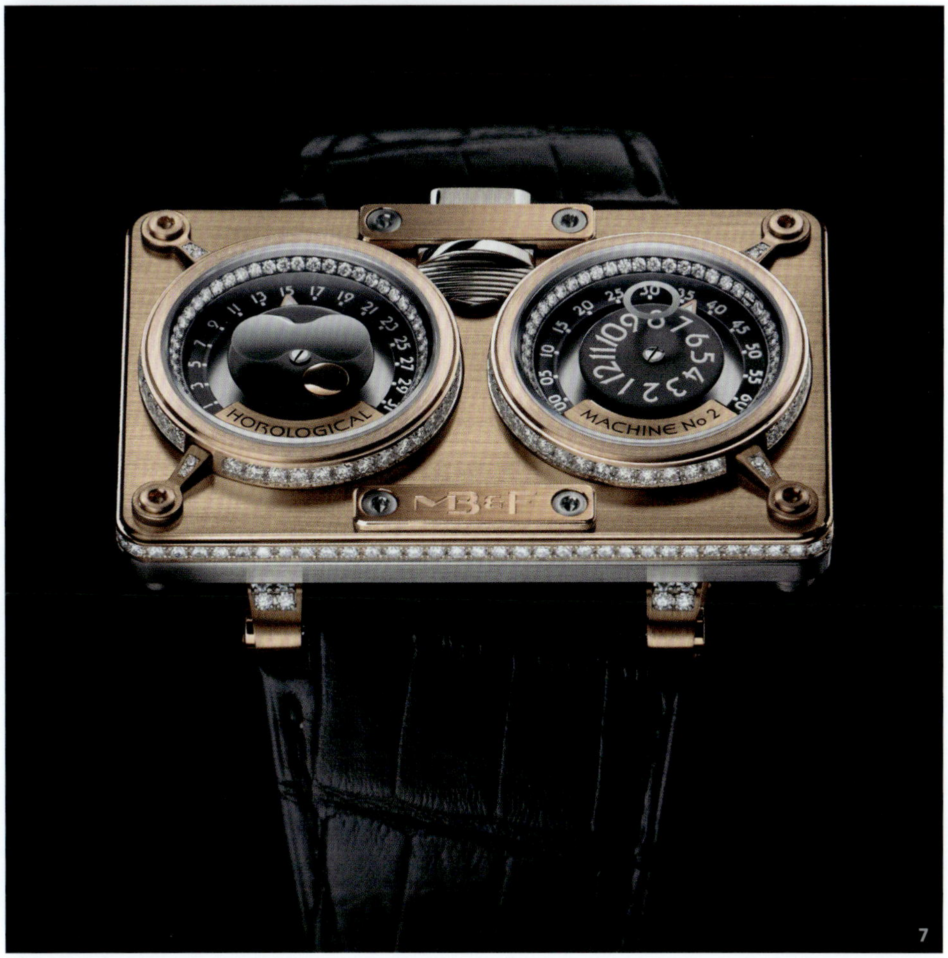

5 **Horological Machine N°2 WT diamonds, grey dials**
Reference: 20.DWWTL.R/D1
Material: 18K white gold and grade 5 titanium full diamond set
Limitation: –
Introduction: 2008
Year in production: 2008
Quantity produced: 6

6 **Horological Machine N°2 WT diamonds, black dials**
Reference: 20.DWWTL.B/D1
Material: 18K white gold and grade 5 titanium full diamond set
Limitation: –
Introduction: 2009
Year in production: 2009
Quantity produced: 2

7 **Horological Machine N°2 RT diamonds**
Reference: 20.DRRTL.R/S1
Material: 18K red gold, grade 5 titanium and partially diamond set
Limitation: 2
Introduction: 2008
Year in production: 2008
Quantity produced: 2

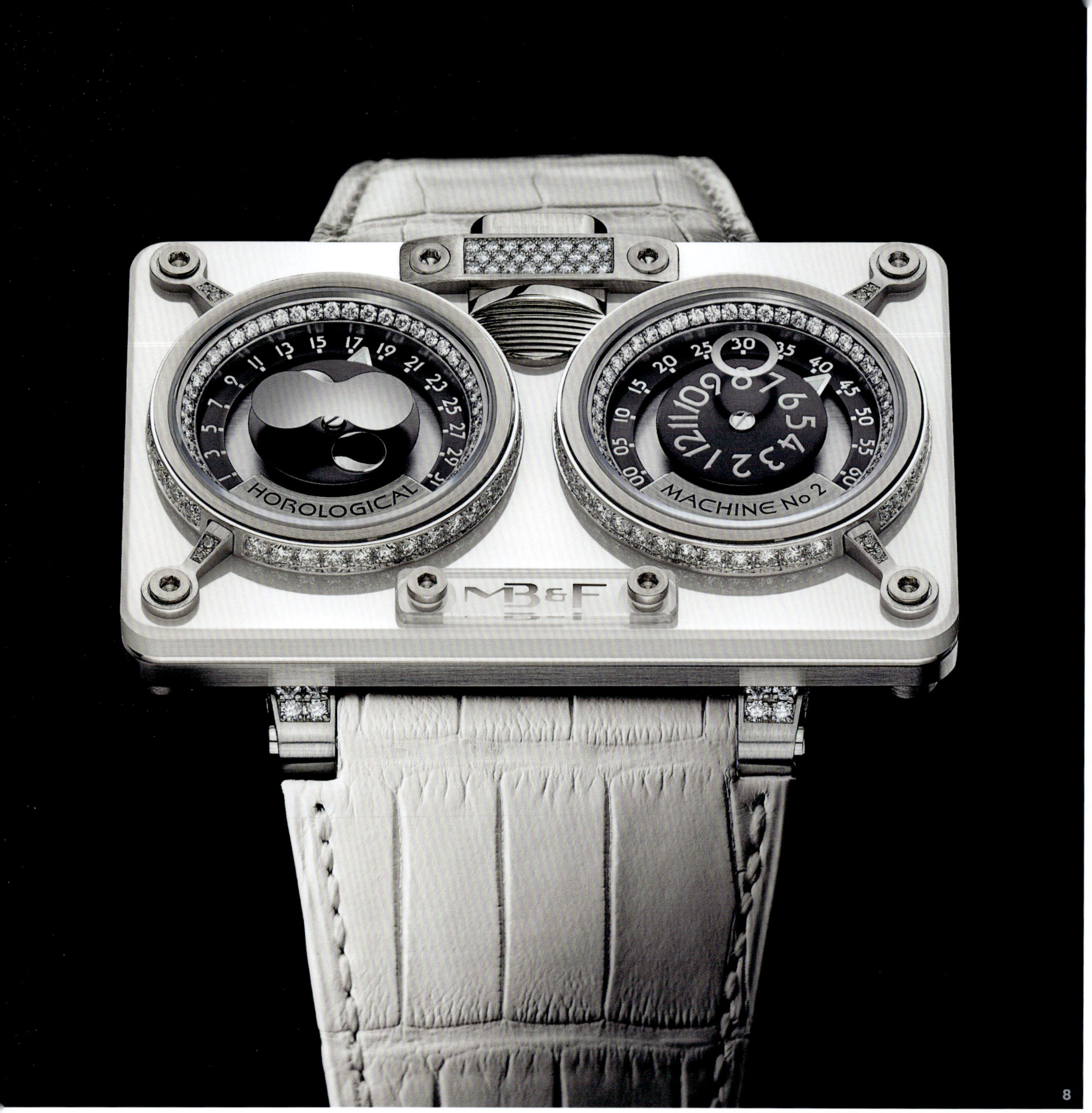

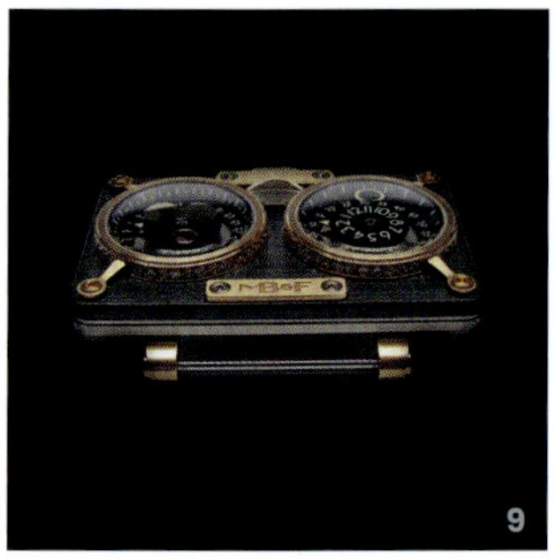

8 **Horological Machine N°2 White CTi diamonds**
Reference: 20.DC1WTL.B/S1
Material: White ceramic and grade 5 titanium partially diamond set
Limitation: –
Introduction: 2009
Year in production: 2009
Quantity produced: 4

9 **Horological Machine N°2 CR Ti diamonds**
Reference: 20.DRC1T.SU
Material: 18K red gold and grade 5 titanium partially diamond set
Limitation: unique piece
Introduction: 2009
Year in production: 2009
Quantity produced: 1

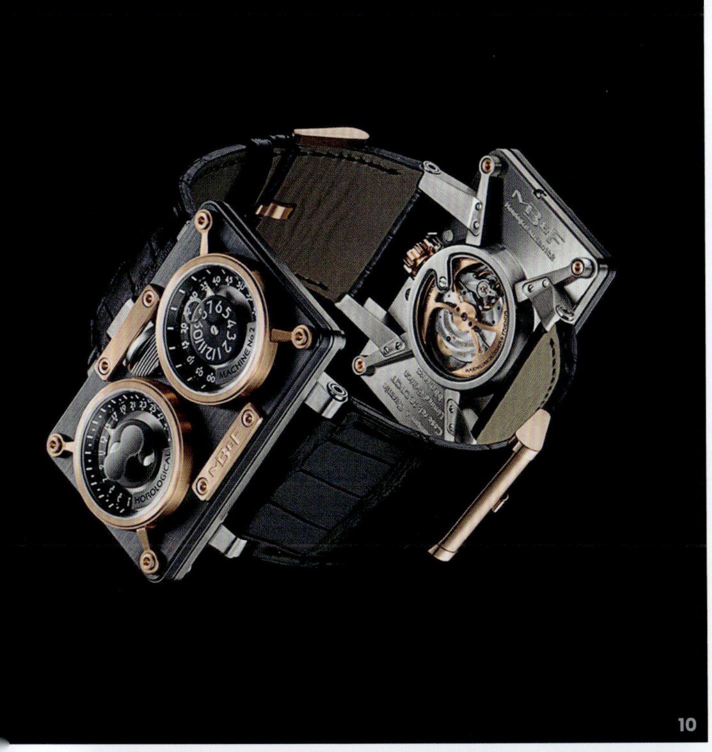

10

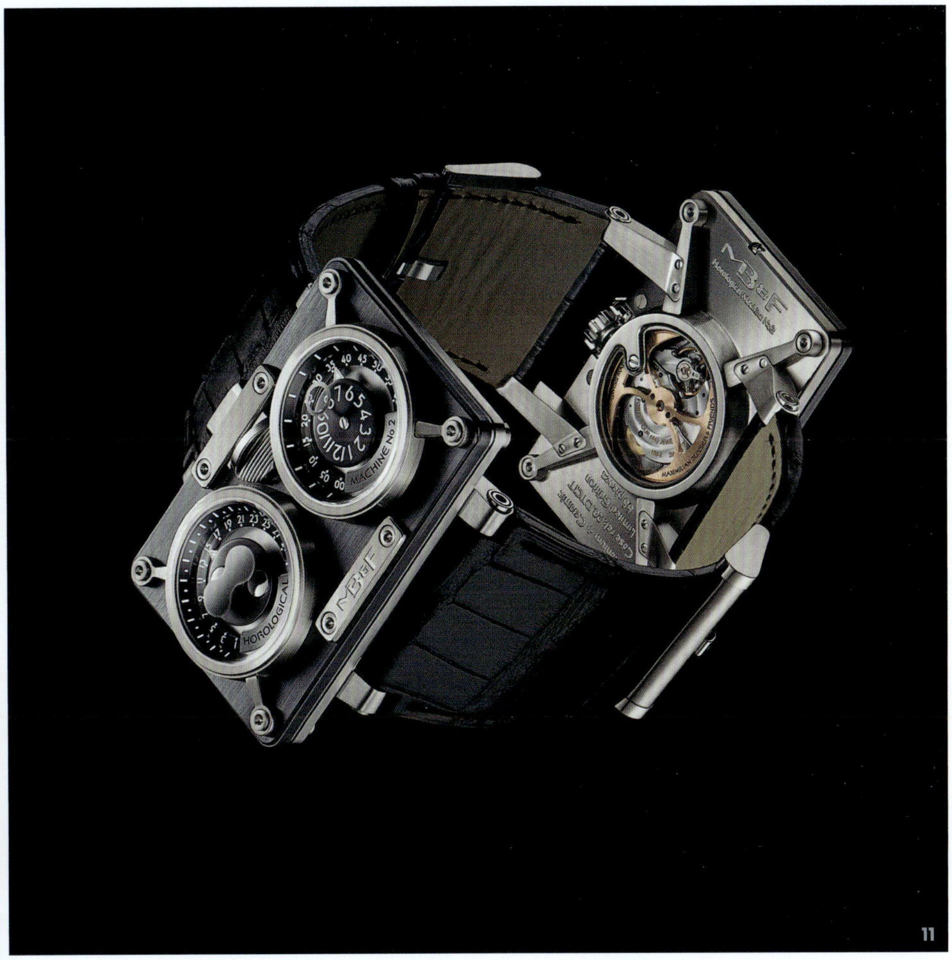

11

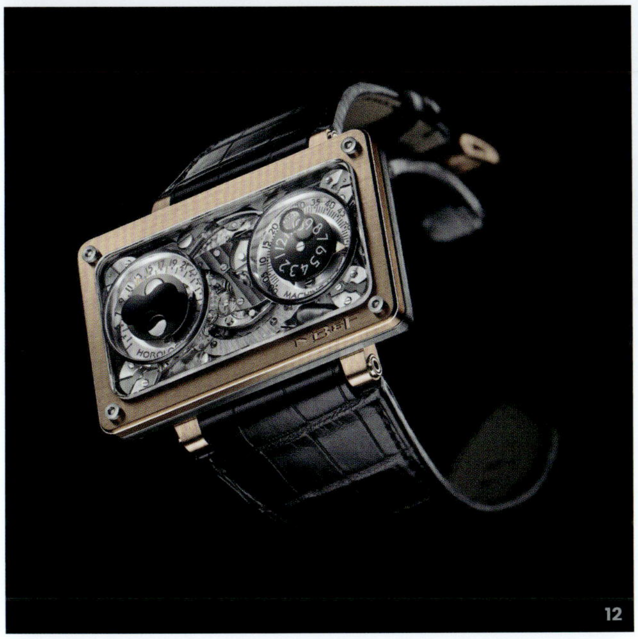

12

10 Horological Machine N°2 CR
Reference: 20.DCRTL.B
Material: black ceramic, titanium and 18K red gold
Limitation: 33
Introduction: 2009
Years in production: 2009–2010
Quantity produced: 21

11 Horological Machine N°2 CTi
Reference: 20.DCTTL.B
Material: black ceramic and grade 5 titanium
Limitation: 66
Introduction: 2009
Years in production: 2009–2010
Quantity produced: 26

12 Horological Machine N°2 RT unique piece
Reference: 20.DRSTL.B
Material: 18K red gold, sapphire crystal and grade 5 titanium
Limitation: unique piece
Introduction: 2020
Year in production: 2020
Quantity produced: 1

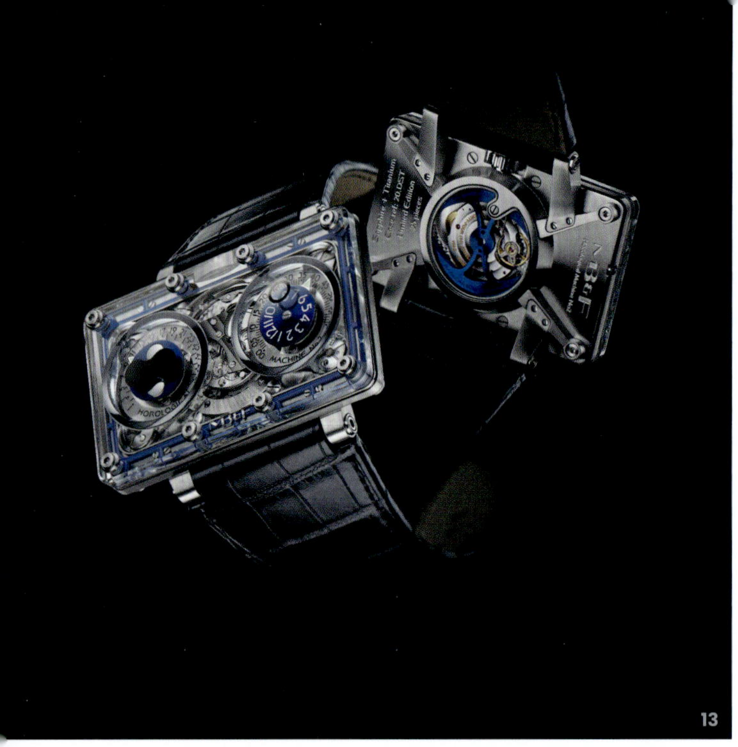

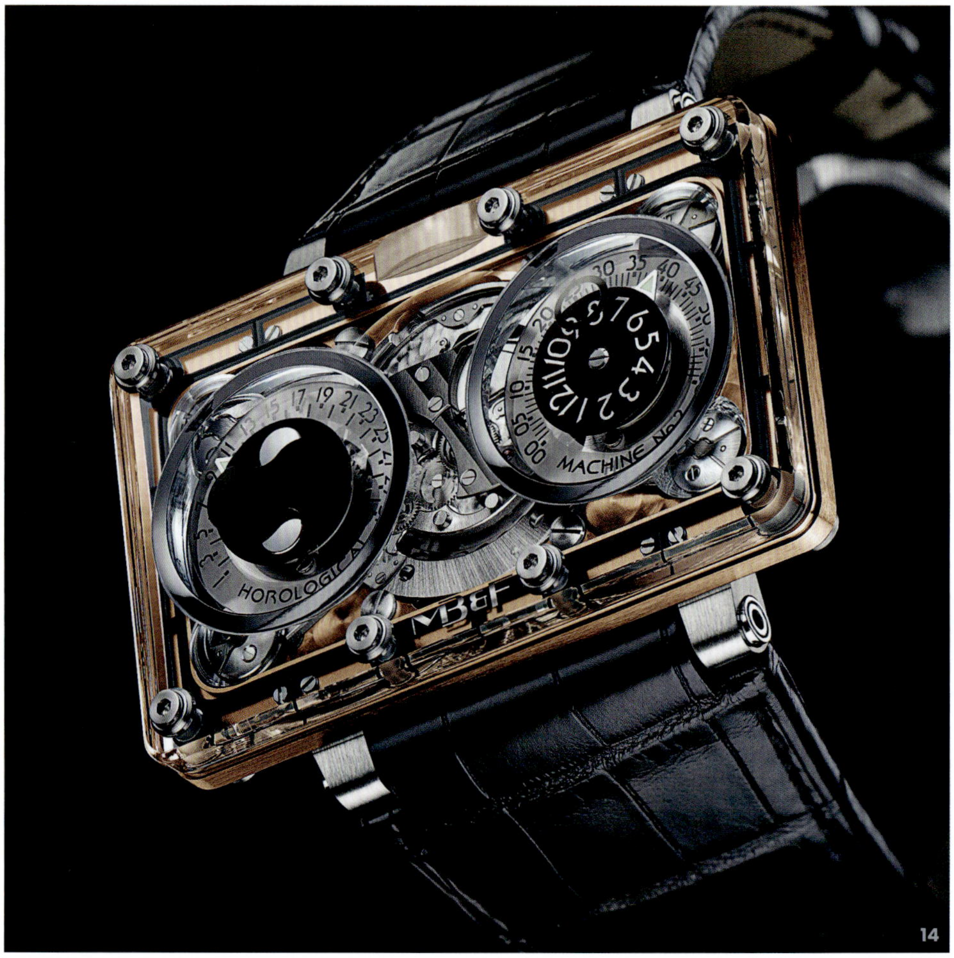

13 Horological Machine N°2 SV
Reference: 20.DSTL.BL
Material: sapphire crystal and grade 5 titanium
Limitation: 25
Introduction: 2010
Year in production: 2010
Quantity produced: 24

14 Horological Machine N°2 Red-SV
Reference: 20.DSRL.B
Material: sapphire crystal and 18K red gold
Limitation: 18
Introduction: 2011
Years in production: 2011-2013
Quantity produced: 17 + 1 prototype

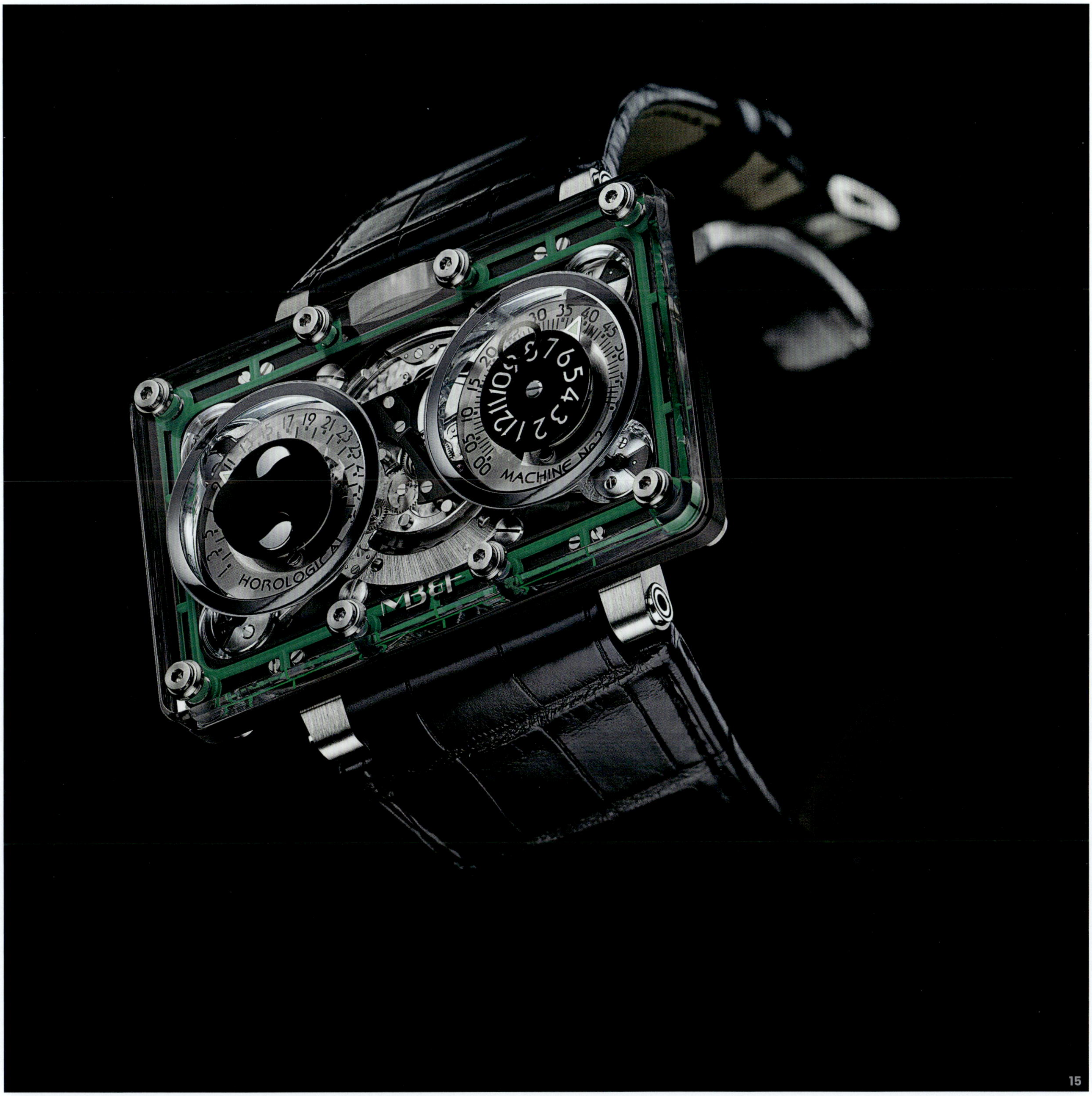

15

15 Horological Machine N°2
Black-SV
Reference: 20.DSTBL.B
Material: sapphire crystal
and black PVD coated
grade 5 titanium
Limitation: 18
Introduction: 2011
Year in production: 2011
Quantity produced: 16

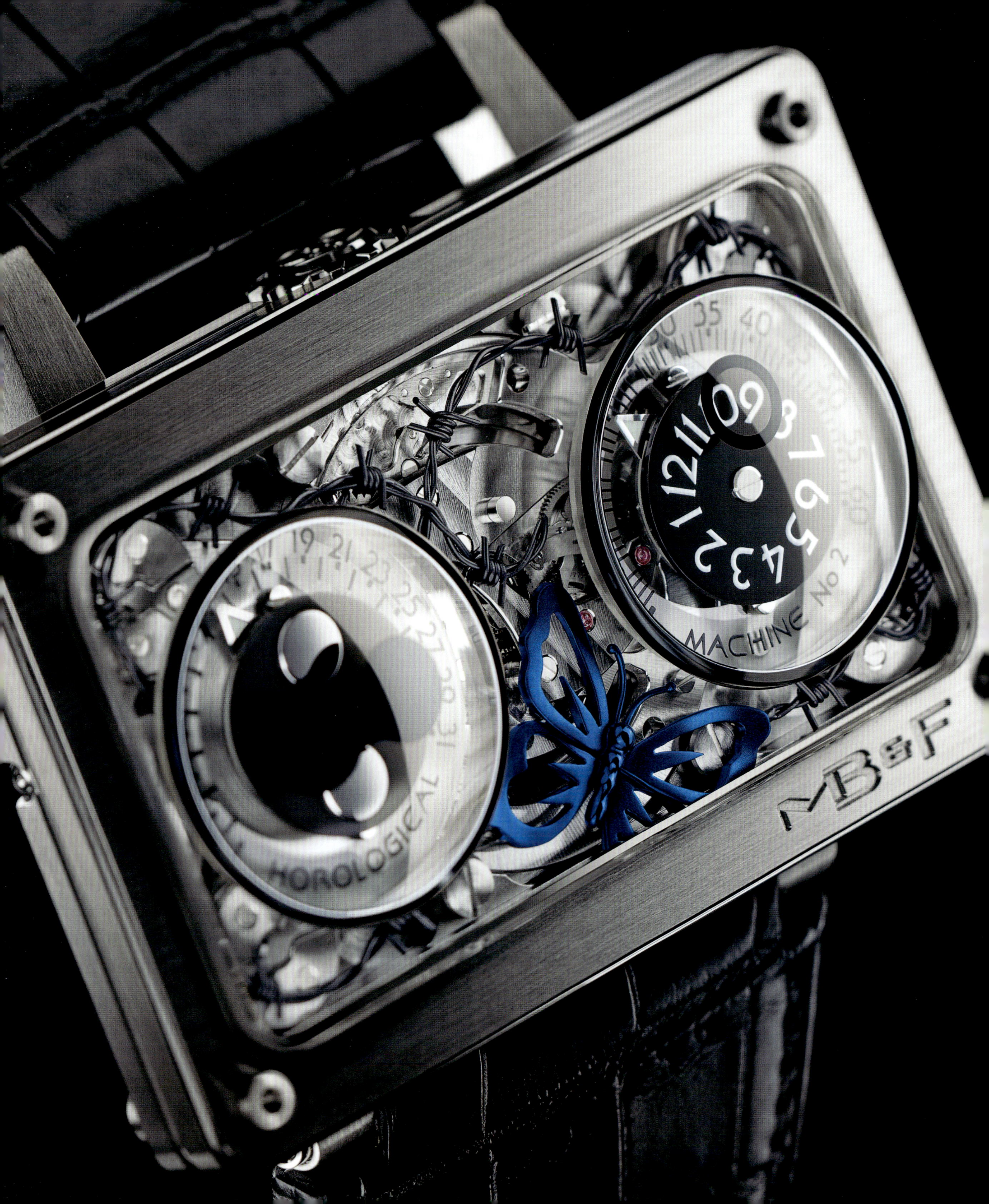
HOROLOGICAL
MACHINE No 2
MB&F

HM2 ONLY WATCH MB&F × SAGE VAUGHN

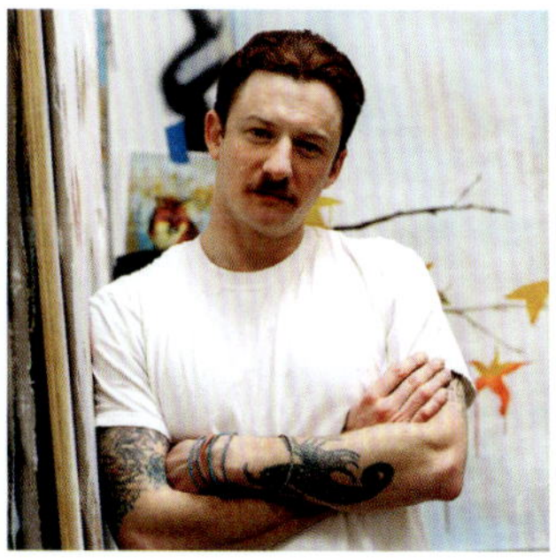

SAGE VAUGHN
American artist Sage Vaughn is a Los-Angeles-based painter and graffiti artist, born in 1976 in Oregon, United States. His work often features colorful depictions of children, birds and butterflies, set against a grey, bleak urban background. Vaughn has spoken publicly of his own seven-year struggle with addiction, the psychological weight of which can be felt throughout his work. When a gallery in Geneva dedicated a solo show to Vaughn's work, Maximilian Büsser paid a visit and bought one of his pieces, featuring a collage of butterflies pinned down by nails. Büsser was intrigued by this tension between innocence and something darker, and he quickly saw the opportunity for a charity collaboration.

The Only Watch auction is held every two years in Monaco or Geneva under the patronage of HSH Prince Albert II. The event features the work of Switzerland's Haute Horlogerie watchmakers, each contributing a unique piece, or the first in a limited series, with no reserve price set by the owner. Proceeds of the auction go to the Monegasque Association against Duchenne Muscular Dystrophy (DMD), which funds research to fight the disease. A genetic disorder affecting children, DMD is characterized by the progressive weakening of muscles, causing respiratory and cardiac problems as the child gets older.
MB&F was immediately interested in participating, not least because its creations often draw inspiration from childhood memories of comic books or science-fiction movies. Sage Vaughn was also keen to participate when he heard about a potential collaboration with MB&F for the event.
The Sage Vaughn HM2 Only Watch is a unique piece based on Horological Machine N°2. It features a butterfly in blue gold, captured behind swirling masses of sharp black barbed wire. The butterfly, a delicate symbol of childhood innocence, is captured and held prisoner by the constraints of the physical world, such as sickness, which is symbolized by the barbed wire. The surface of the case, along with the two porthole dials, is crafted from polished sapphire crystal, revealing the engine of the machine and making Vaughn's vision plainly visible.

Horological Machine
N°2 Only Watch
Material: 18K white gold, sapphire crystal and grade 5 titanium, barbed wire handcrafted in blackened gold and butterfly in blued gold
Limitation: unique piece
Introduction: 2009
Year in production: 2009
Quantity produced: 1

“With Max, I am a little lost. He can’t be so easily described; businessman, entrepreneur, creator, visionary? What I can say: he is an individual who has helped shape a new generation of watchmaking and has helped many people along the way.”

Peter Speake
Independent watchmaker

"I have been a long-time admirer of MB&F. Creating so many dramatically different watches, so consistently, over so long a period of time is nearly impossible to do well, and I can't think of any other company large or small which has managed to produce such a wide variety of timepieces to such a high technical and aesthetic standard."

Jack Forster
Editor-in-chief, Hodinkee

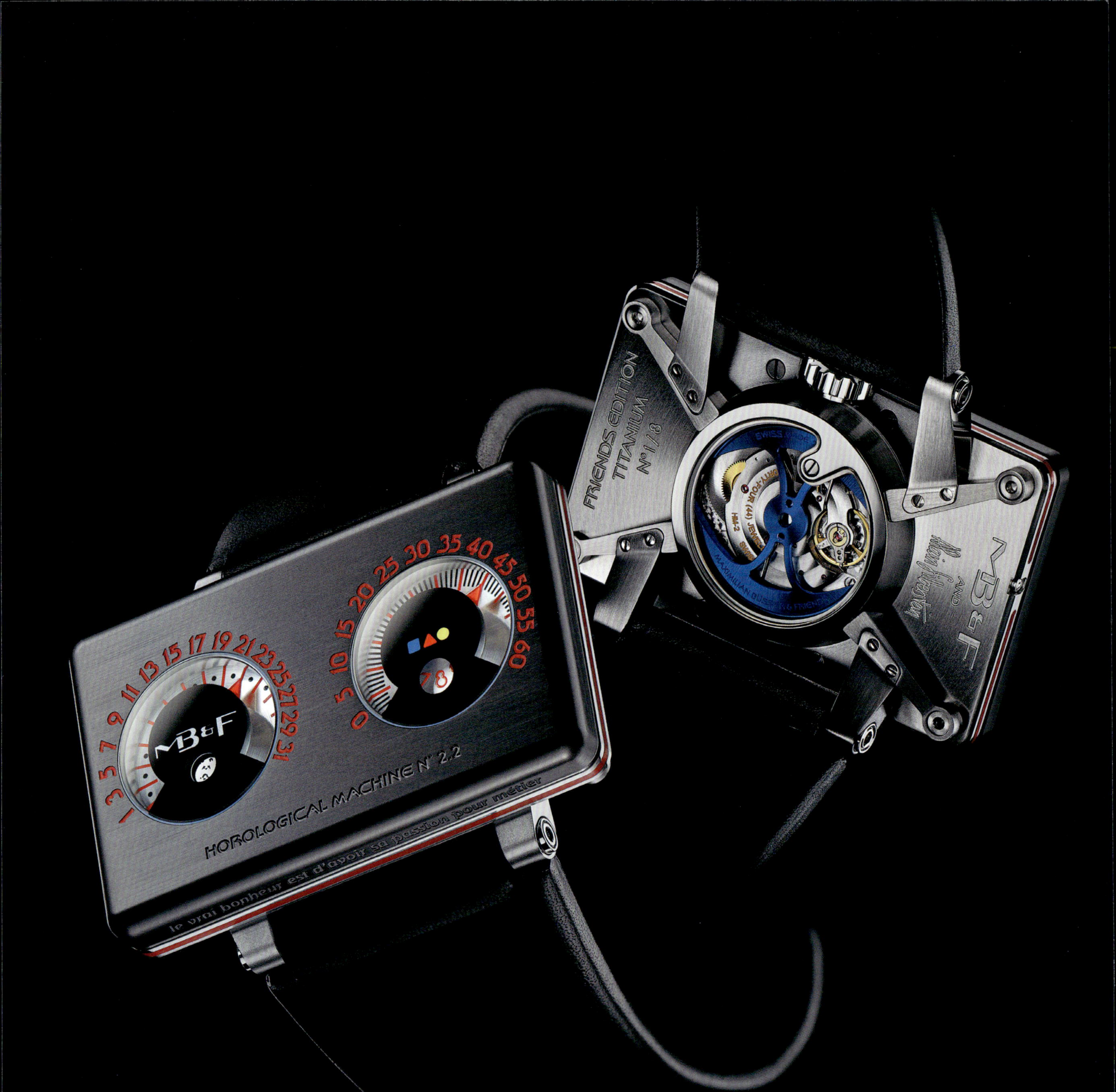
MB&F
HOROLOGICAL MACHINE N° 2.2
le vrai bonheur est d'avoir sa passion pour métier
FRIENDS EDITION
TITANIUM
N°1/8
SWISS MADE
MB&F
AND

HM2.2 BLACK BOX MB&F × ALAIN SILBERSTEIN

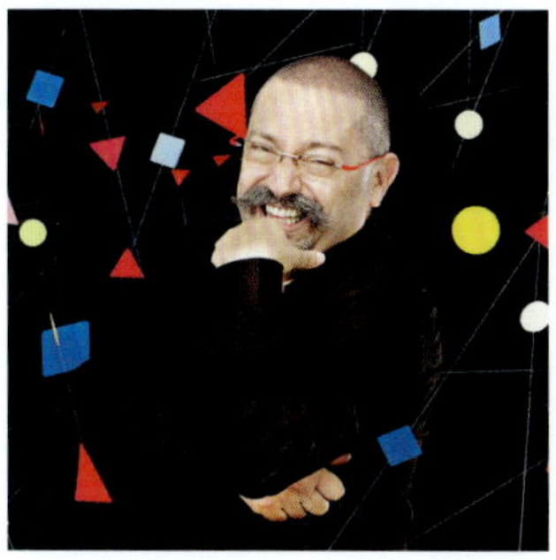

ALAIN SILBERSTEIN

Alain Silberstein was born in Paris, France, in 1950. A trained interior architect and model maker, he worked in Paris as an interior designer before moving to Besançon, the heart of French watchmaking, in 1979.

In 1990, Silberstein created his own brand, Alain Silberstein Créations, which quickly developed a reputation for its bright, playful use of primary colours and geometric shapes, taking inspiration from the German Bauhaus movement.

Following the release of HM2, Büsser asked his friend Silberstein to redesign the machine, for a limited edition of just eight examples, with the simple instruction: "Enjoy yourself!"

When MB&F finally saw what Alain Silberstein – a creator known for his bold use of colour and pattern – had envisaged for his machine, they were shocked. Horological Machine 2.2, affectionately nicknamed "Black Box", was a surprisingly restrained, minimal and almost completely black re-edit of the HM2.

Silberstein's HM2.2 retains the twin dials from the original machine with their sapphire portholes and the flying-buttress lugs. But otherwise, the machine appears almost entirely rethought and recreated, drawing inspiration from the square lines and large lenses that characterize the miniature box cameras of the 1940s. The MB&F HM2.2 mimics the shape of these cameras, with a case carved from a solid block of titanium with black PVD coating.

The right-side dial is dotted with three geometric shapes (a Silberstein staple) in his signature primary colours of blue, red and yellow. Another key Silberstein motif is his use of Bauhaus red Super-LumiNova on all indicators for legibility. One whimsical touch is a subtle change to the moon phase indicator on the left-hand dial, inspired by the "Little Nemo" cartoons of the Art Nouveau period.

Almost hidden on the side of the HM2.2 is an inscription from Silberstein, which reads: "Le vrai bonheur est d'avoir sa passion pour métier"; or "true happiness is having one's passion for a profession" – a sentiment that works well for MB&F and Silberstein alike.

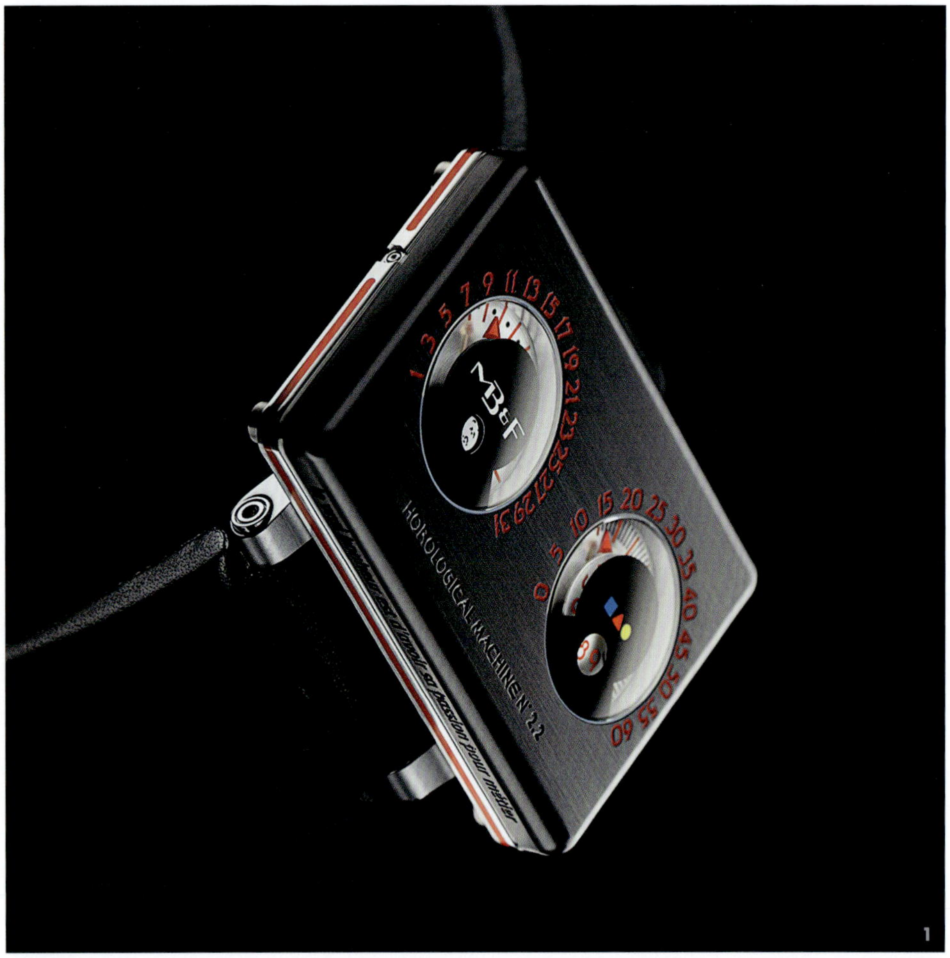

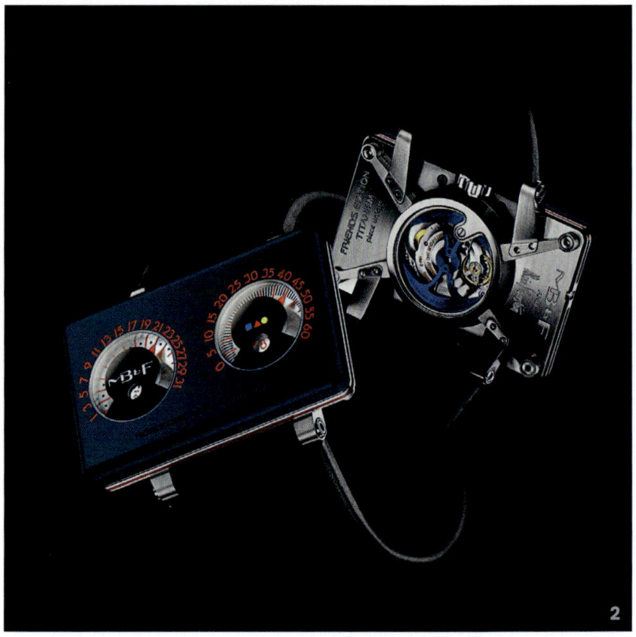

1 Horological Machine N°2.2
Black Box
Reference: 20.DASTTL.S
Material: black PVD coated grade 5 titanium
Limitation: 8
Introduction 2009
Years in production: 2009 - 2010
Quantity produced 8 + 1 prototype

2 Horological Machine N°2.2
Blue Box
Reference: 20.DASTTL.S "Blue"
Material: blue PVD coated grade 5 titanium
Limitation: unique piece
Introduction: 2010
Year in production: 2010
Quantity produced: 1

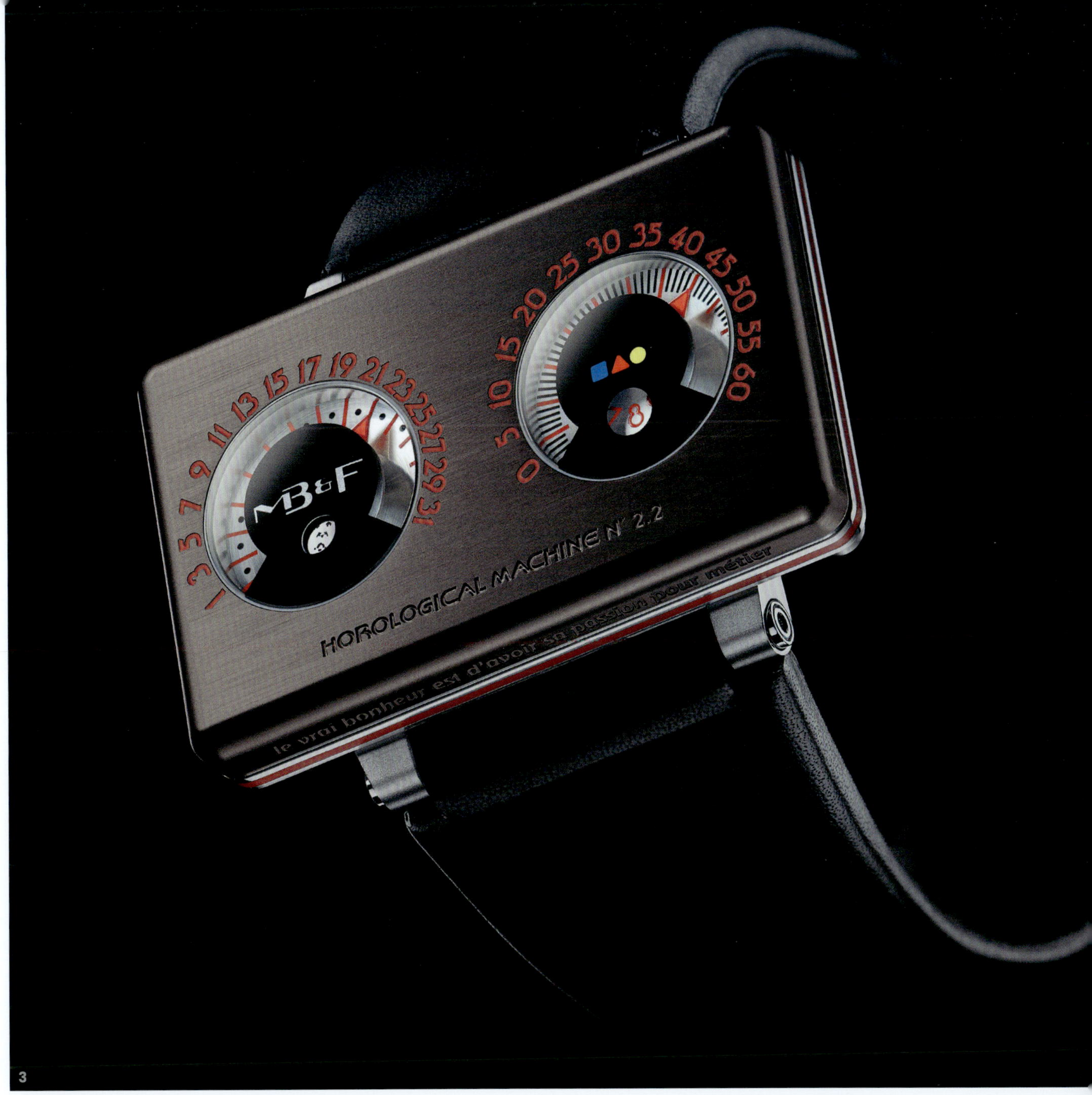

3

3 Horological Machine N°2.2
Chocolate Box
Reference: 20.DASTTL.S
"Chocolat"
Material: chocolate PVD
coated grade 5 titanium
Limitation: unique piece
Introduction: 2010
Year in production: 2010
Quantity produced: 1

HM 3

Released in 2009, Horological Machine N°3 quite literally turned the concept of a horological machine on its physical head. Switching the position of the automatic winding rotor from the base of the movement to the top of the dial puts the solid-gold MB&F battle-axe rotor in full view, displayed to advantage, under crystal-clear sapphire surrounded by a large metal ring of Arabic numerals (the machine's date wheel). Alongside the dial, twin cones or pods, also in sapphire crystal, rise from the three-dimensional base to display the hours and minutes.

Just as the design of HM2 was inspired by the shape and profile of a space station, so HM3 represents its space ship. At the back of the space ship are its reactors, in the form of two ceramic bearings, transmitting power to the time indicators and date wheel.

This was the first time a Horological Machine was offered in two versions: "Sidewinder", with its cones lined perpendicular to the wrist; and "Starcruiser", with cones following the line of the wrist.

HOROLOGICAL MACHINE Nº3

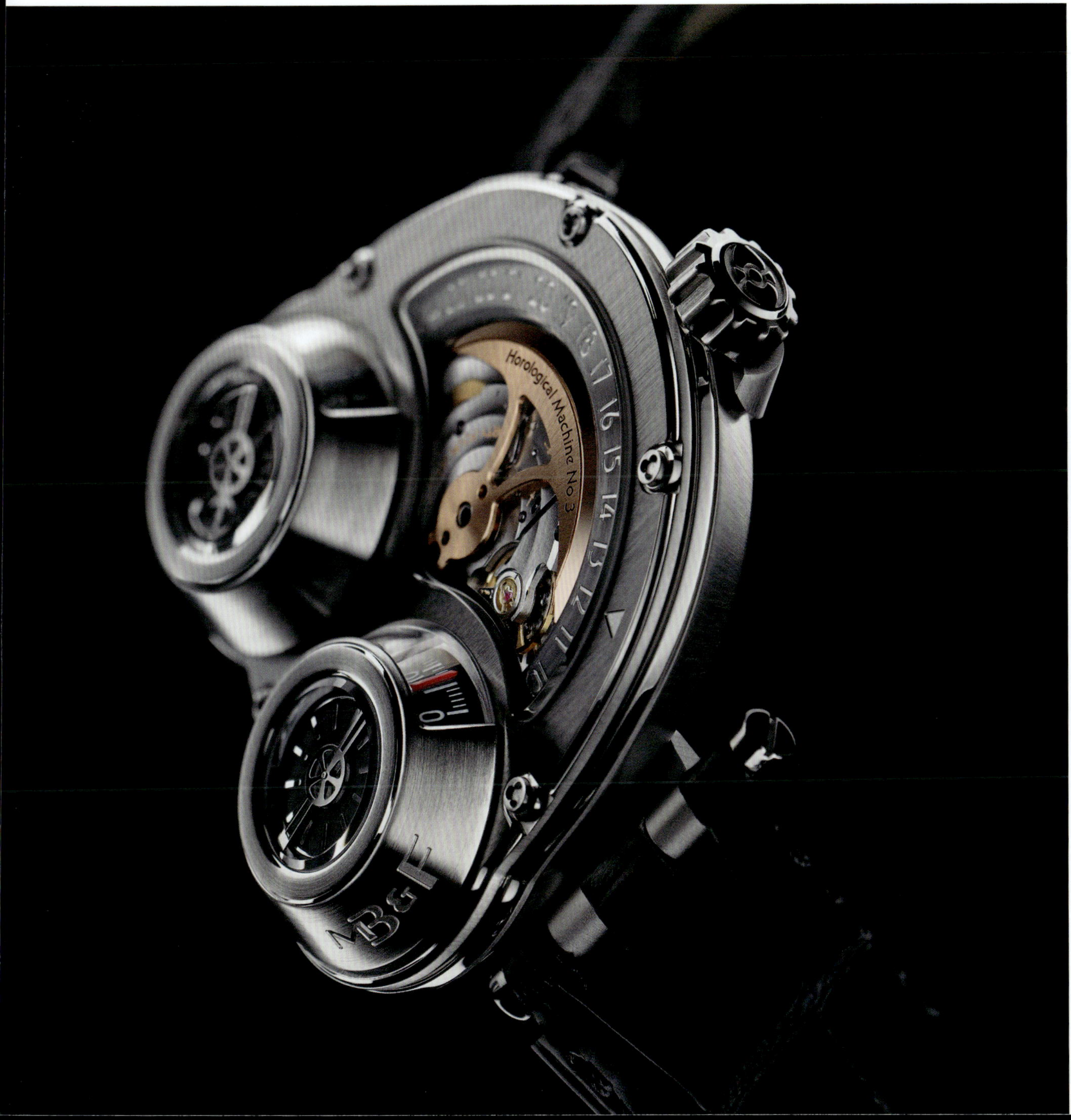

HM3 CALIBRE

The HM3 was designed by Maximilian Büsser and Eric Giroud; their drawings were then made mechanical reality by Jean-Marc Wiederrecht, winner of the 2007 award for Best Watchmaker at the Grand Prix d'Horlogerie de Genève.

The most striking features of the HM3 are probably its raised, twin cones for time display, which were also among the most difficult details to realise. Three-dimensional cones had never before been employed to display time. In fact, they were initially thought impossible to manufacture due to the considerable polishing required to make the sapphire transparent.

The design of the oversized date wheel also required significant ingenuity. A pusher was needed for the manual adjustment of the date, but a traditional pusher can only move a date wheel by about 1mm – which would not be enough for a date wheel of this size. So an amplifier was developed to increase the pusher distance by a factor of four.

Collection
Horological Machines
Series
Horological Machine N°3
Common name
HM3
Movement
Developed for MB&F by Jean-Marc Wiederrecht(Agenhor), powered by a Girard-Perregaux base
Winding
Automatic
Power reserve
45 hours
Components
304/36 jewels
Functions
Hours and day/night indicator on one cone; minutes on second cone; date around the movement
Add. features
22K gold "battle-axe" winding rotor
Hour and minutes information transmitted via ceramic ball bearings to laser-cut hands
Case
18K white gold/grade 5 titanium or 18K red gold/grade 5 titanium
Dimensions
47 × 50 × 16 mm
Case components
53 for Starcruiser editions, 57 for Sidewinder editions
Water resistance
30 m/3 ATM
Strap
Alligator
Buckle
Folding buckle, 18K white or red gold matching the case
Introduction
2009
Years in production
2009–2012
Quantity produced
231 + 2 prototypes
Comments
HM3 is available in 3 versions: 'Sidewinder', with cones lined perpendicular to the arm, 'Starcruiser', with cones in line with the arm and a special version for people who wear their watch on the right wrist: the HM3 ReBel.

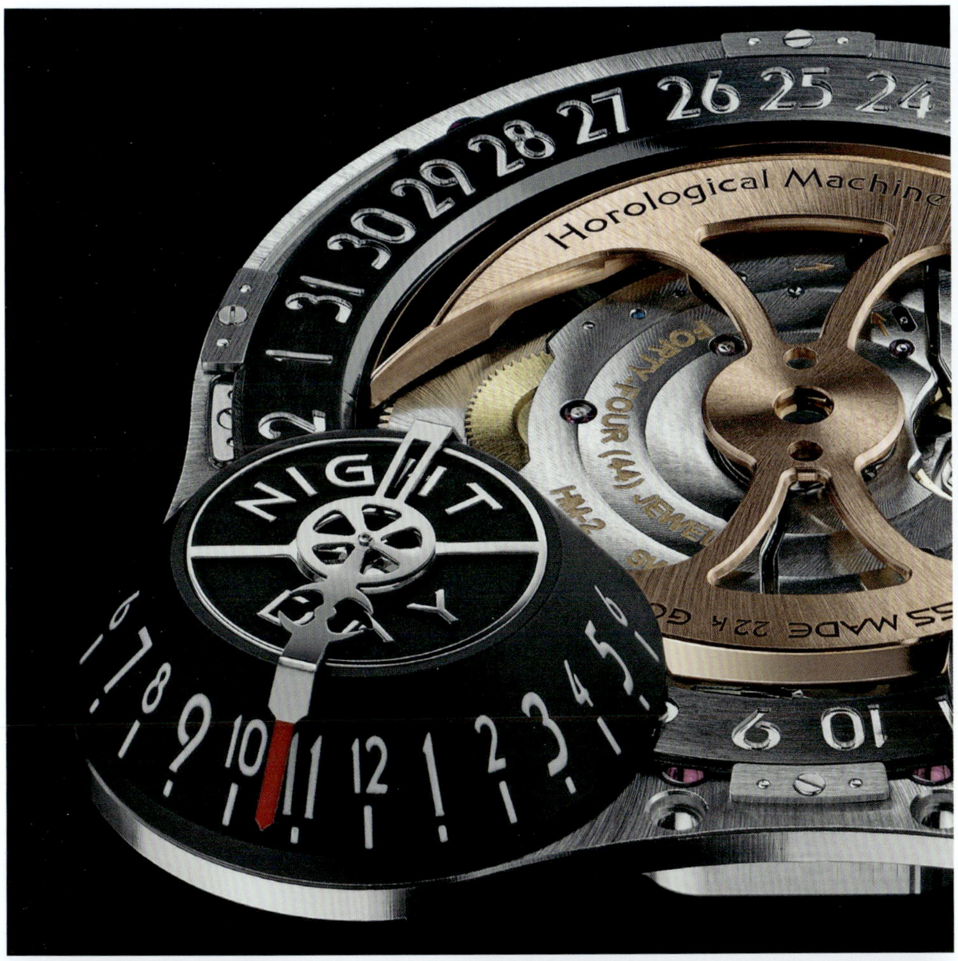
Horological Machine

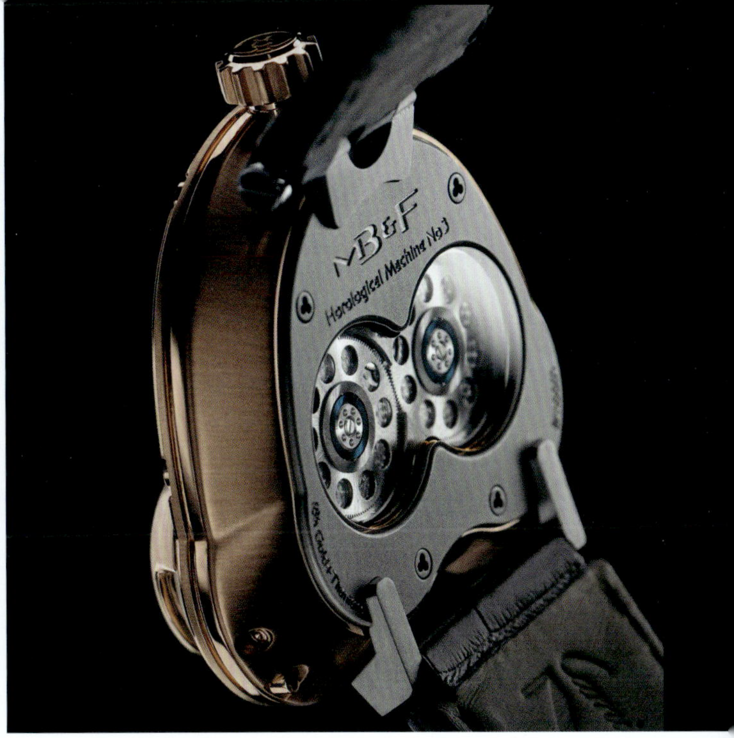
MB&F
Horological Machine No3

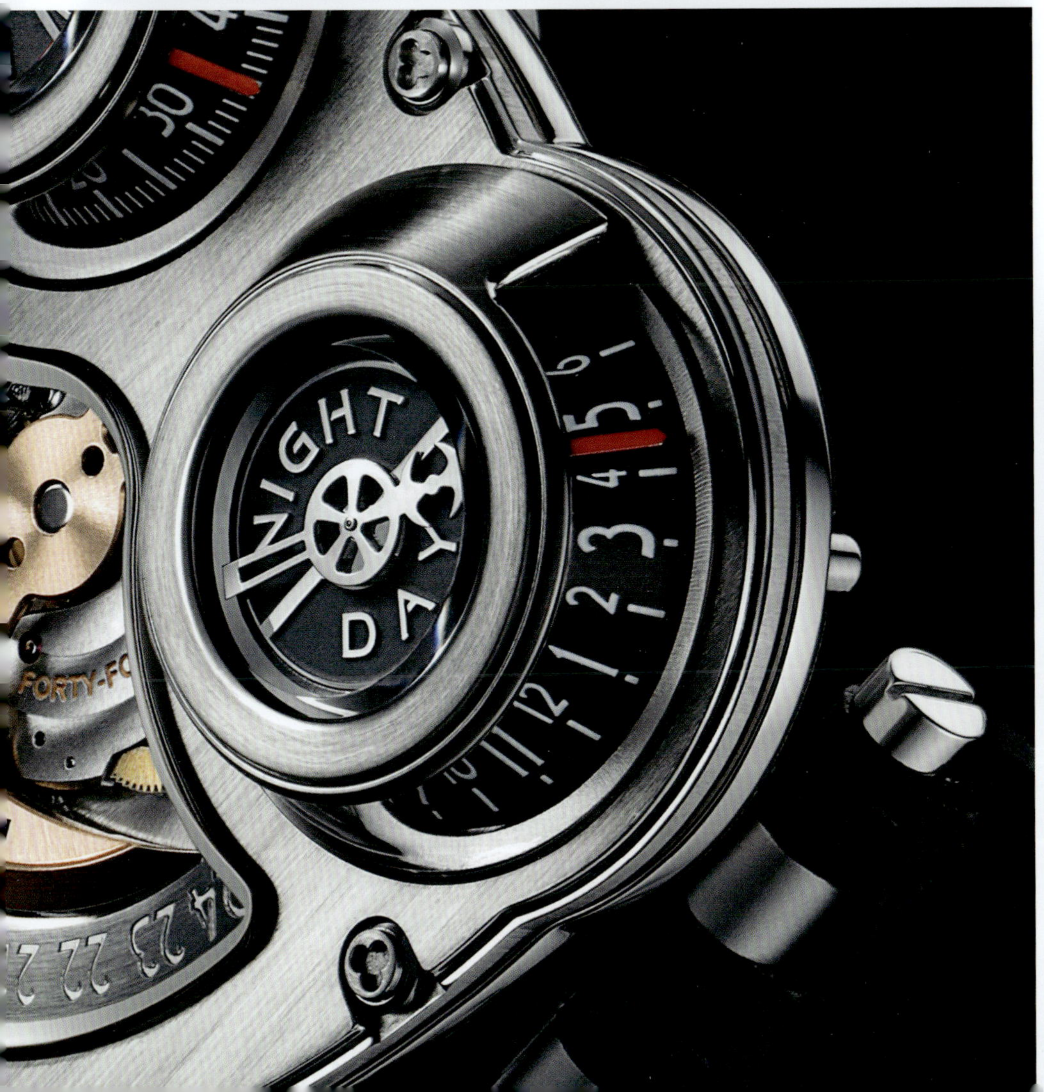
NIGHT
DAY

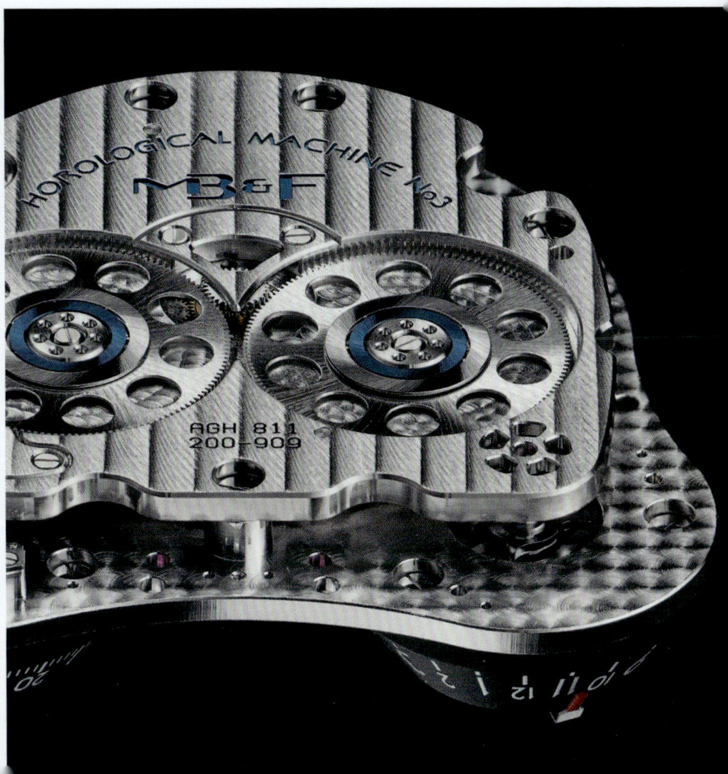
HOROLOGICAL MACHINE No3
MB&F
AGH 811
200-909

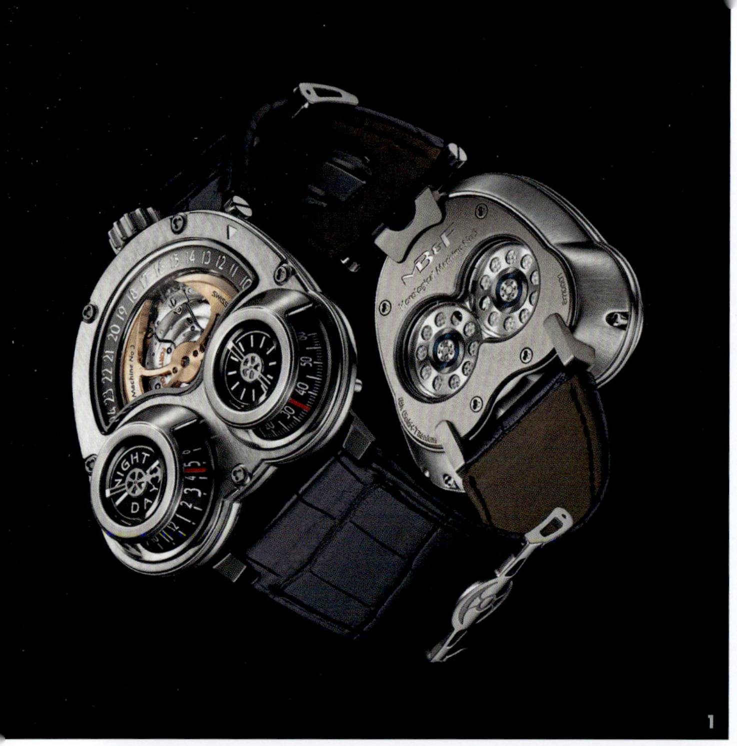

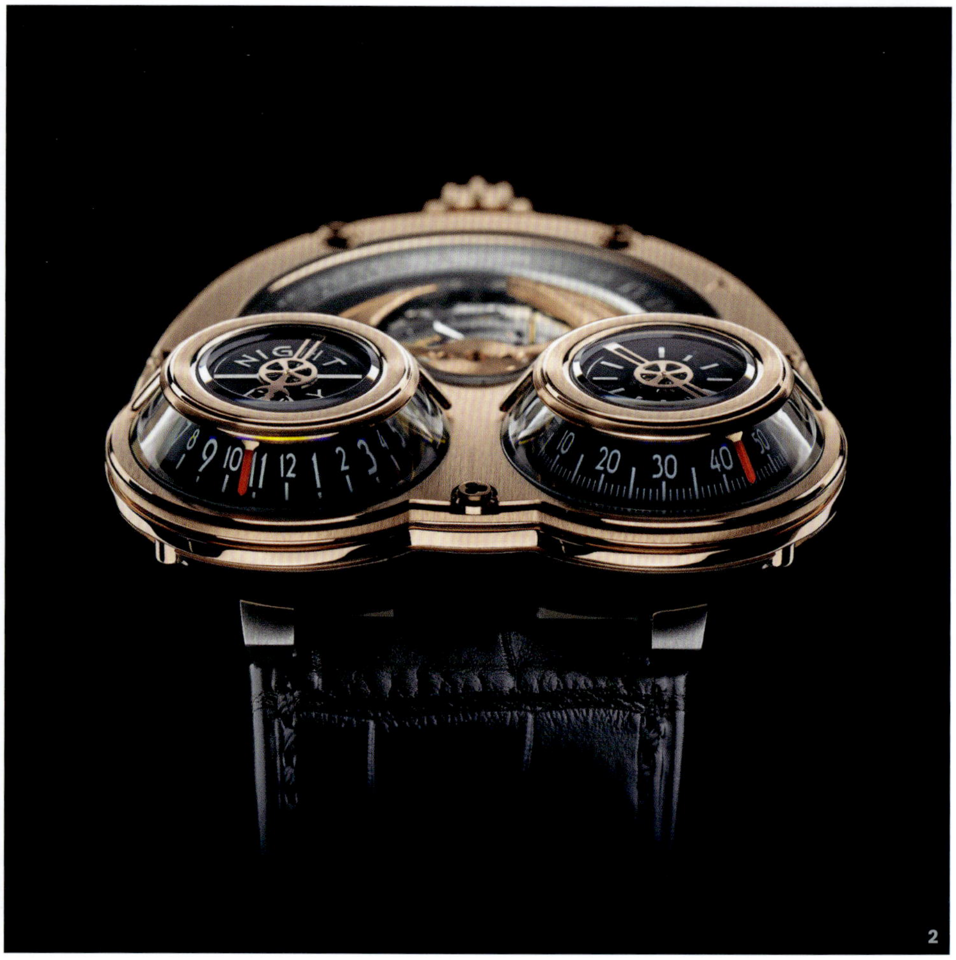

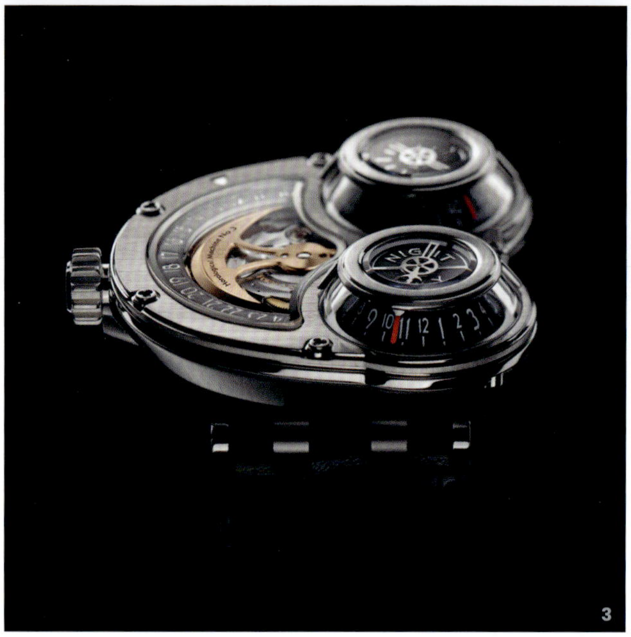

1 Horological Machine N°3
Starcruiser WG
Reference: 30.WTL.B
Material: 18K white gold and grade 5 titanium
Limitation: –
Introduction: 2009
Years in production: 2009–2011
Quantity produced: 36

2 Horological Machine N°3
Starcruiser RG
Reference: 30.RTL.B
Material: 18K red gold and grade 5 titanium
Limitation: –
Introduction: 2009
Years in production: 2009–2011
Quantity produced: 57 + 1 prototype

3 Horological Machine N°3
Sidewinder WG
Reference: 31.WTL.B
Material: 18K white gold and grade 5 titanium
Limitation: –
Introduction: 2009
Years in production: 2009–2011
Quantity produced: 38

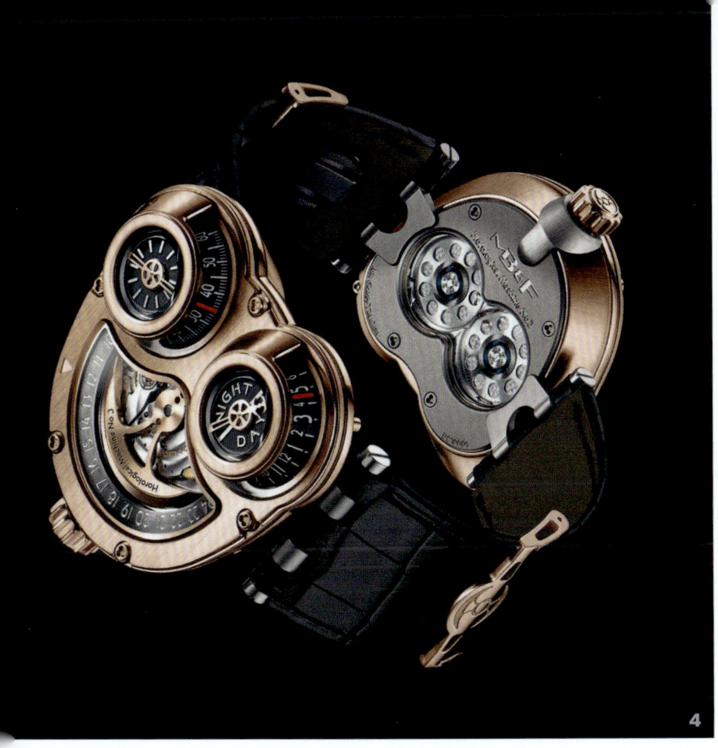

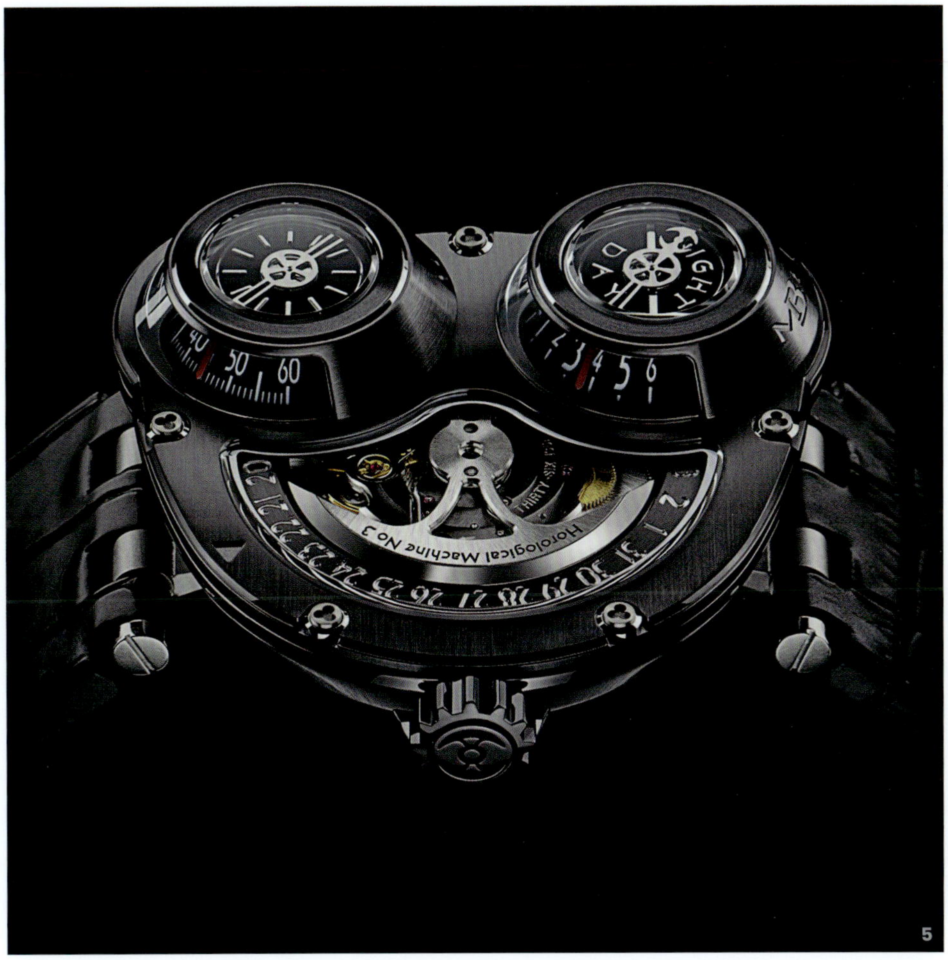

4 Horological Machine N°3
Sidewinder RG
Reference: 31.RTL.B
Material: 18K red gold and
grade 5 titanium
Limitation: –
Introduction: 2009
Years in production:
2009–2011
Quantity produced: 55

5 Horological Machine
N°3 ReBel
Reference: 31.GWBL.B
Material: black PVD coated
18K white gold
Limitation: 18
Introduction: 2011
Year in production: 2011
Quantity produced:
18 + 1 prototype

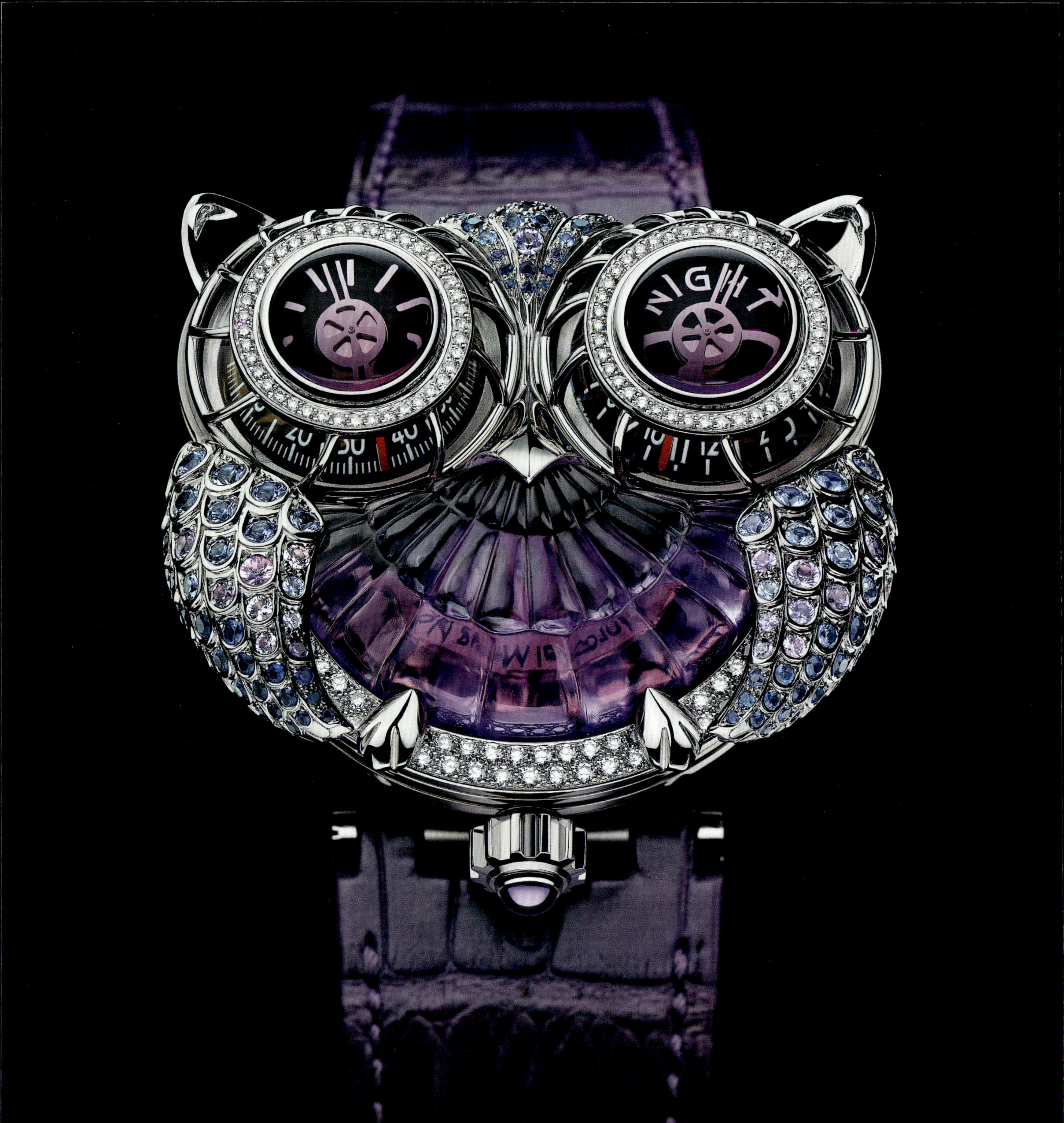
NIGHT

JWLRYMACHINE MB&F × BOUCHERON

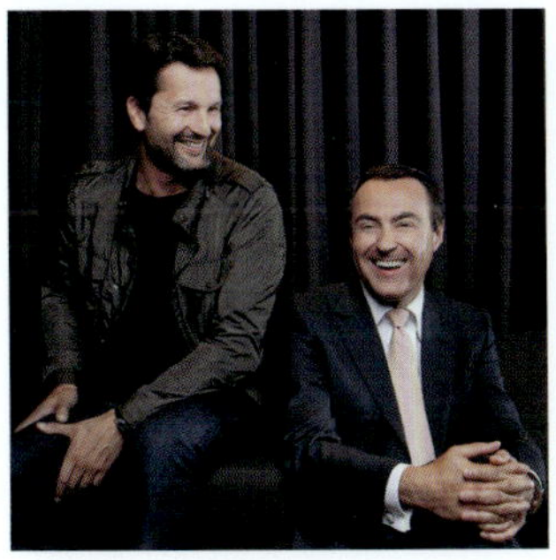

Portrait of Maximillian Büsser and Jean-Christophe Bedos, Former CEO of Boucheron

BOUCHERON

French High Jewellery House Boucheron was founded in 1858 by Frédéric Boucheron, and has long been known for its international clientele of aristocrats, royalty, authors, artists, film stars and billionaires. The House is famous worldwide for its innovative and free-spirited approach, making bold use of rare and flamboyantly colored stones to create an exquisite collection of glittering designs.

A love and appreciation of the natural world is a frequent source of inspiration for Boucheron master jewellers. Flowers, birds, insects, and other animal forms are recurrent motifs, always presented in an eclectic, luxurious, and fascinating style.

Some might be surprised by the pairing of a historic High Jewellery house located in Paris' prestigious Place Vendôme with MB&F, a young and rebellious Swiss watch brand. As it turned out, Boucheron and MB&F's shared creative passion, dedication to quality, and avant-garde attitude made this an especially rewarding collaboration.

What would become the JwlryMachine was selected from more than 60 Boucheron concepts presented to MB&F: a brilliant, bejewelled modification of the HM3 featuring a glowing, glittering owl with wide, gleaming eyes, and a trembling, beating heart beneath its feathered breast.

The eyes of the owl are formed by two large cabochons encircled by diamonds, set over the twin cones of the HM3. Brilliant-cut stones give the owl's wings their pave-set sparkle, hugging and protecting the movement.

The spirit and heritage of Boucheron come shining through in the playful but careful use of colour, expressed in violet amethyst, diamonds, and blue and violet sapphires set in 18-karat white gold; or 18-karat red gold set with pink tourmaline, rose quartz, diamonds and pink sapphires.

In the centre of the JwlryMachine is a translucent breastplate made from either carved amethyst or rose quartz, with the swinging battle-axe rotor of the HM3 just visible beneath. Breathing life into this little Boucheron creature is the movement of the winder, cleverly employed to mimic the beating of the owl's heart.

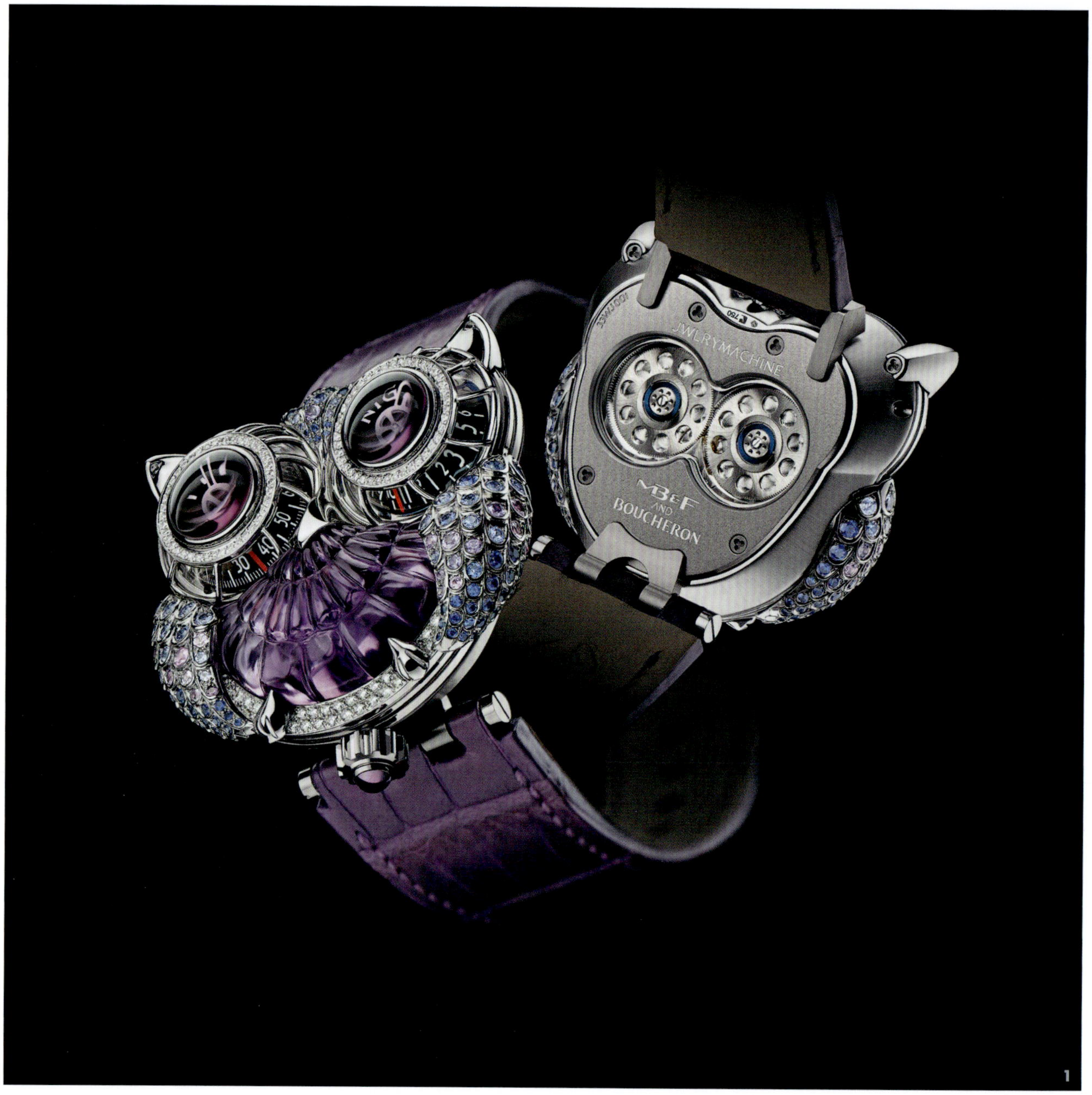

1 **JwlryMachine Purple**
Reference: 33.WATL.B
Material: 18K white gold with amethyst, diamonds, blue and violet sapphires
Limitation: –
Introduction: 2010
Years in production: 2010–2012
Quantity produced: 19

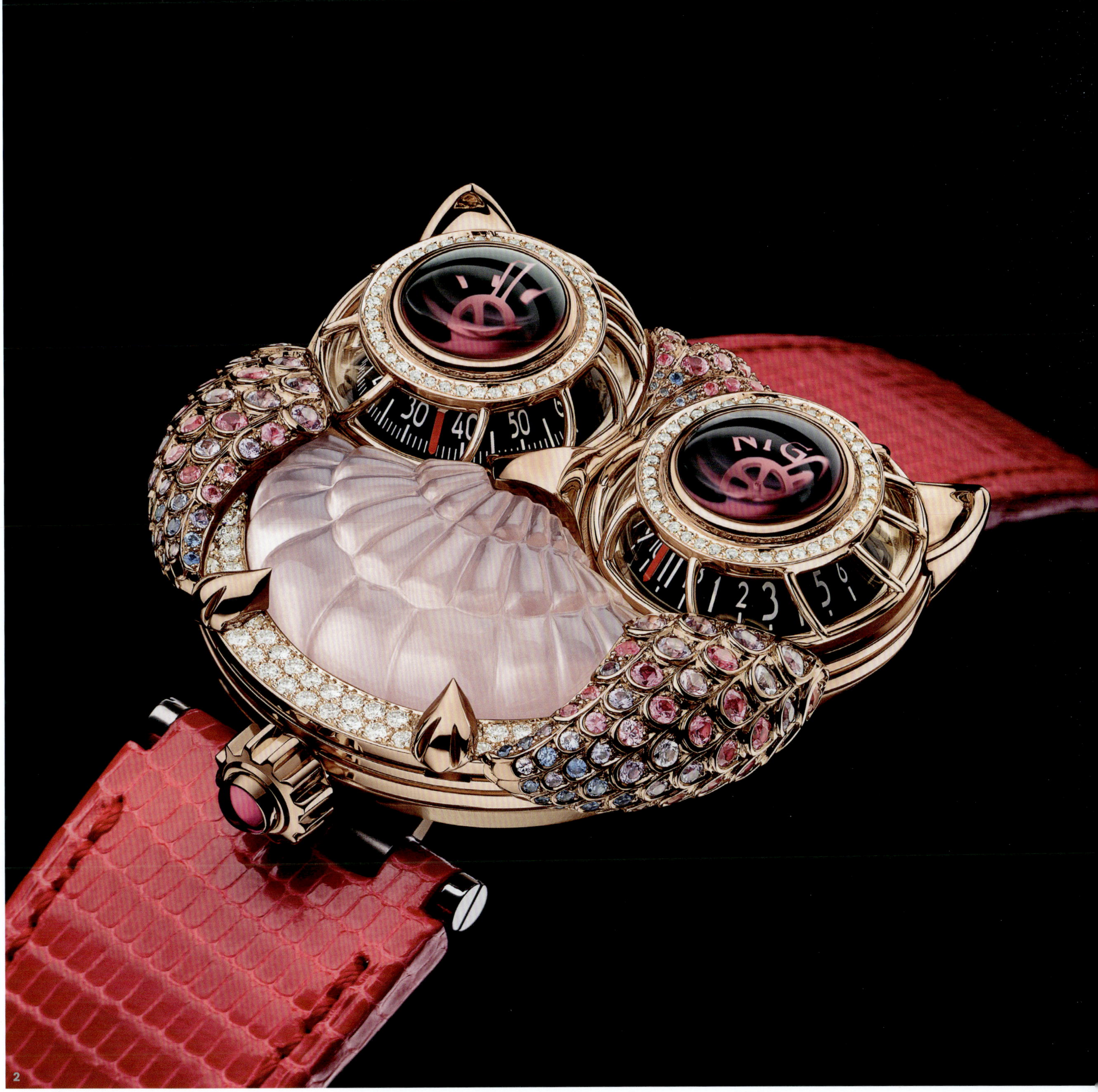

2 JwlryMachine Pink
Reference: 33.RQTL.B
Material: 18K red gold with pink tourmaline, rose quartz, diamonds and pink sapphires
Limitation: –
Introduction: 2010
Years in production: 2010–2011
Quantity produced: 8

“Maximilian Büsser is to independent high-end watchmaking something like Karl Lagerfeld was to haute couture. He has reinvented the profession of artistic director and is revisiting all the codes of tradition… Above all, he is determined to defend his creative freedom and let his childlike soul wander.”

Stéphane Gachet
Co-Editor-in-chief, Watch Around

“MB&F is arguably the most unconventional, imaginative and complex luxury watch, clock and music box creator in the world. Inspired by science fiction with some retro-futuristic vehicle and aquatic life influences, the timepieces from MB&F showcase a complex, playful sophistication that has really shaken up the industry.”

Brice Goulard
Managing editor, Monochrome Watches

HM3 FROG

The MB&F Horological Machine N°3 Frog is a playful but seriously intended re-edit of the Horological Machine N°3. It was one of Maximilian Büsser's main principles when he founded MB&F to bring joy and a sense of childlike awe to the otherwise restrained world of high-end watchmaking. The HM3 Frog is a perfect expression of this ambition.

Rising prominently from the dial are two rotating sapphire eyes, which twist and turn like the eyes of a real-life frog. Taking their cue from Nature, the semi-spherical sapphire domes of the HM3 offer a wide field of view that allows the wearer to read the time without turning the wrist, in the same way that a frog can see in many directions without turning its head.

The dial of the HM3 Frog is dominated by the iconic MB&F battle-axe, the machine's automatic winding rotor. Traditionally located at the bottom of the movement, the HM3 instead places the rotor in prime position at the top of the movement, on display under a pane of clear sapphire.

Surrounding the battle-axe is an oversized date wheel that rotates slowly around the movement (a feature not present on the later model, the Frog X).

HOROLOGICAL MACHINE Nº3 FROG

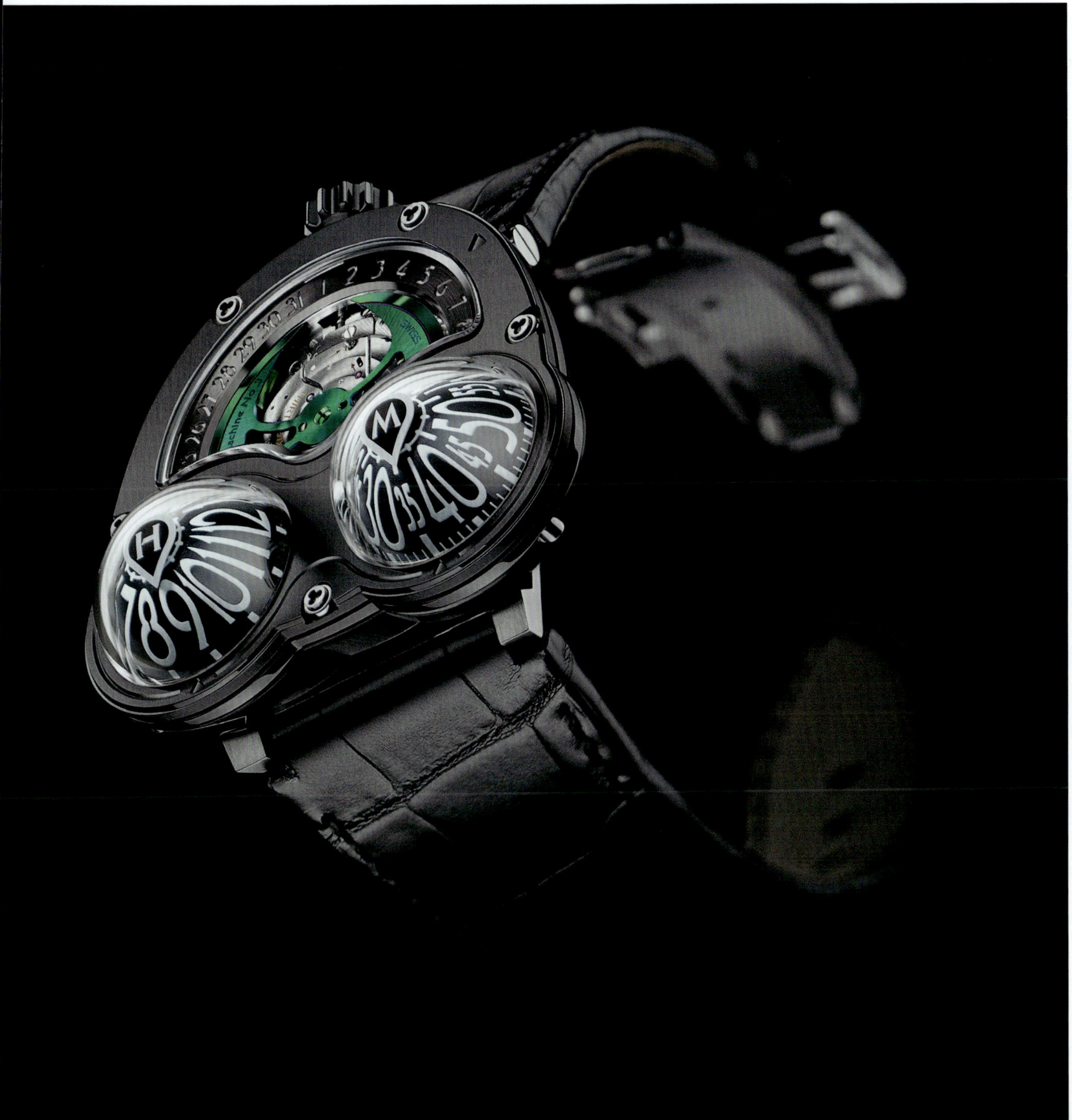

HM3 FROG CALIBRE

Twin semi-spherical, rotating domes topped with sapphire form the "eyes" of the HM3 Frog and serve to indicate hours and minutes. The fabrication of these domes is a feat of engineering in its own right, the sapphire worked and shaped with great care then uniformly polished to perfection. Any slight imperfection at this stage can produce a disconcerting magnification effect on a portion of the crystal. Inside the crystal are the rotating domes that display the hours and minutes. Aluminium was chosen as the base material for its favorable strength-to-mass ratio and energy-saving potential. The finished domes are paper-thin and weigh only 0.5g, reducing energy consumption to the very minimum.

Collection
Horological Machines
Series
Horological Machine N°3 Frog
Common name
HM3 Frog
Movement
Movement developed for MB&F by Jean-Marc Wiederrecht (Agenhor), powered by a Girard-Perregaux base
Winding
Automatic
Power reserve
45 hours
Components
HM3 Frog: 304/36 jewels
MoonMachine: 319/36 jewels
HM3 FrogX: 277/36 jewels
Functions
Hour on one dome (aluminium dome rotating in 12 hours) and minutes on second dome (aluminium dome rotating in 60 minutes); date around the movement for Frog editions (not for FrogX and MoonMachine); moon phase for MoonMachine editions
Add. features
22K gold "battle-axe" winding rotor
FrogX: removal of the date, enlarged 22K gold and titanium rotor and "secret message" crown reflects light rays to project the MB&F battle-axe icon (Rayform technology)
MoonMachine: date replaced by moon phase, winding rotor illustrating northern sky constellations
Case
HM3 Frog and MoonMachine: grade 5 titanium, 18K red gold, black PVD coated zirconium
HM3 FrogX: full sapphire crystal with gasket matching the battle-axe rotor
Dimensions
HM3 Frog: 47 × 50 × 18mm
MoonMachine: 47 × 50 × 19mm
HM3 FrogX: 48.3 × 52.7 × 17.5mm
Case components
HM3 Frog: 53
MoonMachine: 55
HM3 FrogX: 59
Water resistance
30 m/3 ATM
Strap
Alligator
Buckle
Folding buckle, titanium or 18K red gold matching the case
Introduction
HM3 Frog: 2010
HM3 FrogX: 2020
MoonMachine: 2012
Years in production
HM3 Frog: 2010-2014
HM3 FrogX: 2020 – in production
MoonMachine: 2012-2014
Quantity produced
HM3 Frog: 110 + 3 prototypes
HM3 FrogX (at the end of 2020): 10
MoonMachine: 54 + 3 prototypes
Comments
The FrogX celebrates the 10th anniversary of the HM3 Frog in a full sapphire crystal case, with a hidden message in its crown.

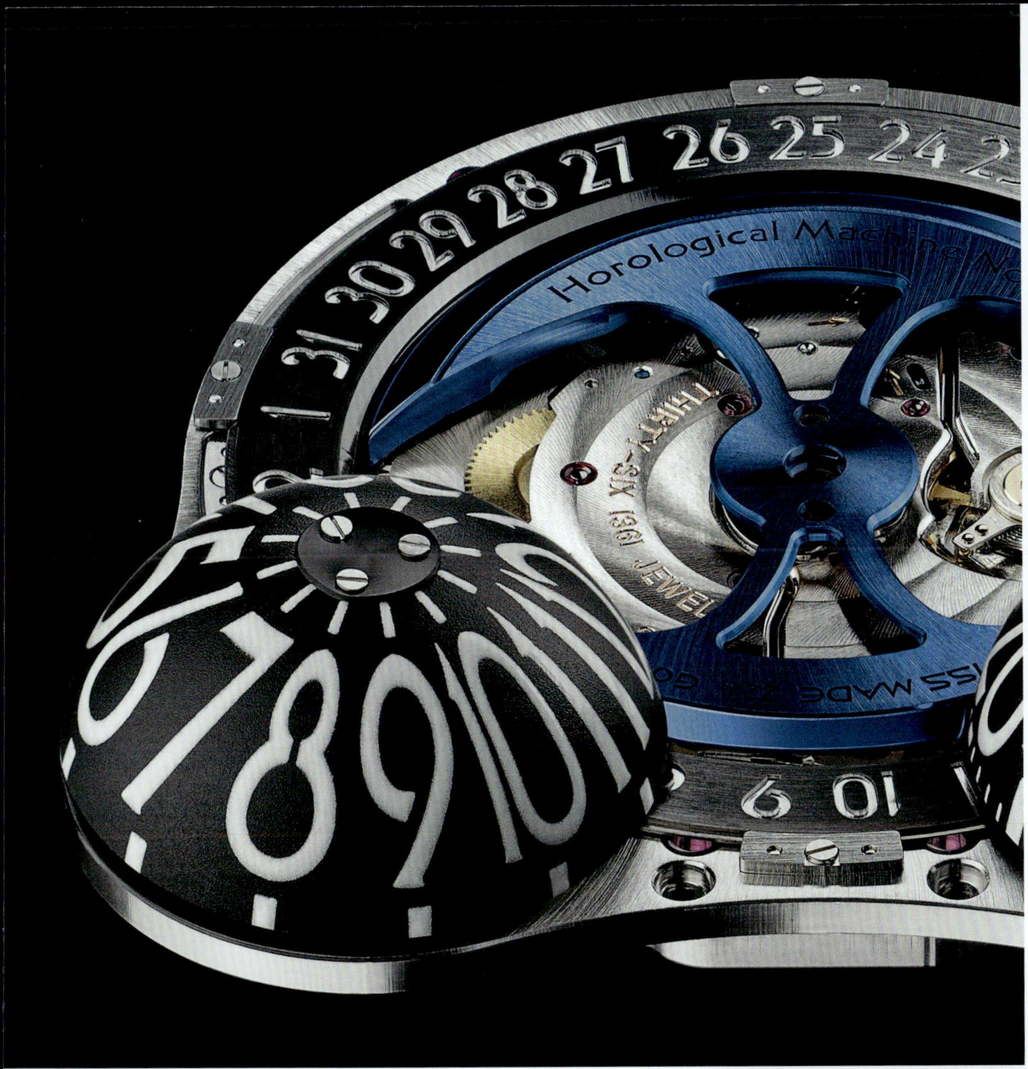
Horological Machine No
THIRTY-SIX (36) JEWELS

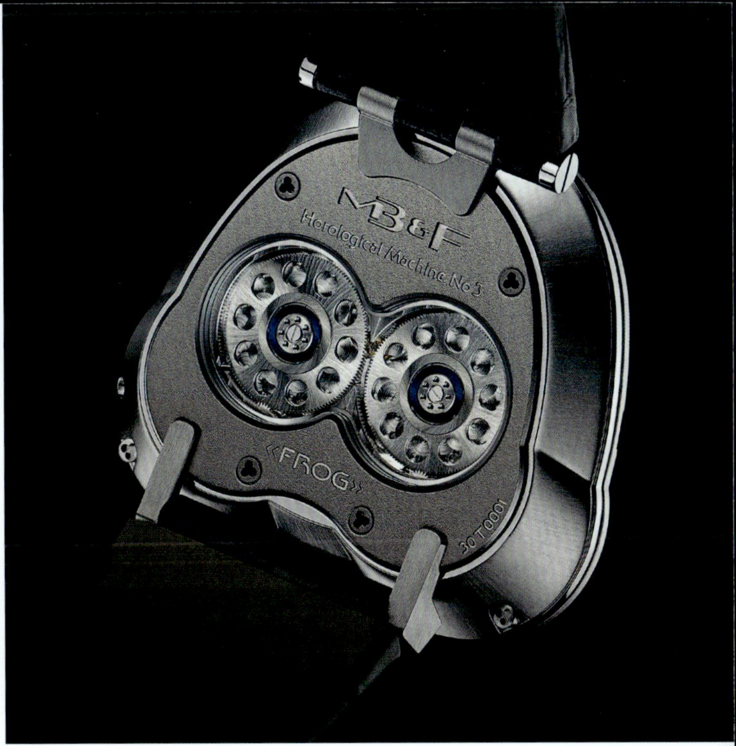
MB&F
Horological Machine No3
«FROG»

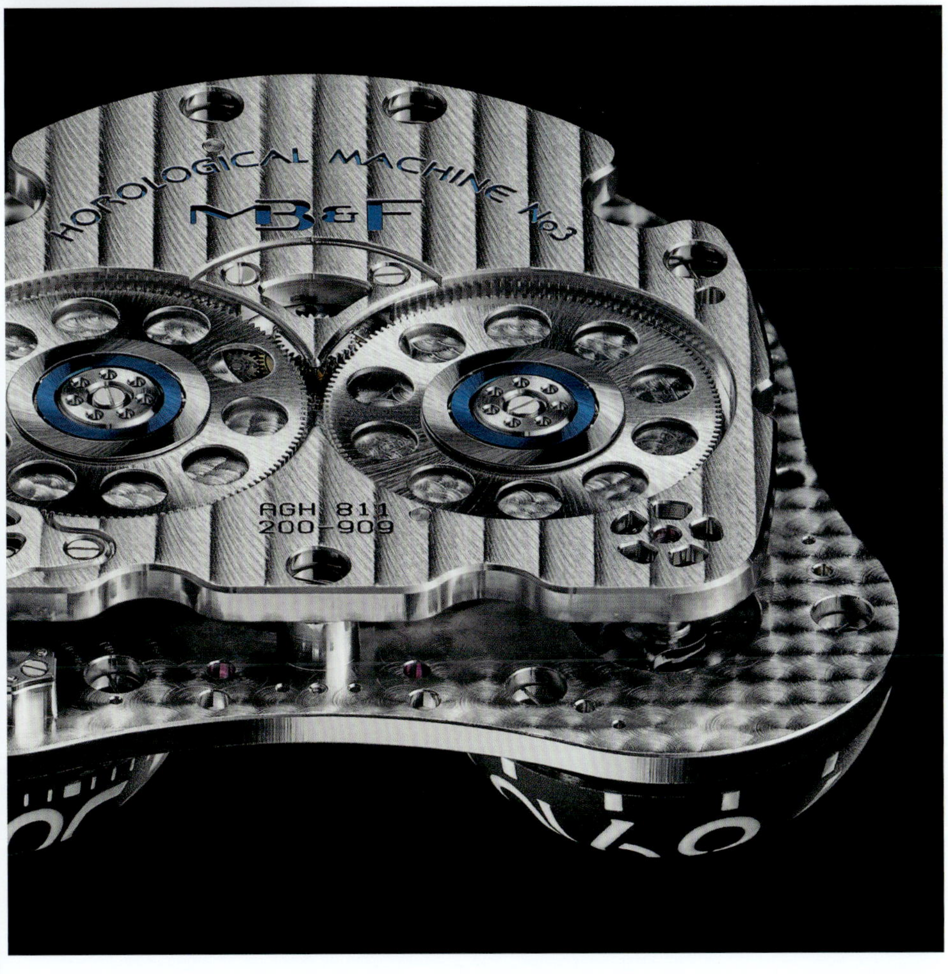
HOROLOGICAL MACHINE No3
MB&F
AGH 811
200-909

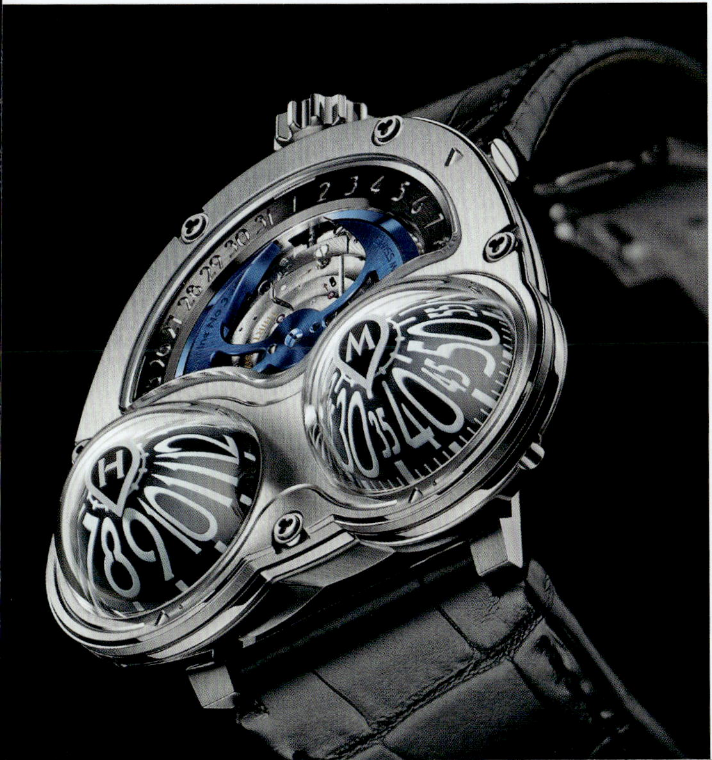

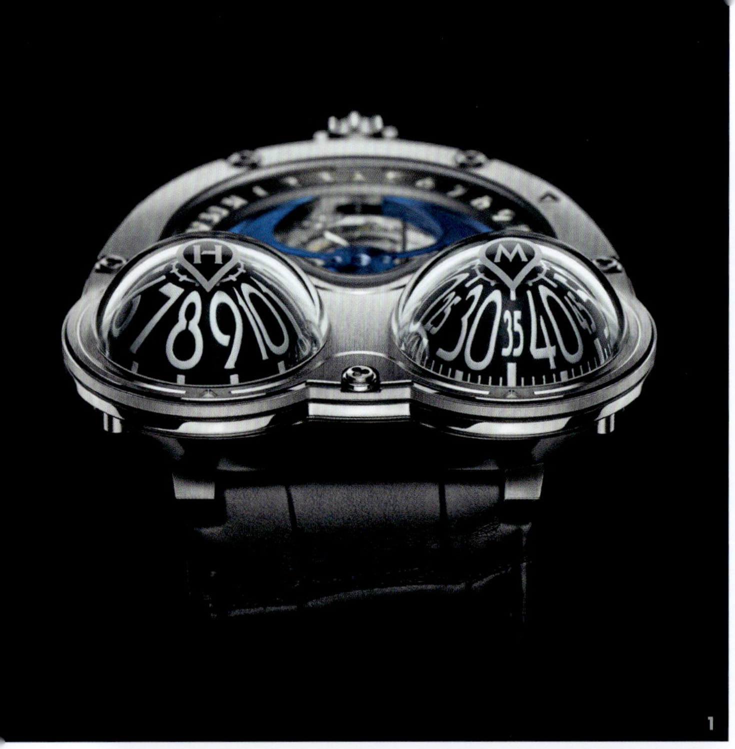

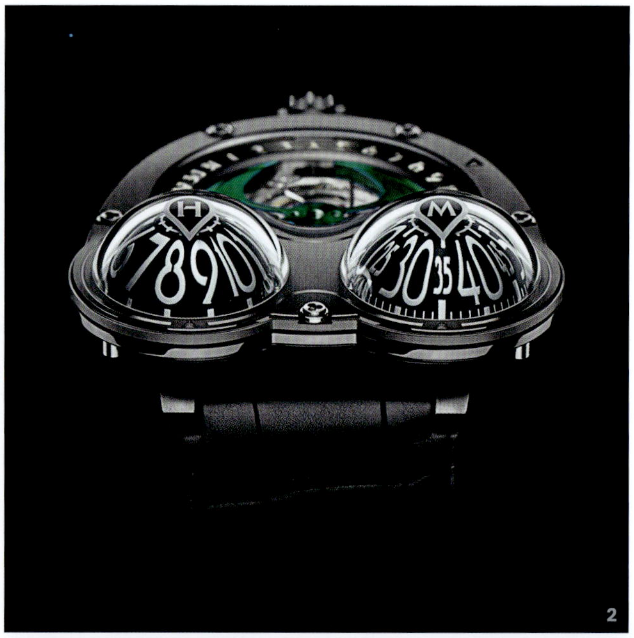

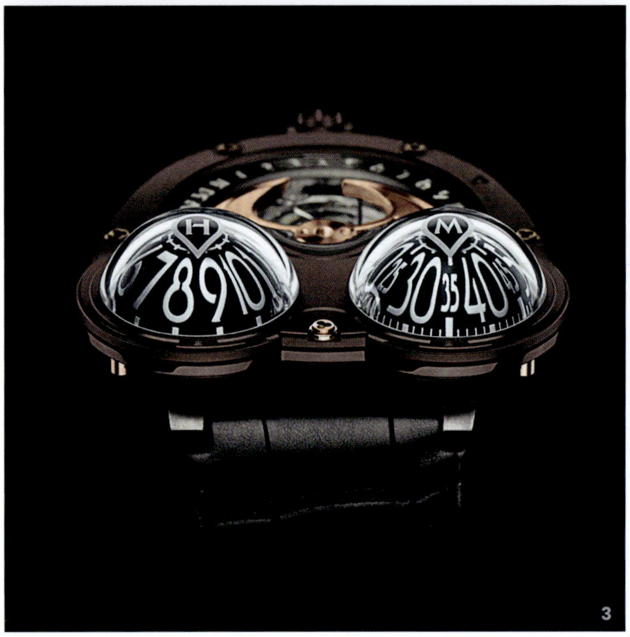

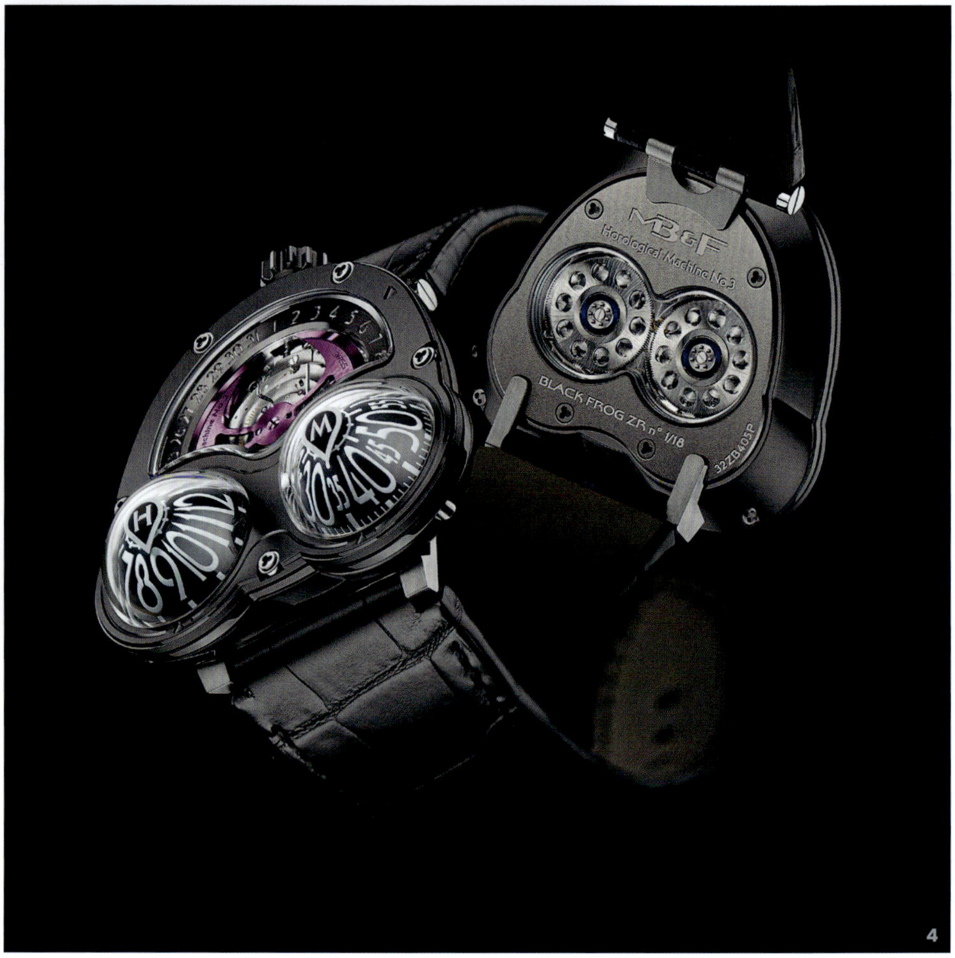

1 Horological Machine N°3
Frog Ti
Reference: 32.TL.B
Material: grade 5 titanium with blue "battle-axe" rotor
Limitation: –
Introduction: 2010
Years in production: 2010-2012
Quantity produced: 50 + 1 prototype

2 Horological Machine N°3
Frog Ti Black
Reference: 32.BTL.B
Material: black PVD coated grade 5 titanium with green "battle-axe" rotor
Limitation: 12
Introduction: 2010
Year in production: 2010
Quantity produced: 12

3 Horological Machine N°3
Frog Chocolate (Purists 10th Anniversary edition)
Reference: 32.TC.B
Material: chocolate PVD coated grade 5 titanium with 22K red gold "battle-axe" rotor
Limitation: 10
Introduction: 2011
Year in production: 2011
Quantity produced: 10 + 1 prototype

4 Horological Machine N°3
Frog ZR
Reference: 32.ZRBL.B
Material: black PVD coated zirconium with purple "battle-axe" rotor
Limitation: 18
Introduction: 2011
Year in production: 2011
Quantity produced: 18

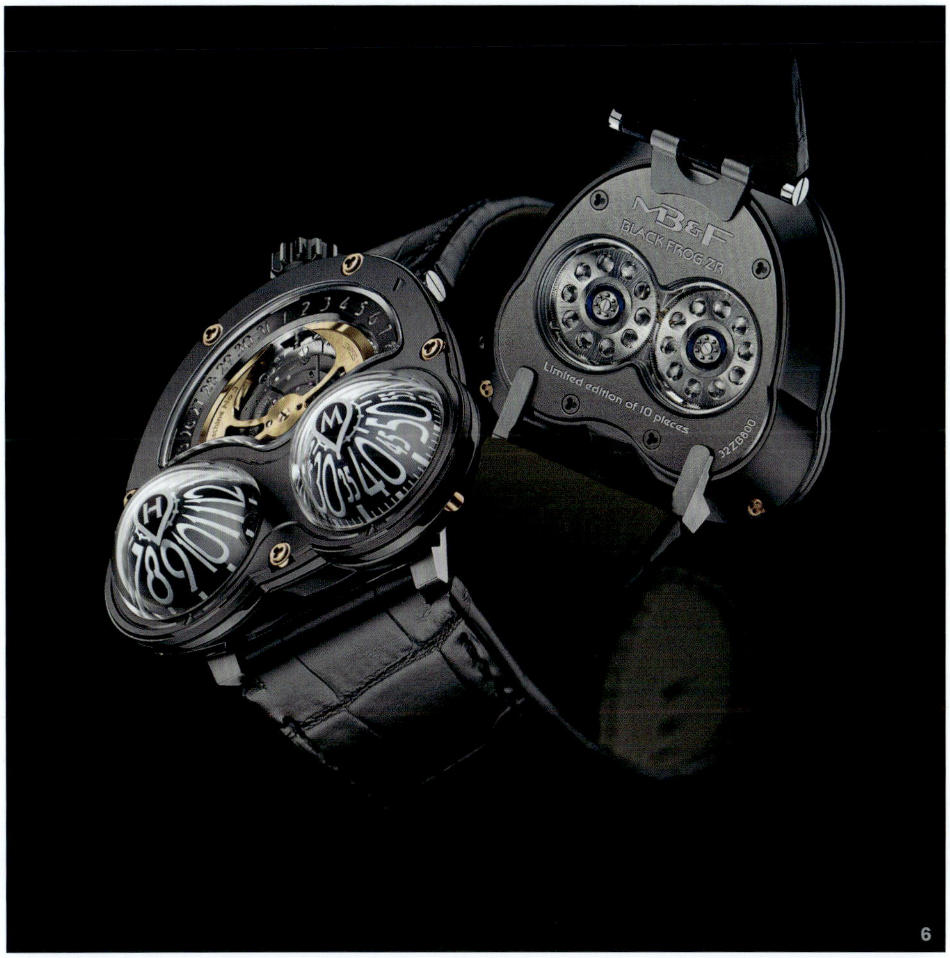

6

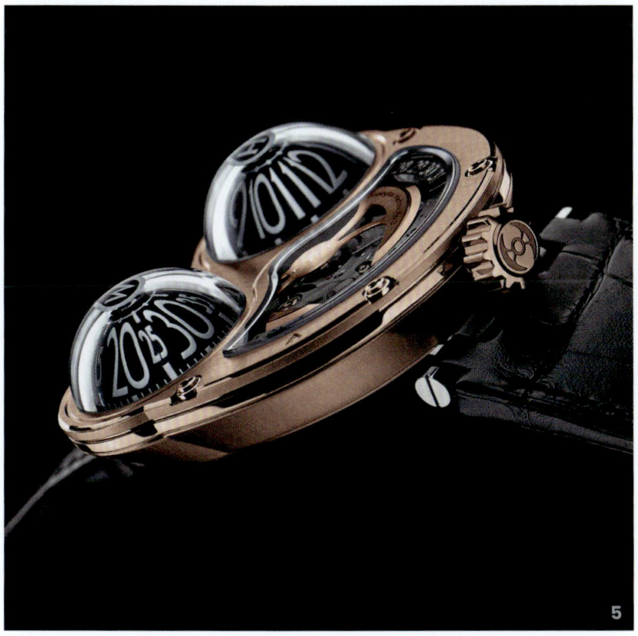

5

5 Horological Machine N°3
Fire Frog
Reference: 32.RL.B
Material: 18K red gold with
22K red gold "battle-axe" rotor
Limitation: 10
Introduction: 2012
Years in production:
2012–2014
Quantity produced: 10

6 Horological Machine N°3
Poison Dart Frog
Reference: 32.ZRBRL.B
Material: black PVD coated
zirconium and yellow gold, with
yellow gold "battle-axe" rotor
Limitation: 10
Introduction: 2012
Years in production:
2012–2013
Quantity produced: 10 + 1
prototype

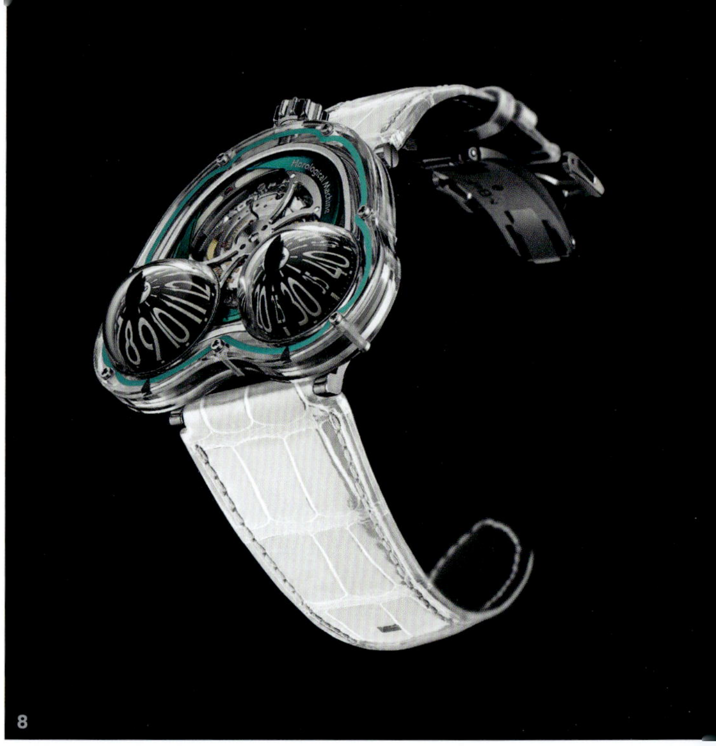

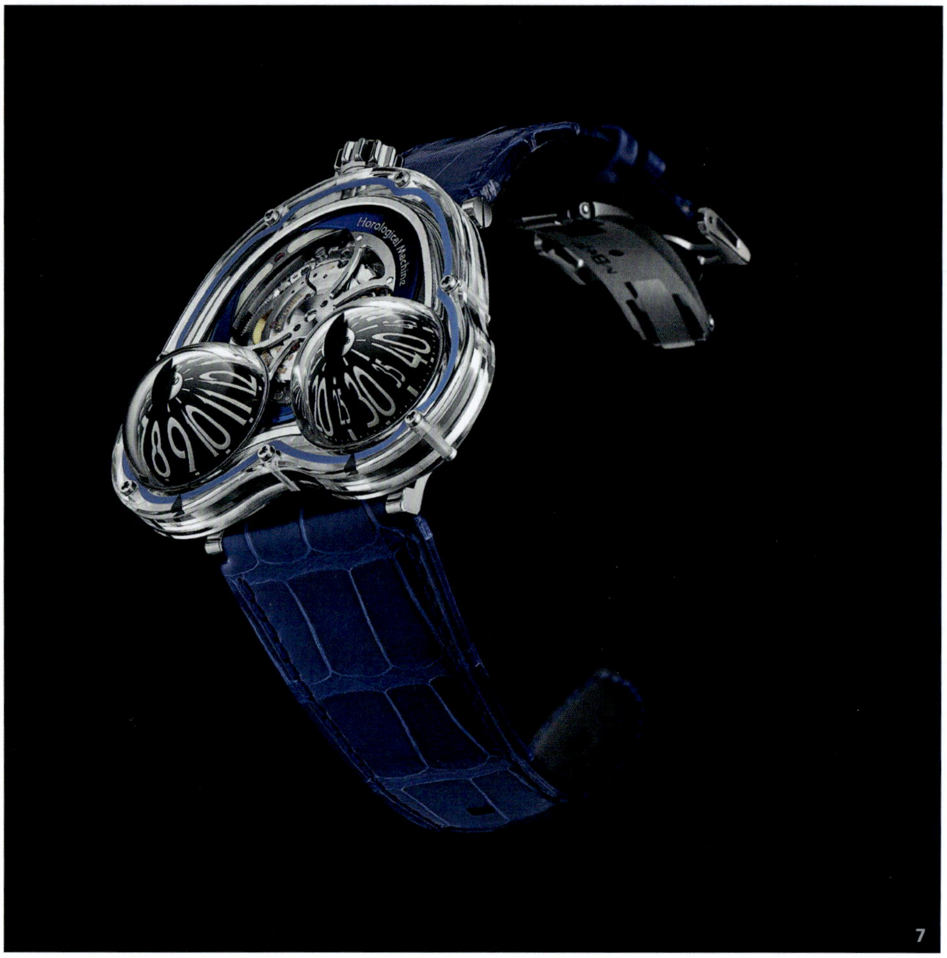

7 Horological Machine N°3
FrogX Blue
Reference: 36.SVL.BU
Material: sapphire crystal
with blue "battle-axe" rotor
Limitation: 10
Introduction: 2020
Years in production:
2020 – in production
Quantity produced (at the
end of 2020): 3

8 Horological Machine N°3
FrogX Turquoise
Reference: 36.SVL.GR
Material: sapphire crystal
with turquoise "battle-axe" rotor
Limitation: 10
Introduction: 2020
Years in production:
2020 – in production
Quantity produced (at the
end of 2020): 3

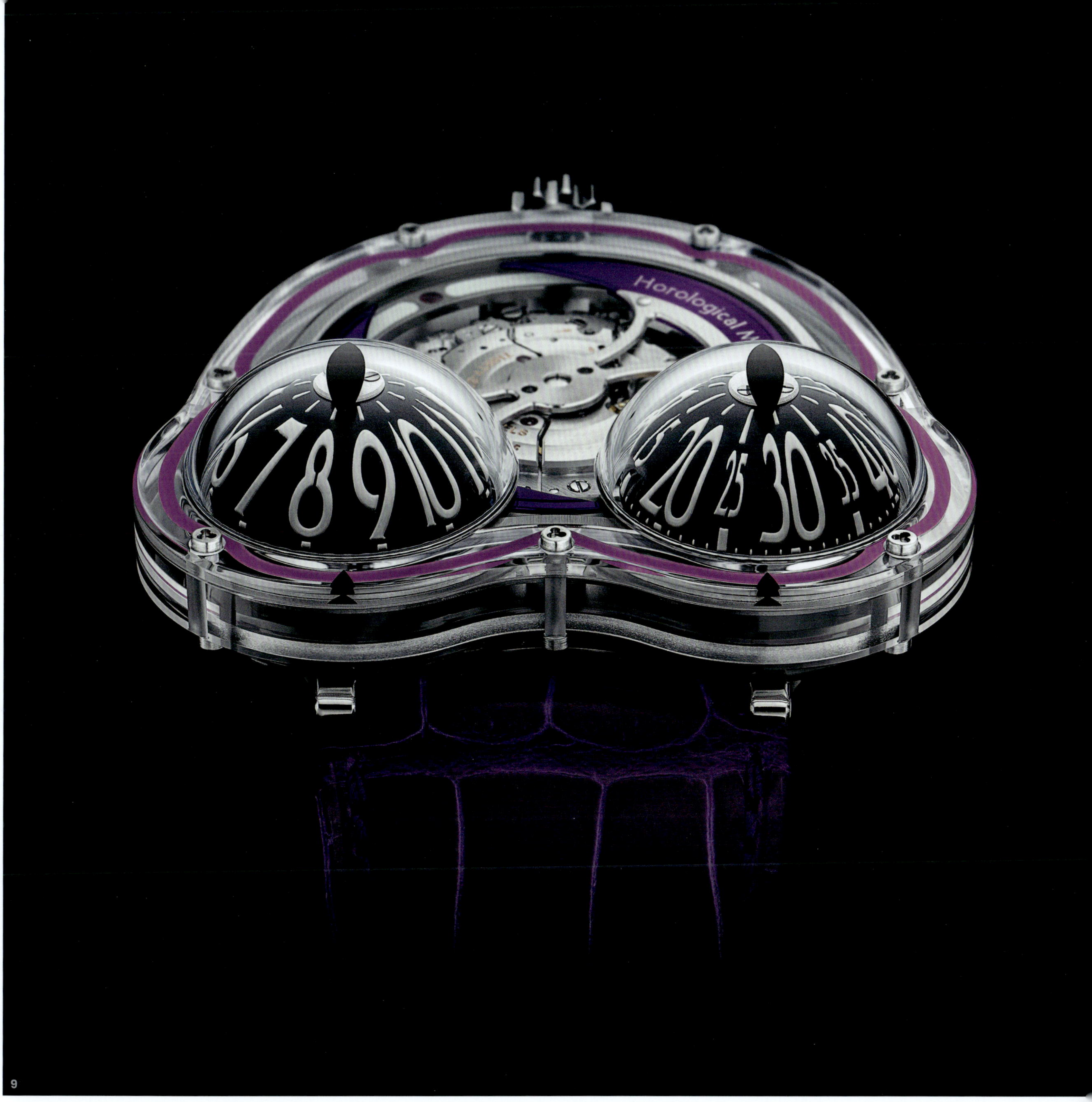

9

9 **Horological Machine N°3**
FrogX Purple
Reference: 36.SVL.PU
Material: sapphire crystal with purple "battle-axe" rotor
Limitation: 10
Introduction: 2020
Years in production: 2020 – in production
Quantity produced (at the end of 2020): 4

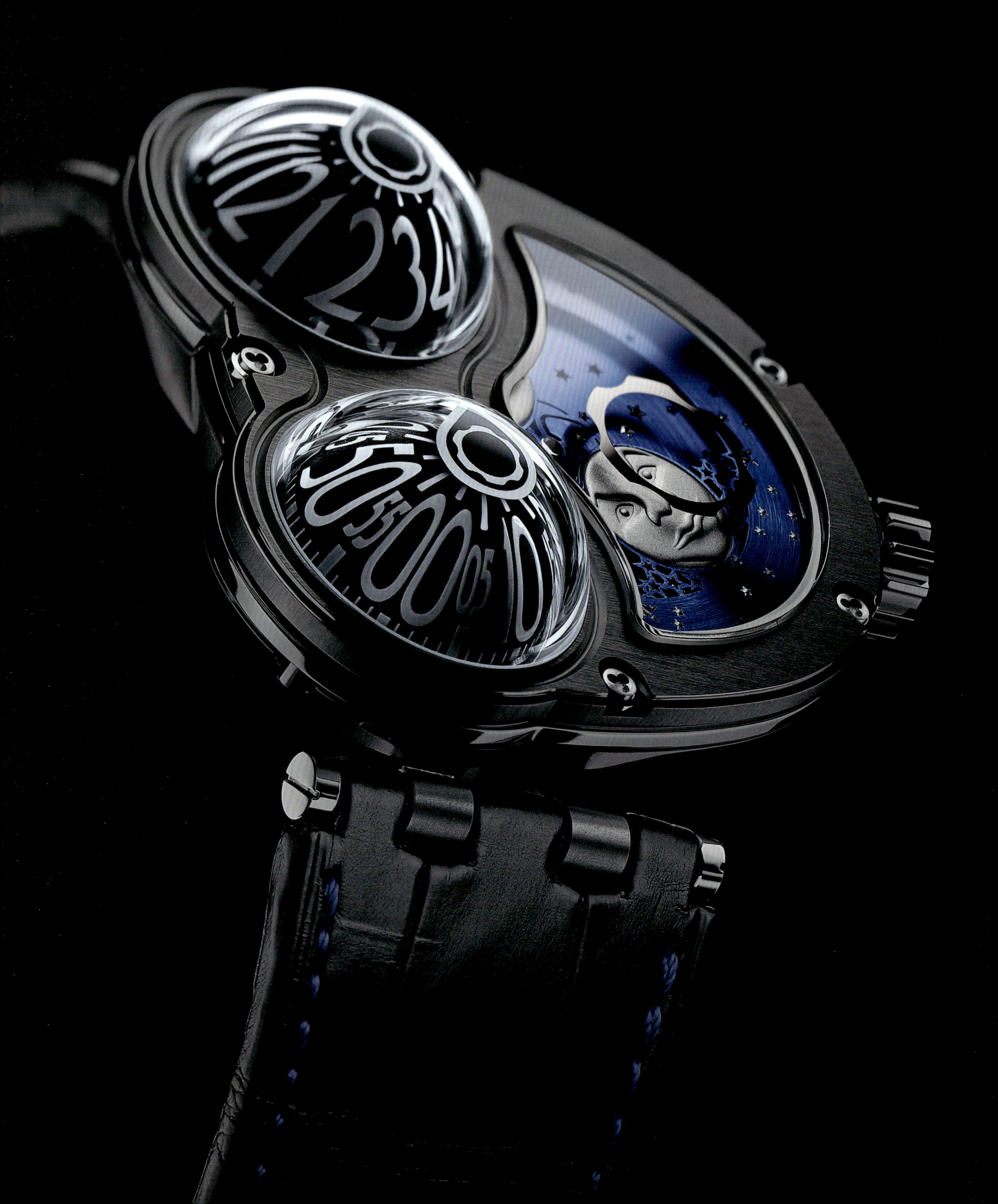

MOONMACHINE MB&F × STEPAN SARPANEVA

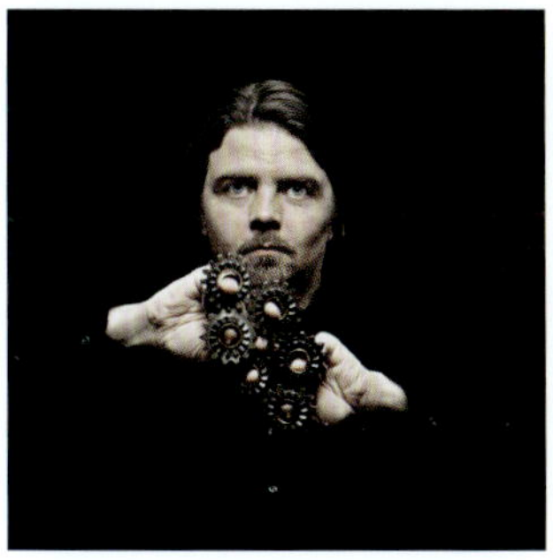

STEPAN SARPANEVA

Stepan Sarpaneva was born in 1970 in Turku, Finland. Prior to launching his own brand Sarpaneva Watches in 2003, he worked for several prestigious Swiss brands including Piaget and Parmigiani, alongside fellow Finnish MB&F "Friend" Kari Voutilainen.

Three themes are central to Sarpaneva's creations: a distinctive moon face, based on his own likeness; northern stars and constellations as seen from his native Finland; and his "Korona" case, inspired by the glowing corona around the Sun. Growing up in Finland, where the winter months are characterized by long, dark nights and limited sunlight, Sarpaneva found inspiration in the lingering moon and stars of the Finnish night sky. He would draw on these celestial images in his collaboration with MB&F. Taking the HM3 Frog as his starting point, chosen for its large, open dial, he set to work on what would become the MoonMachine.

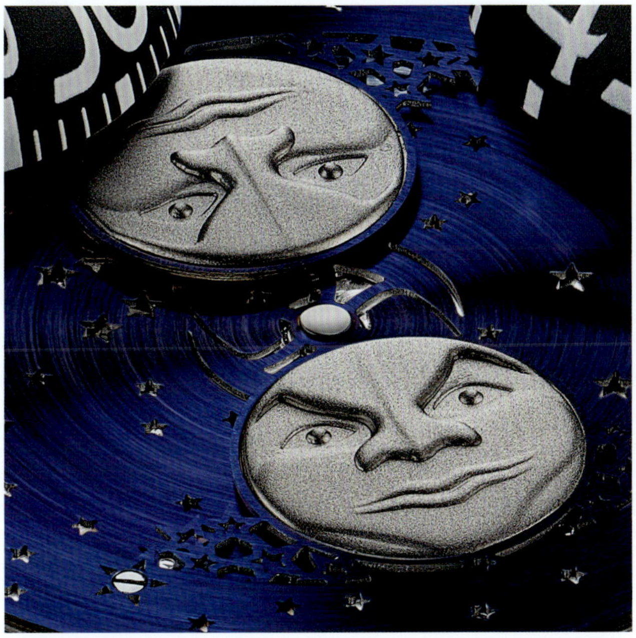

The MoonMachine is MB&F's first performance art project to feature another watchmaker, and the first performance art piece to employ a new complication – the moon phase.

Here the exposed winding rotor of the HM3 Frog is replaced by a new, multi-layered rotor: a mystery rotor, featuring a northern sky full of laser-cut constellations. The stars are carefully chosen to display Ursa Major and Ursa Minor, including Polaris, the North Star, and ingeniously placed so that one of the stars provides access to a servicing tool. Whirring beneath the stars, the machine's movement adds shine and the occasional twinkle.

The sapphire domes that were formerly the "eyes" of the HM3 Frog (indicating hours and minutes) now resemble something closer to a moon colony under a starry sky.

Twin, hand-finished moon faces move according to the Lunar cycle, displaying the phases of the moon through the bright, crenellated aperture of Sarpaneva's "Korona" ring.

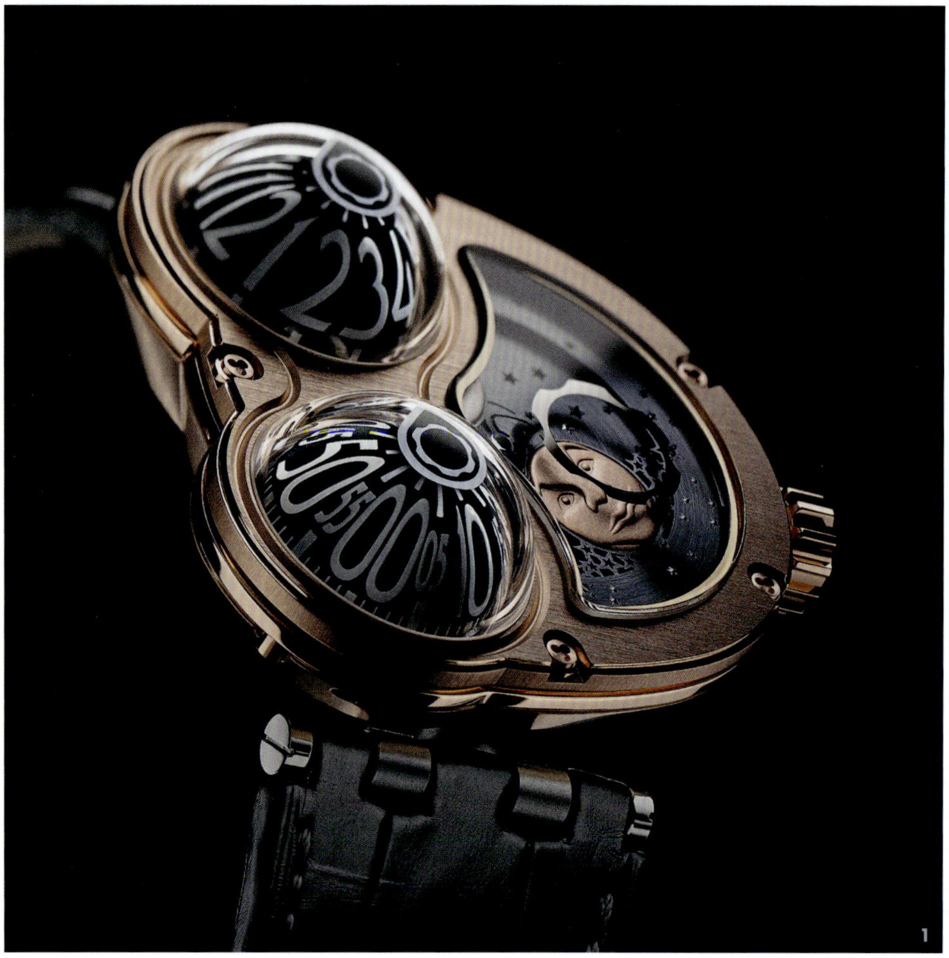

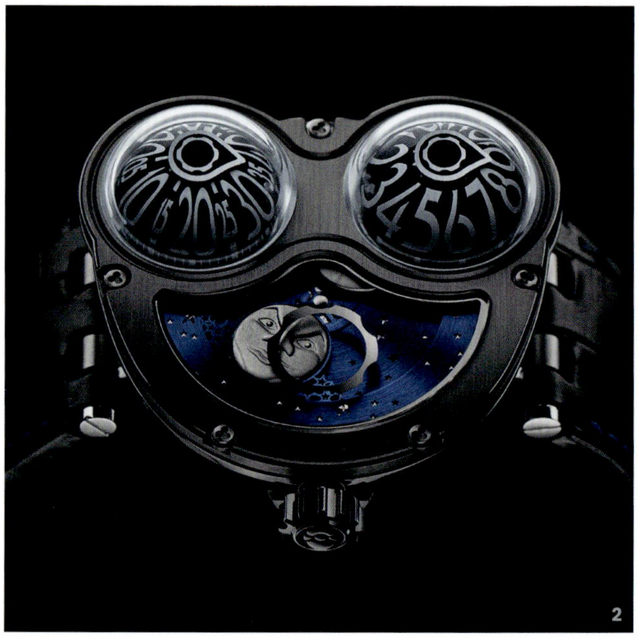

1 MoonMachine RG
Reference: 34.RL.B
Material: 18K red gold case with red gold moon face on anthracite sky
Limitation: 18
Introduction: 2012
Years in production: 2012–2013
Quantity produced: 18 + 1 prototype

2 MoonMachine Black Ti
Reference: 34.BTL.B
Material: black PVD coated grade 5 titanium case with white gold moon face on dark blue sky
Limitation: 18
Introduction: 2012
Years in production: 2012–2014
Quantity produced: 18 + 1 prototype

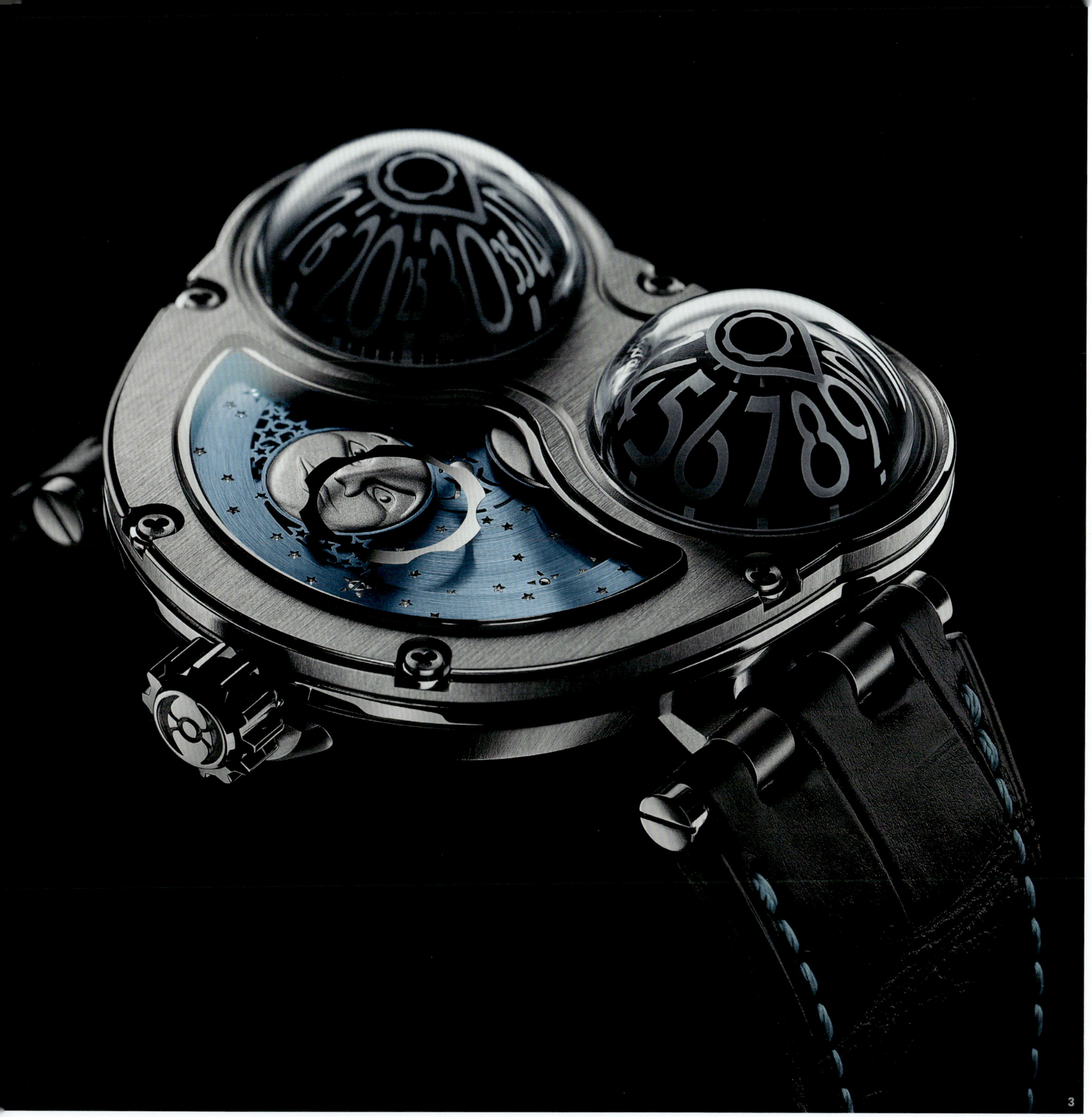

3

3 MoonMachine Ti
Reference: 34.TL.B
Material: grade 5 titanium
case with white gold moon face
on light blue sky
Limitation: 18
Introduction: 2012
Years in production:
2012–2014
Quantity produced:
18 + 1 prototype

HM3

MEGAWIND

The concept behind Horological Machine N°3 MegaWind, together with the name, came from MB&F "Friend" Stepan Sarpaneva. Released in 2013, the timepiece follows the principles established by Horological Machine N°3, with a few important changes.

Like the HM3, the MegaWind flips the construction of a traditional automatic winding movement on its head – literally. The winding rotor, for which the machine is named, is an oversized version of the iconic MB&F battle-axe winding rotor, displayed under a dial of crystal-clear sapphire. To showcase the winding rotor in all its glory, the date wheel of the HM3 is removed, revealing more of the movement on the dial side. Twin cones wrapped in clear sapphire stand clear of the dial to show the hours and minutes. At the top of the cones, taking the place of the day/night indicator on the HM3, are now mirror indicators that enhance the legibility of the numerals.

HOROLOGICAL MACHINE Nº3 MEGAWIND

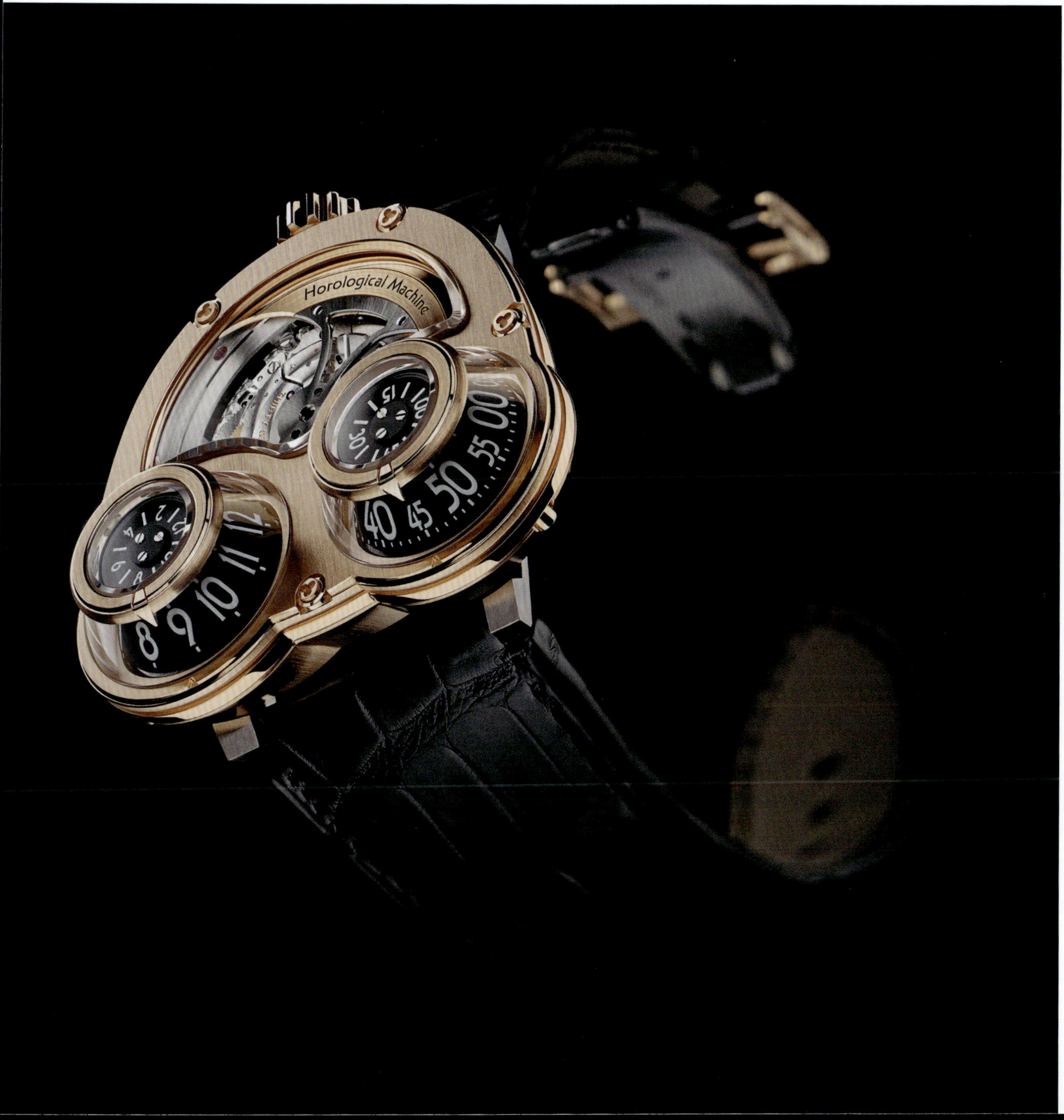

HM3 MEGAWIND CALIBRE

The MegaWind movement was designed by Jean-Marc Wiederrecht, winner of the 2007 award for Best Watchmaker at the Grand Prix d'Horlogerie de Genève, based on a drawing by Maximilian Büsser and accomplished designer Eric Giroud.

The MegaWind's battle-axe winding rotor is dubbed a "mystery" rotor because unlike a conventional rotor it appears to be symmetrically balanced. In fact, the so-called "mystery" is all down to engineering: machining away the underside of one half of the battle-axe produces a razor-sharp edge, thus reducing its mass relative to the opposing side.

Two large, high-tech ceramic bearings at the bottom of the MegaWind transfer the power from the rotor back up the dial side to drive the hours and minutes. A traditional system of pinions would have caused increased friction and required support from both top and bottom. Using the optimised ceramic bearings reduces friction and requires less support, allowing for a thinner movement.

Collection
Horological Machines
Series
Horological Machine N°3 MegaWind
Common name
HM3 MegaWind, MegaWind
Movement
Developed for MB&F by Jean-Marc Wiederrecht (Agenhor), powered by a Girard-Perregaux base
Winding
Automatic
Power reserve
45 hours
Components
270/36 jewels
Functions
Hours (on one cone) and minutes (on second cone)
Add. features
22K gold large "battle-axe" winding rotor. Time indications transmitted via ceramic ball bearings Final edition: Super-LumiNova numerals and band under the rotor
Case
18K red or white gold with grade 5 titanium
Dimensions
47 × 50 × 17 mm
Case components
52
Water resistance
30 m/3 ATM
Strap
Alligator
Buckle
Folding buckle, 18K red or white gold with titanium matching the case
Introduction
2013
Years in production
2013–2015
Quantity produced
91 + 5 prototypes

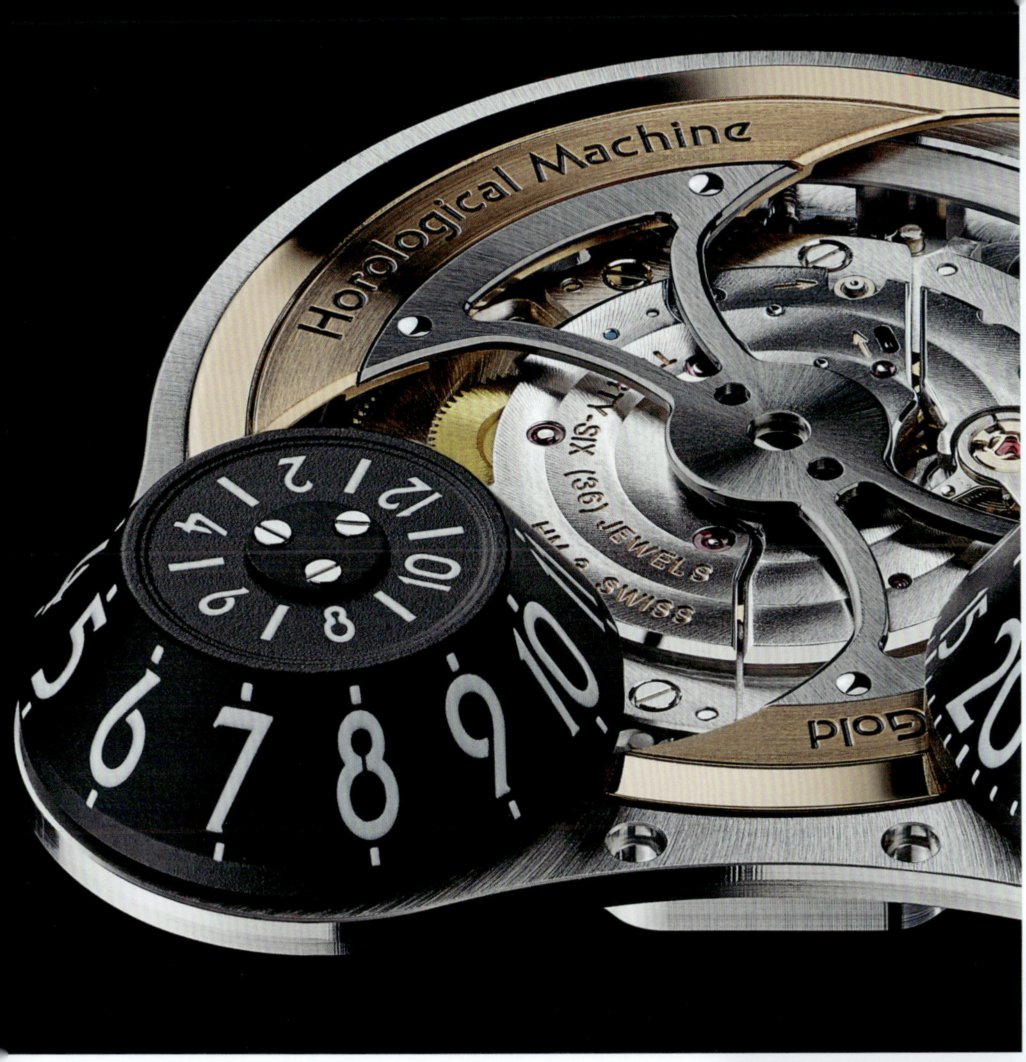
Horological Machine
JEWELS
SWISS
Gold

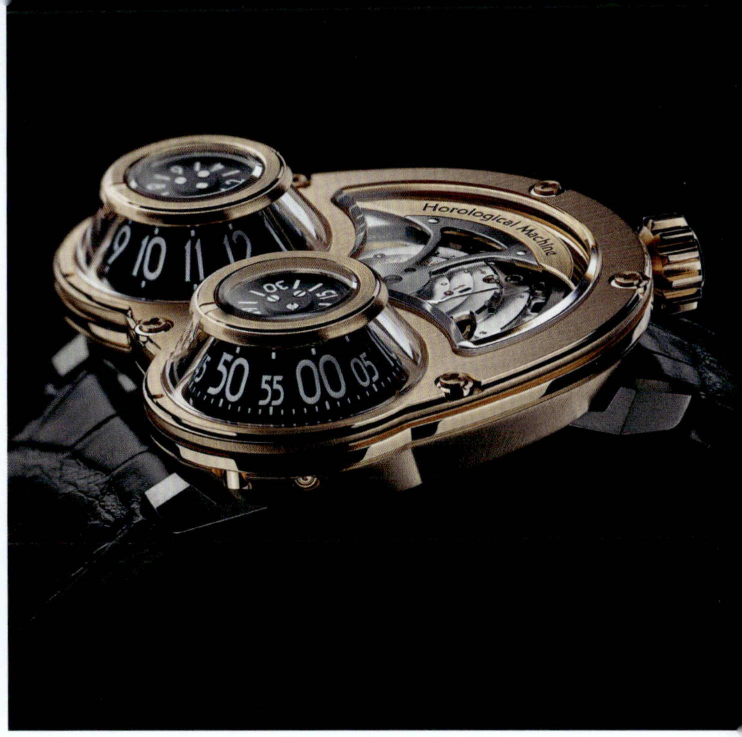
Horological Machine

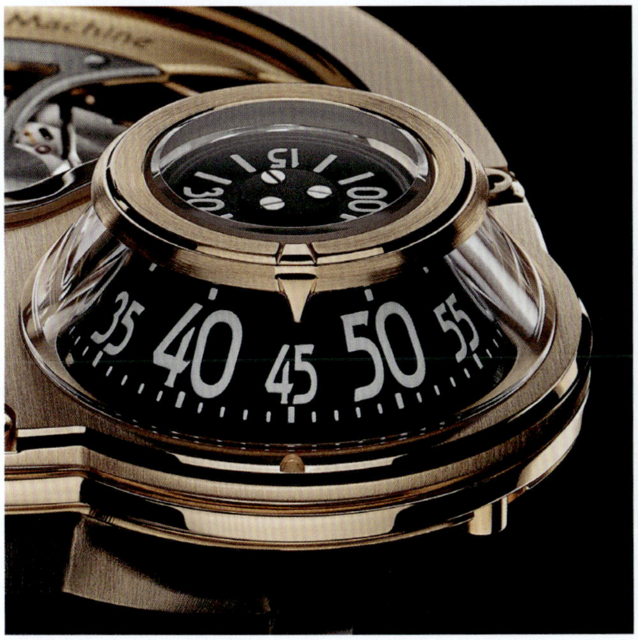

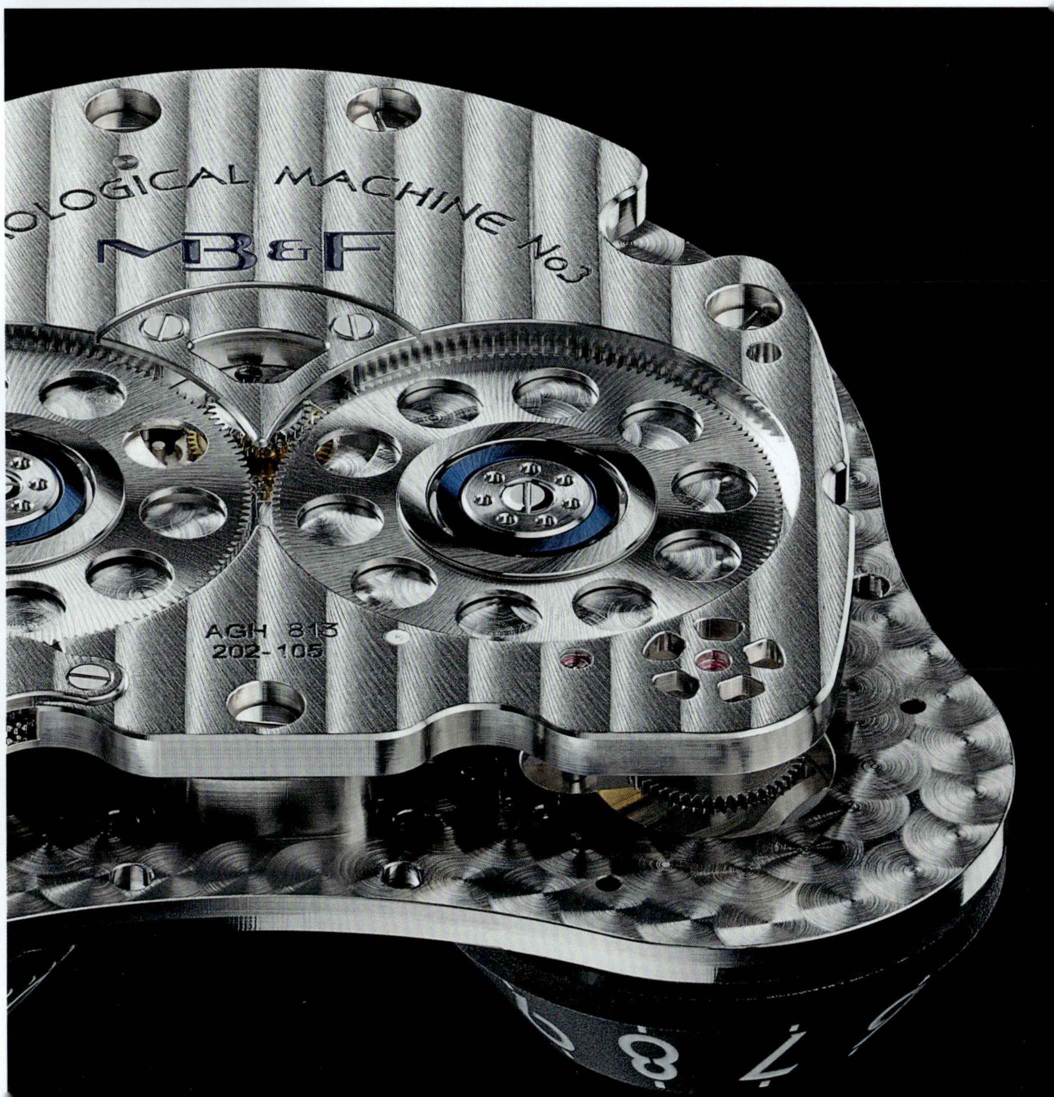
MB&F
AGH 813
202-105

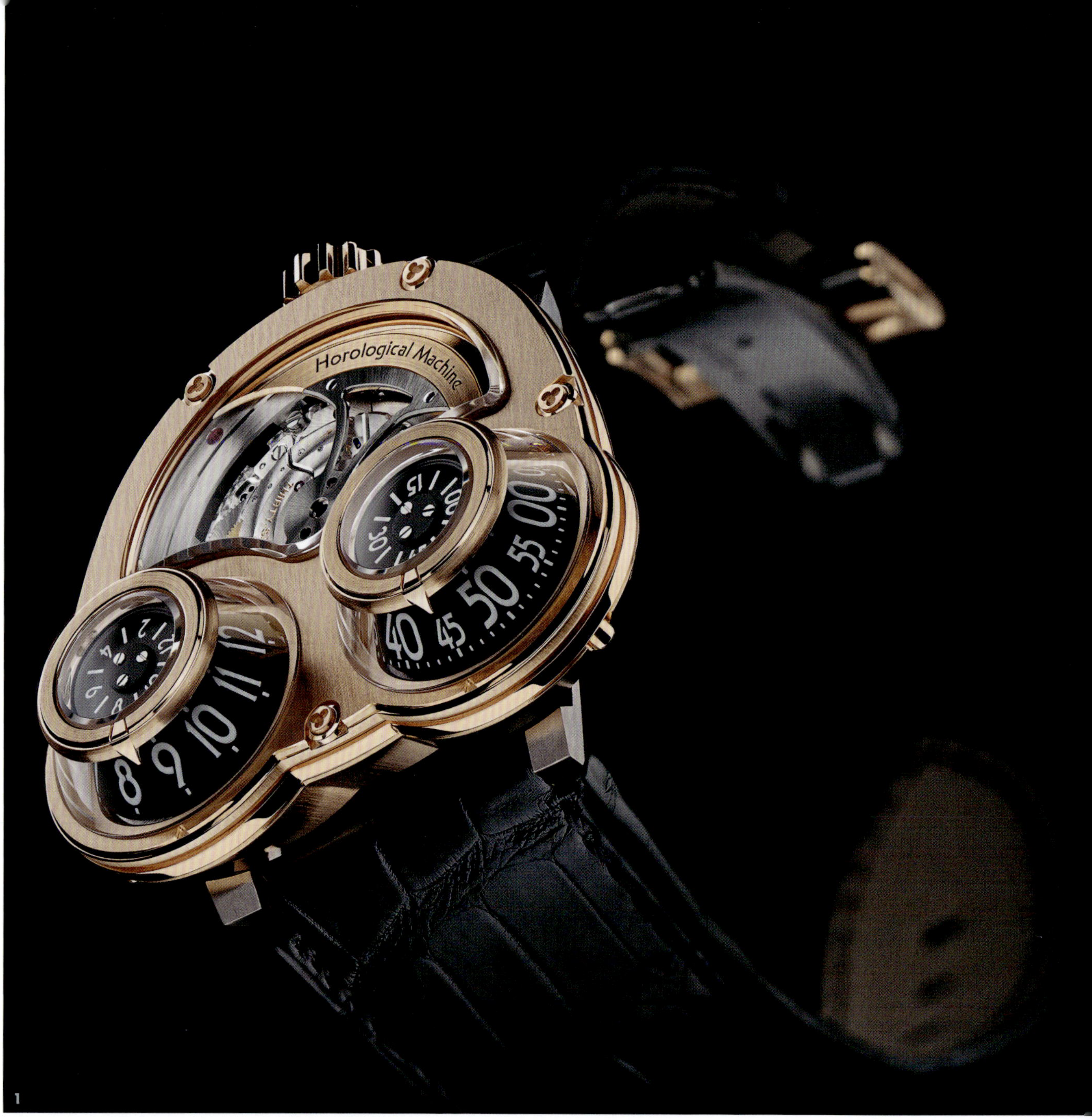

1

1 Horological Machine N°3
MegaWind RG
Reference: 35.RTL.B
Material: 18K red gold and
grade 5 titanium
Limitation: –
Introduction: 2013
Years in production:
2013–2015
Quantity produced:
46 + 2 prototypes

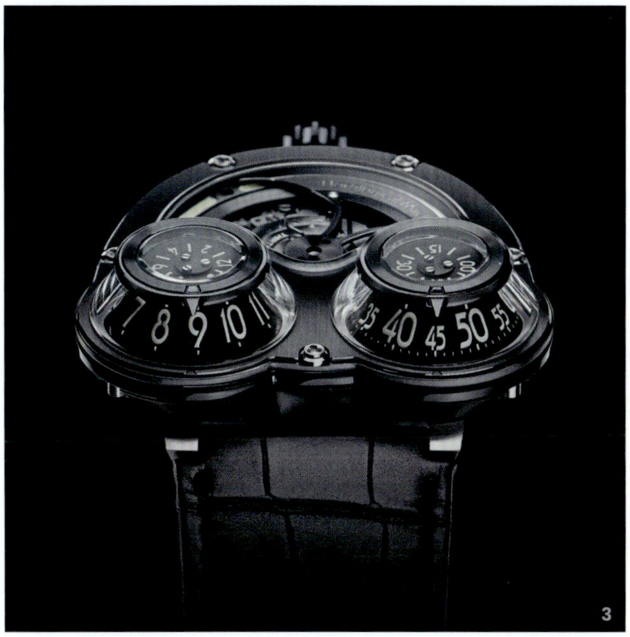

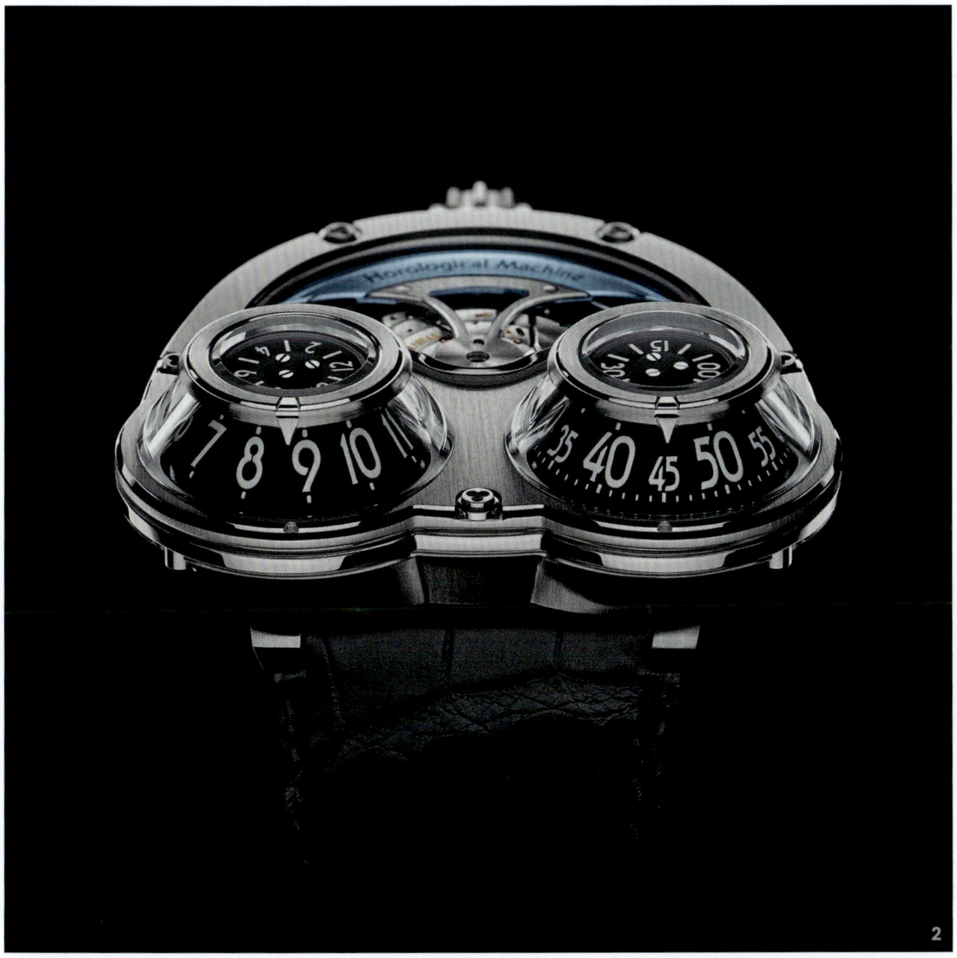

2 **Horological Machine N°3 MegaWind WG**
Reference: 35.WTL.B
Material: 18K white gold and grade 5 titanium
Limitation: –
Introduction: 2013
Years in production: 2013–2015
Quantity produced: 20 + 2 prototypes

3 **HM3 MegaWind Final Edition**
Reference: 35.WBTL.B
Material: black PVD coated 18K white gold and grade 5 titanium; Super-LumiNova numerals and band under the rotor
Limitation: 25
Introduction: 2015
Years in production: 2015
Quantity produced: 25 + 1 prototype

HM4

THUNDERBOLT

Horological Machine N°4 is not a traditional wristwatch. First released in 2010, the machine was dubbed the "Thunderbolt" for its resemblance to the A-10 Thunderbolt aircraft, a reflection of Maximilian Büsser's childhood enthusiasm for model airplane kits. Büsser recalls his childhood bedroom hung from floor to ceiling with assembled planes. Model aircraft were the first thing he saw when waking up in the morning, and the last thing he saw at night. But the A-10 was not a typical aircraft, and neither is the HM4 a typical timepiece. Twin shafts, capped by two displays showing power reserve on the left, and hours and minutes on the right, reach out across the case of the HM4, dominating the physical space. At the centre of the dial, a transparent sapphire cover exposes the machine's whirring balance wheel.

The case looks almost impossibly complicated, but it is in fact constructed of only three sections: two blocks of metal, interrupted by a single block of sapphire. The sapphire block alone required 185 hours of machining and polishing. The two crowns, one for winding and the other for adjusting the time, are located at the end of each shaft, whose sides are decorated with long, functional screws that seal the case shut and make it water resistant.

The first HM4 models, nicknamed "Thunderbolt", were followed by additional iterations including 2011's "Razzle Dazzle" and "Double Trouble" limited editions of 8 pieces each featuring WWII-inspired, hand-painted nose art and real rivets in their titanium fuselages. The HM4 RT was released in 2012, in a limited edition of 18 pieces in red gold, titanium and sapphire. The Final Edition in 2013 was a limited edition of 8 pieces in blackened titanium and sapphire.

HOROLOGICAL MACHINE Nº4 THUNDERBOLT

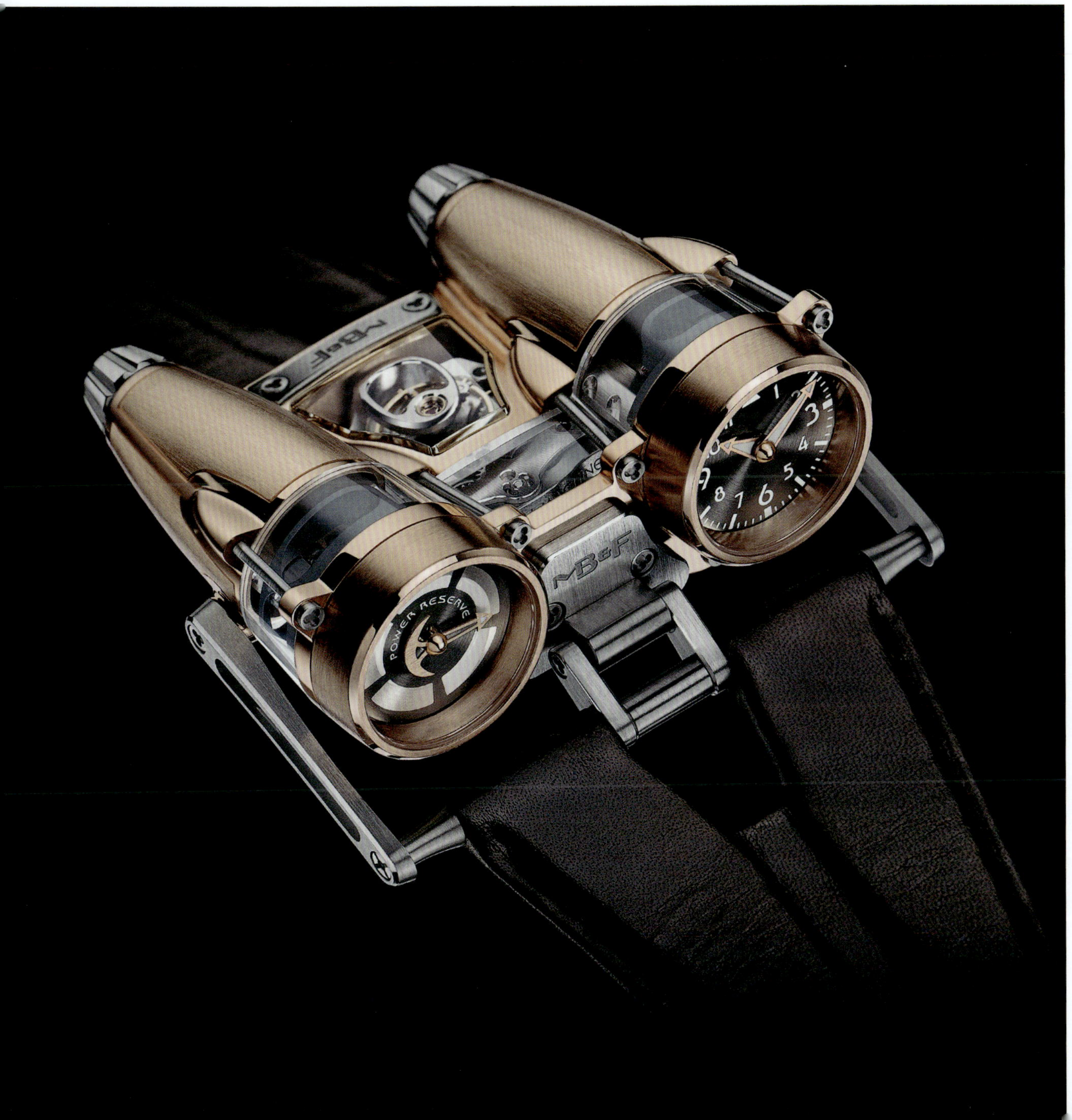

HM4 THUNDERBOLT CALIBRE

The HM4 Calibre was designed and developed by MB&F with Laurent Besse and Beranger Reynard. All of the 311 components for the movement were developed specifically to suit this machine and its radical, never-before-seen three-dimensional case structure. The movement itself mimics the shape of the case, which it fills completely. No off-the-shelf parts were used at all due to the revolutionary nature of the case architecture.

The machine is powered by two mainspring barrels, connected in parallel to supply 72 hours of energy, transferred upwards via vertical gear trains to the twin dials that show power reserve and time. This was the first flat movement to transfer power vertically, applying a technique called renvoi d'angle.

Collection
Horological Machines
Series
Horological Machine N°4 Thunderbolt
Common name
HM4 Thunderbolt, HM4
Movement
Developed for MB&F by Laurent Besse and Beranger Reynard
Winding
Manual
Power reserve
72 hours
Components
311/50 jewels
Functions
Hours and minutes (right dial), power reserve indicator (left dial)
Add. features
Separate crowns for time setting and winding
Case
Grade 5 titanium and sapphire crystal, 18K red gold and sapphire crystal, 18K yellow gold and sapphire crystal or black PVD coated titanium and sapphire crystal
Dimensions
54 × 52 × 24 mm
Components
65 (67 for Final Edition)
Water resistance
30 m/3 ATM
Strap
Calfskin or authentic vintage leather strap crafted from Swiss military bags for Razzle Dazzle and Double Trouble editions
Buckle
Folding buckle, titanium or 18K red gold matching the case
Introduction
2010
Years in production
2010–2013 and 2020 (HM4 Kittyhawk anniversary piece)
Quantity produced
99 + 4 prototypes

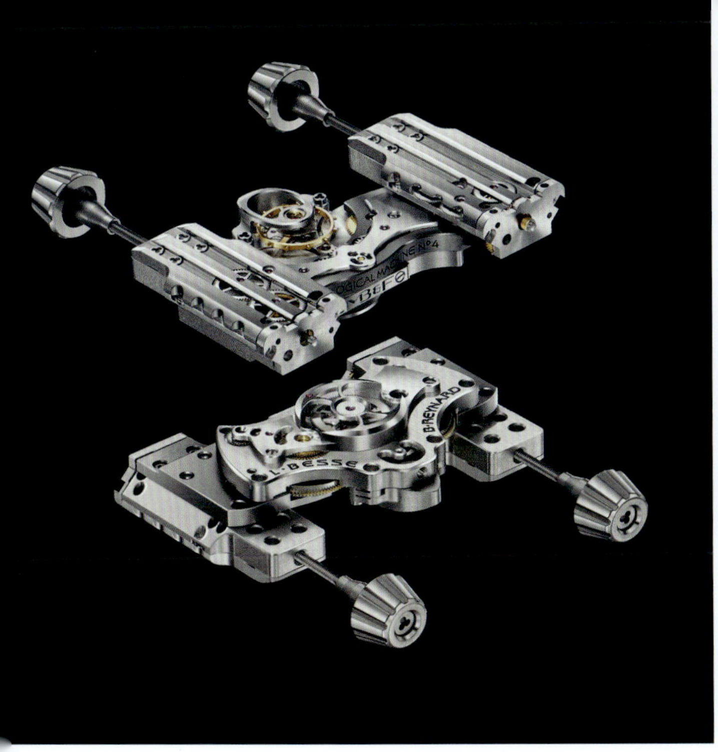

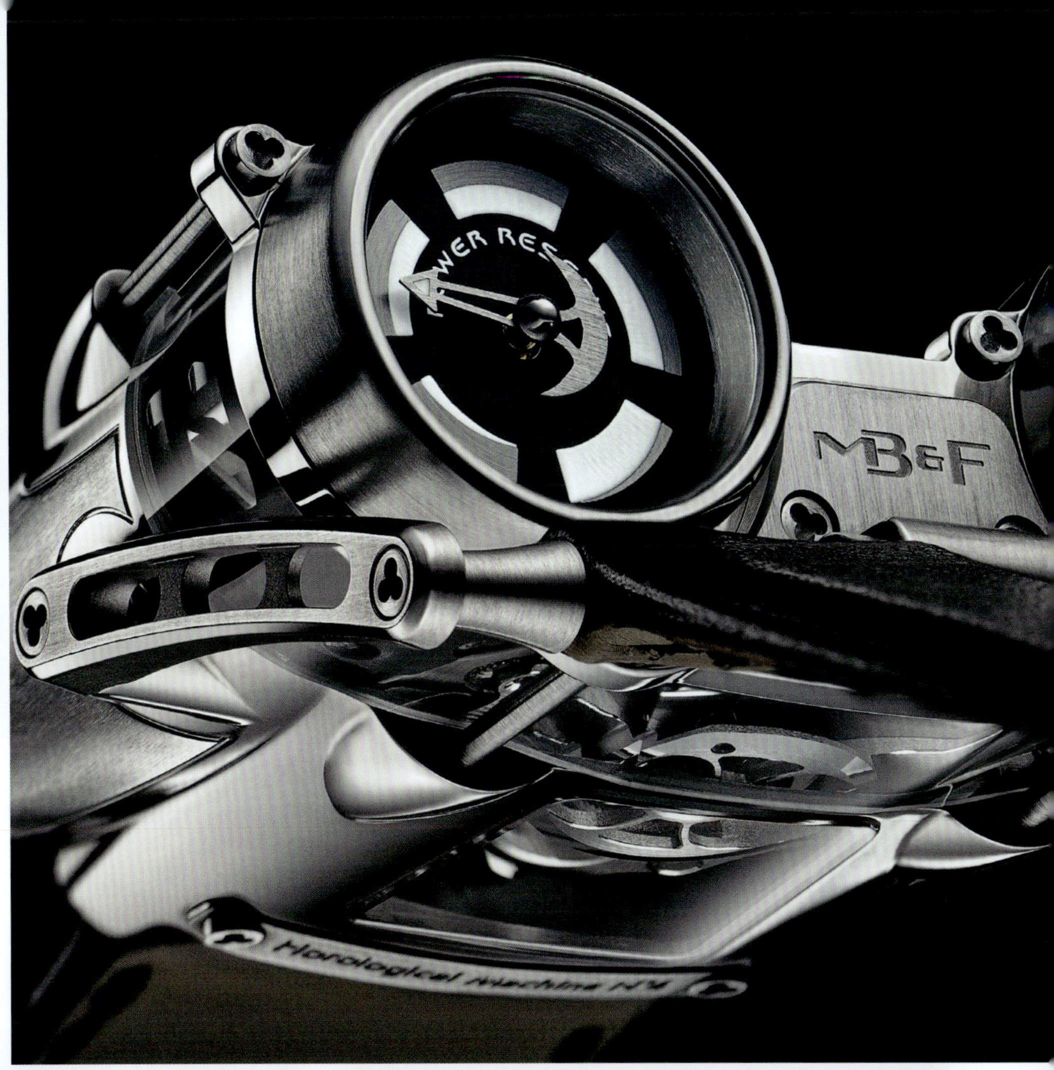
MB&F
Horological machine N°4

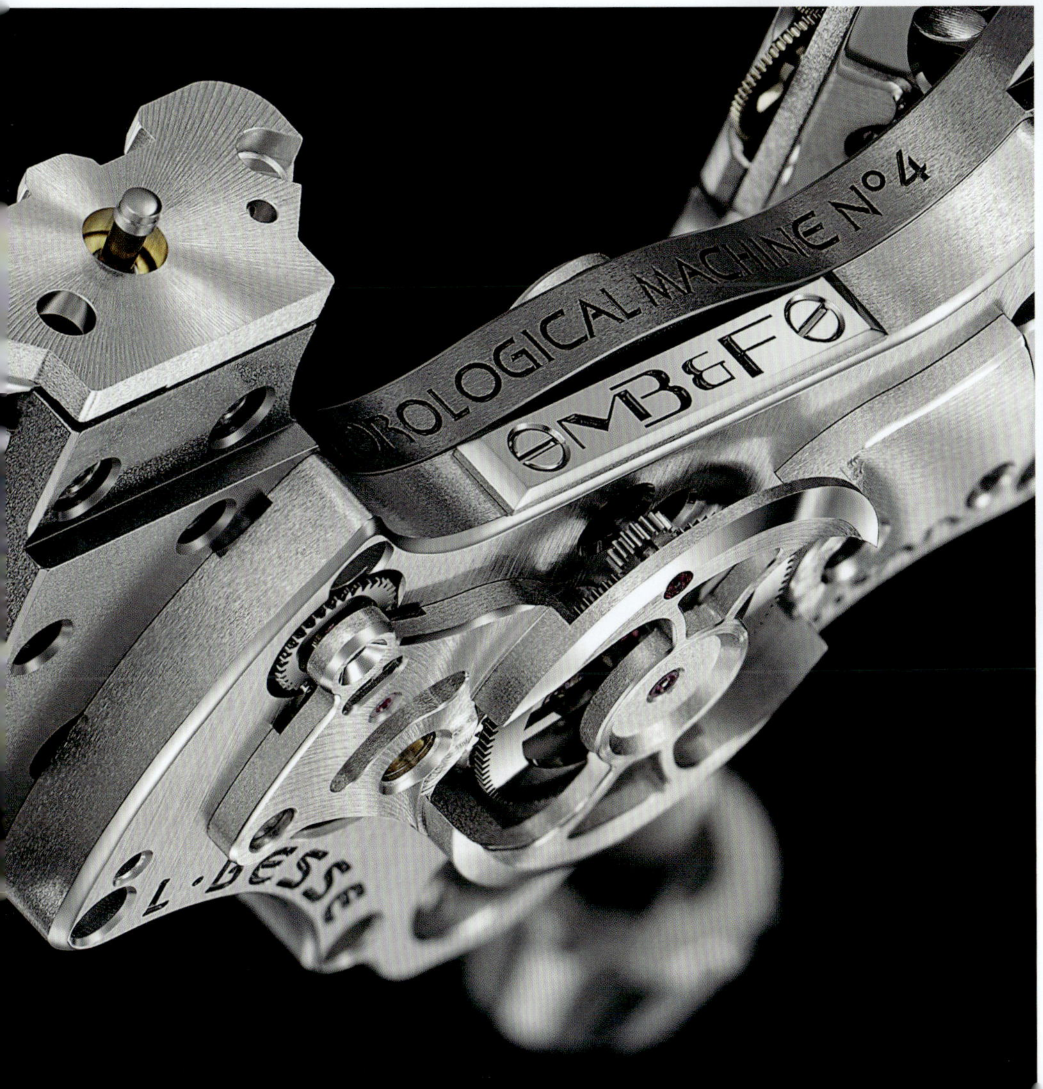
OROLOGICAL MACHINE N°4
MB&F
L.BESSE

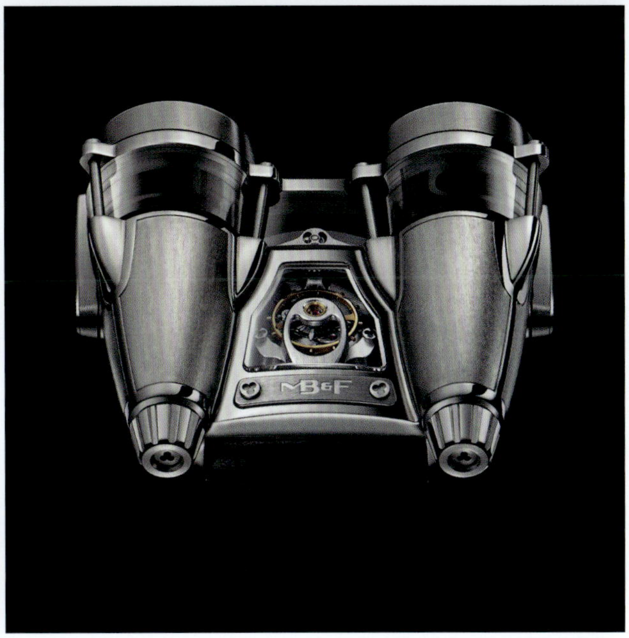
MB&F

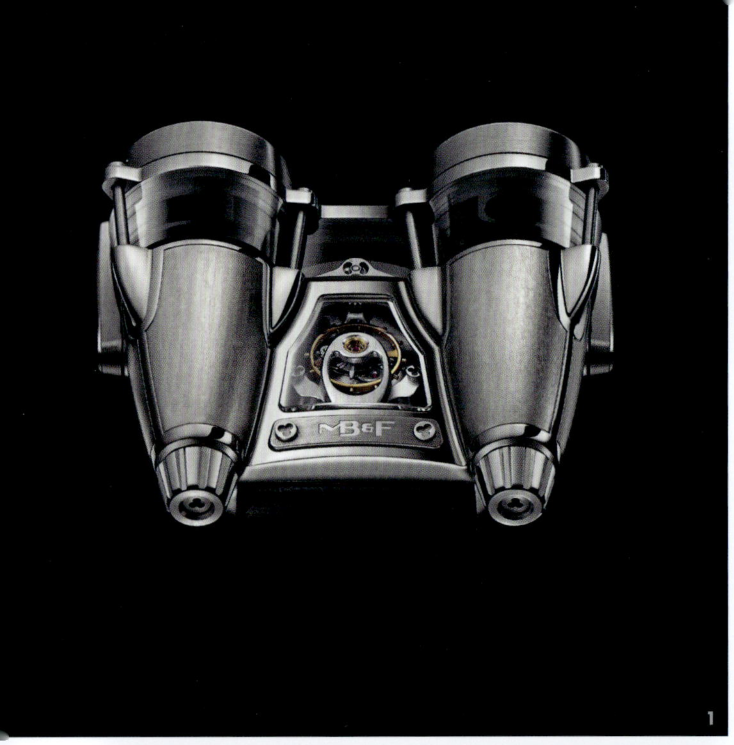

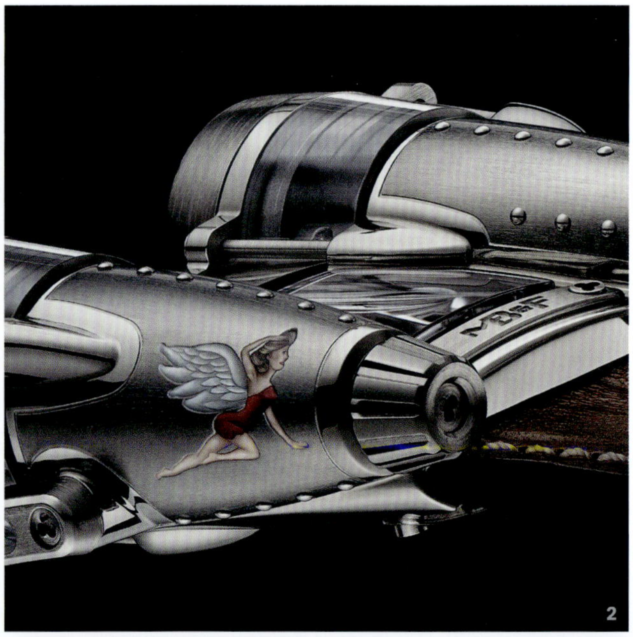

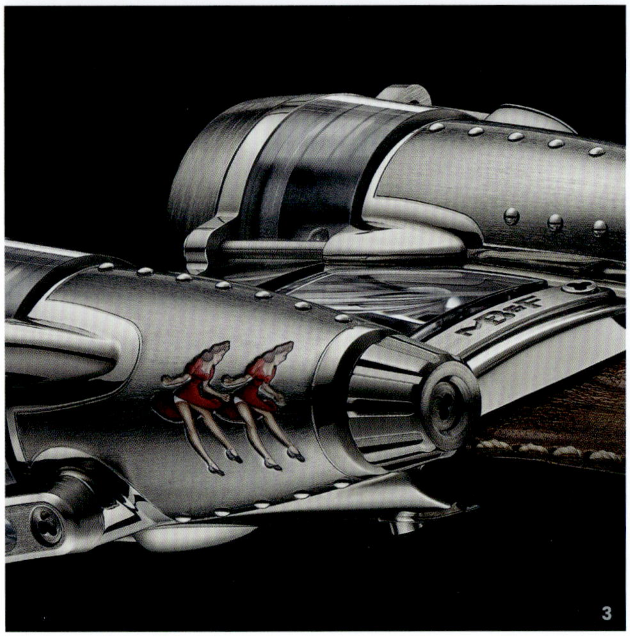

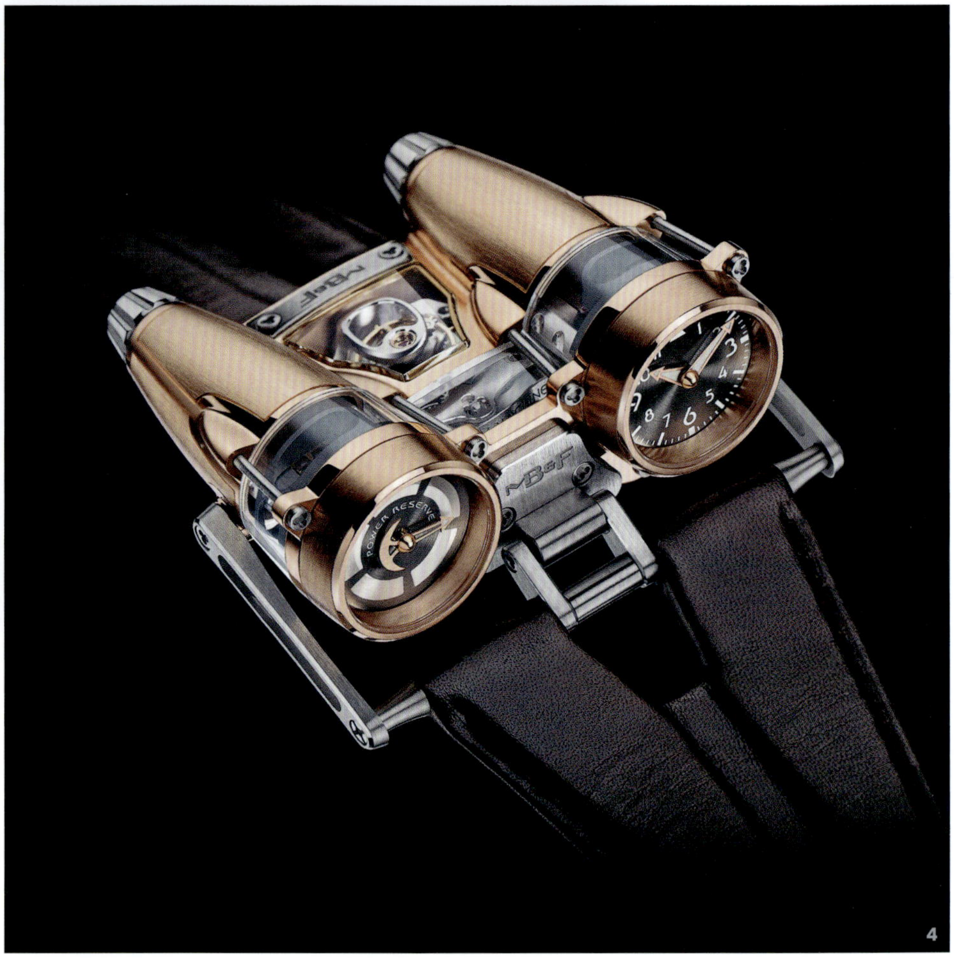

1 Horological Machine N°4
Thunderbolt Ti
Reference: 40.TSL.B
Material: grade 5 titanium and sapphire crystal
Limitation: –
Introduction: 2010
Years in production: 2010–2012
Quantity produced: 53

2 Horological Machine N°4
Razzle Dazzle
Reference: 41.TSL.BRD
Material: grade 5 titanium and sapphire crystal with miniature painting
Limitation: 8
Introduction: 2011
Years in production: 2011–2012
Quantity produced: 8 + 1 prototype

3 Horological Machine N°4
Double Trouble
Reference: 41.TSL.BDT
Material: grade 5 titanium and sapphire crystal with miniature painting
Limitation: 8
Introduction: 2011
Years in production: 2011–2013
Quantity produced: 8 + 1 prototype

4 Horological Machine N°4 RT
Reference: 40.RSL.R
Material: 18K red gold, grade 5 titanium and sapphire crystal
Limitation: 18
Introduction: 2012
Years in production: 2012–2013
Quantity produced: 18 + 1 prototype

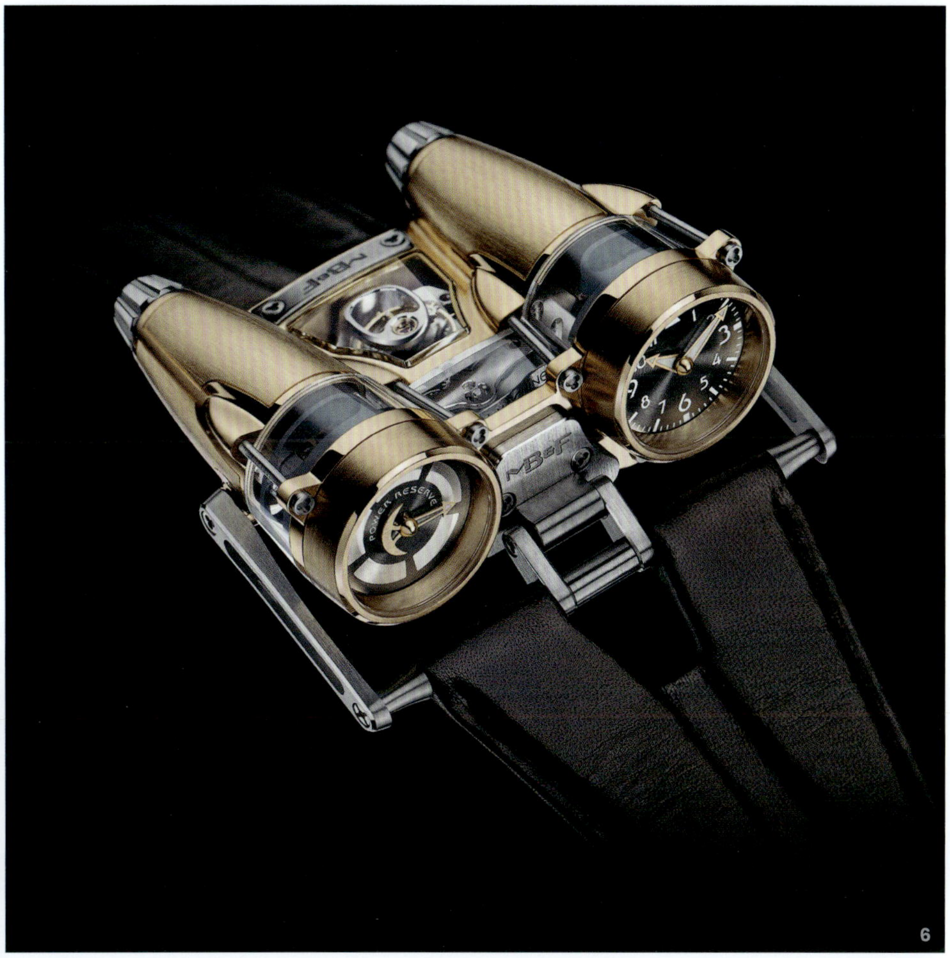

6

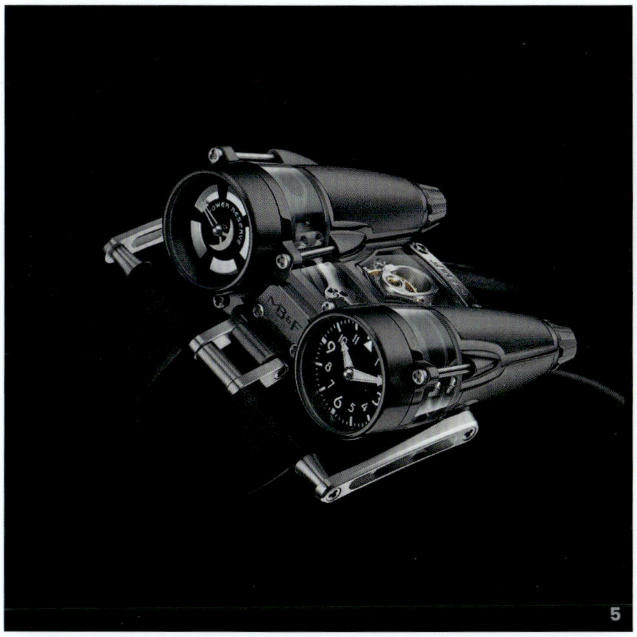

5

5 Horological Machine N°4 Black
Reference: 40.BTSL.B
Material: black PVD coated grade 5 titanium and sapphire crystal
Limitation: unique piece
Introduction: 2012
Year in production: 2012
Quantity produced: 1

6 Horological Machine N°4 YG
Reference: 40.YSL.R
Material: sapphire crystal and 18K yellow gold
Limitation: unique piece
Introduction: 2012
Year in production: 2012
Quantity produced: 1

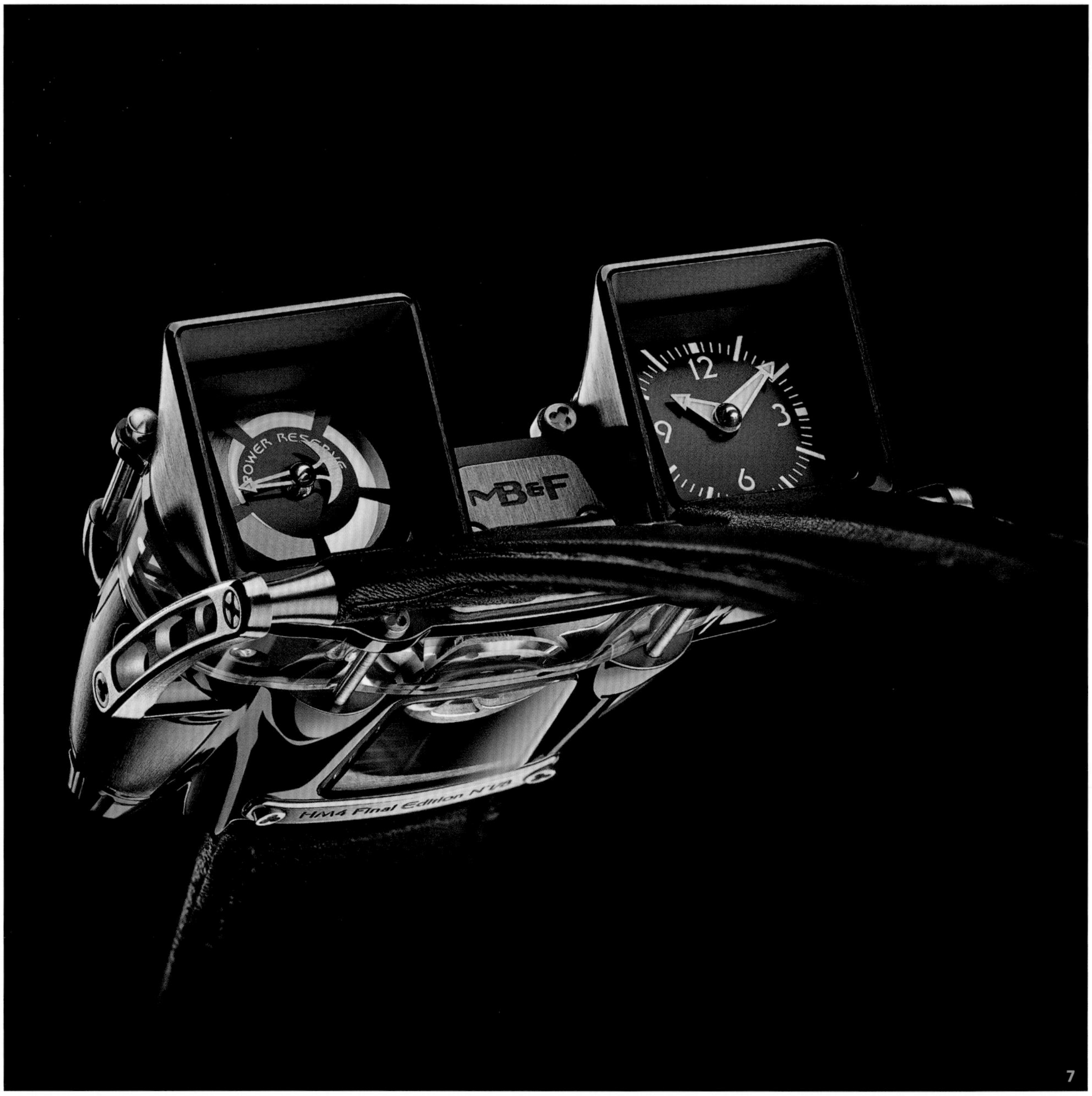

7 **Horological Machine N°4 Final Edition**
Reference: 42.BTSL.B
Material: sapphire crystal and black PVD coated grade 5 titanium
Limitation: 8
Introduction: 2013
Year in production: 2013
Quantity produced: 8 + 1 prototype

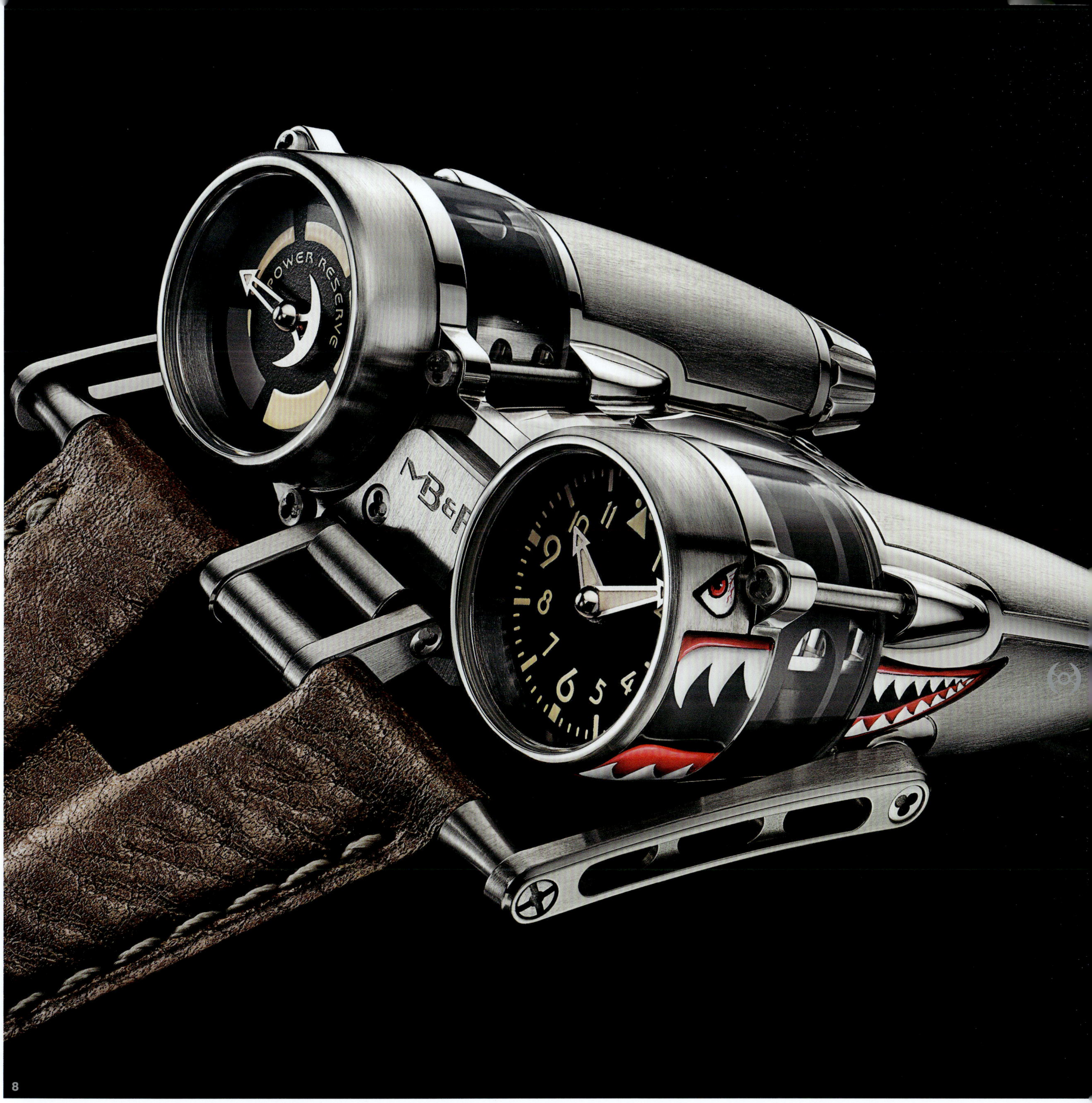

8

8 Horological Machine N°4
Kittyhawk
Reference: 40.TSL.B1
Material: sapphire crystal
and grade 5 titanium with
miniature painting
Limitation: unique piece
Introduction: 2020
Year in production: 2020
Quantity produced: 1

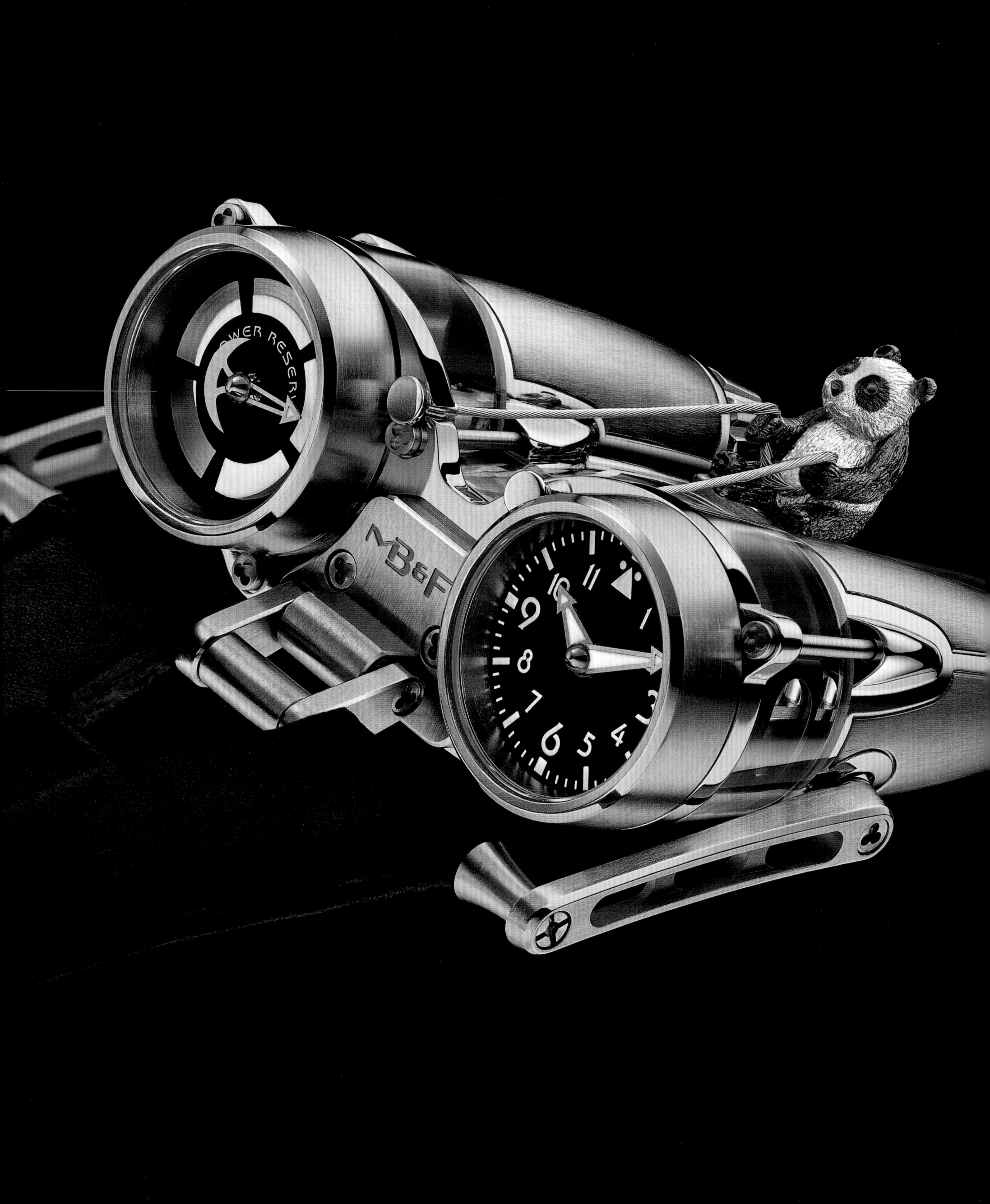
MB&F

HM4 ONLY WATCH MB&F × HUANG HANKANG

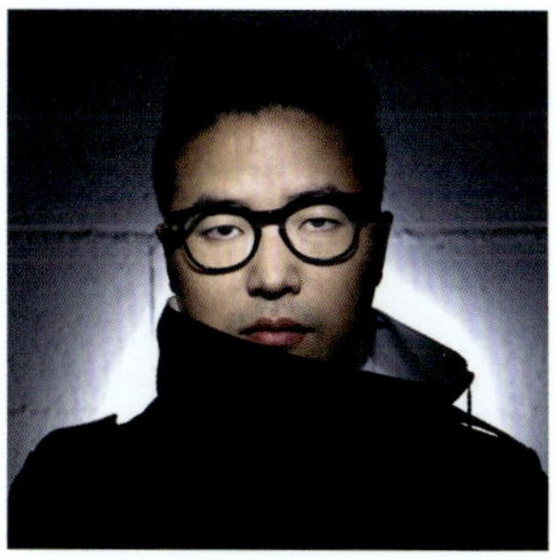

HUANG HANKANG

Chinese artist Huang Hankang was born in Suzhou, China, in 1977. He started to paint at the young age of seven then obtained an art degree in China before moving to Paris to pursue his studies. In 2007, he won international recognition for his work at his first exhibition in New York, followed by subsequent exhibitions in Europe.

Painting mainly in watercolors, Huang frequently depicts animal life, often placed in unusual contexts. The soft subtlety of watercolors gives the resulting images a dreamlike and ethereal quality.

It was exactly this aspect of Huang's work that attracted the interest of Maximilian Büsser, specifically one painting that he would eventually purchase: a watercolor of a panda bear straddling a rocket on take-off. When Büsser told Huang Hankang about the Only Watch charity auction, Huang jumped at the opportunity to loan his talent for the 2011 event, and create a unique piece based on the HM4.

The Only Watch auction is held every two years in Monaco or Geneva under the patronage of HSH Prince Albert II. The event features the work of Switzerland's Haute Horlogerie watchmakers, each contributing a unique piece, or the first in a limited series, with no reserve price set by the owner. Proceeds of the auction go to the Monegasque Association against Duchenne Muscular Dystrophy (DMD), which funds research to fight the disease. A genetic disorder affecting children, DMD is characterized by the progressive weakening of muscles, causing respiratory and cardiac problems as the child gets older.

Childhood memories of model-making or collecting science-fiction comic books are a frequent source of inspiration for MB&F designers. Huang Hankang's fantastical, imaginative creations were thus perfectly in tune with MB&F's philosophy and its commitment to support a charitable cause fighting a disease that affects children.

Huang Hankang's HM4 Only Watch features a miniature flying panda, sitting upright on the aerodynamic case of the HM4 that it appears to be piloting with long, woven reins in its paws. It's a wonderfully carefree image, much like a fleeting memory of a happy dream. The figurine itself is a unique piece, crafted from solid gold hand-carved and cast by the lost wax process.

The HM4 was accompanied by the original Huang Hankang watercolor presenting his vision for the project.

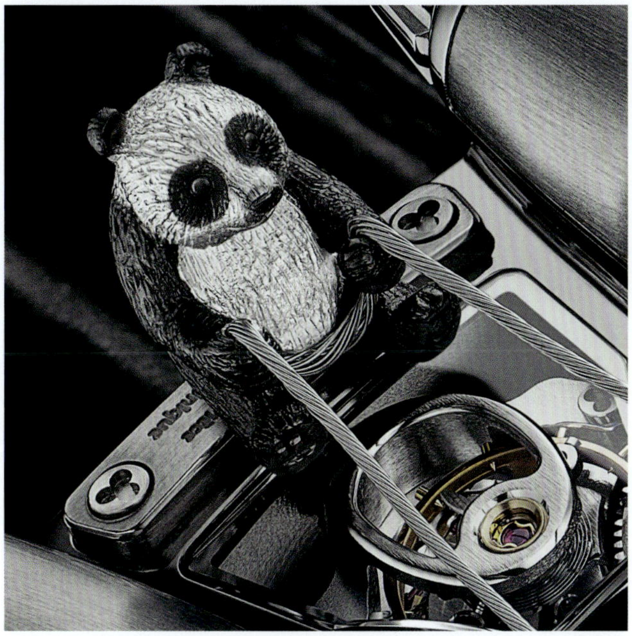

Horological Machine N°4 Only Watch,
Reference: 40.TSL.B
Material: sapphire crystal and grade 5 titanium with 18K white gold panda
Limitation: unique piece
Introduction: 2011
Year in production: 2011
Quantity produced: 1

"I first met Max in Singapore in 2010, when he was presenting Horological Machine N°4 'Thunderbolt'. I knew of the brand before, but this was the first time I had actually held a Machine in my hand and spoken to Max in person. It's appropriate that the Machine in this encounter was the Thunderbolt, because that's how it hit me. A coup de foudre. It was a St-Paul-on-the-road-to-Damascus kind of experience, and it converted me for life."

Suzanne Wong
Editor-in-chief, WorldTempus

“One of the aspects of MB&F I appreciate most in what they do and who they are: even though they are extremely competent and professional, they have really kept a child’s soul.”

Laurent Picciotto
Founder, Chronopassion

HM5

ON THE ROAD AGAIN

Horological Machine N°5, released in 2012, is a mechanical timepiece inspired by the 1970s when quartz was king. Everything about the HM5 recalls the futuristic muscle cars of the 1970s, from its sleek, aerodynamic wedge shape, to the folding rear louvres and the "exhaust pipes". This is the first car-inspired timepiece from MB&F, and it is very much a driver's watch, its time indicators visible from the side so that drivers with their two hands on the steering wheel can easily read the time without turning the wrist (or lifting a hand from the steering wheel).
A mechanical digital watch, the HM5 employs a reflective prism made of machined and polished sapphire, which is also curved for magnification, to display the jumping hours and minutes. The case shape together with the mechanical digital time display is inspired by the Amida Digitrend, one of the most iconic driving watches of the 1970s.
As with a car, the movement is located inside an interior watertight chassis, and the exterior case then screwed on. Unlike a car however, the louvres do not serve to block the rays of the sun to prevent overheating. On the HM5 they do the opposite, opening and closing to admit sunlight for the re-charging of the Super-LumiNova numerals on its dials. Should any water enter the case through the louvres, the two "exhaust pipes" at the back of the case serve as a draining system.
The HM5 was cased in zirconium (66 pieces), red gold (66 pieces), and finally in black Carbon-Macrolon (a final 66 pieces).

HOROLOGICAL MACHINE Nº5 ON THE ROAD AGAIN

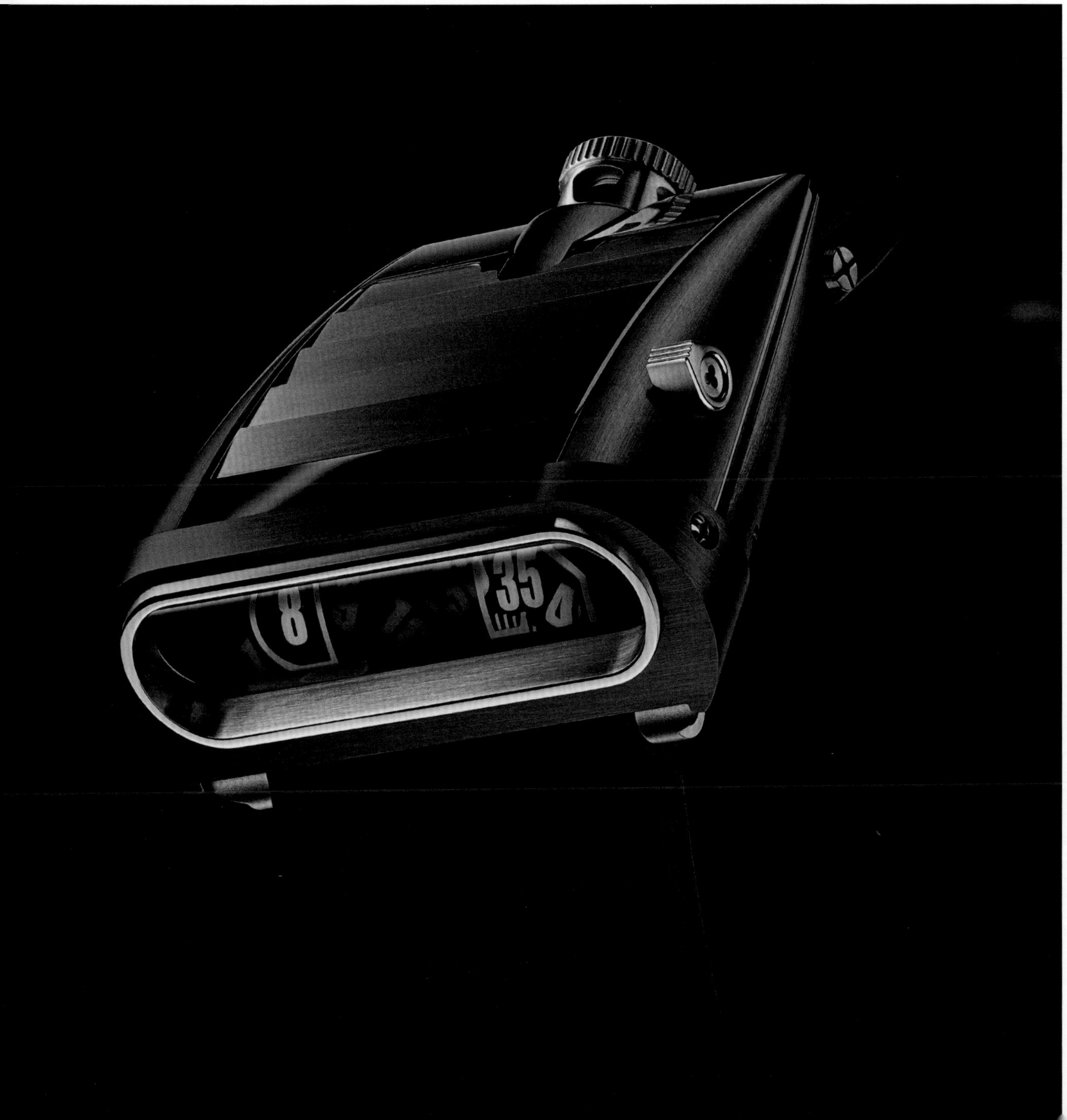

HM5
CALIBRE
ON THE ROAD AGAIN

The movement of the HM5 – this muscle car's engine – is located inside its interior case, or chassis. Developed by Jean-François Mojon, Vincent Boucard and the team at Chronode, the watch movement provides for bi-directional jumping hours, so the time can safely be set either forward or backward.
The mechanical digital display of the HM5 is made possible by overlapping discs that rotate flat on top of the movement (to suit the slim case design) each numbered to display the time (one for hours, another for minutes). The flat numbers are reflected upward and forward, and then magnified 20% by a triangular sapphire prism. For this reason, the numerals are painted onto their dials in mirror image so as to read the right way round.
The HM5 is powered by a 22-karat gold battle-axe "mystery" winding rotor (a mystery because there is no visible imbalance in mass that would otherwise cause the rotor to swing). Power reserve is 42 hours.

Collection
Horological Machines
Series
Horological Machine N°5 "On the Road Again"
Common name
HM5 On the Road Again, HM5
Movement
Developed for MB&F by Jean-François Mojon and Vincent Boucard of Chronode, powered by a Girard-Perregaux base
Winding
Automatic
Power reserve
42 hours
Case components
224/30 jewels
Functions
Minutes and bi-directional jumping hours displayed by reflective sapphire crystal prism with integrated 20% magnifying lens
Add. features
22K gold "battle-axe" winding rotor; slide button to open/close flaps; smoked optical grade sapphire crystal with anti-reflective coating
Case
Zirconium, 18K red gold or Black CarbonMacrolon
Dimensions
51.5 × 49 × 22.5 mm
Case components
80
Water resistance
30 m/3 ATM
Strap
Rubber, moulded in aircraft-grade Fluorocarbon FKM 70 Shores A elastomer
Buckle
Titanium pin buckle
Introduction
2012
Years in production
2012–2016
Quantity produced
170 + 4 prototypes
Comments
A water resistant inner container houses the engine; exhaust ports to drain water

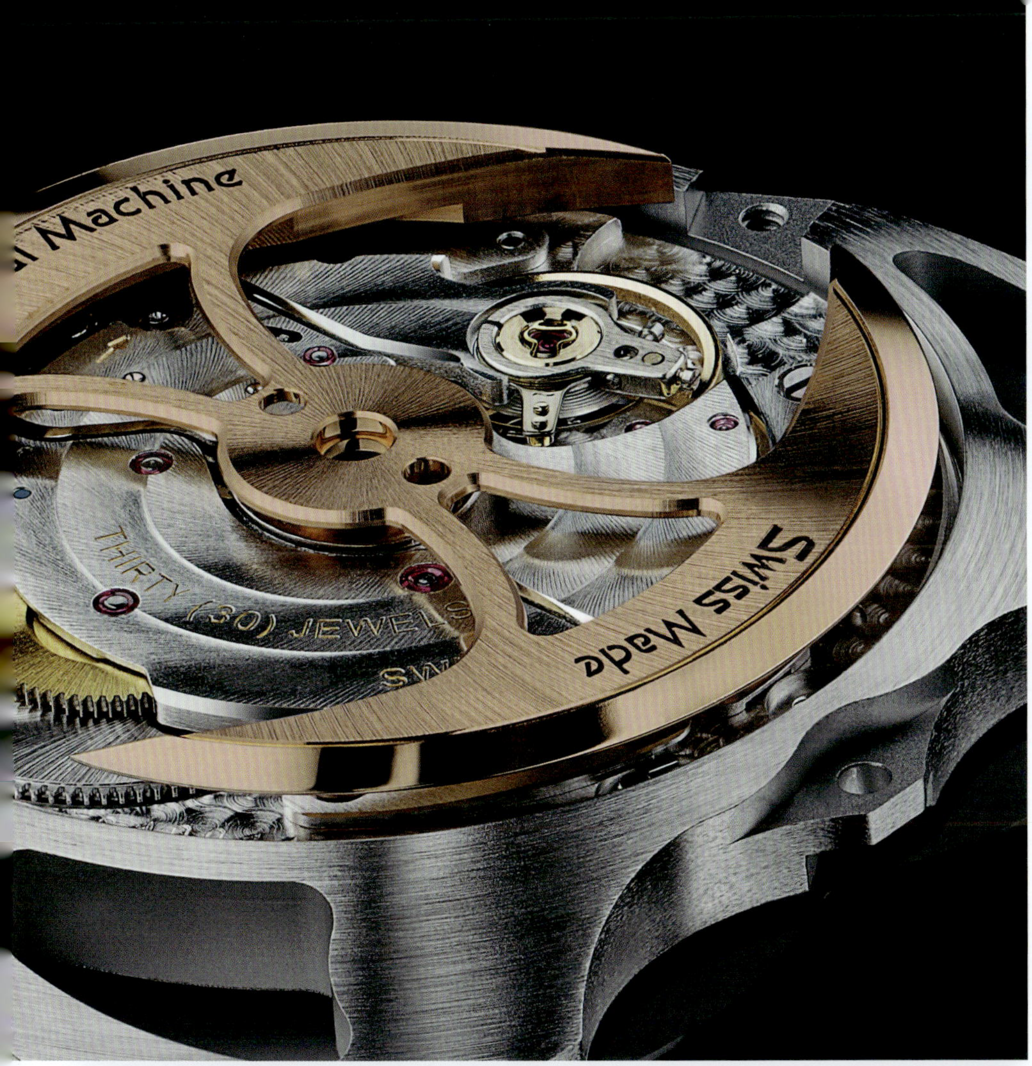
Machine
Swiss Made
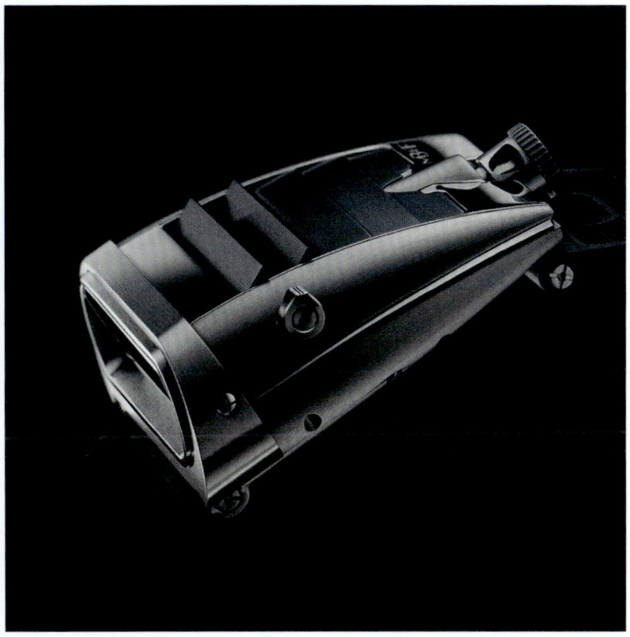
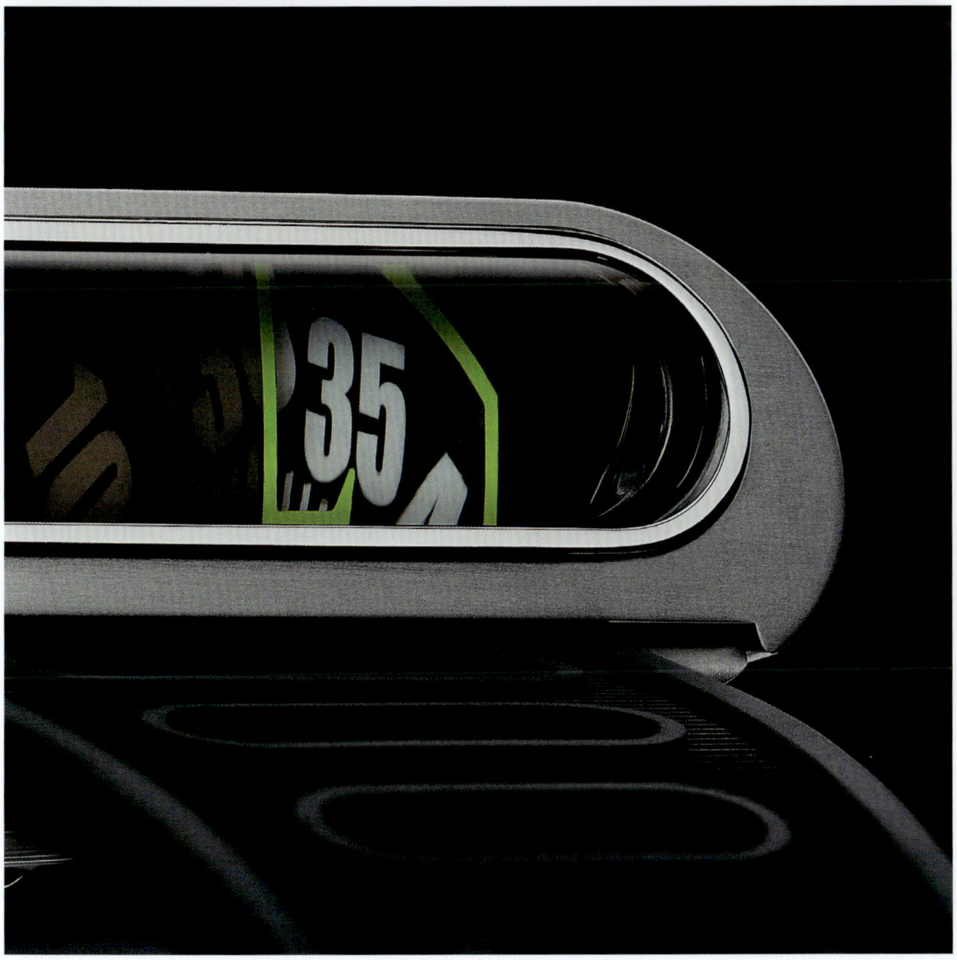
35
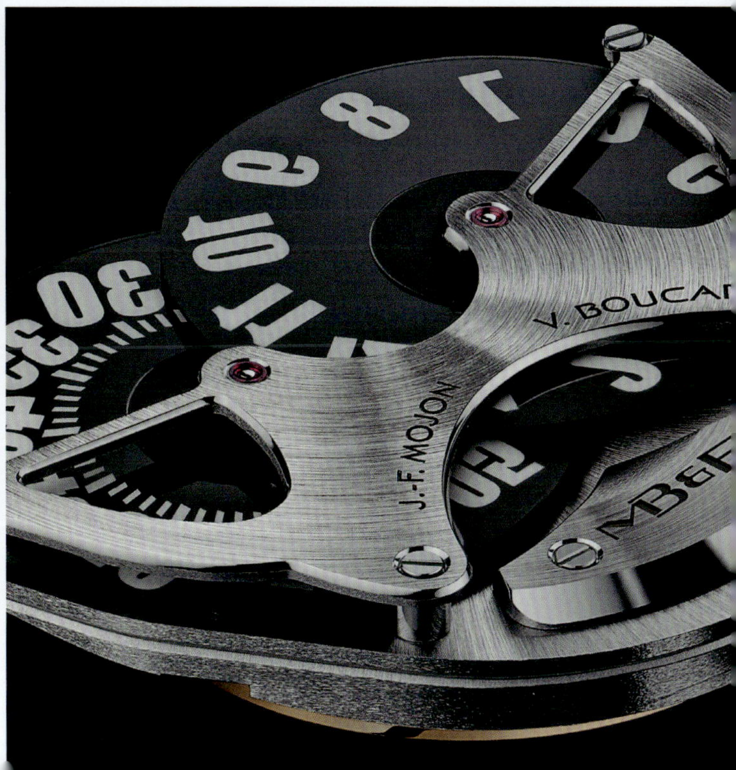
J.F. MOJON

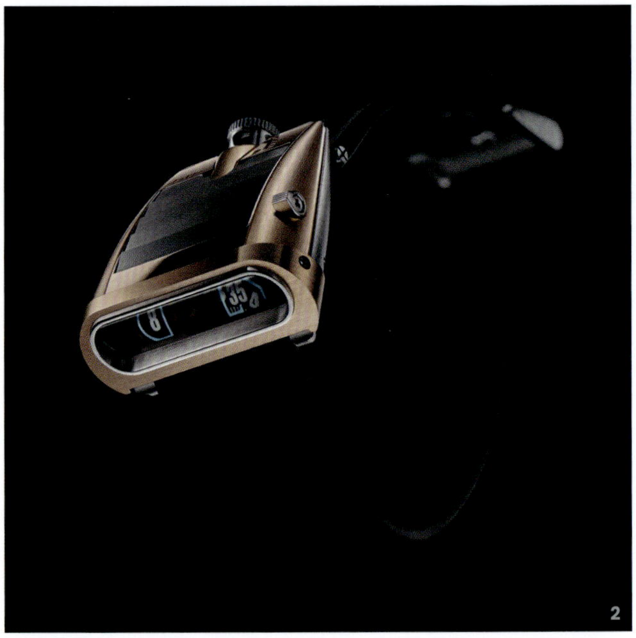

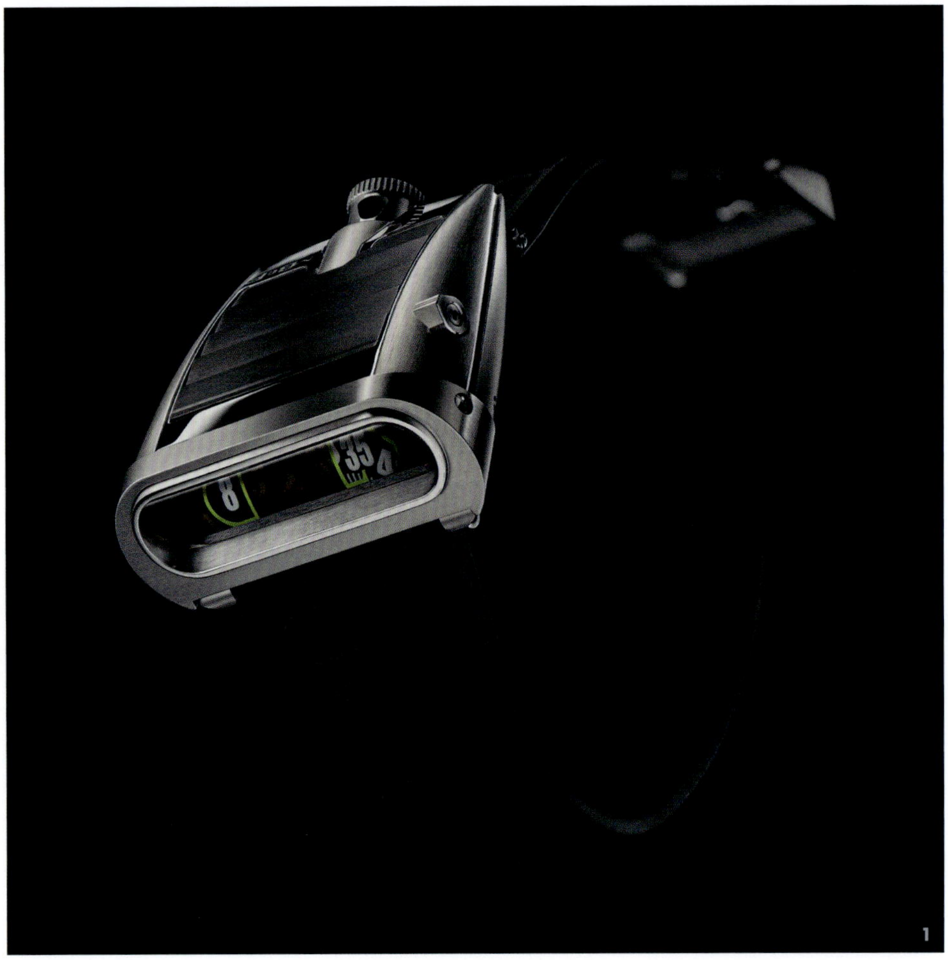

1 Horological Machine N°5 ZR
Reference: 55.ZL.B
Material: zirconium with stainless steel container
Limitation: 66
Introduction: 2012
Years in production: 2012–2014
Quantity produced: 66 + 2 prototypes

2 Horological Machine N°5 RT
Reference: 55.RL.B
Material: 18K red gold and grade 5 titanium with titanium container
Limitation: 66
Introduction: 2013
Years in production: 2013–2016
Quantity produced: 57 + 1 prototype

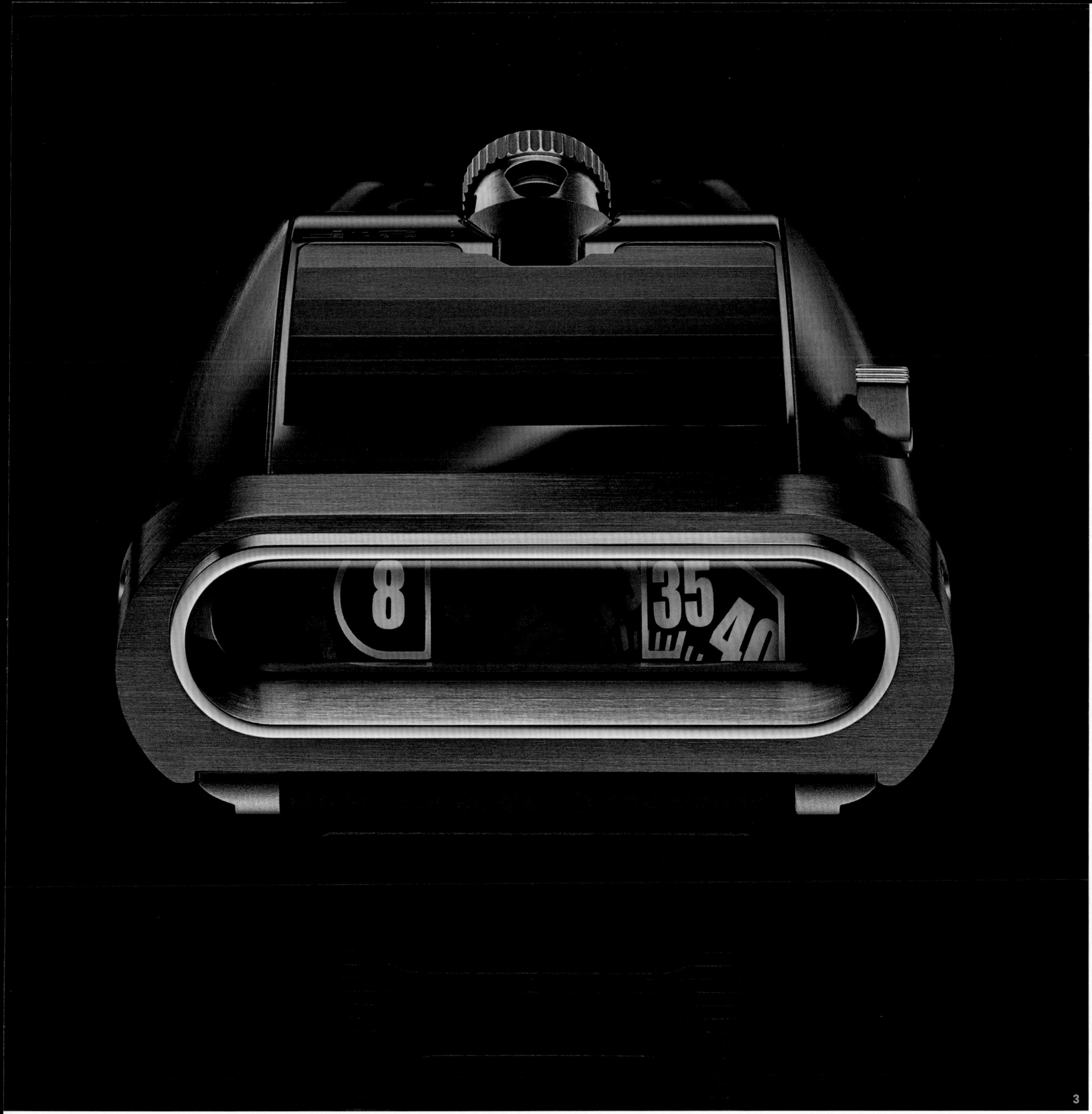

3 Horological Machine N°5 CarbonMacrolon
Reference: 55.CMT.B
Material: CarbonMacrolon with stainless steel container
Limitation: 66
Introduction: 2014
Years in production: 2014-2016
Quantity produced: 47 + 1 prototype

HM 6

SPACE PIRATE

Horological Machine N°6, nicknamed the "Space Pirate", was first released in 2014 and draws inspiration from 1970s and 1980s science fiction, especially the Japanese animated TV series, Captain Future, and his bulbous spacecraft. Other influences are the engineering projects and designs of the "Streamline Era" — but most importantly, the work of 1970s industrial designer Luigi Colani, a pioneer of biomorphic design who embraced soft, rounded shapes and curved lines at a time when the design world was dominated by quite the opposite aesthetic.

Time is displayed across two spherical dials cased in sapphire, with hours on the left, and minutes on the right. At the opposite end of the case are twin turbines, also under sapphire domes, which create air resistance to protect the machine's automatic winding system. In the middle of the HM6 between its dials and turbines is its beating heart, a flying tourbillon under a sapphire dome protected by a retractable shield, operated by turning one of two crowns.

HM6 Space Pirate was released as a limited edition of 50 pieces in titanium plus a further 18 in red gold. Then came HM6-SV (Sapphire Vision), 10 pieces in platinum and 10 pieces in red gold, followed by four unique, all-sapphire pieces in the Alien Nation edition, each featuring white gold figurines engraved by Olivier Kuhn. A final edition of 8 pieces in stainless steel was released in 2019.

HOROLOGICAL MACHINE Nº6 SPACE PIRATE

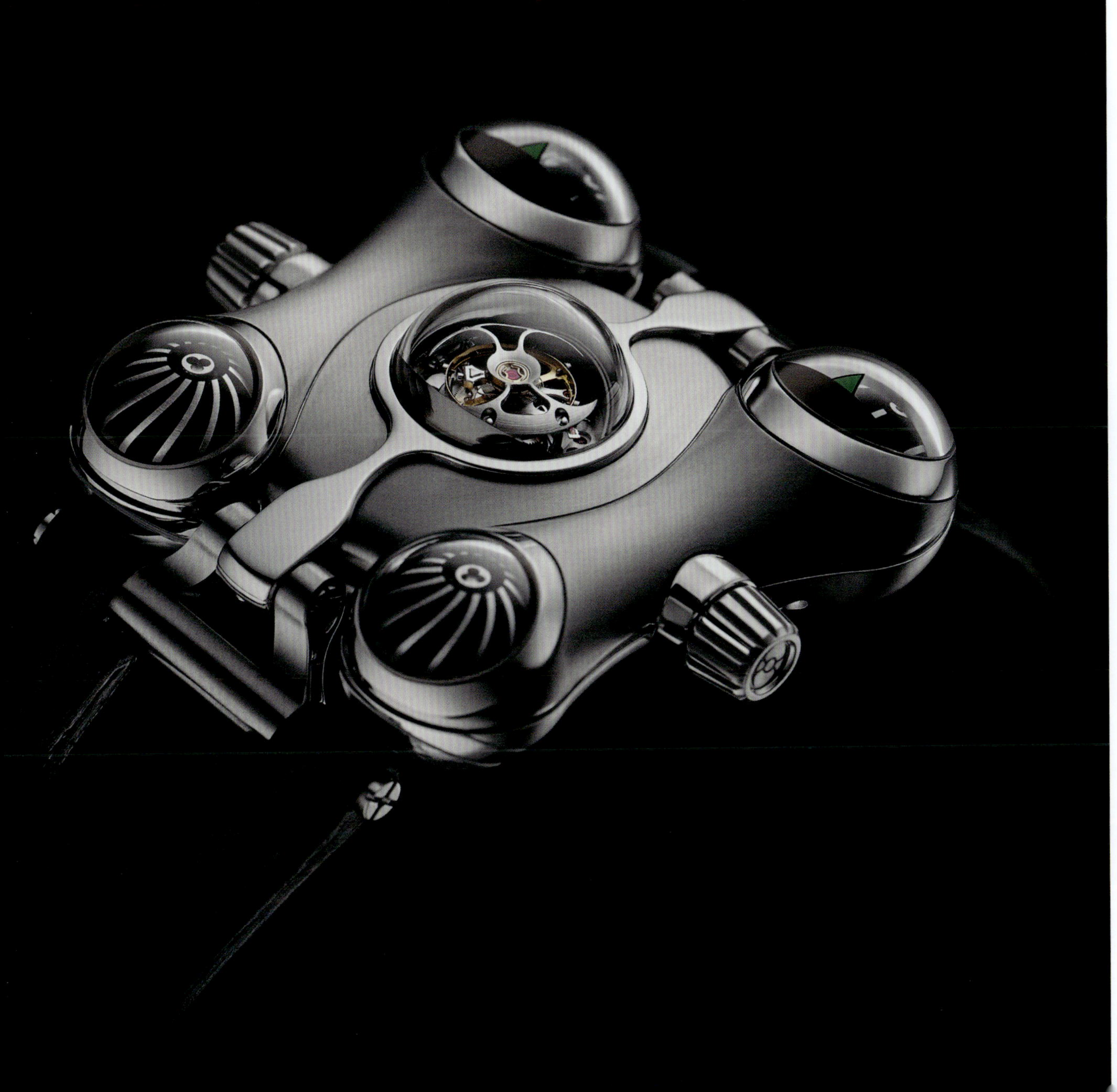

HM6 SPACE PIRATE CALIBRE

The engine of the HM6 is one of the most complicated ever created by MB&F, requiring more than three years of research and development to produce its 475 components (496 for the Alien Nation edition). This was MB&F's second movement with a tourbillon, and its first ever with a flying tourbillon – whose extreme height relative to the movement seems to defy the laws of engineering. The only concession to those laws is the retractable titanium shield, which guards the tourbillon's lubricating oils against the sun's radiation.

Hours and minutes are each indicated by two rotating hemispheres, which are milled to near paper-thinness to reduce the torque load on the mainspring barrels and extend the power reserve to its maximum capacity (72 hours). Conical gears are employed to transfer energy vertically within the machine without compromising precision.

The turbines flanking the machine's twin spherical dials are not for aesthetics alone. Their curved fins act as a winding buffer, creating air resistance to keep the winding rotor within a safe range of oscillation speed, thus extending the life of this machine's complex 475-component engine.

Collection
Horological Machines
Series
Horological Machine N°6 "Space Pirate"
Common name
HM6 Space Pirate, HM6
Movement
Automatic movement developed for MB&F
Winding
Automatic
Power reserve
72 hours
Components
475/68 jewels (496 components/68 jewels for Alien Nation)
Functions
Hours and minutes on separate semi-spherical aluminium indications
Add. features
Flying tourbillon with retractable shield and twin aluminium turbines driven by winding rotor; crown to open/close tourbillon shield
Case
Grade 5 titanium, 18K red gold and titanium, sapphire crystal with red gold or platinum, full sapphire crystal or stainless steel
Dimensions
49.5 × 52.3 × 20.4 mm; 51 × 50 × 22.7 mm for HM6 SV and Alien Nation editions
Case components
HM6 Ti and RT: 80
HM6-SV: 78
HM6 Alien Nation: 95
HM6 FE: 85
Water resistance
30 m/3 ATM
Strap
Hand-stitched calfskin strap for the HM6 Ti and RT; alligator strap for the HM6-SV, HM6 Alien Nation and HM6 Final Edition
Buckle
Folding buckle matching the case material
Introduction
2014
Years in production
2014–2019
Quantity
100 + 5 prototypes

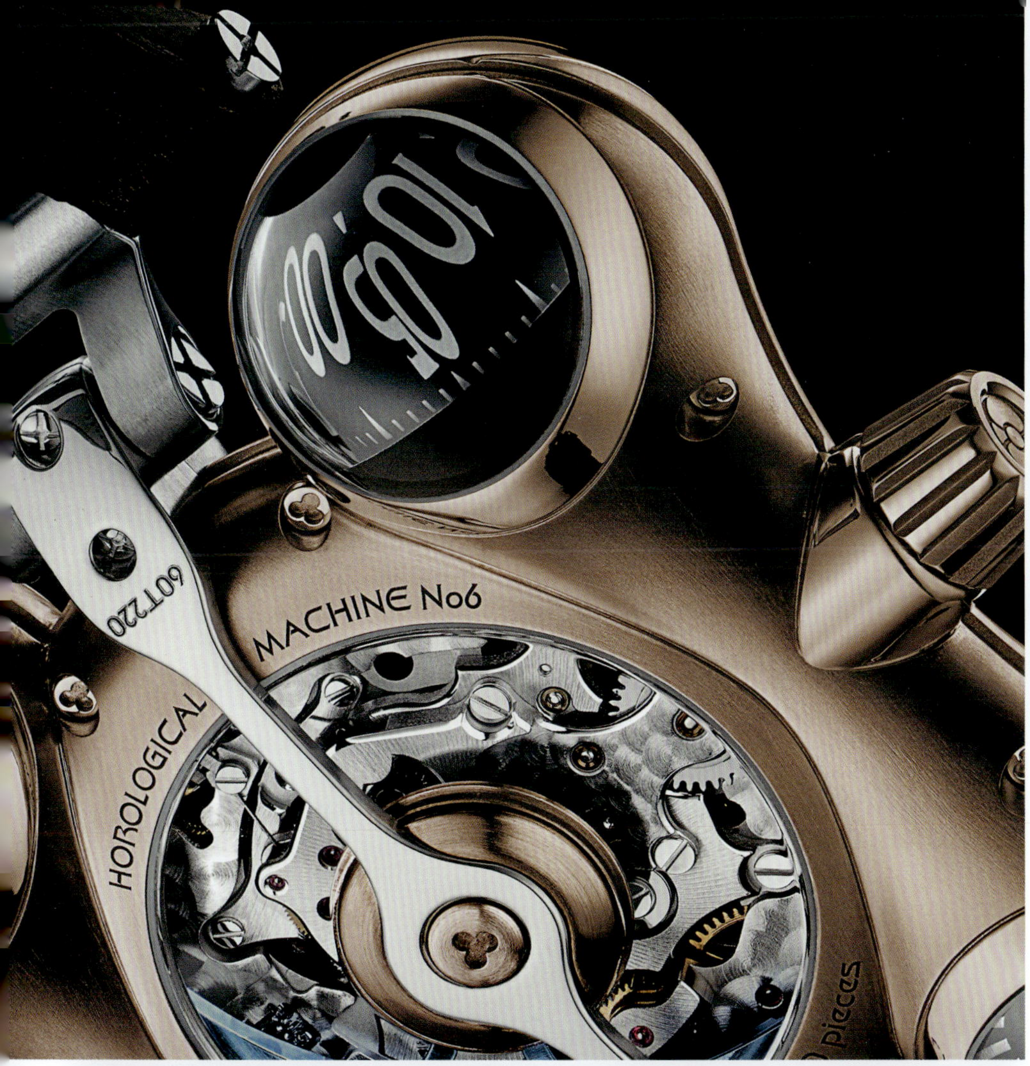
HOROLOGICAL MACHINE No6
60T220
pieces

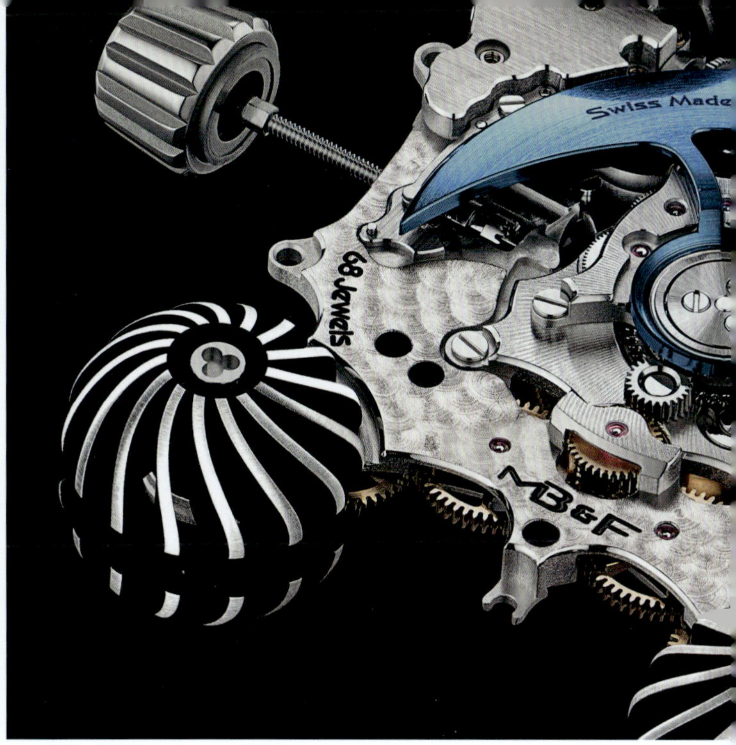
Swiss Made
68 Jewels
MB&F

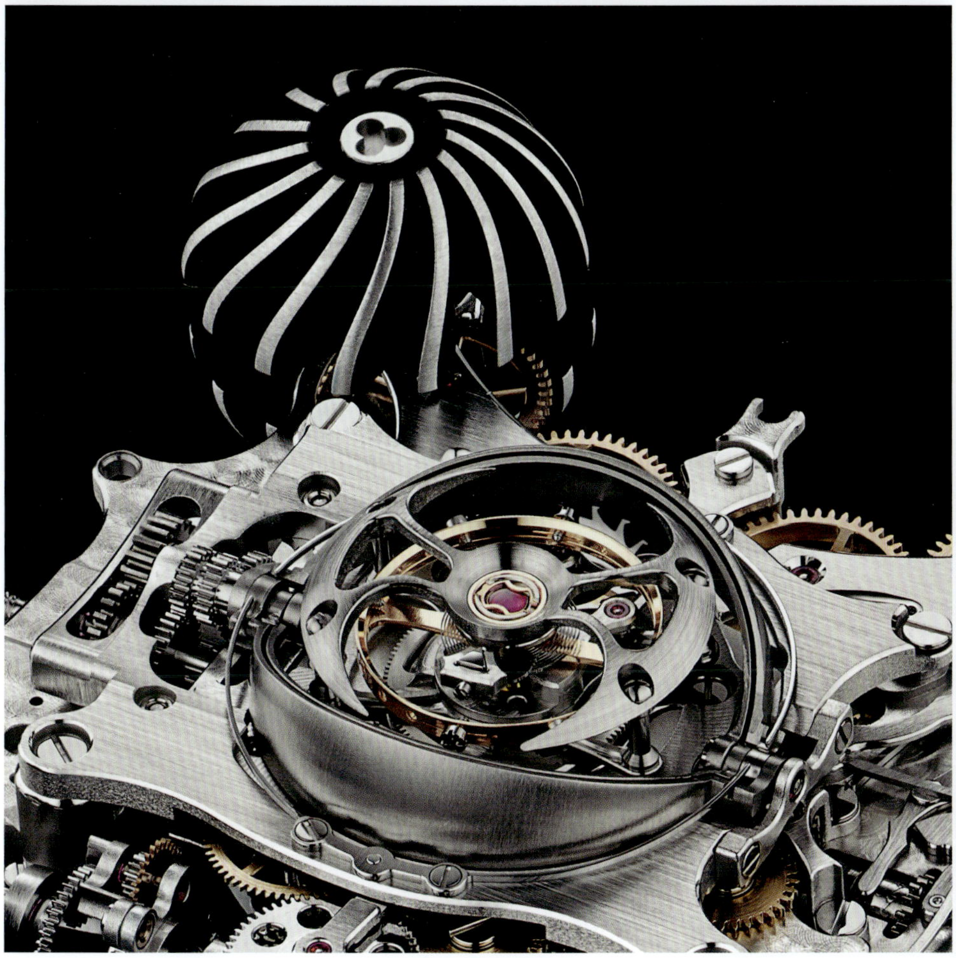

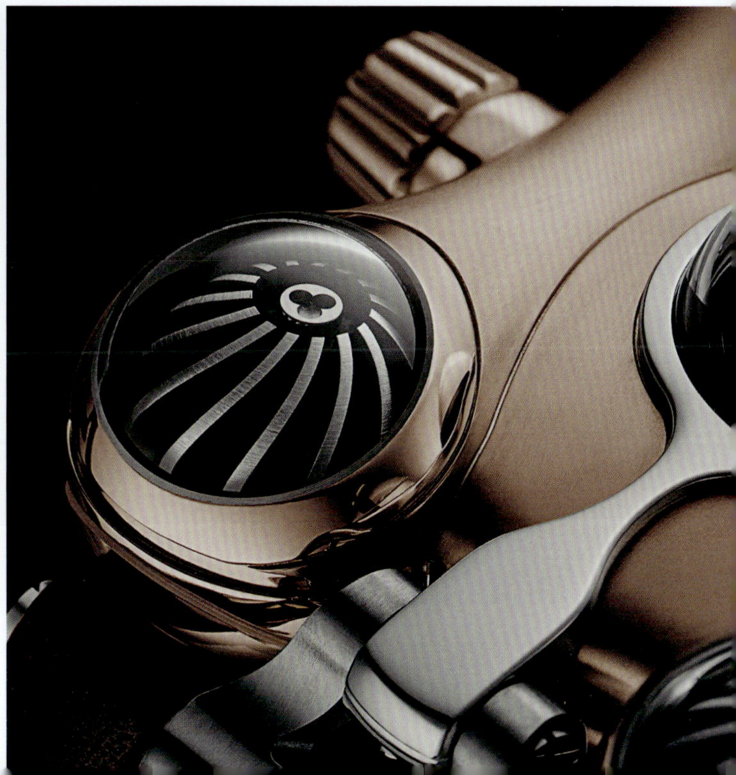

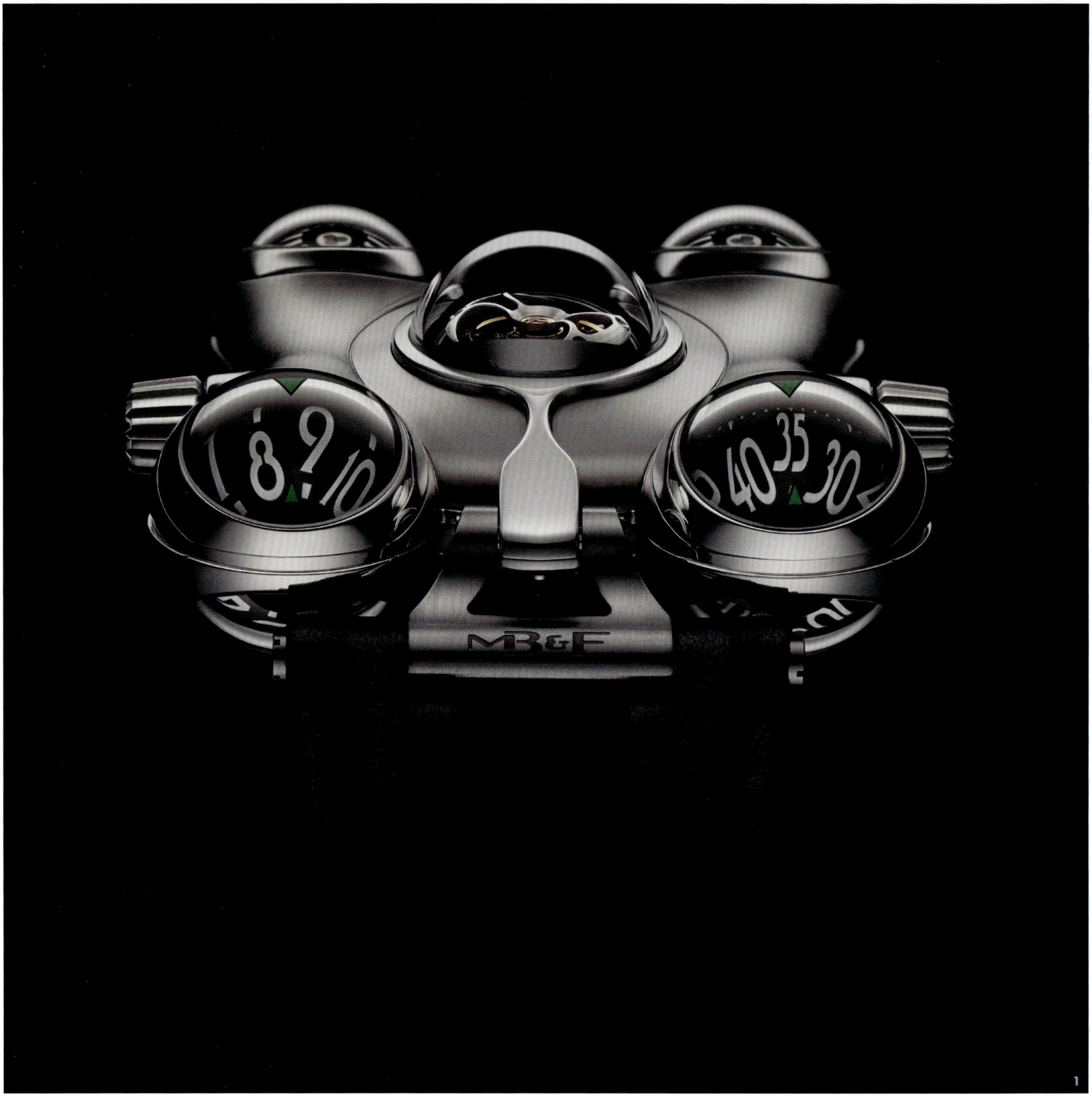

1 Horological Machine N°6 Ti
Reference: 60.TL.B
Material: grade 5 titanium
Limitation: 50
Introduction: 2014
Years in production:
2014–2016
Quantity produced:
50 + 1 prototype

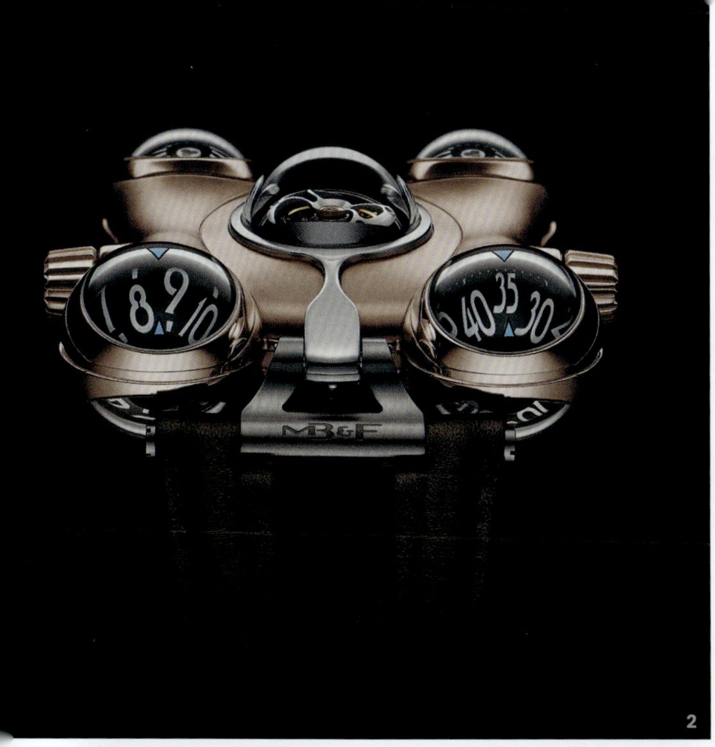

2

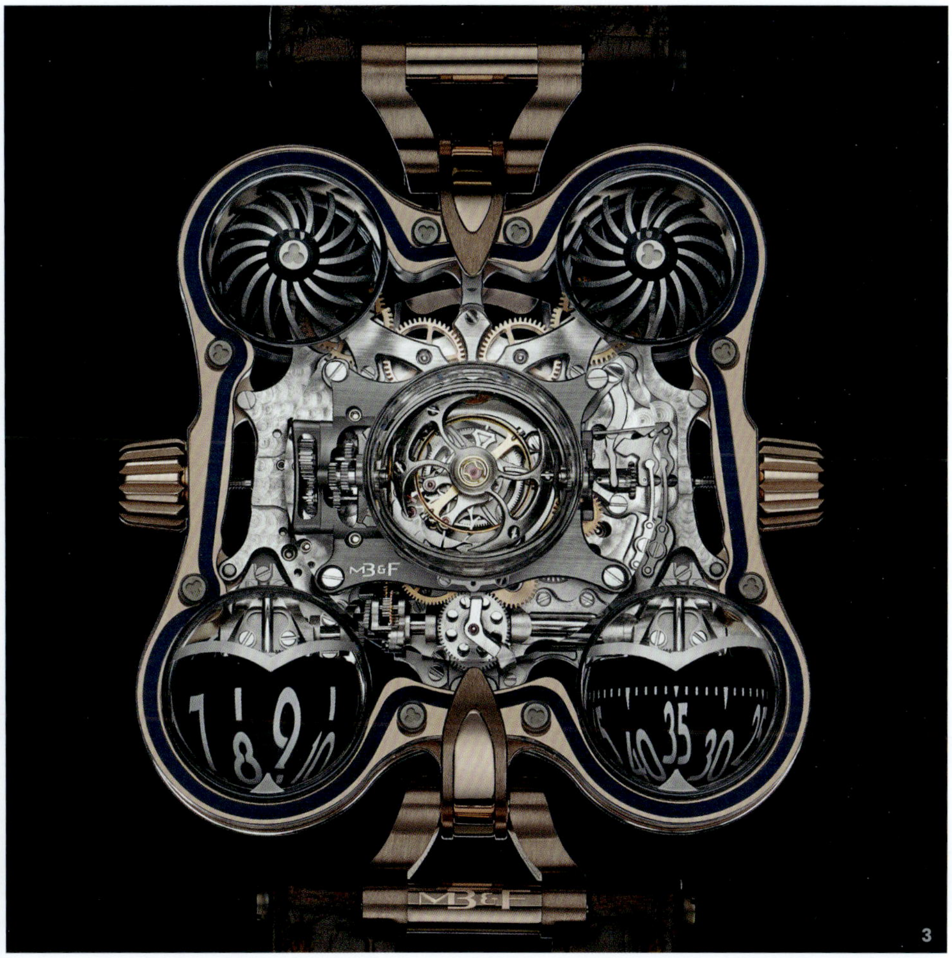

3

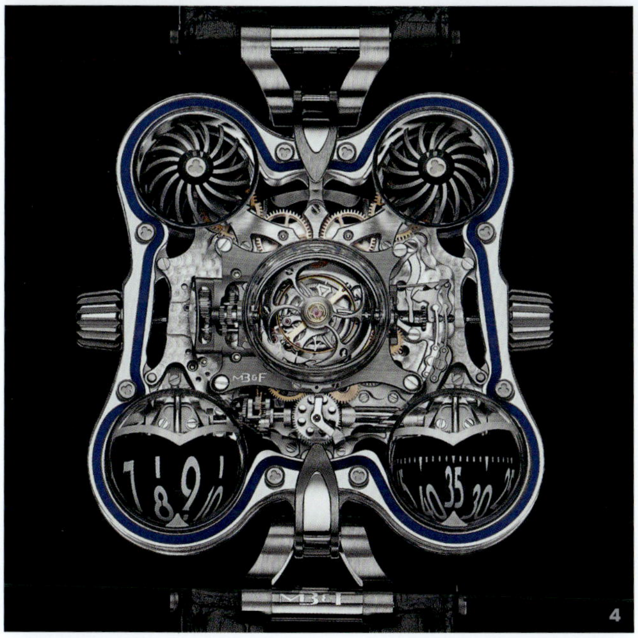

4

2 Horological Machine N°6 RT
Reference: 60.RL.B
Material: 18K red gold and grade 5 titanium
Limitation: 18
Introduction: 2015
Years in production: 2015–2016
Quantity produced: 18 + 1 prototype

3 Horological Machine N°6 SV RG
Reference: 60.SRL.B
Material: sapphire crystal and 18K red gold
Limitation: 10
Introduction: 2016
Years in production: 2016–2017
Quantity produced: 10 + 1 prototype

4 Horological Machine N°6 SV PT
Reference: 60.SPL.B
Material: sapphire crystal and platinum 950
Limitation: 10
Introduction: 2016
Years in production: 2016–2017
Quantity produced: 10 + 1 prototype

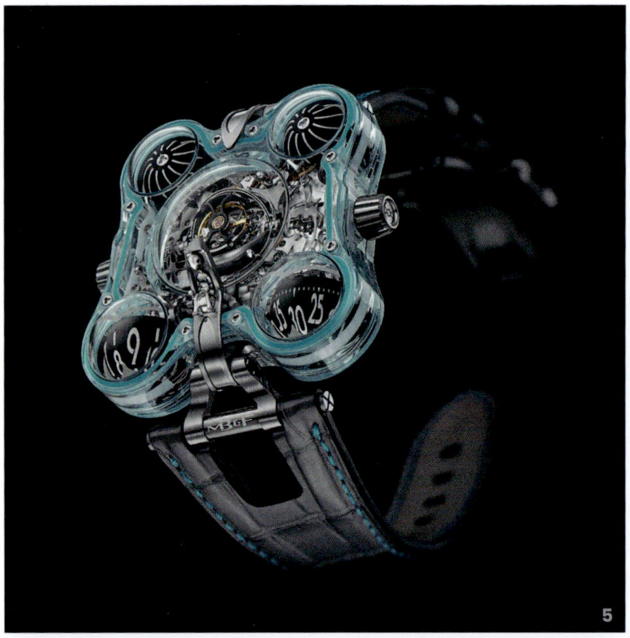

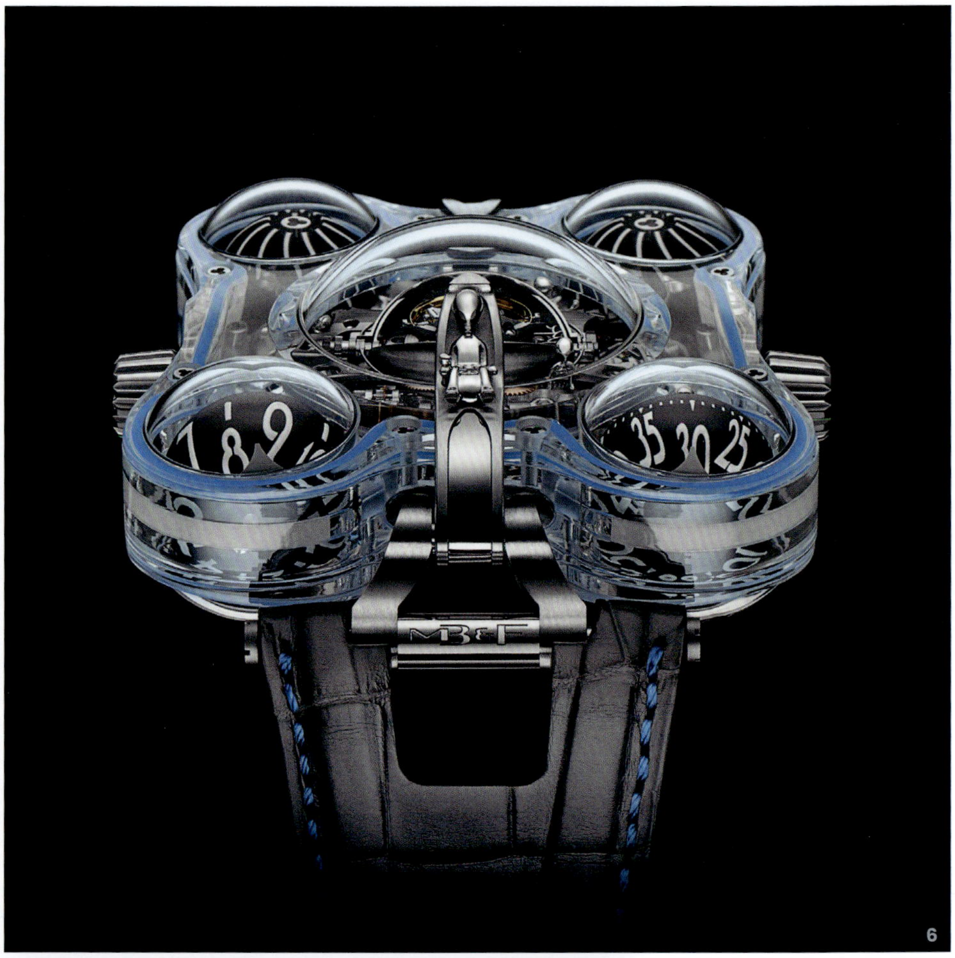

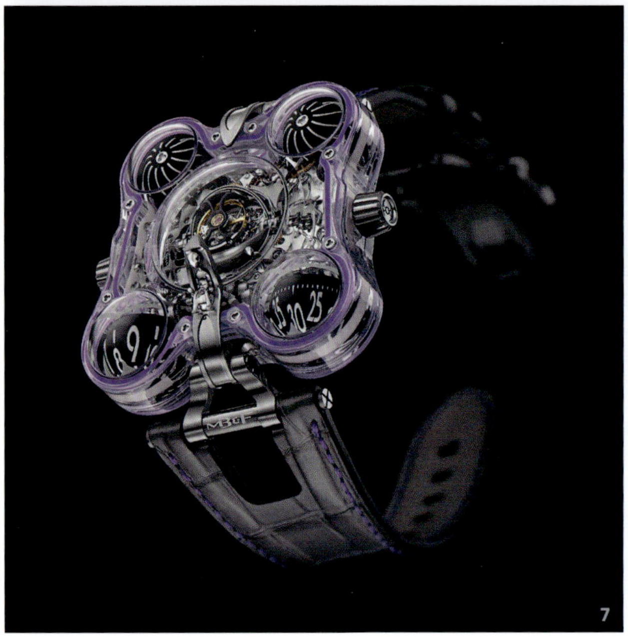

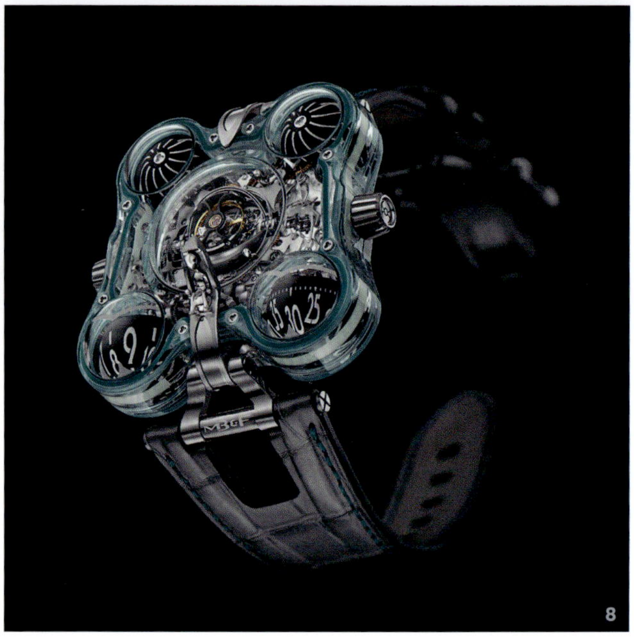

5 Horological Machine N°6
Alien Nation Turquoise
Reference: 60.SVTL.B
Material: sapphire crystal with turquoise luminescent details
Limitation: unique piece
Introduction: 2017
Year in production: 2018
Quantity produced: 1

6 Horological Machine N°6
Alien Nation Blue
Reference: 60.SVBL.B
Material: sapphire crystal with blue luminescent details
Limitation: unique piece
Introduction: 2017
Year in production: 2018
Quantity produced: 1

7 Horological Machine N°6
Alien Nation Purple
Reference: 60.SVPL.B
Material: sapphire crystal with purple luminescent details
Limitation: unique piece
Introduction: 2017
Year in production: 2017
Quantity produced: 1

8 Horological Machine N°6
Alien Nation Green
Reference: 60.SVGL.B
Material: sapphire crystal with green luminescent details
Limitation: unique piece
Introduction: 2017
Year in production: 2017
Quantity produced: 1

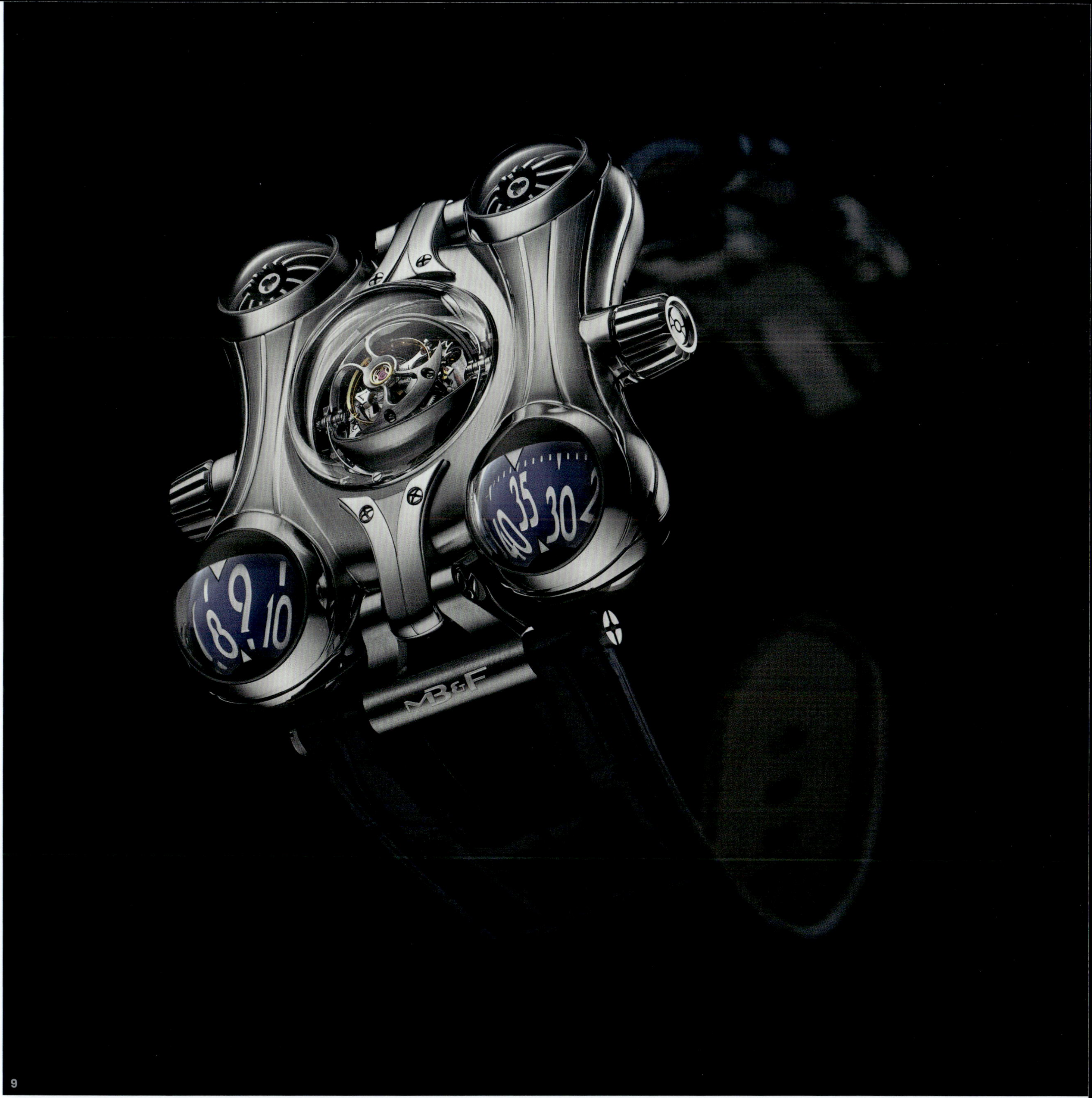

9

9 Horological Machine N°6
Final Edition
Reference: 60.SL.B
Material: stainless steel
Limitation: 8
Introduction: 2018
Years in production:
2018–2019
Quantity produced:
8 + 1 prototype

HMX

HMX was conceived to celebrate the 10th anniversary of MB&F in 2015, and as a thank-you to all those "Friends" who had supported the young brand in its first ten years. Anniversaries in the watch industry are typically treated as an opportunity to develop an ultra-expensive timepiece. Instead, MB&F chose to make this their most affordable Horological Machine to date, as you might expect from a company that readily breaks all the rules.

As a child, Maximilian Büsser dreamed of becoming a car designer, and there is no better proof of this than the design of HMX. Inspired by the world of automobiles and supercars, the HMX is a driver's watch, its time indicators visible from the side so that drivers can easily read the time without needing to turn the wrist (or lift a hand from the steering wheel). Bi-directional jumping hours and trailing minutes are projected thanks to a digital mechanical display, the numerals painted onto rotating disks that are reflected through a triangular prism of sapphire crystal. The cover of the HMX is also made of sapphire, allowing light to stream into the machine and illuminate its display for improved legibility.

The HMX was produced as a limited edition of twenty pieces each in four automobile-inspired colours: black, green, blue and red. These were followed by three "Black Badger" editions of eighteen pieces each.

HOROLOGICAL MACHINE X

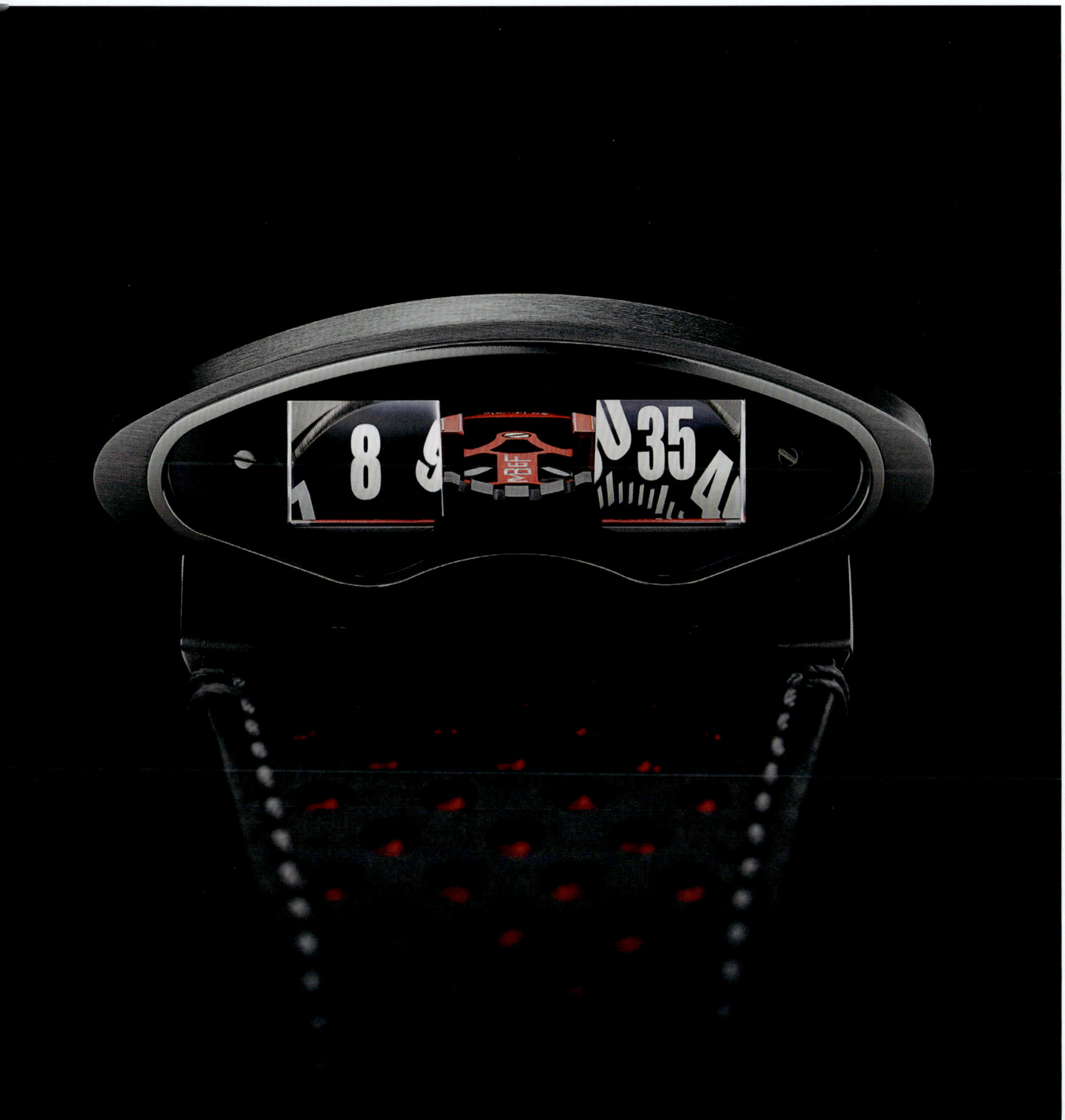

HMX CALIBRE

The three-dimensional HMX movement was developed in-house by MB&F and is powered by a Sellita gear train. The mechanical digital display of the HMX is derived from circular discs that rotate horizontally (to fit its slim case), each numbered to display the time (one for hours, another for minutes). Because the numerals are projected forward by a set of two triangular sapphire prisms, they are painted in mirror image so as to read the right way round. On display through the sapphire crystal engine cover are supercar inspired twin "rocker covers" on top of the movement, each with a shiny chrome oil filler cap. These caps are not purely aesthetic – each unscrews so the watchmaker can oil the jeweled bearings of the indication discs.
The crystal display back meanwhile reveals the 22-karat gold rotor of the HMX engraved with the MB&F motto: "A Creative Adult Is a Child Who Survived."

Collection
Horological Machines
Series
HMX
Common name
HMX
Movement
Trailing minutes and jumping hours module developed in-house by MB&F, powered by a Sellita gear train
Winding
Automatic
Power reserve
42 hours
Components
223/29 jewels
Functions
Bi-directional jumping hours and trailing minutes
Add. features
The time is displayed by dual reflective sapphire crystal prisms with integrated magnifying lens; high-efficiency lume components for the Black Badger editions
Case
Grade 5 titanium and stainless steel with detailing in black, green, red or blue (purple, green and blue for the Black Badger editions)
Dimensions
46.8 × 44.3 × 20.7 mm
Case components
44
Water resistance
30 m/3 ATM
Strap
Partially perforated calfskin, black and colour matching with the engine
Buckle
Titanium pin buckle
Introduction
2015
Years in production
2015–2016
Quantity produced
134 + 7 prototypes + 1 artist's proof (Black Badger)

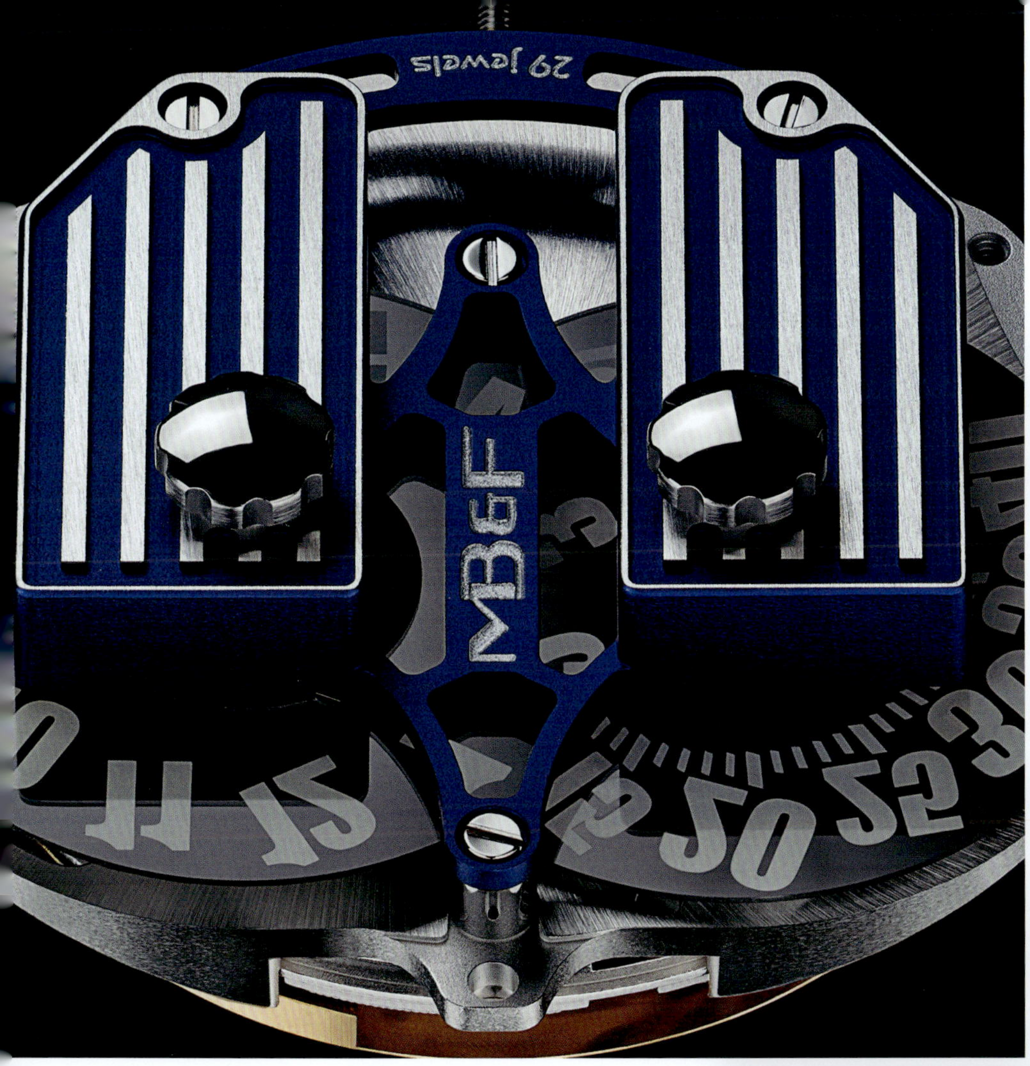
29 jewels
MB&F

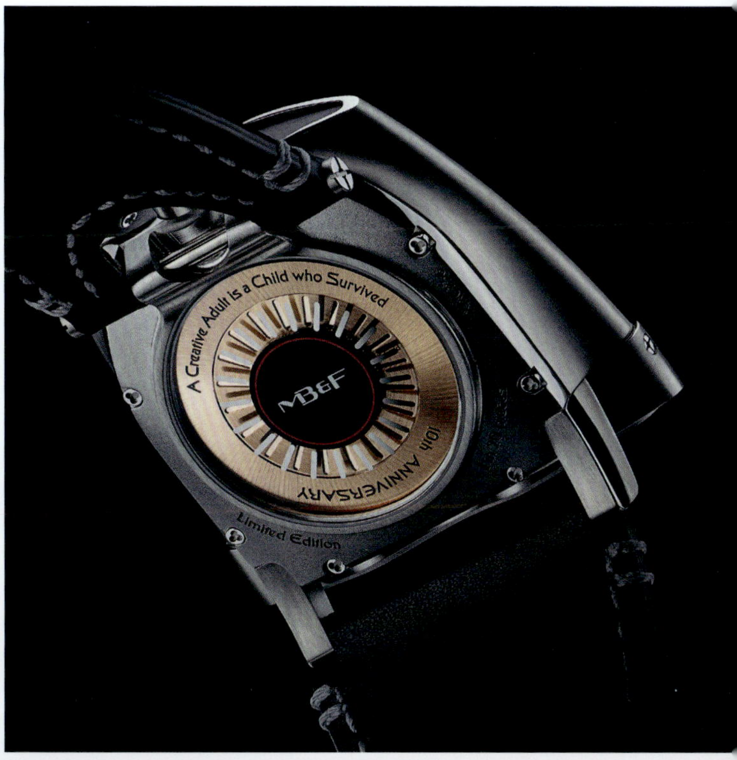
A Creative Adult is a Child who Survived
MB&F
10th ANNIVERSARY
Limited Edition

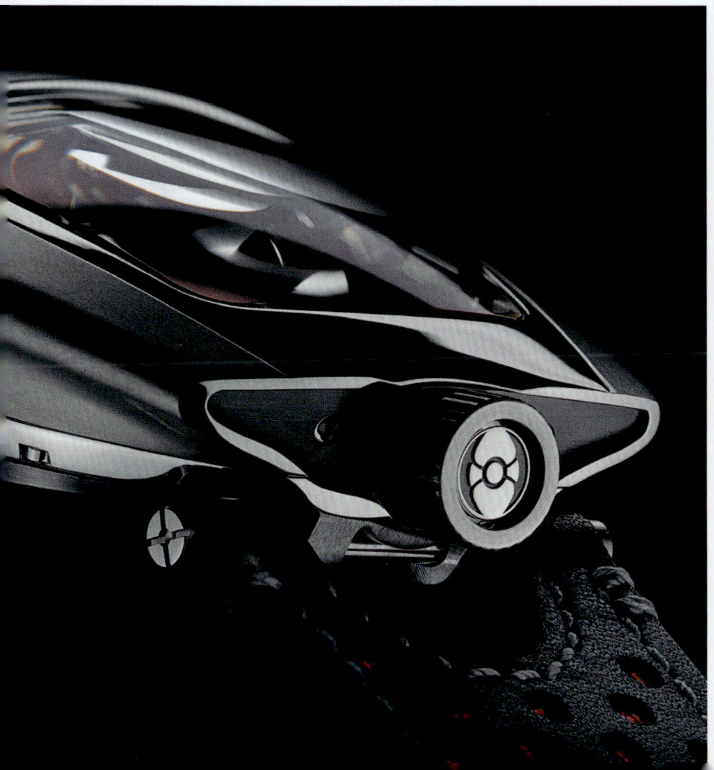

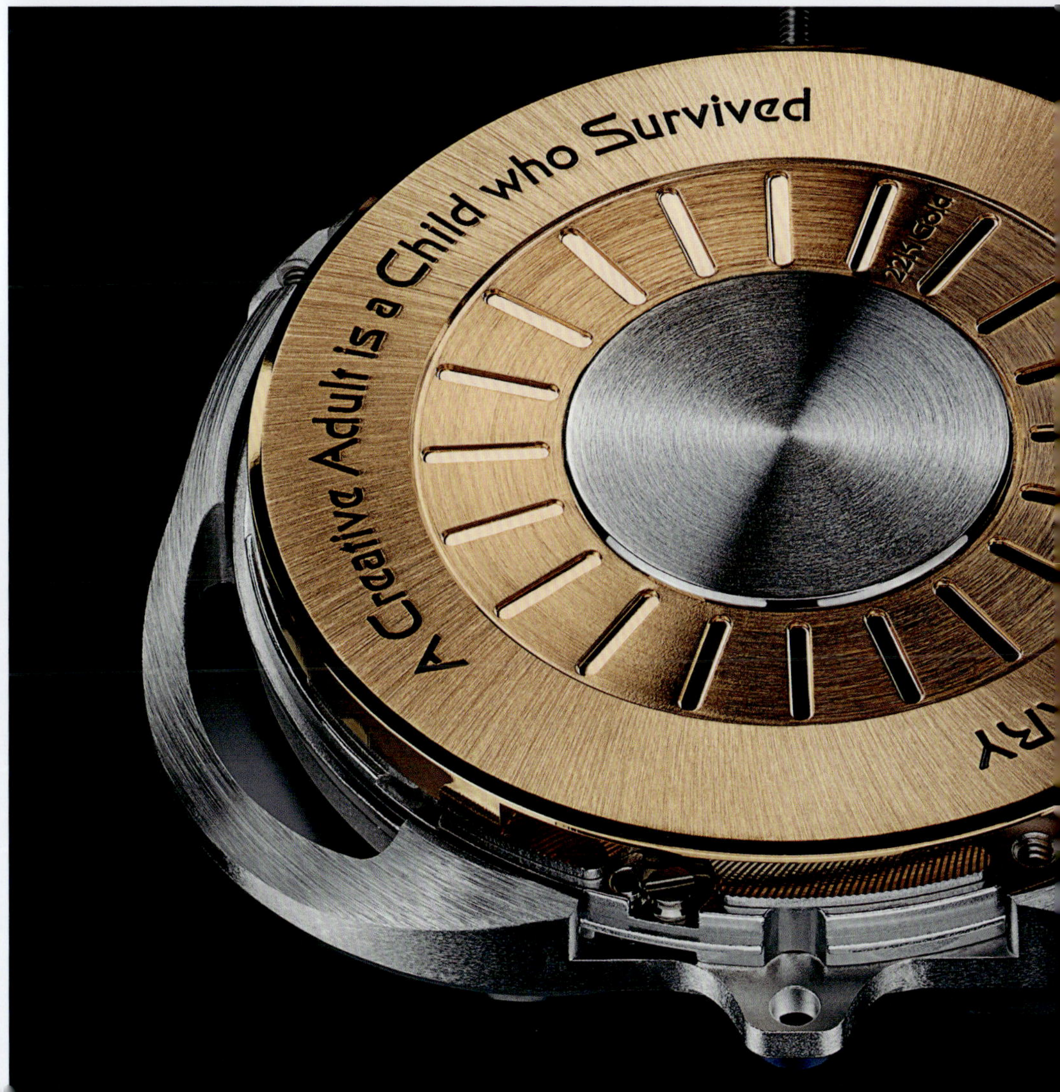
A Creative Adult is a Child who Survived

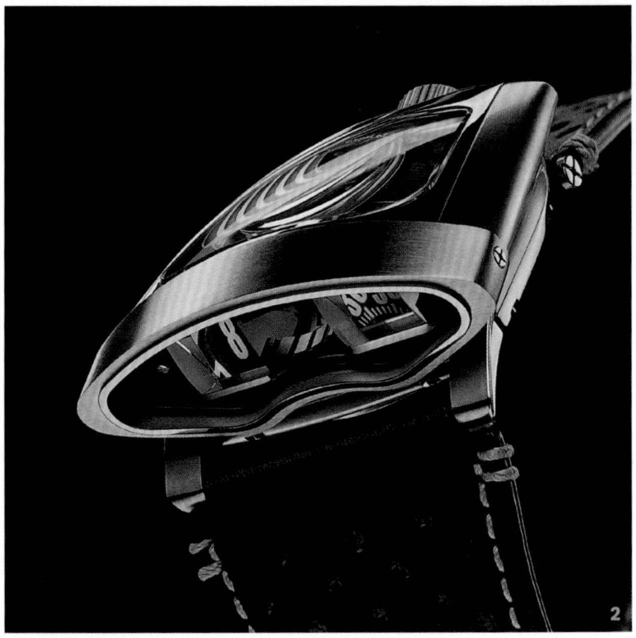

2

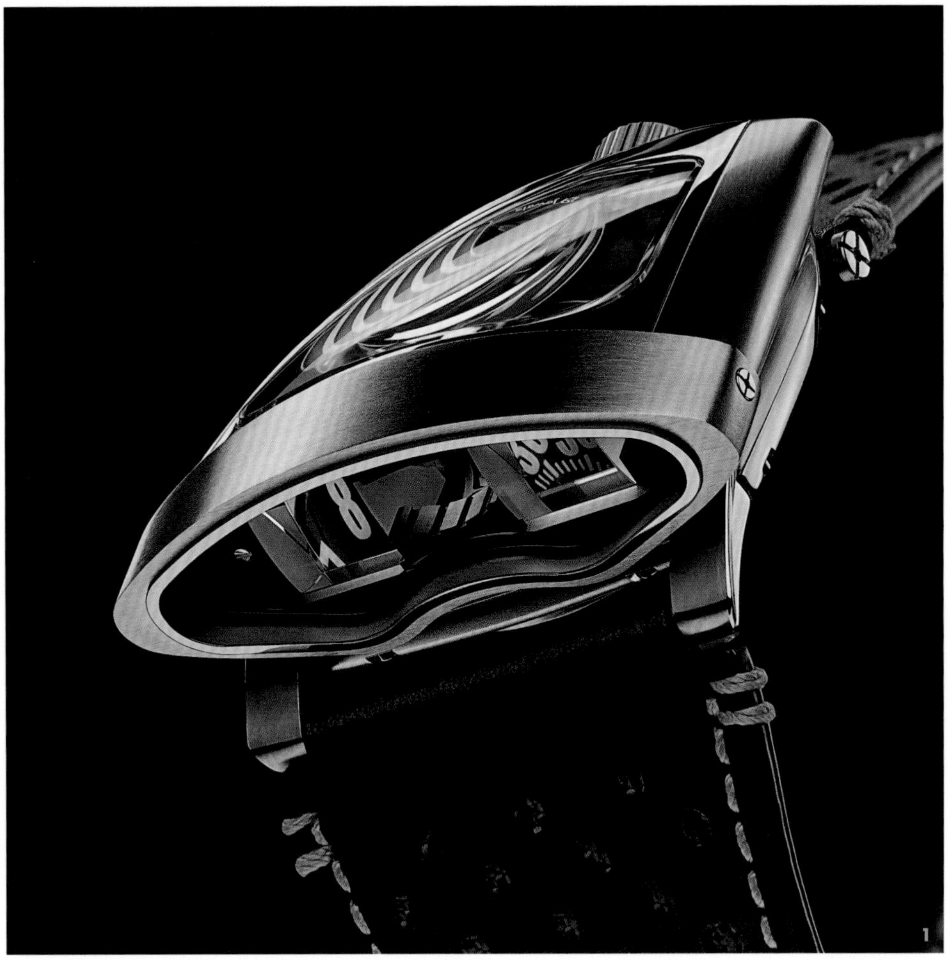

1

1 HMX Black
Reference: 56.STDL.B
Material: grade 5 titanium and stainless steel with black details
Limitation: 20
Introduction: 2015
Year in production: 2015
Quantity produced:
20 + 1 prototype

2 HMX Green
Reference: 56.STGL.B
Material: grade 5 titanium and stainless steel with green details
Limitation: 20
Introduction: 2015
Year in production: 2015
Quantity produced:
20 + 1 prototype

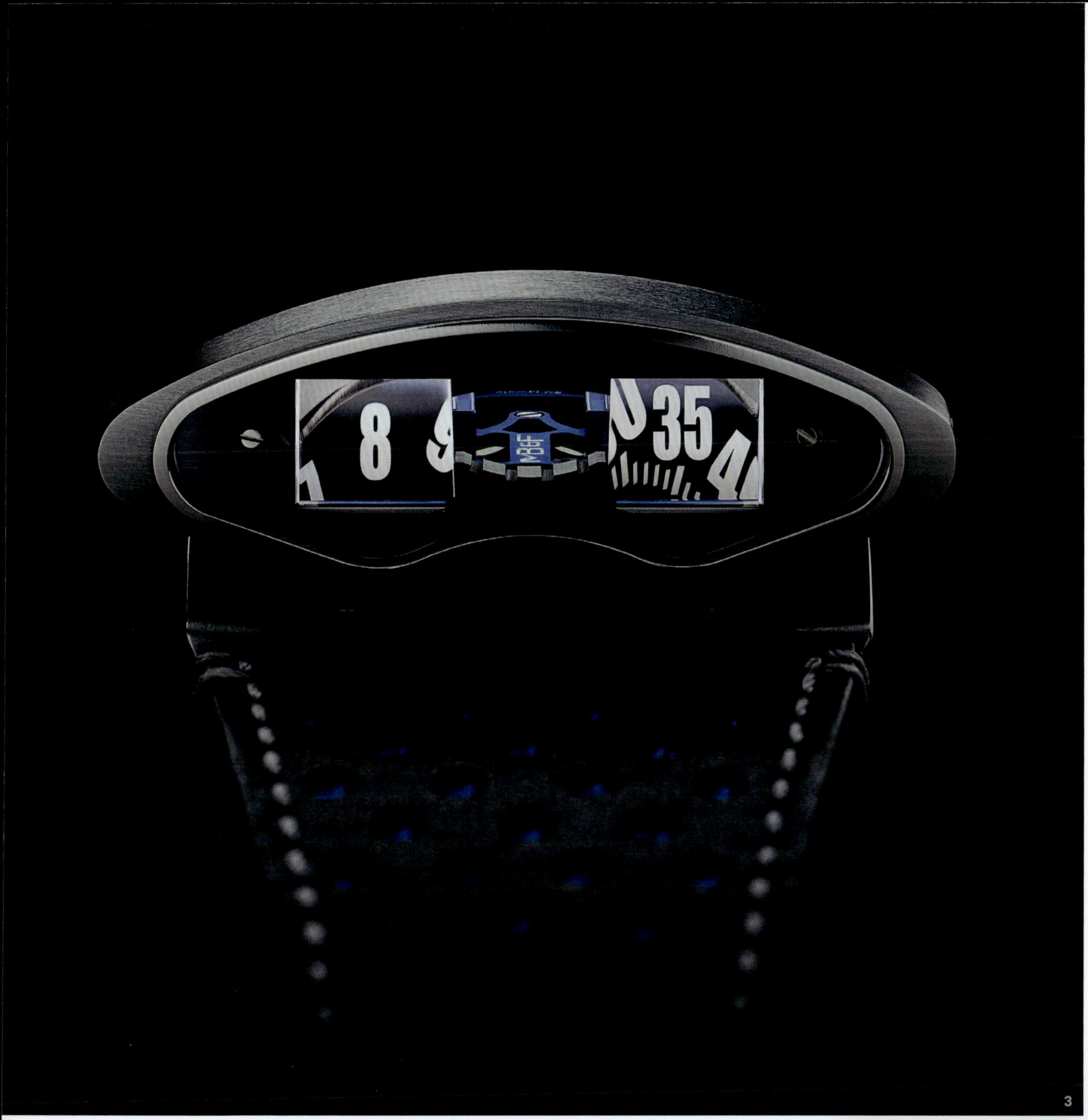

3

4
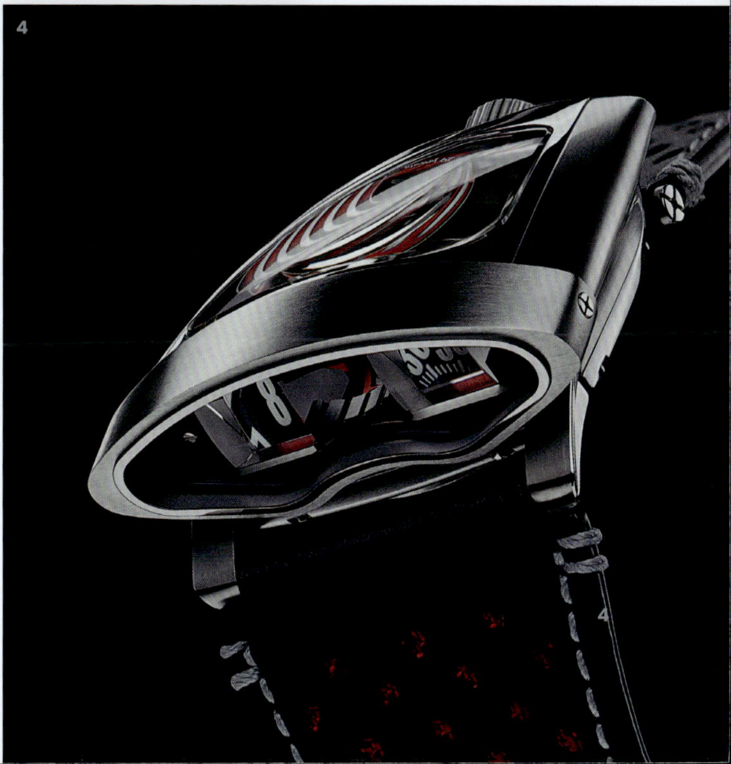

3 **HMX Blue**
Reference: 56.STBL.B
Material: Grade 5 titanium and stainless steel with blue details
Limitation: 20
Introduction: 2015
Year in production: 2015
Quantity produced: 20 + 1 prototype

4 **HMX Red**
Reference: 56.STRL.B
Material: grade 5 titanium and stainless steel with red details
Limitation: 20
Introduction: 2015
Year in production: 2015
Quantity produced: 20 + 1 prototype

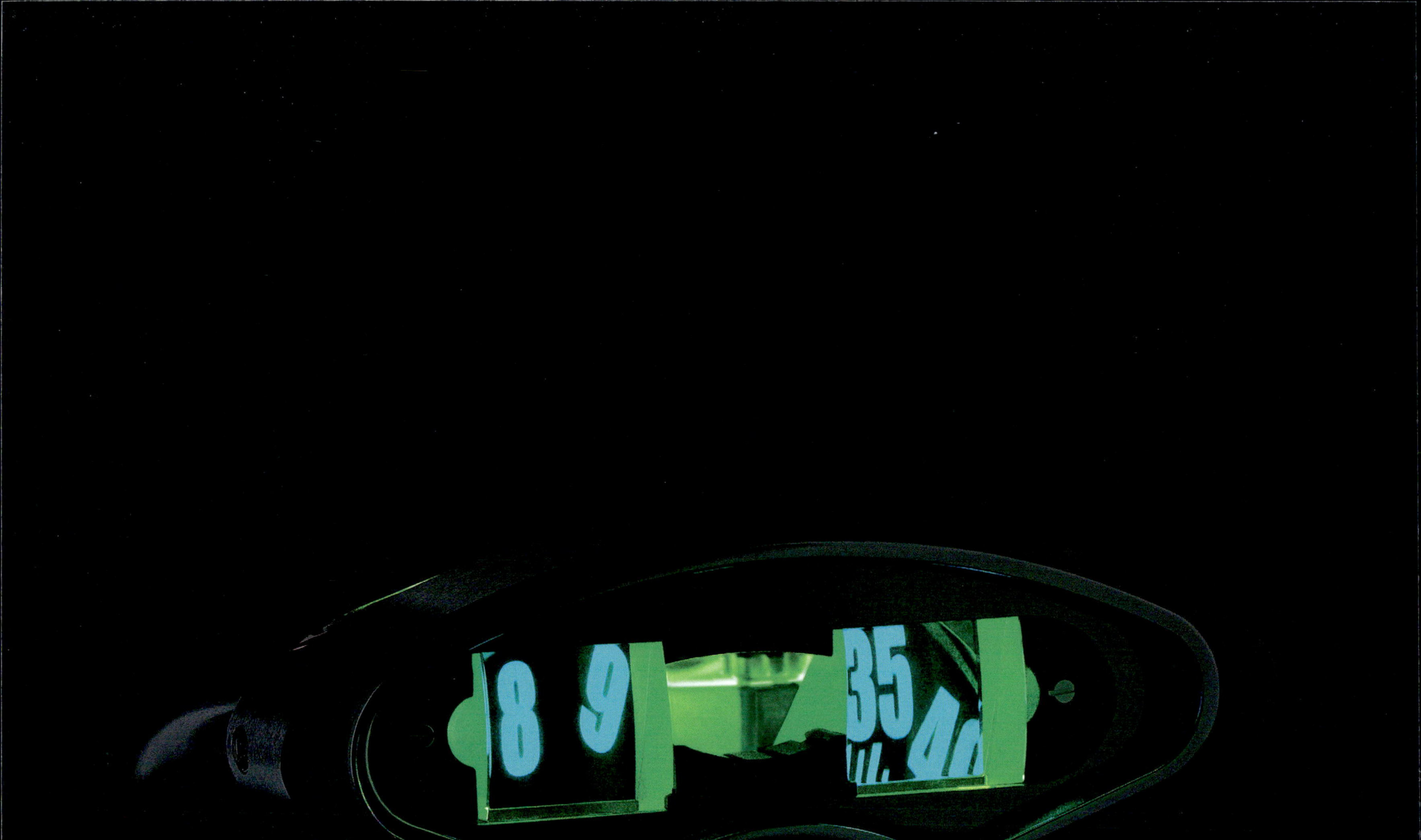
8
9
35

HMX MB&F × BLACK BADGER

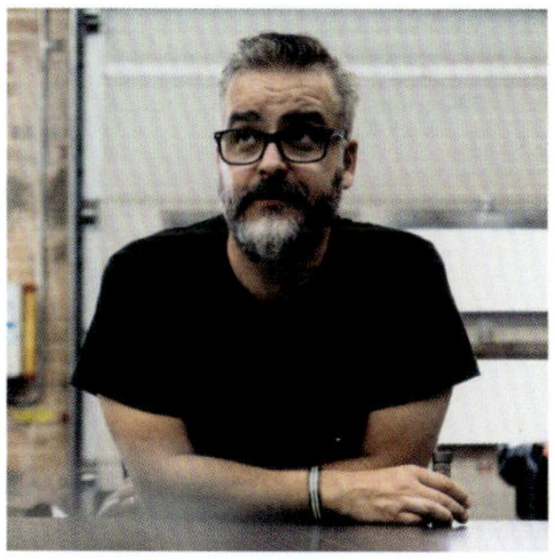

BLACK BADGER

Canadian artist James Thompson, better-known under his professional name "Black Badger", was born in Ottawa, Canada in 1976 and raised in Vancouver. Though he attended an industrial design school, Thompson did not graduate due to the mismatch between his own interests and the courses offered at his university. Instead he pursued his studies abroad, enrolling in a master's degree course in industrial design in Lund, Sweden in 2002. Eighteen months later he was forced to leave the program following administrative changes affecting international students.

Thompson took it very personally. He describes the moment as one of feeling brushed aside – abandoned through no fault of his own, and the experience filled him with creative rage. It was then that he set up his "Black Badger" design studio, named after an animal that never backs down from a fight, no matter the size or strength of its adversary.

He quickly earned a reputation for his use of solid lume as a material for making jewelry, having fashioned a ring out of the material as an experiment. His glowing, almost hypnotizing creations soon attracted a loyal group of fans, Maximilian Büsser among them.

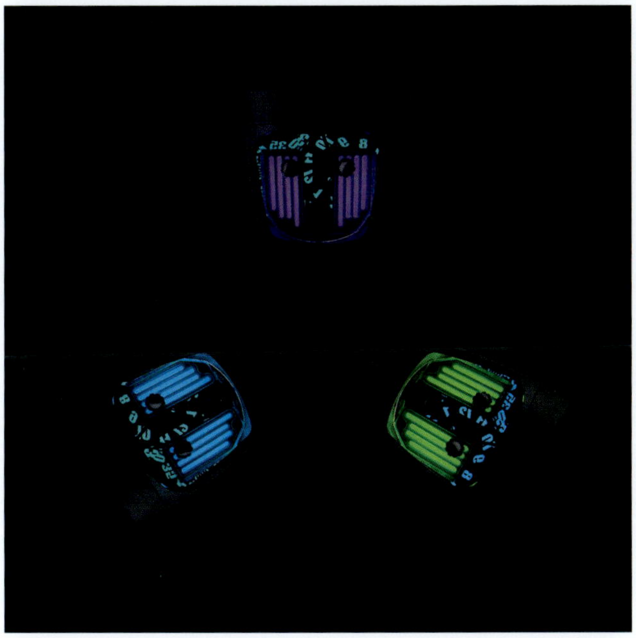

For his collaboration with MB&F, Thompson used his signature lume with great flair and to great effect for a reinterpretation of the HMX. His solid, highly efficient lume is applied to the "rocker cover" under the sapphire hood of the HMX, in Radar Green, Phantom Blue, and Purple Reign.

Watchmakers' lume typically refers to a paint-like liquid used for coating surfaces. Black Badger's lume is instead a solid, which is worked by hand or machined into the desired form. Lume in solid form shines brighter and for longer compared to painted lume, bringing out the colours of the HMX Black Badger.

What we call "white light," which includes sunlight, is composed of red, green and blue wavelengths. Other wavelengths are known to exist but they lie beyond the visible spectrum. Ultraviolet light is a good example. Blue and green wavelengths, on the other hand, are found in sunlight. This means that Radar Green and Phantom Blue can be easily charged during normal wearing, as the sapphire hood of the HMX exposes Black Badger's blocks of lume to natural, white light. In the case of the Purple Reign model, however, charging the lume depends on ultraviolet light.

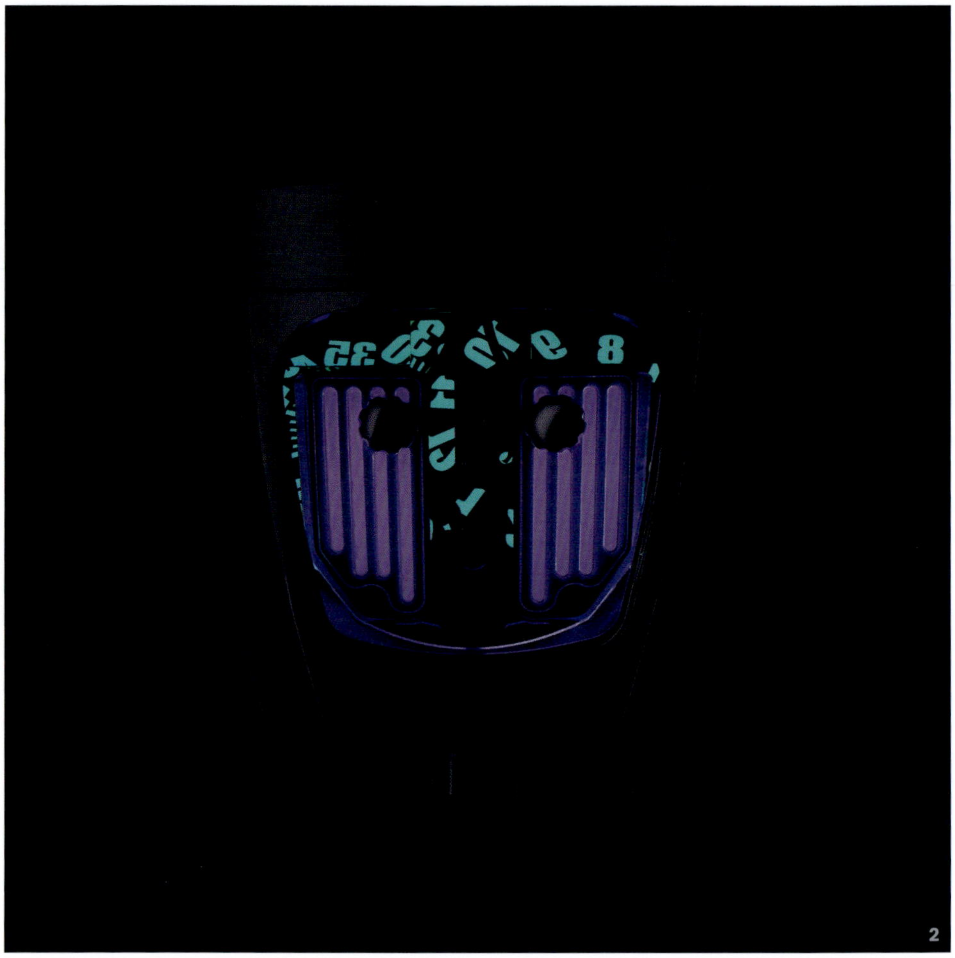

1 HMX Black Badger
Radar Green
Reference: 57.STGL.B
Material: grade 5 titanium and stainless steel with green luminescent details
Limitation: 18
Introduction: 2016
Year in production: 2016
Quantity produced:
18 + 1 prototype

2 HMX Black Badger
Purple Reign
Reference: 57.STPL.B
Material: grade 5 titanium and stainless steel with purple luminescent details
Limitation: 18
Introduction: 2016
Year in production: 2016
Quantity produced:
18 + 1 prototype

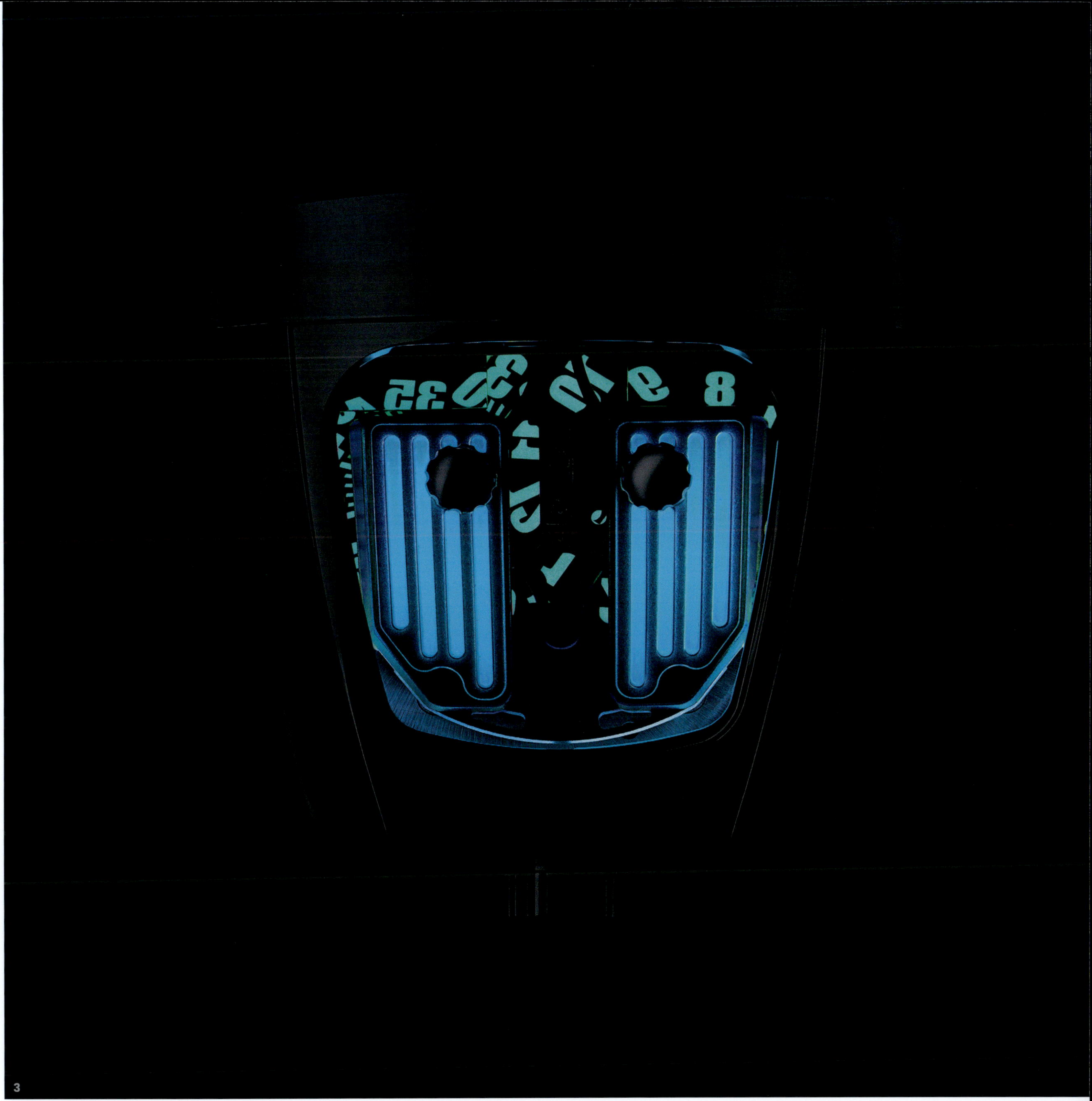
3

3 HMX Black Badger
Phantom Blue
Reference: 57.STBL.B
Material: grade 5 titanium
and stainless steel with
blue luminescent details
Limitation: 18
Introduction: 2016
Year in production: 2016
Quantity produced:
18 + 1 prototype + 1 artist's proof

“Max Büsser is not a watchmaker, but he’s got an uncanny ability to make friends with one. An engineer by trade, who never imagined himself playing a role in the watchmaking industry, Büsser is now one of its leading actors. Under MB&F – an independent outfit that hires and collaborates with talented artists – Büsser has created some of the bravest and most provocative timepieces ever made. Not that he cares if you disagree.”

Arthur Touchot
Head of Digital Strategy,
Phillips in Association with Bacs & Russo

"The coherence behind MB&F's Machines always stems from an idea: one product, one idea. The creative challenge is straightforward: maintaining that idea throughout the entire process. The energy powering the process is teamwork, and a sort of team culture has settled in — or rather, a counter-culture, as opposed to what you find in a typical watch brand, which bases everything on its DNA or on dogma. MB&F is instead driven by a way of thinking, a vision."

Eric Giroud
Designer

HM7

AQUAPOD

Horological Machine N°7, dubbed the "Aquapod", is the first round-shaped Horological Machine and, as its name implies, the first to draw inspiration from aquatic life. Maximilian Büsser had the idea for the machine following a close encounter between his wife and a jellyfish while on holiday in Ras Al Khaimah in 2013. So was born the mechanical jellyfish that eventually became the HM7.

Symmetrical like its aquatic counterpart, the HM7 is constructed around concentric rings that mimic those of a jellyfish. Just as a jellyfish brain consists of radially symmetrical neuron rings, so the brain of the Horological Machine – the flying tourbillon at the top of the machine – is surrounded by rotating rings displaying hours and minutes.

On the underside are the tentacles (the winding rotor) of the HM7. It relies on these as a source of energy, much like a jellyfish uses its tentacles to trap and feed upon its prey.

The outermost ring of the Aquapod is its rotating bezel, ingeniously positioned beyond the boundary of the machine's two crowns so it appears to float on air or water, like a lifebuoy.

HM7 was produced in four limited editions: first in red gold with a black ceramic bezel and blue lume (66 pieces), and in titanium with a blue ceramic bezel and blue lume (33 pieces); then in titanium with a green sapphire crystal bezel and green lume (50 pieces); and later in platinum with a red sapphire crystal bezel.

HOROLOGICAL MACHINE N°7 AQUAPOD

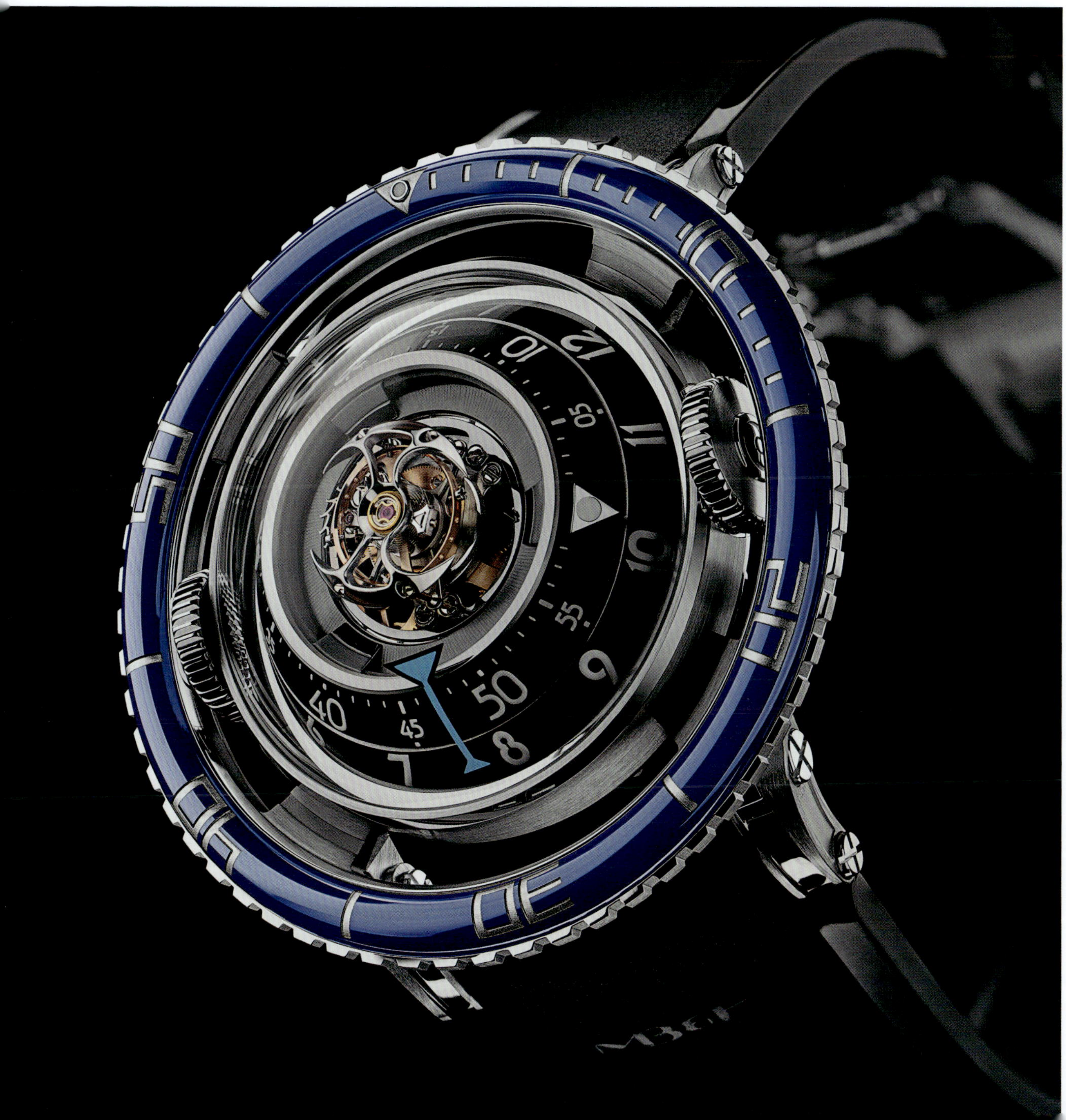

HM7 AQUAPOD CALIBRE

The 303-component engine of the HM7 was developed entirely in-house by MB&F and is three dimensional throughout, from the central flying 60-second tourbillon to the far reaching, curved tentacles of its winding rotor. Energy travels from the rotor at the bottom of the machine to the flying tourbillon at the top via a set of gears that serve as stairs, allowing power to transition from one level to the next.
The large diameter of its display rings posed a particular challenge to the team's engineers. To minimize friction, extra-large diameter ceramic ball bearings were developed to support the hour and minute displays.
The curved, three-dimensional tentacles on the underside of the HM7 were painstakingly crafted from a single block of titanium. Hidden beneath them is a section in heavy platinum to ensure that the HM7 is properly and efficiently wound.

Collection
Horological Machines
Series
Horological Machine N°7 Aquapod
Common name
HM7 Aquapod, HM7
Movement
Developed in-house by MB&F
Winding
Automatic
Power reserve
72 hours
Components
303 (391 components for the Platinum Red edition) / 35 jewels
Functions
Hours and minutes displayed by two aluminium / titanium spherical segment discs rotating on oversized central ceramic bearings (for the Platinum Red edition: three-dimensional "floating" hour and minute numerals)
Add. features
Central 60-second flying tourbillon; unidirectional rotating bezel for elapsed time; 3 panels of AGT (Ambient Glow Technology) lume around the flying tourbillon and numerals, markers and segments along the rotor in Super-LumiNova. Two crowns for winding and time-setting
Case
Grade 5 titanium, 18K red gold or platinum 950
Dimensions
53.8 × 21.3 mm
Case components
95 (83 for the red edition)
Water resistance
50 m/5 ATM
Strap
Rubber, moulded in aircraft-grade Fluorocarbon FKM 70 Shores A elastomer
Buckle
Folding buckle, titanium, 18K red gold or platinum matching the case
Introduction
2017
Years in production
2017 – in production
Quantity produced
(at the end of 2020) 142 + 5 prototypes

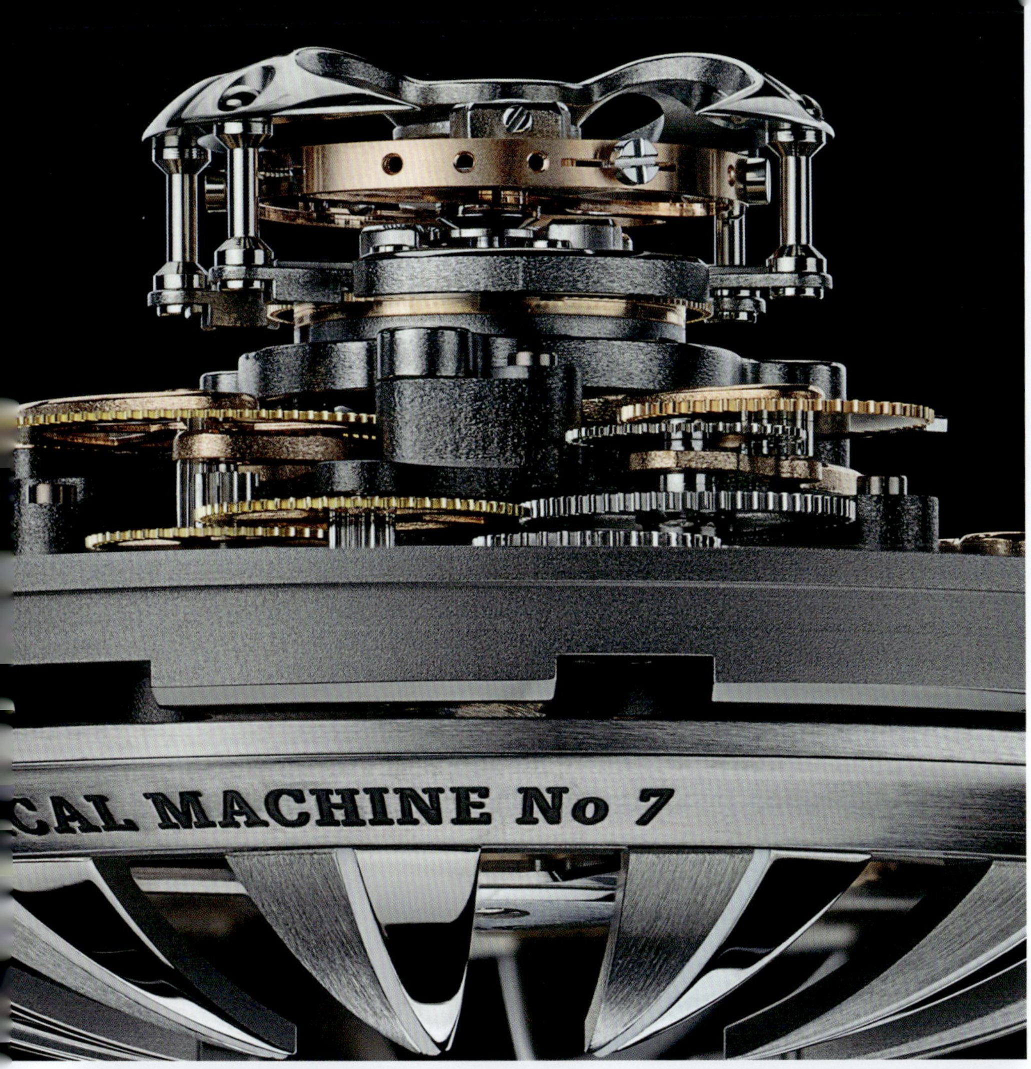
CAL MACHINE No 7

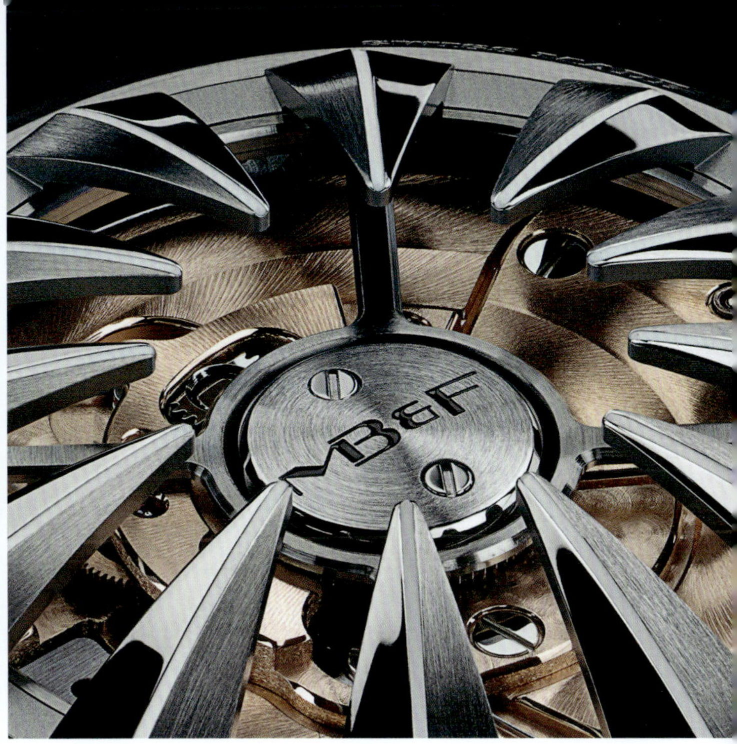
MB&F

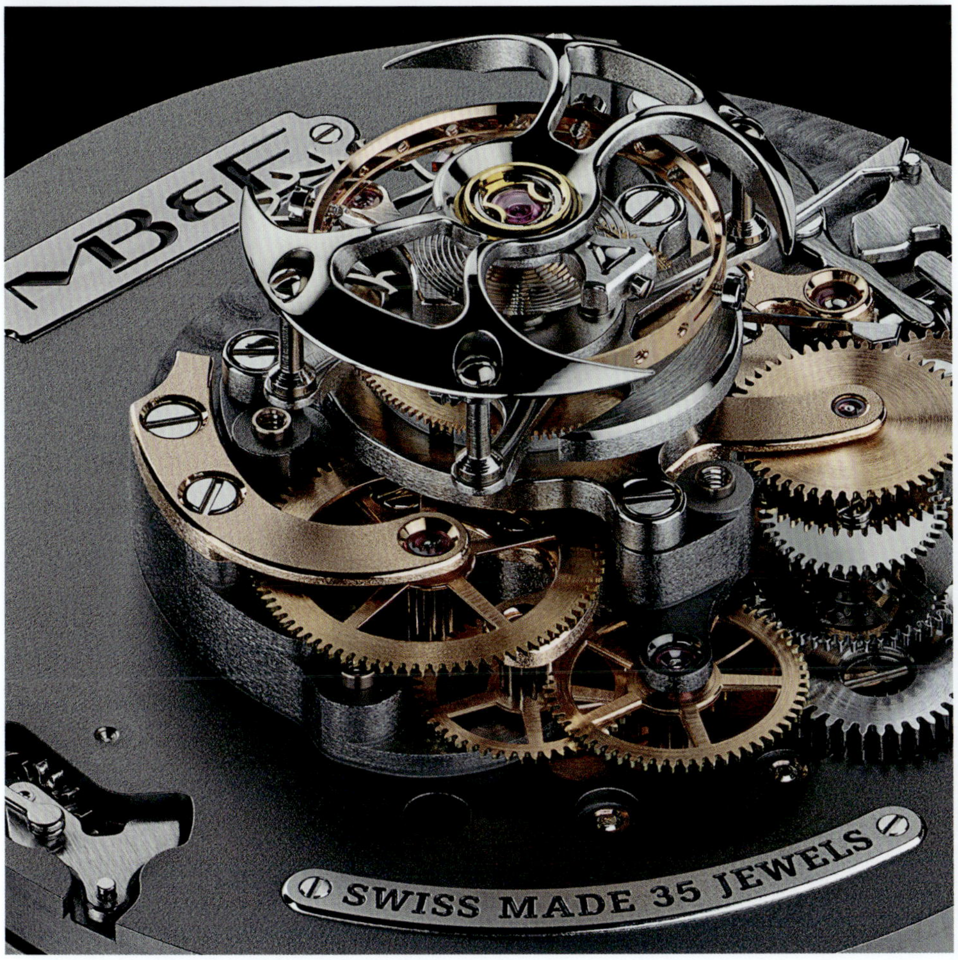
MB&F
SWISS MADE 35 JEWELS

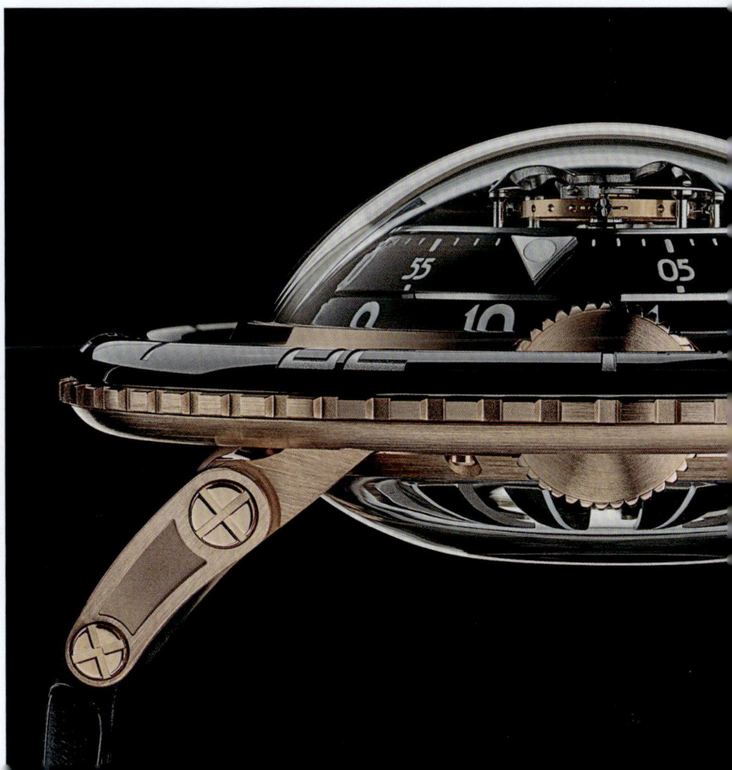
55
05
10

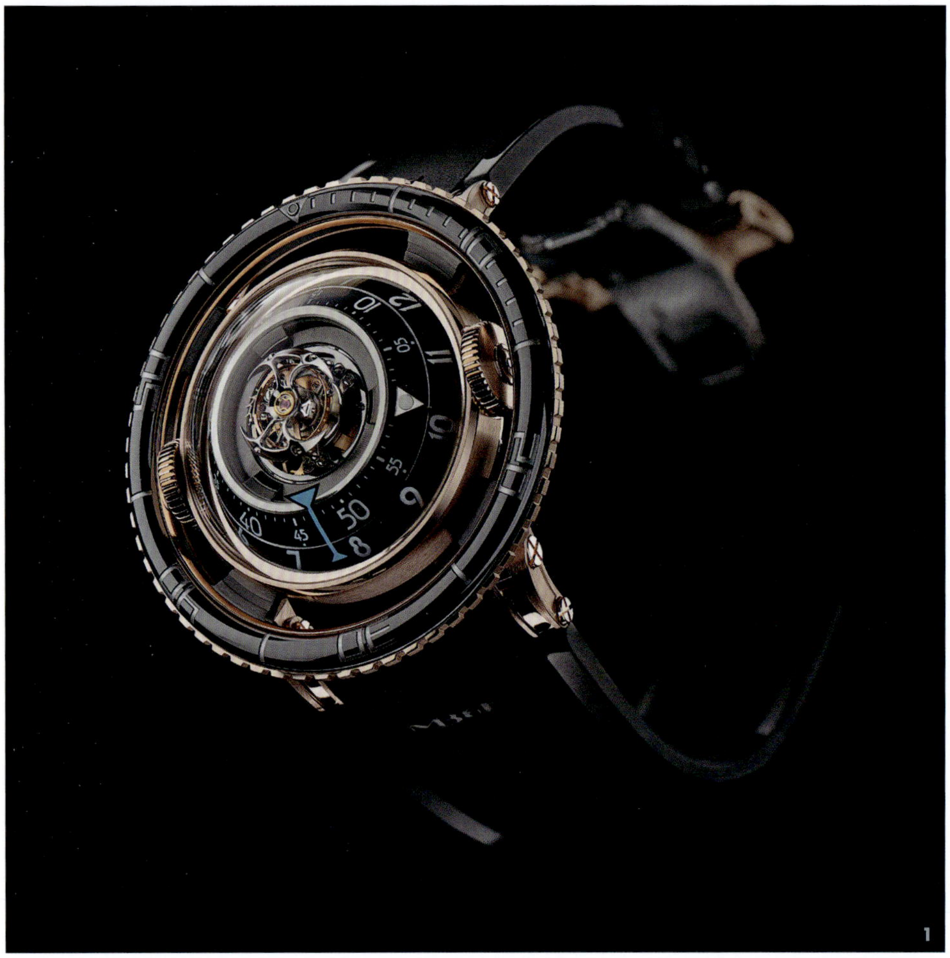

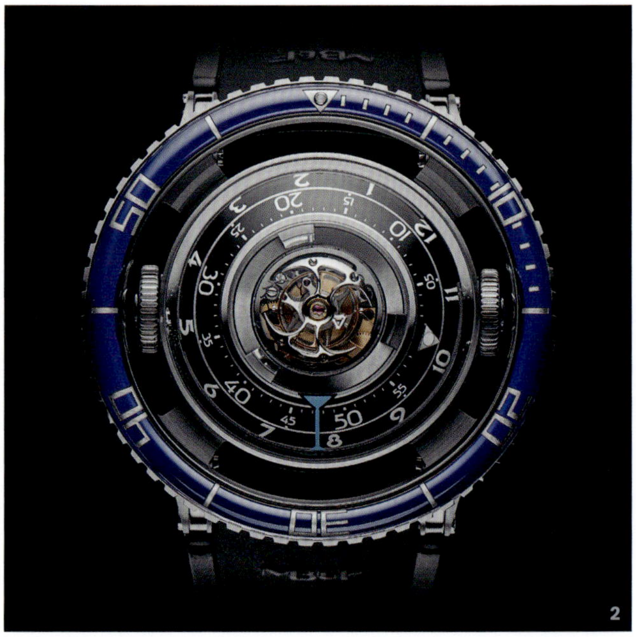

1 Horological Machine N°7 RG
Reference: 70.RBL.B
Material: 18K red gold with black ceramic bezel
Limitation: 66
Introduction: 2017
Years in production: 2017–2019
Quantity produced: 61 + 2 prototypes

2 Horological Machine N°7 Ti Blue
Reference: 70.TSL.B
Material: grade 5 titanium with blue ceramic bezel
Limitation: 33
Introduction: 2017
Years in production: 2017–2018
Quantity produced: 33 + 1 prototype

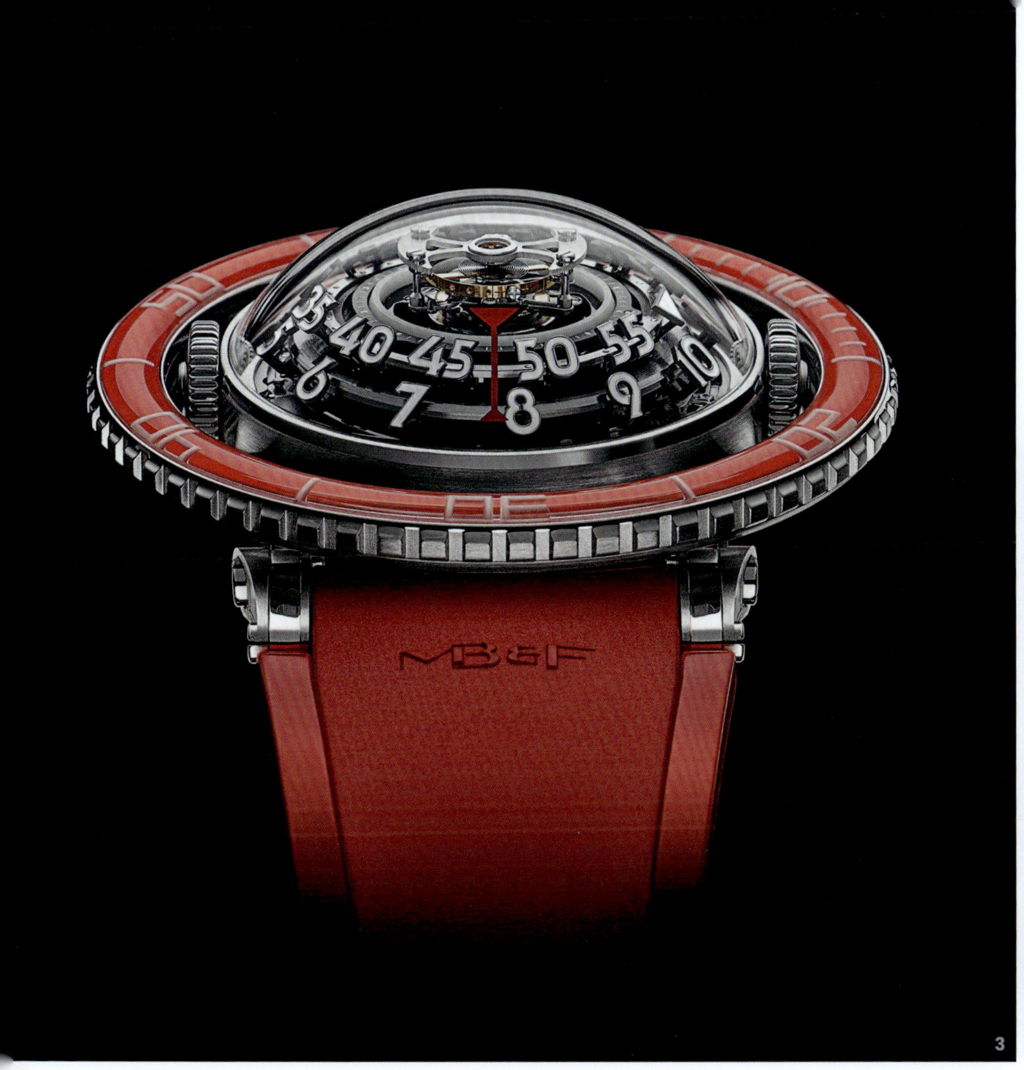

3

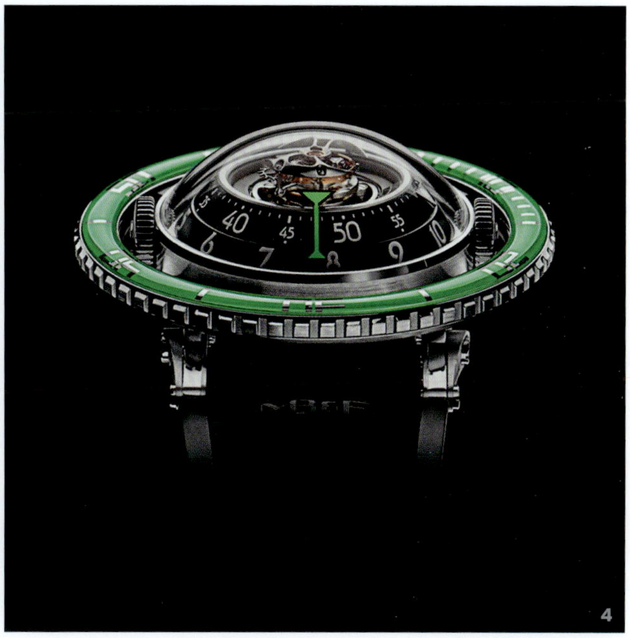

4

3 Horological Machine N°7
PT Red
Reference: 70.PBL.B
Material: platinum 950
with red sapphire crystal bezel
Limitation: 25
Introduction: 2019
Years in production:
2019 – in production
Quantity produced
(at the end of 2020):
14 + 1 prototype

4 Horological Machine N°7
Ti Green
Reference: 70.TGL.B
Material: grade 5 titanium
with green sapphire
crystal bezel
Limitation: 50
Introduction: 2018
Years in production:
2018–2019
Quantity produced:
34 + 1 prototype

HM8

CAN-AM

Horological Machine N°8 follows HM5 and HMX as the last in a trilogy of homages to traditional drivers' watches (which display time vertically, on the side so drivers can read time without turning the wrist). This one bears the nickname "Can-Am", a reference to the Canadian-American Challenge Cup sports car racing series that ran from 1966 to 1987, and was known for its lax safety rules and "anything goes" attitude. Racing a car at 240 miles per hour has its risks, which is why Can-Am racing cars were fitted with their distinctive "roll bars", to protect the driver if the vehicle should roll over. This was the inspiration for one of HM8's most recognizable features, the polished titanium bars that sweep from the front to the back of the machine.

Whirring away below these bars, under a sleek slab of polished sapphire, is the famous MB&F battle-axe winding rotor in blued gold that powers the HM8 "muscle car".

Bi-directional jumping hours and trailing minute displays are projected vertically thanks to a set of two sapphire prisms that reflect the numerals upward and forward.

HOROLOGICAL MACHINE Nº8 CAN-AM

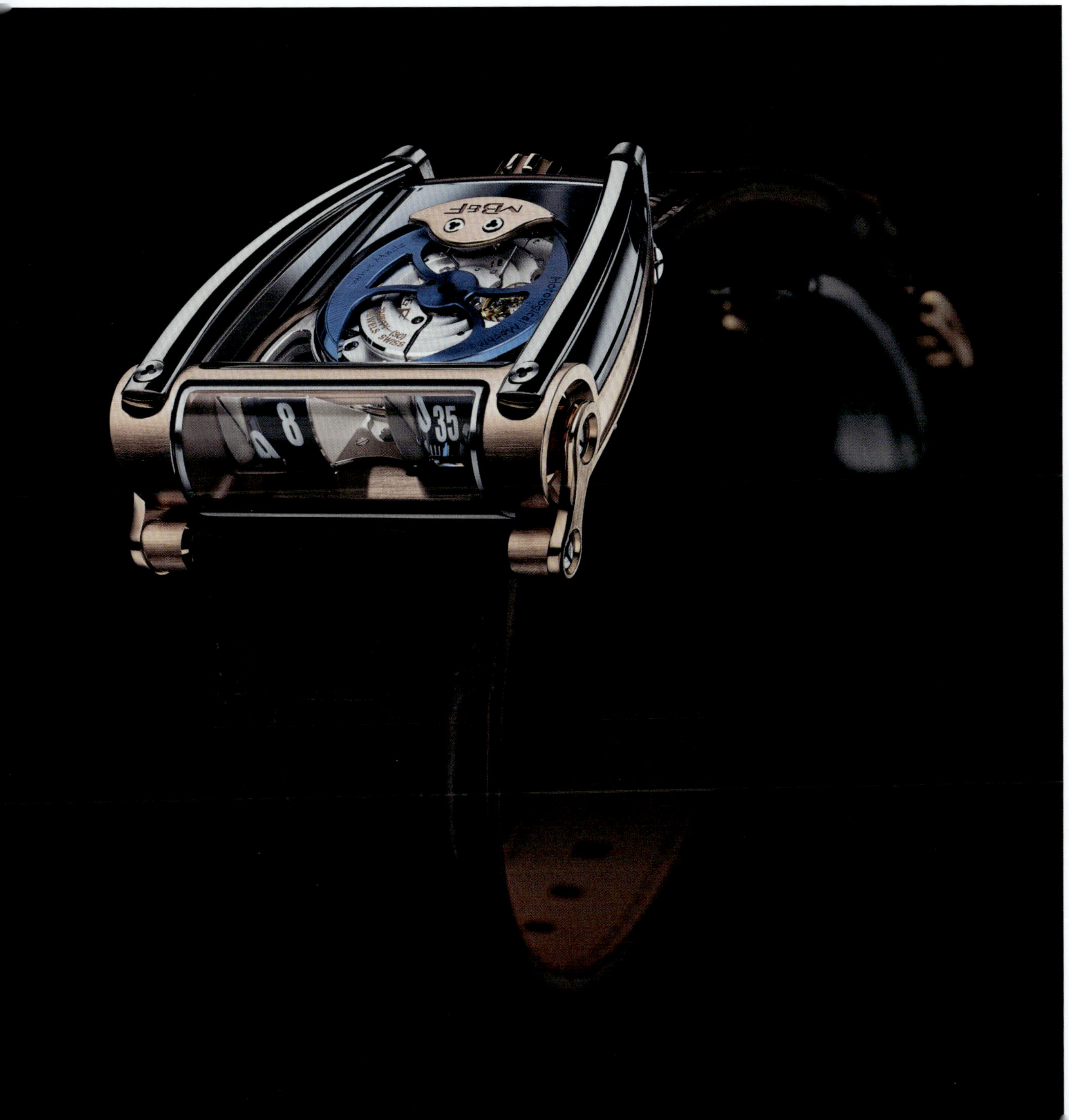

HM8 CAN-AM CALIBRE

The engine of the supercar HM8 was developed in-house, built on a Girard-Perregaux base movement that has been inverted so that the 22-karat battle-axe winding rotor faces the wearer.
The battle-axe rotor is dubbed a "mystery" rotor because unlike a conventional rotor that has an apparently unbalanced mass, it appears to be symmetrically balanced. In fact, the so-called "mystery" is all down to engineering: machining away the underside of one half of the battle-axe produces a razor-sharp edge, thus reducing its mass relative to the opposing side.
Overlapping discs (one for hours, another for minutes) are painted with mirror image numerals for the hours and minutes (so the prisms then project them right way round). The prisms' reflective sapphire crystal also magnifies the time display by 20% for increased legibility.

Collection
Horological Machines
Series
Horological Machine N°8 Can-Am
Common name
HM8 Can-Am, HM8
Movement
Developed by MB&F, powered by a Girard-Perregaux base; moon phase complication by Stepan Sarpaneva for the MoonMachine 2
Winding
Automatic
Power reserve
42 hours
Components
HM8: 247/30 jewels
MoonMachine 2: 293/30 jewels
Functions
Bi-directional jumping hours and trailing minutes; moon phase indicator for the MoonMachine 2 editions
Add. features
Automatic 22K gold "battle-axe" winding rotor; hours and minutes displayed by two optical prisms that both reflect and magnify
Case
Grade 5 titanium, 18K red gold or white gold
Dimensions
HM8: 49 × 51.5 × 19 mm
MoonMachine 2: 49 × 51.5 × 19.5 mm
Case components
HM8: 60
MoonMachine 2: 59
Water resistance
30 m/3 ATM
Strap
Alligator
Buckle
Folding buckle, 18K red gold or white gold with titanium matching the case
Introduction
2016
Years in production
HM8: 2016–2018
MoonMachine 2: 2018
Quantity produced
(at the end of 2020)
HM8: 83 + 2 prototypes
MoonMachine 2: 36 + 3 prototypes + 1 artist's proof

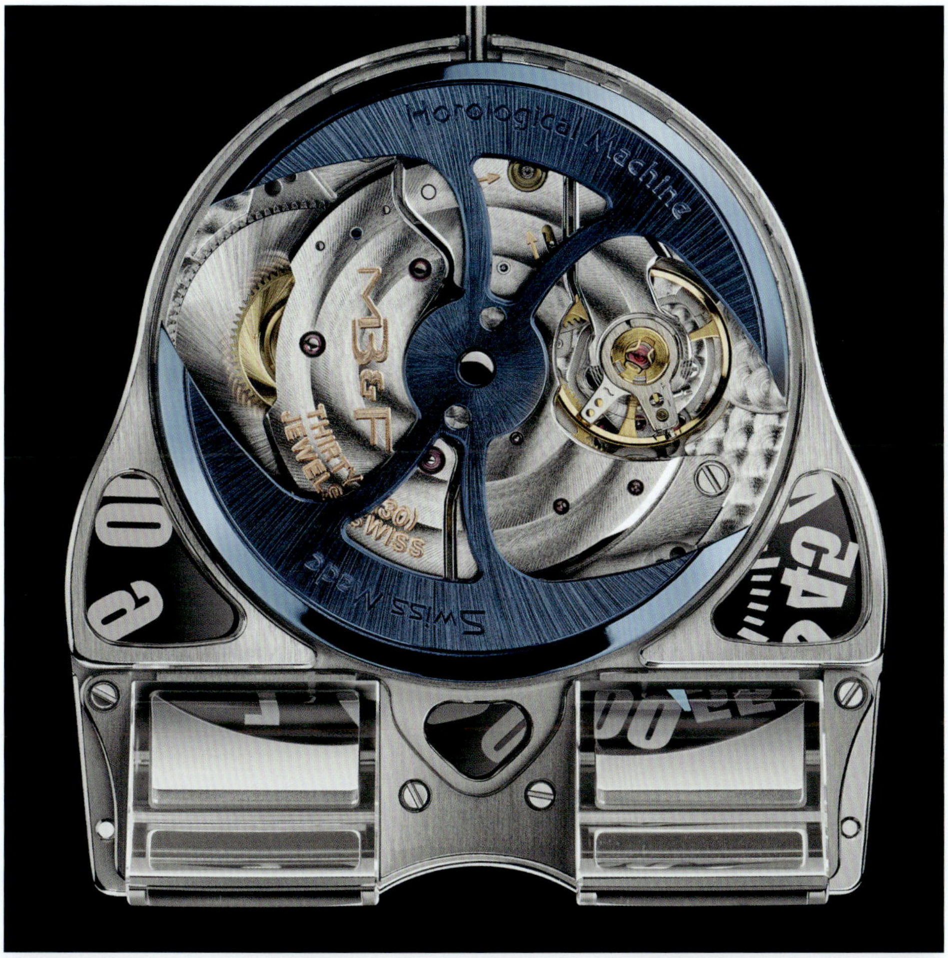
Horological Machine
MB&F
Swiss Made

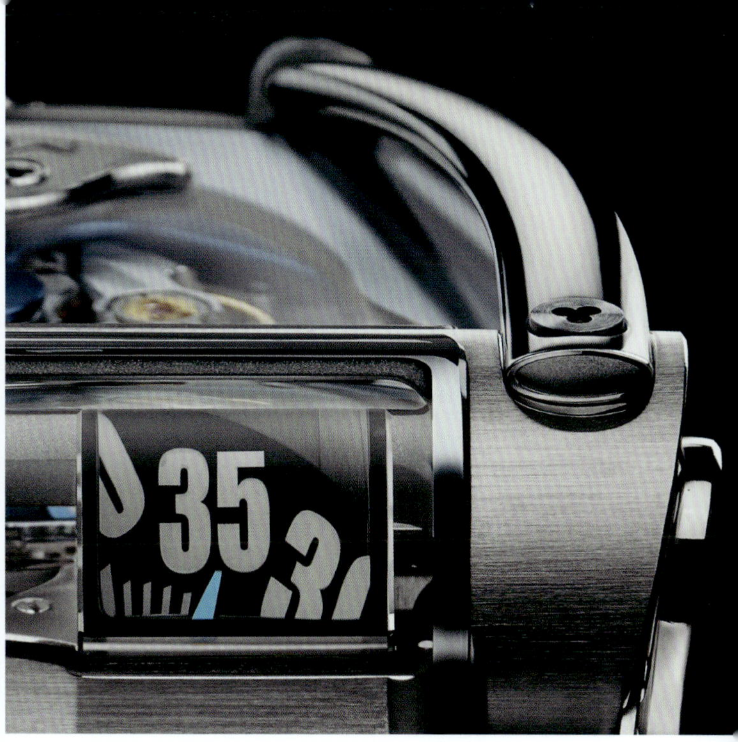
35

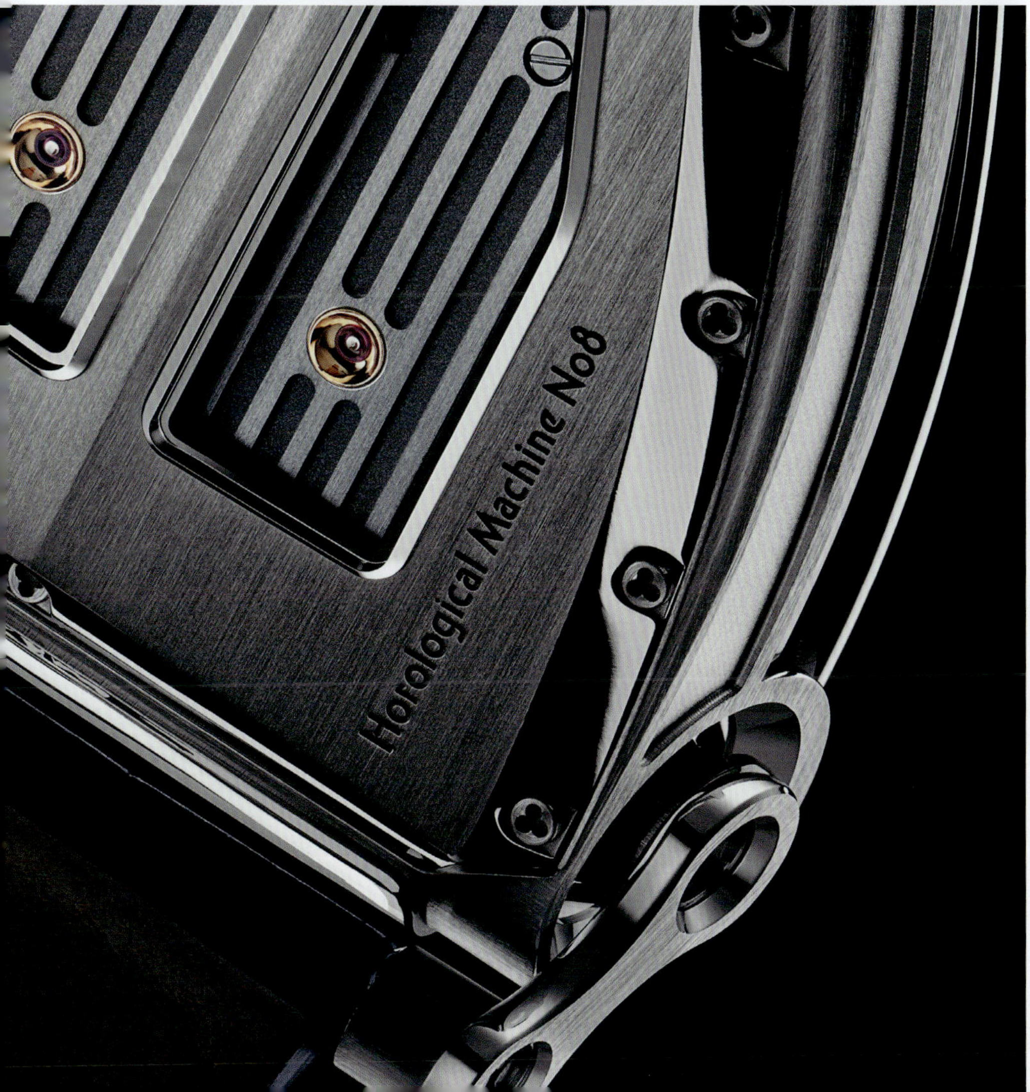
Horological Machine No8

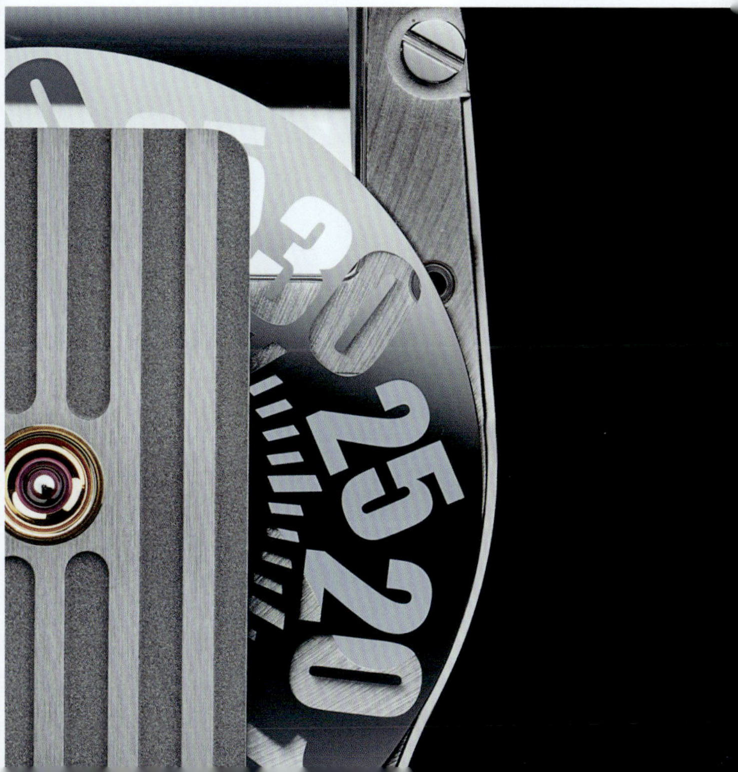
25
20

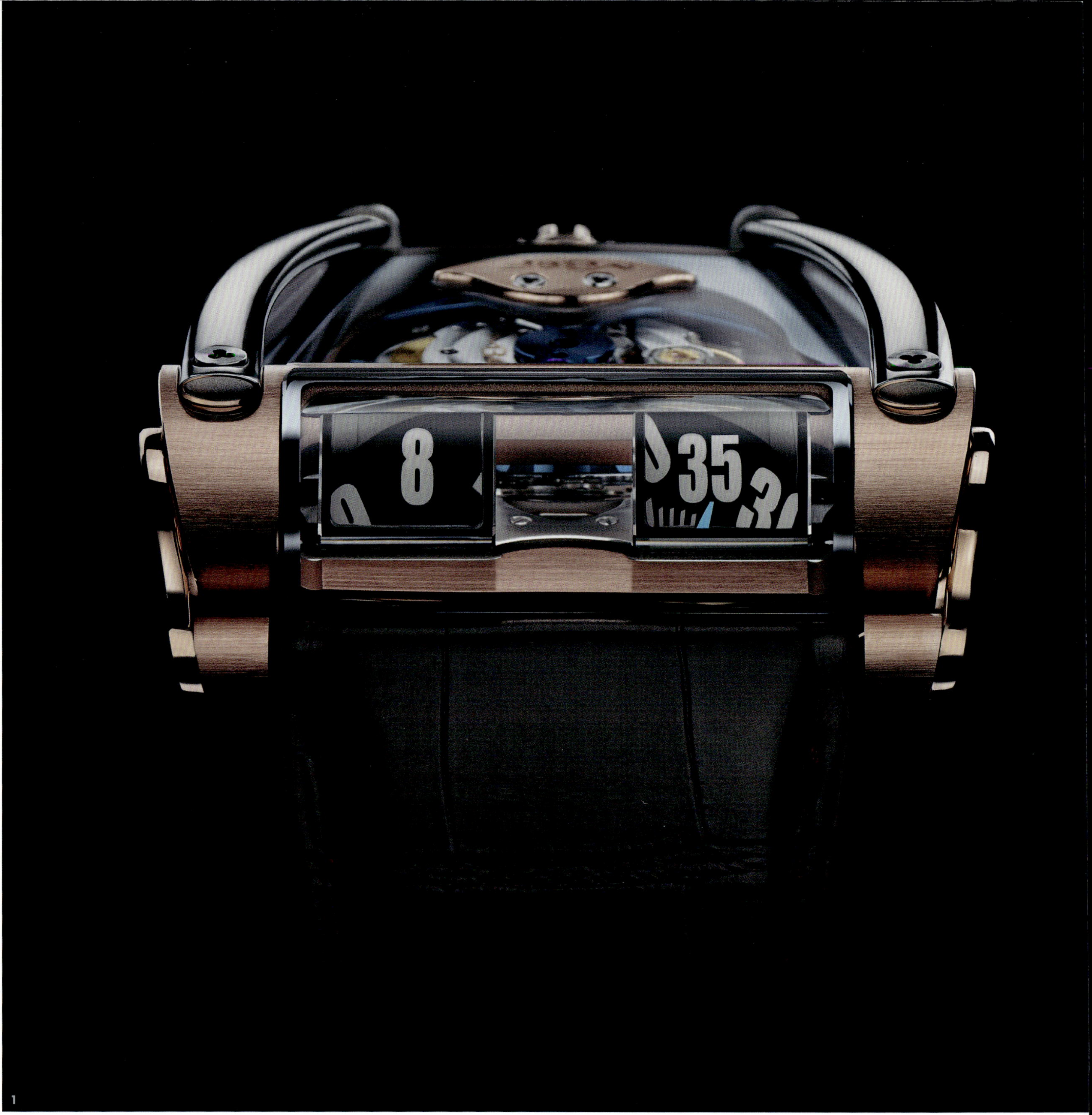

1

1 Horological Machine N°8
Can-Am RT
Reference: 80.RTL.B
Material: 18K red gold and
grade 5 titanium
Limitation: –
Introduction: 2016
Years in production:
2016–2018
Quantity produced:
46 + 1 prototype

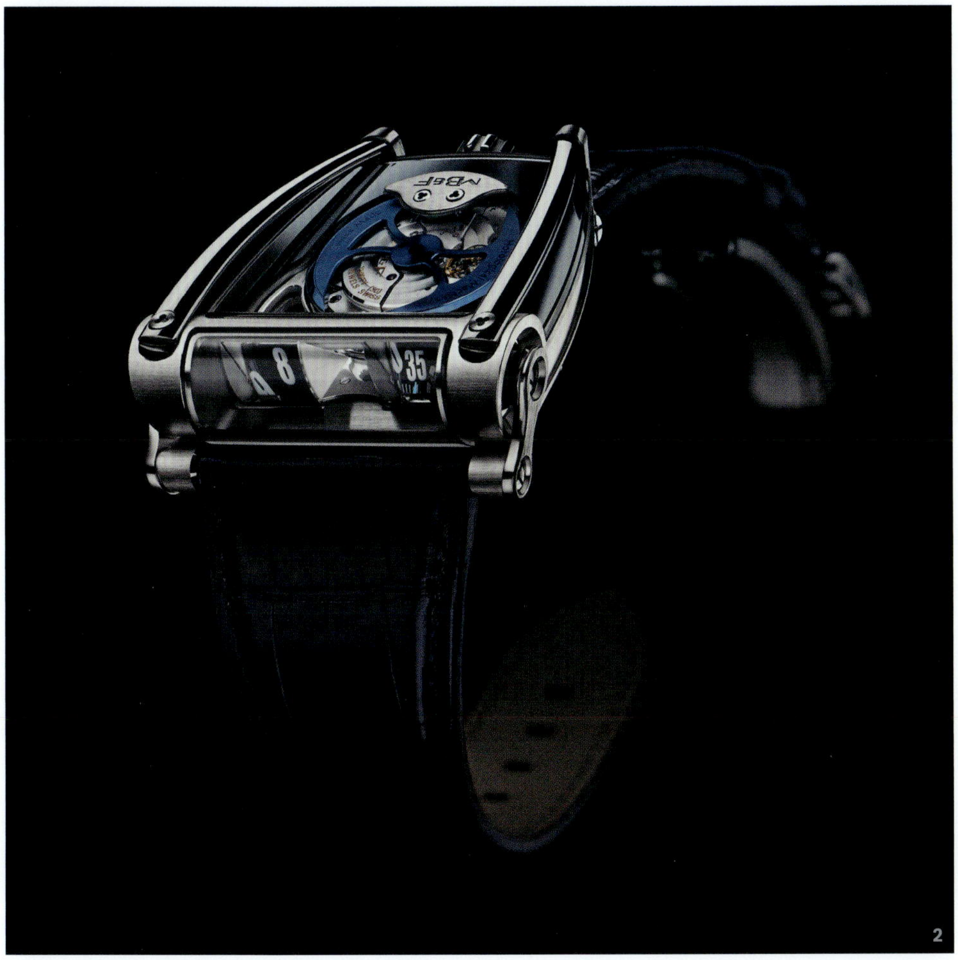

2 Horological Machine N°8
Can-Am WT
Reference: 80.WTL.B
Material: 18K white gold
and grade 5 titanium
Limitation: –
Introduction: 2016
Years in production:
2016–2018
Quantity produced:
36 + 1 prototype

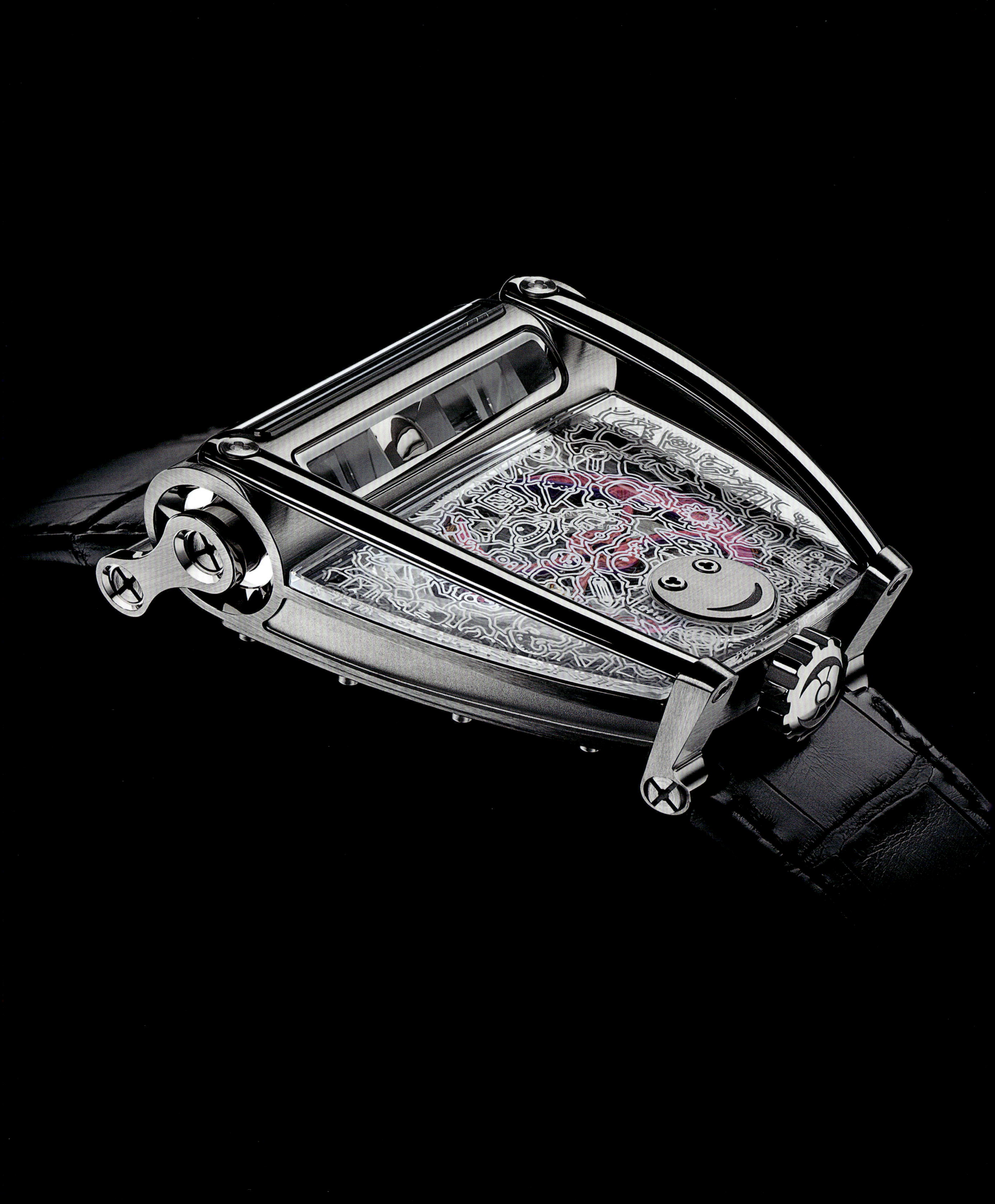

HM8 ONLY WATCH MB&F × CASSANDRA LEGENDRE

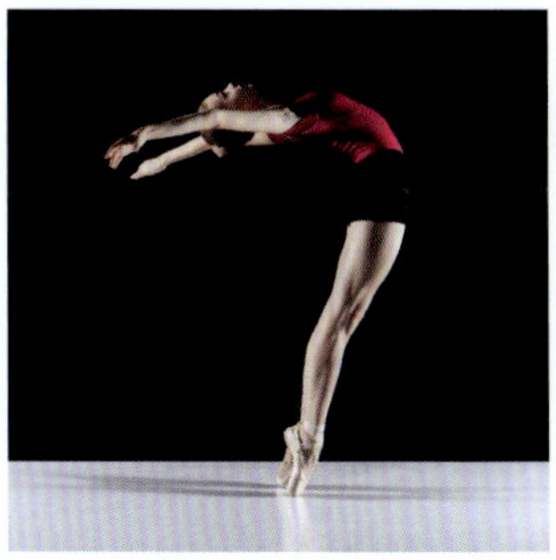

CASSANDRA LEGENDRE

In 2017 Cassandra Legendre was a 15-year old classical ballet student at the Victor Ullate School in Madrid – an unlikely background for an MB&F creative collaborator. She says that she first started drawing at school, just to relieve the tedium of particularly boring math classes. And from that beginning, drawing came to be her preferred means of self-expression. Encouraged by her parents, she presented her work on social media, and it was there that she came to the attention of Maximilian Büsser. What most attracted him was her unashamedly playful style – a whimsical, youthful approach that earned her an invitation to collaborate with MB&F in the creation of a Unique Piece for the 2017 Only Watch Auction.

Specially produced for the 2017 auction, MB&F's rework of its supercar-inspired watch, Horological Machine N°8, takes its inspirational cues from young artist Cassandra Legendre. The bold case provides the perfect foil for her extravagant doodle-like drawing that is engraved into the sapphire crystal pane. Close inspection reveals several of the signature motifs common to the MB&F Machine family, among them a jellyfish, a robot, a toy car and a jet plane. These and other shapes combine together to form an image that shines out with the artist's youthful enthusiasm. To complete the picture, a smiley-faced double-screwed plate replaces the MB&F logo, its two screws representing the eyes. The elephant in the doodle is reproduced in three dimensions between the hour and minute prisms, crafted in white gold by Olivier Kuhn. It took more than 120 hours to make and provides a vivid reminder of the much-loved pachyderm featured most famously in stories by Kipling and Disney. The finished work expressed youthful exuberance and unabashed optimism.

The Only Watch auction is held every two years in Monaco or Geneva under the patronage of HSH Prince Albert II. The event features the work of Switzerland's Haute Horlogerie watchmakers, each contributing a unique piece, or the first in a limited series, with no reserve price set for the auction. Proceeds go to the Monegasque Association against Duchenne Muscular Dystrophy (DMD), which funds research to fight the disease. DMD is a genetic disorder affecting children, particularly young boys, characterized by the progressive weakening of muscles that causes respiratory and cardiac problems as the child gets older.

Max Büsser saw this as the perfect opportunity to feature the work of a young artist whose creations embody the hopes and dreams of childhood – voicing her vision to combat a disease that robs too many children of a normal life.

Horological Machine N°8
Only Watch
Performance art piece
Reference: 80.WTL.O
Material: 18K white gold and grade 5 titanium
Limitation: unique piece
Introduction: 2017
Year in production: 2017
Quantity produced: 1

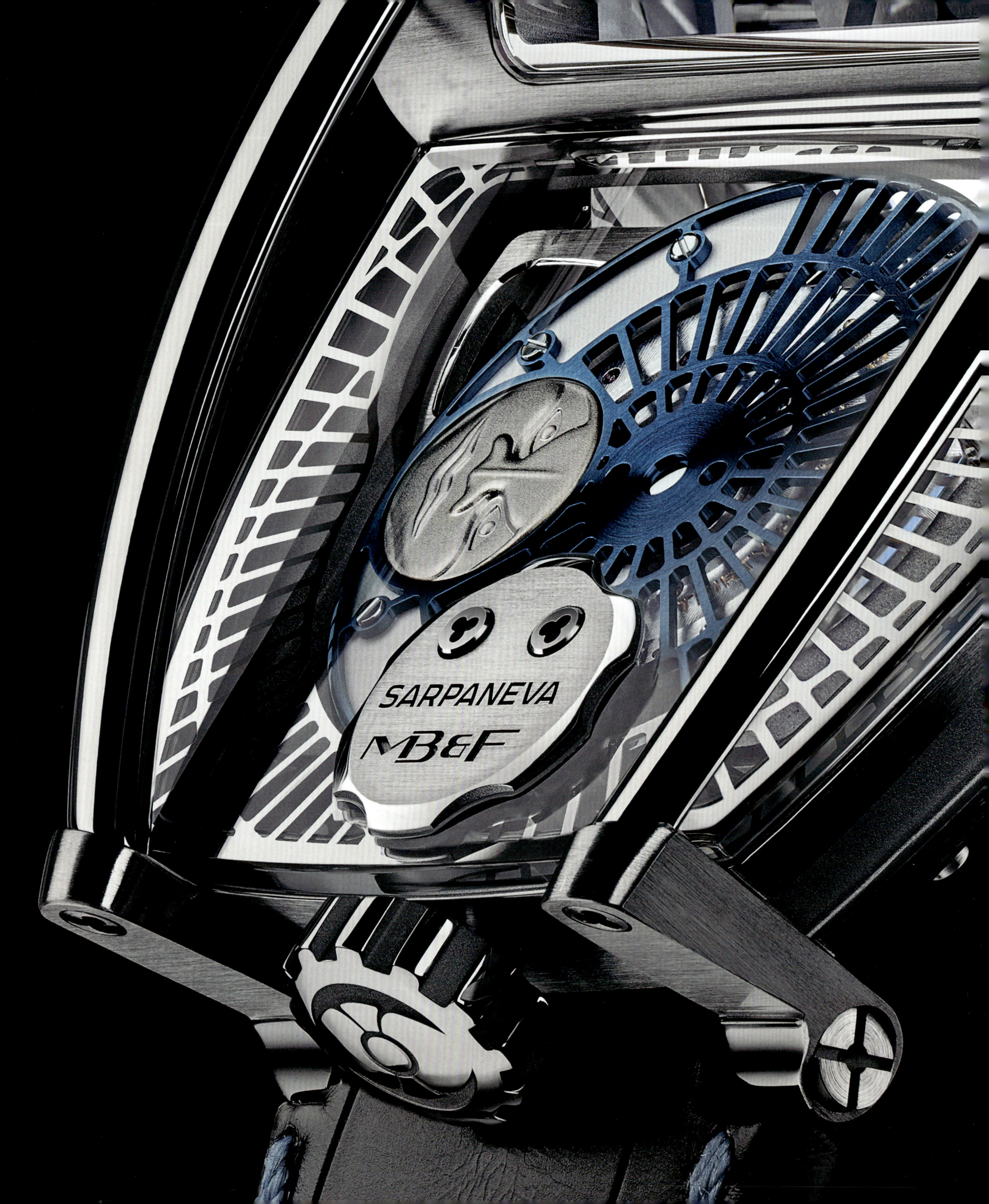
SARPANEVA
MB&F

MOONMACHINE 2 MB&F × STEPAN SARPANEVA

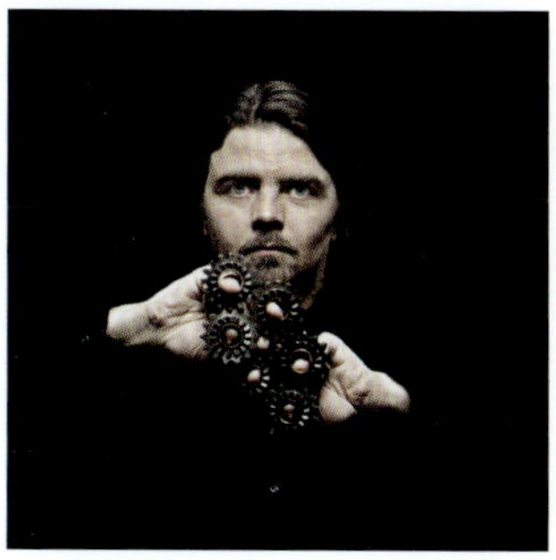

STEPAN SARPANEVA

Stepan Sarpaneva was born in 1970 in Turku, Finland. Before launching his own Sarpaneva Watches brand in 2003, he had worked for several prestigious Swiss brands including Piaget and Parmigiani, the latter alongside fellow Finn and MB&F "Friend" Kari Voutilainen. Three elements are central to Sarpaneva's creations: a distinctive moon face, based on his own likeness; northern stars and constellations as seen from his native Finland; and his "Korona" case, inspired by the glowing corona around the Sun. Growing up in Finland made him familiar with long, dark nights and limited sunlight in the winter months – the lingering moon and stars of the Finnish night sky. He would draw on these celestial images in his collaboration with MB&F. Sarpaneva's first collaboration with MB&F was the 2012 MoonMachine, based on the framework of the HM3 Frog. This time, again inspired by his love of the night sky, Sarpaneva turned MB&F's Horological Machine N°8 into MoonMachine 2.

As its name implies, the MoonMachine 2 is designed around a moon phase complication – the first-ever projected moon-phase display. Like the time indication, the moon phase is projected upward and forward from horizontal discs onto a vertical display, using a specially fashioned optical prism.

The signature Sarpaneva moon face features on three hand-finished moons in gold: one large moon set in the machine's automatic winding rotor; and two smaller, dual moons that display the phases of the moon and rotate under a Korona ring – another Sarpaneva hallmark. Measuring just 4.55 mm in diameter, these dual moons are the smallest ever created by Sarpaneva and proved a particular challenge, partly because of their unusually small diameter but also because of their wafer-thin profiles, almost like gold foil. The battle-axe rotor of the HM8 has meanwhile been transformed into an open-work radial web in titanium, a further expression of Sarpaneva's unmistakable design vocabulary.

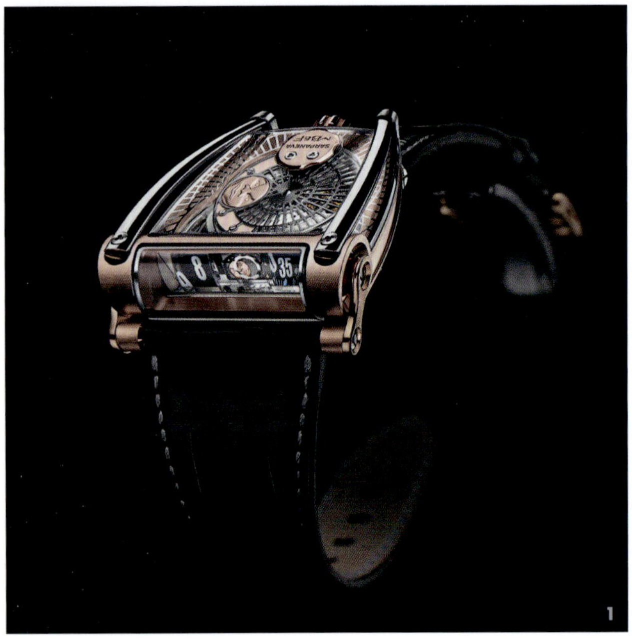

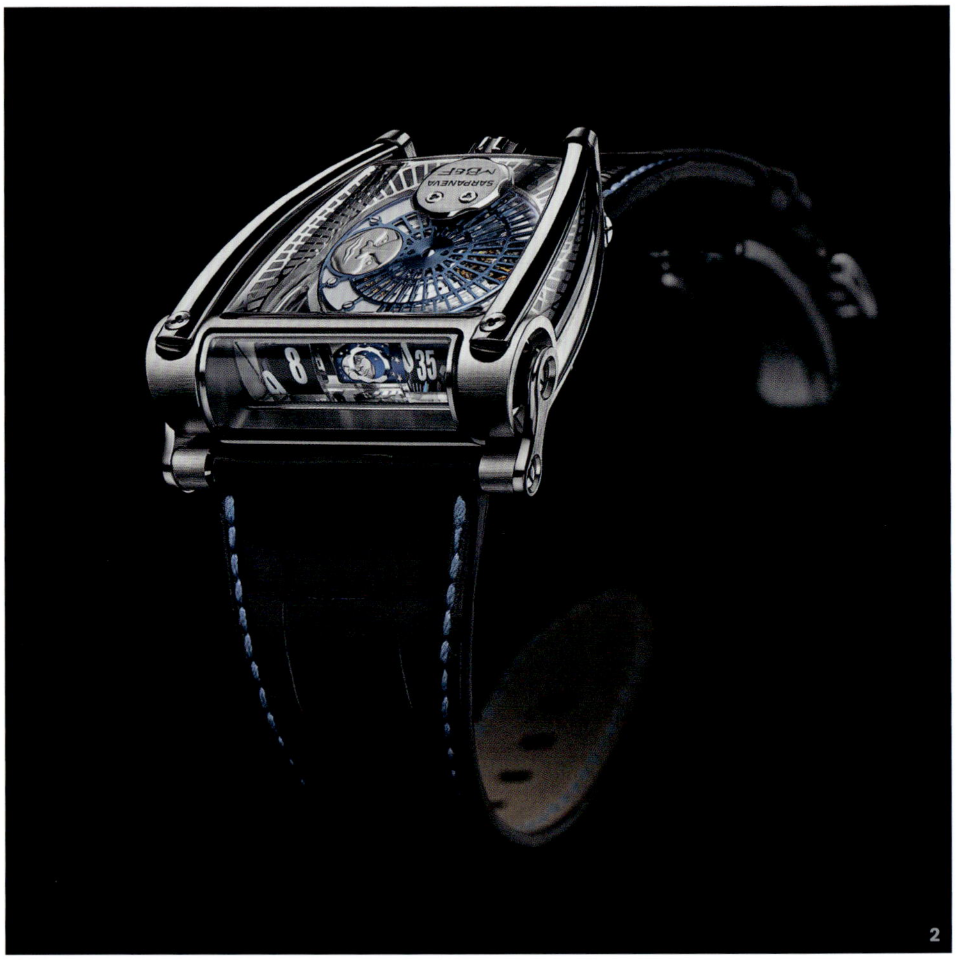

1 MoonMachine 2 RG
Reference: 81.RTL.B
Material: 18K red gold and titanium case, red gold moon faces and anthracite sky
Limitation: 12
Introduction: 2018
Year in production: 2018
Quantity produced: 12 + 1 prototype + 1 artist's proof

2 MoonMachine 2 Ti
Reference: 81.TL.B
Material: grade 5 titanium case, white gold moon faces and light blue sky
Limitation: 12
Introduction: 2018
Year in production: 2018
Quantity produced: 12 + 1 prototype

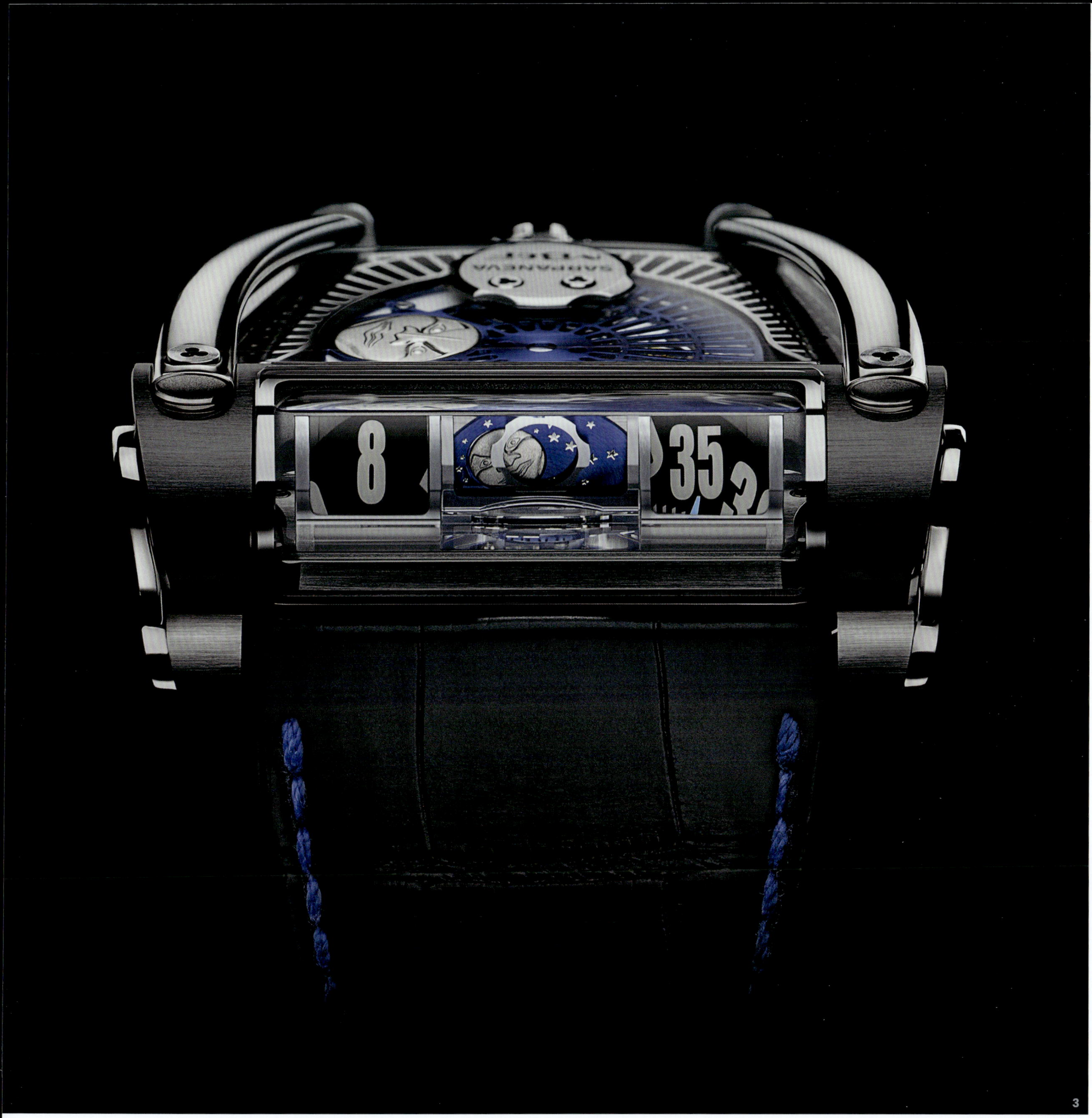

3 MoonMachine 2 Black Ti
Reference: 81.BTL.B
Material: black PVD coated grade 5 titanium, white gold moon faces and dark blue sky
Limitation: 12
Introduction: 2018
Year in production: 2018
Quantity produced:
12 + 1 prototype

HM9 FLOW

Horological Machine N°9 "Flow" was released in 2018 in titanium, in two versions – "Air" and "Road" – taking inspiration from two contrasting sources. The post-war years of the late 1940s and early 1950s marked a new focus for engineering and automotive design, with a shift from angular, boxy shapes to the sleek, streamlined forms of aerodynamic styling. Computer technology and wind tunnels had yet to be invented, so aesthetic considerations took precedence over scientific precepts. So was born the Streamline Era, ushering in some of the most beautiful objects ever made, among them the Mercedes-Benz W196 automobile and the De Havilland Venom aircraft.

Twin balance wheels beat independently on each flank of the HM9, seen through a dome of polished sapphire crystal. Each wheel feeds into the central body: a planetary differential that averages output from both wheels to supply one stable reading of the time.

The sleek, aerodynamic profile of the HM9 takes its lead from the HM4 Thunderbolt and the HM6 Space Pirate. Its three-dimensional case is highly complex, built to look like a jet engine and fitted with a specially designed 3D gasket to make it waterproof – a first from MB&F.

The HM9 "Air" version features an aviation-style dial displaying hours and minutes; the "Road" version has a speedometer-style dial. Each version is available in titanium or red gold.

HOROLOGICAL MACHINE Nº9 FLOW

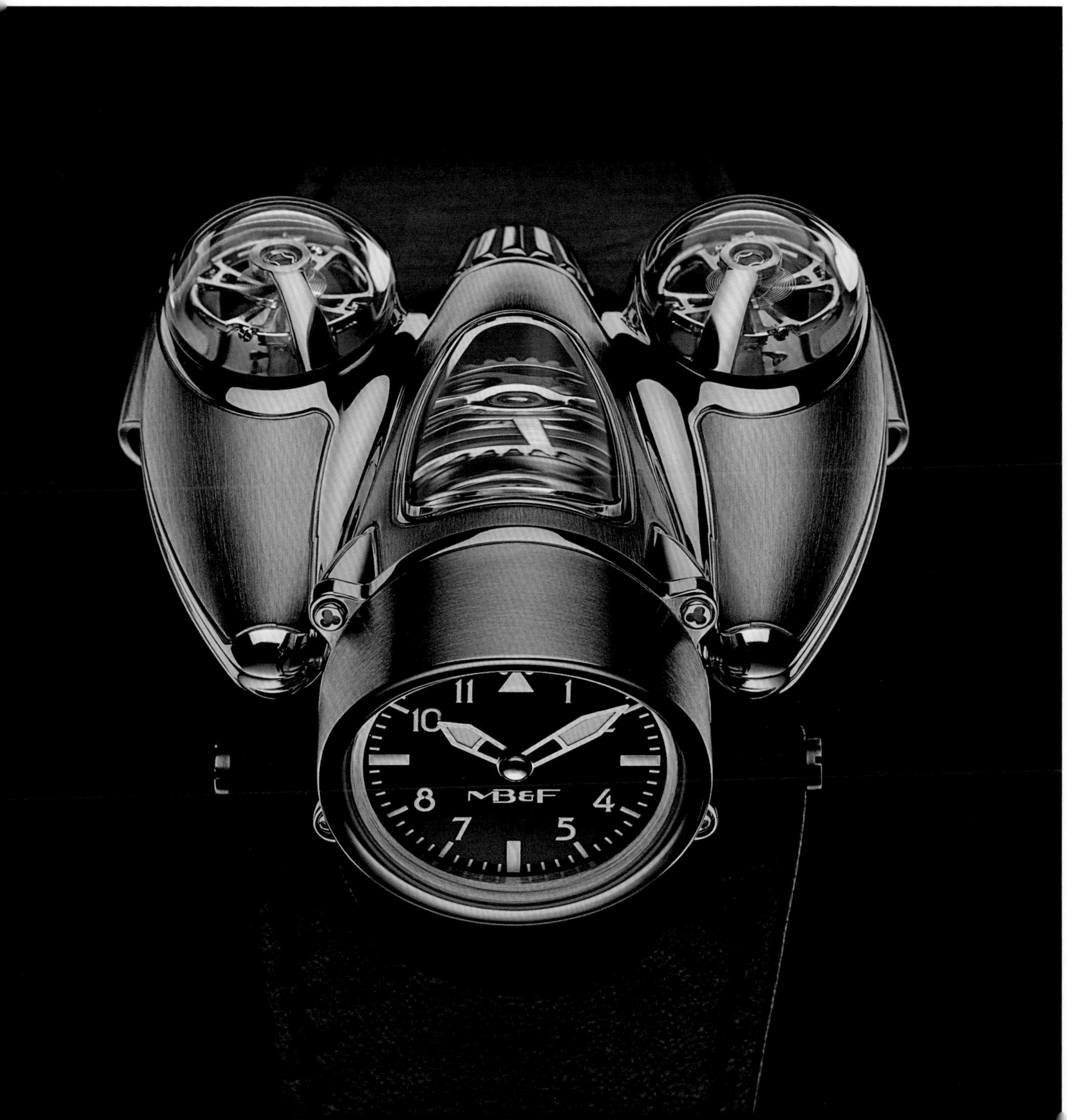

HM9 CALIBRE

Released in 2018, the manual-winding engine of the HM9 was created entirely in-house, drawing on MB&F's 13 years of design experience, with 14 previous movements to its credit. The machine's double balance wheels, together with its planetary differential, are descended from a similar system in Legacy Machine N°2, though with vastly different aesthetics.

The HM9 has two fully independent balance wheels that each beat at 2.5Hz (18,000bph), specifically designed to avoid the phenomenon of resonance – important to ensure that each wheel obtains its own distinct chronometric data, which is then fed to the differential for an averaged reading.

Collection
Horological Machines
Series
Horological Machine N°9 Flow
Common name
HM9 Flow, HM9
Movement
Developed in-house by MB&F
Winding
Manual
Power reserve
45 hours (single barrel)
Components
301/52 jewels

Functions
Hours and minutes on vertical display
Add. features
Two fully independent balance wheels with planetary differential
Case
Grade 5 titanium or 18K red gold
Dimensions
57 × 47 × 23 mm
Case components
43 for Ti editions; 49 for RG editions

Water resistance
30 m/3 ATM
Strap
Calf leather
Buckle
Folding buckle, titanium or 18K red gold matching the case
Introduction
2018
Years in production
2018 – in production
Quantity produced
(at the end of 2020) 79 + 2 prototypes

Comments
The HM9 comes in 2 versions: "Air" version with aviation-style dial and "Road" version with speedometer dial

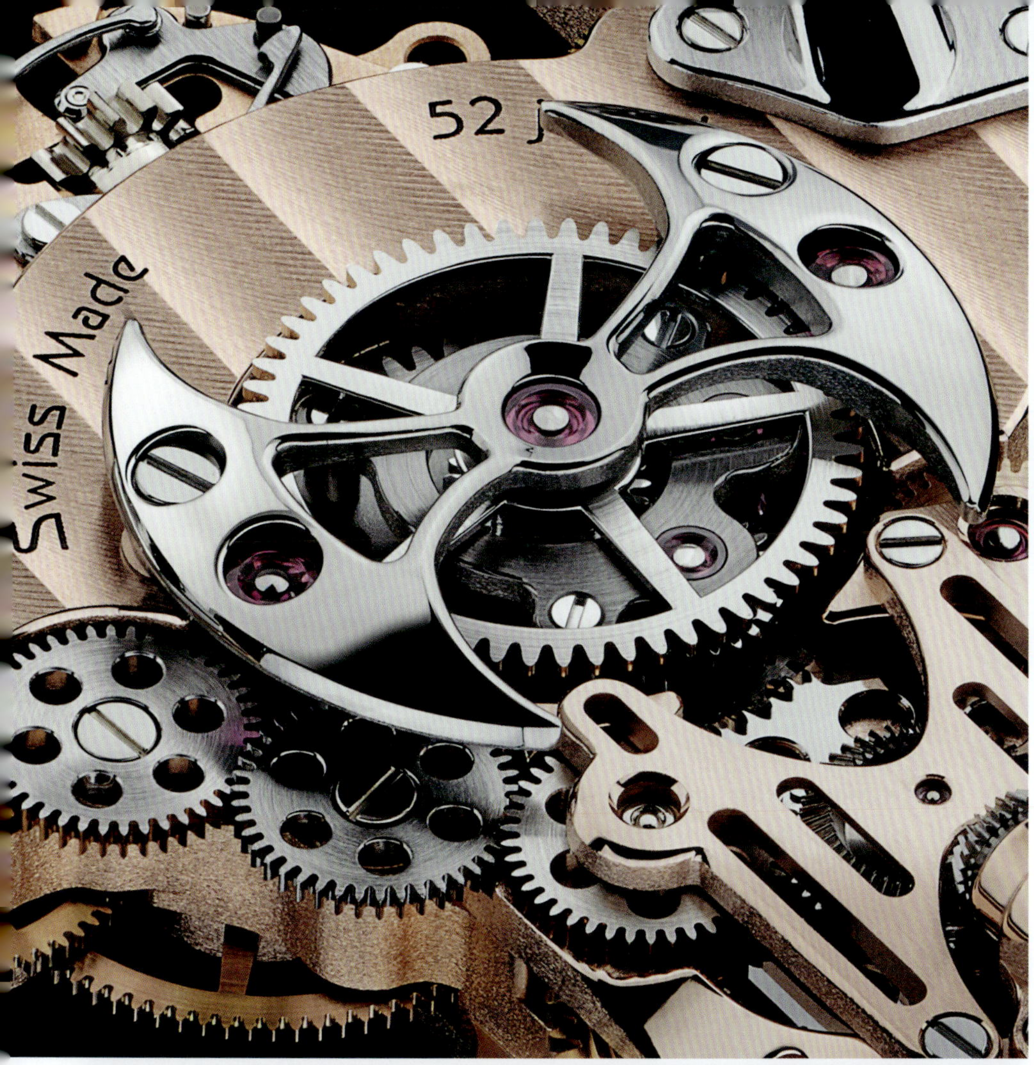
Swiss Made

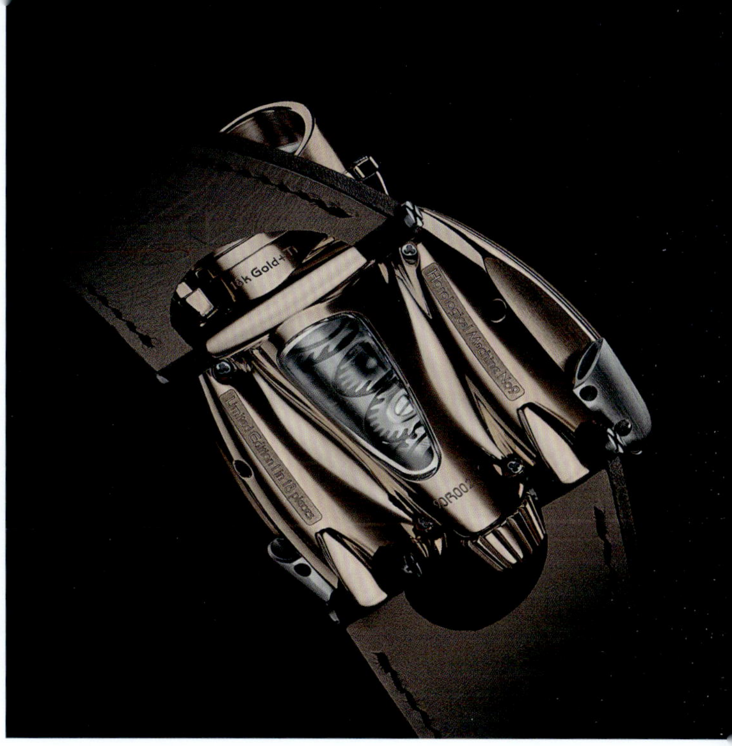

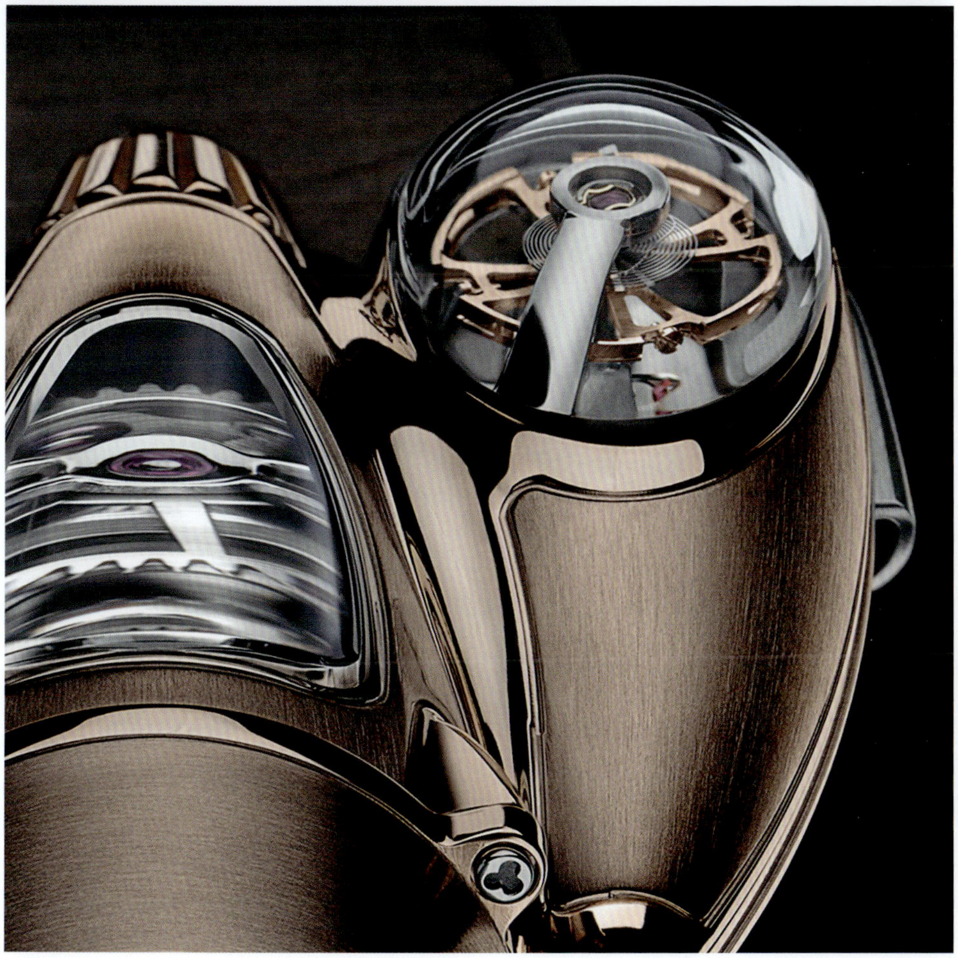

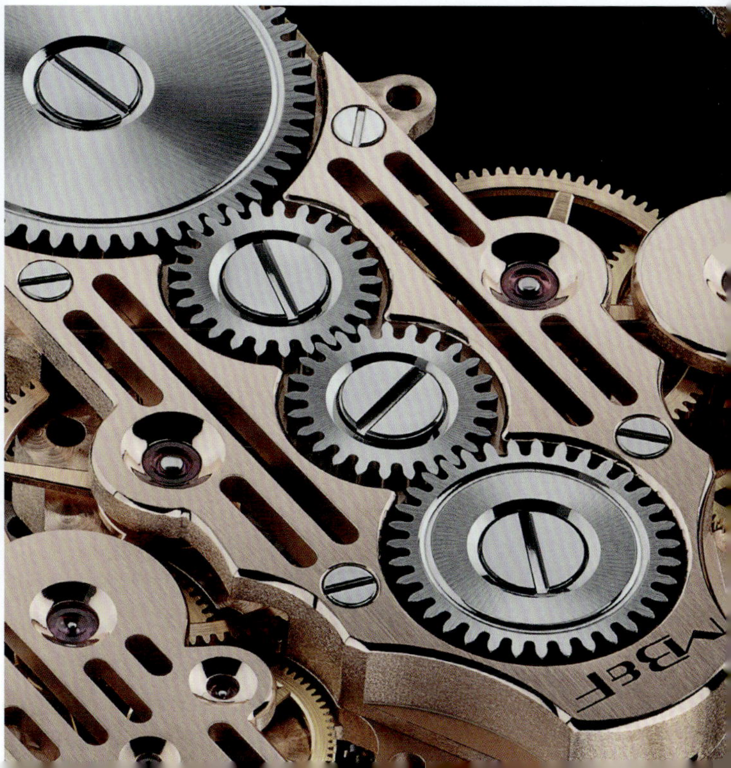

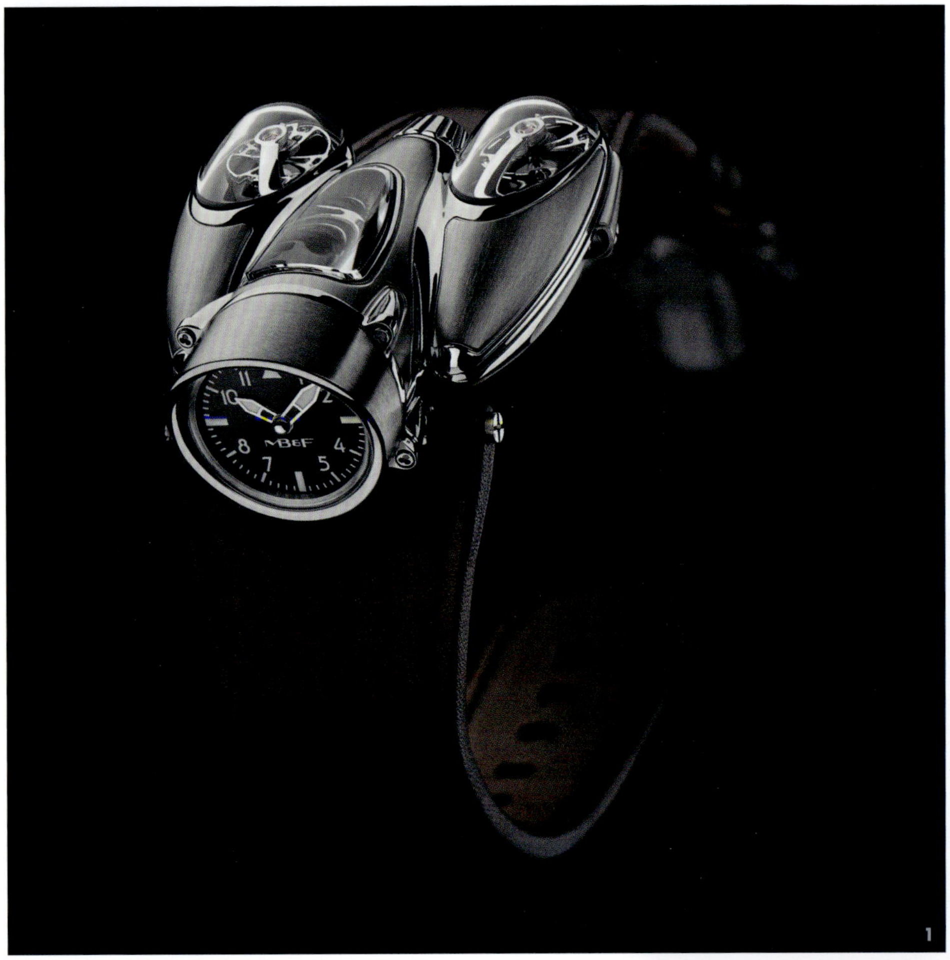

1

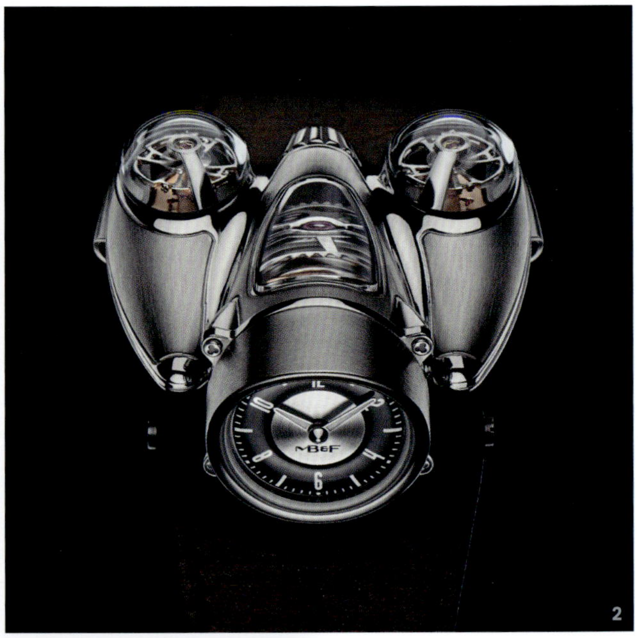

2

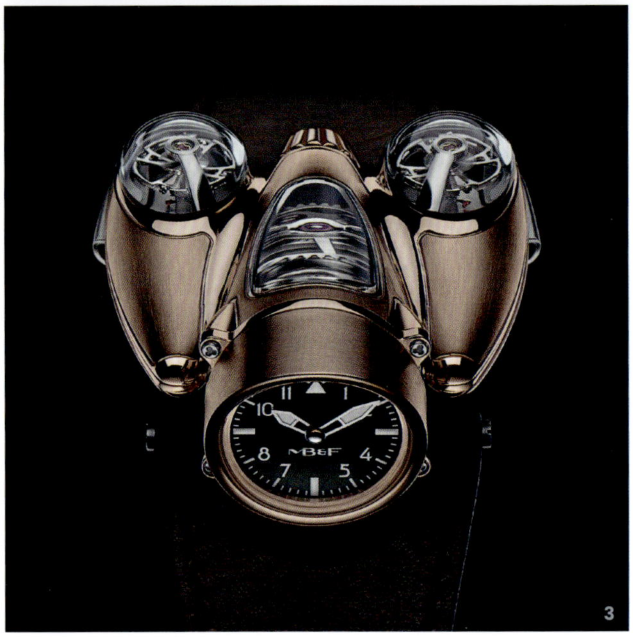

3

1 Horological Machine N°9
Ti Air Edition
Reference: 90.TL.AB
Material: grade 5 titanium
Limitation: 33
Introduction: 2018
Years in production:
2018 – in production
Quantity produced
(at the end of 2020):
28 + 1 prototype

2 Horological Machine N°9
Ti Road Edition
Reference: 90.TL.RB
Material: grade 5 titanium
Limitation: 33
Introduction: 2018
Years in production:
2018 – 2020
Quantity produced :
33 + 1 prototype

3 Horological Machine N°9
RG Air Edition
Reference: 90.RL.AB
Material: 18K red gold
Limitation: 18
Introduction: 2019
Years in production:
2019 – in production
Quantity produced
(at the end of 2020): 11

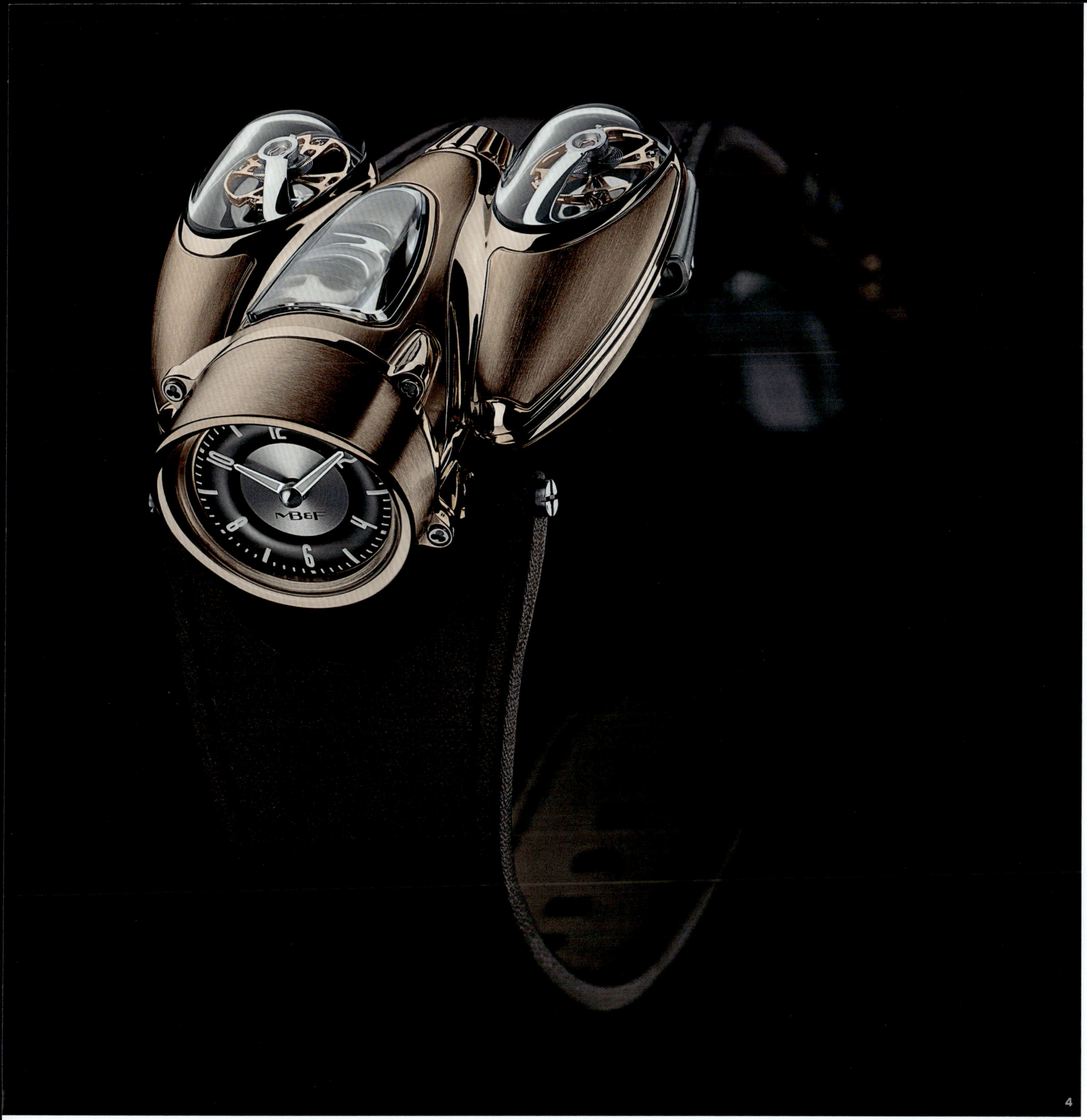

4

4 Horological Machine N°9
RG Road Edition
Reference: 90.RL.RB
Material: 18K red gold
Limitation: 18
Introduction: 2019
Years in production:
2019 – in production
Quantity produced
(at the end of 2020): 7

HM10
BULLDOG

Horological Machine N°10, nicknamed the Bulldog, was released in 2020 and is the first horological machine inspired by man's best friend. With bulging eyes, short legs, round body, studded collar and toothy grin, the machine bears a striking resemblance to its namesake.
The hour and minute indicators are similar to those of the HM3 Frog. They are displayed on paper-thin rotating aluminium domes protected by a sapphire crystal shield. Above them, a flying balance wheel – signature feature of the Legacy Machine collection – spins hypnotically, powering the time indications below.
The crowns of the HM10 are dotted with spikes and can supply 45 hours of mainspring energy when fully wound. The level of barrel wind at any moment is displayed by the Bulldog's jaws, which are located immediately below the time display. They open and shut according to the amount of wind remaining in the mainspring: a broad, open grin indicates a full power reserve; clenched jaws tell you that the Bulldog is run down and ready for a nap.
The articulated lugs give the Bulldog its legs. Despite the Machine's large dimensions, the lugs make the HM10 surprisingly wearable, across a range of wrist sizes.
HM10 was released in two editions: grade 5 titanium, with blue eyes; and 18-karat red gold, with black eyes.

HOROLOGICAL MACHINE Nº10 BULLDOG

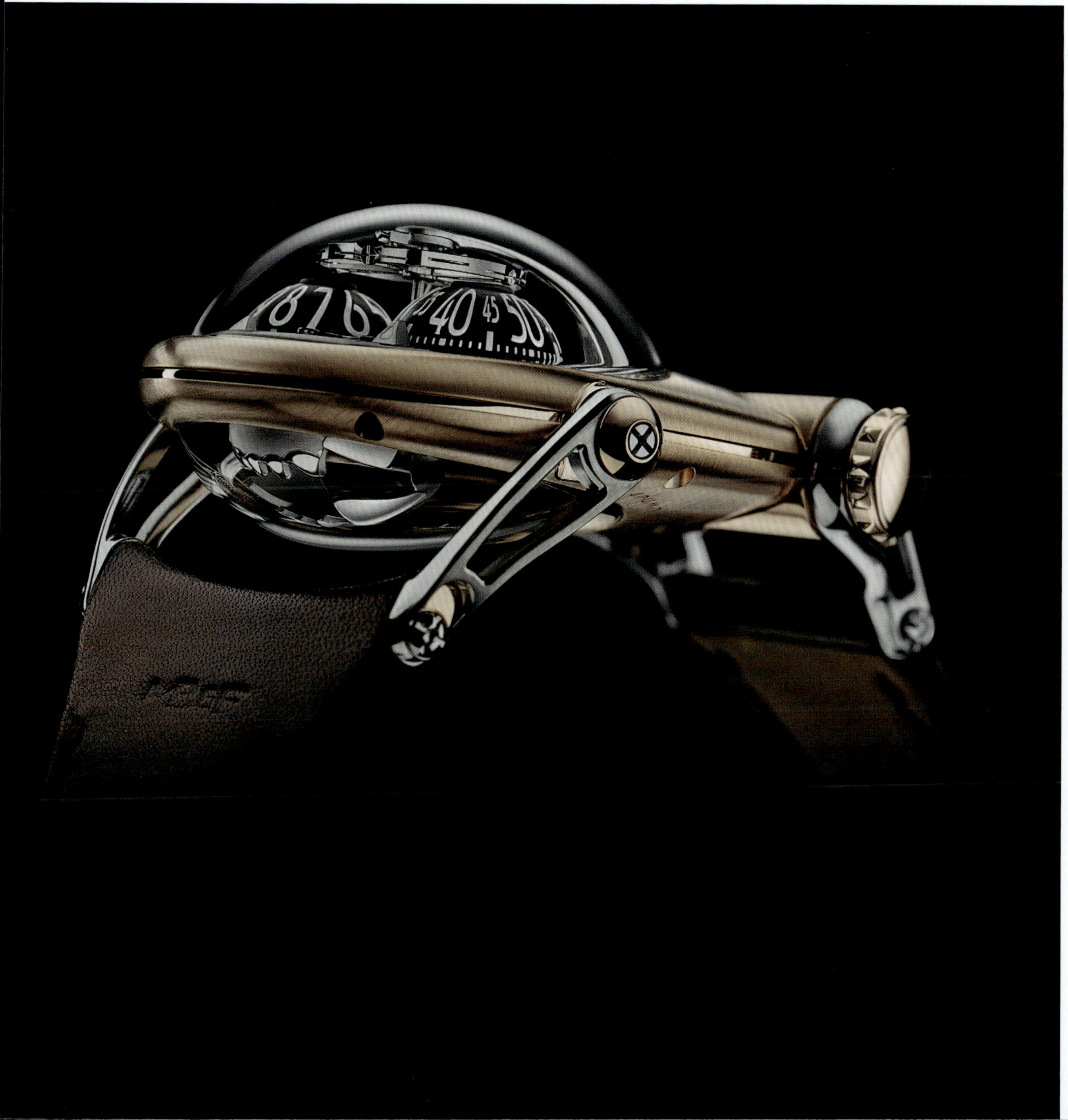

HM10 CALIBRE

The manual-winding engine of the HM10 was created entirely in-house, drawing on MB&F's 15 years of design expertise.
The machine's flying balance wheel first made its entrance in 2011 in MB&F's Legacy Machine N°1. It beats at a cadence of 2.5Hz (18,000bph) and has since featured in most other Legacy Machines, as well as the HM9, and most recently the HM10 Bulldog. Other features serve to remind the wearer of MB&F's journey to date and the adventures still to come. The time-telling domes are reminiscent of the HM3 Frog, while the kinetic power display recalls the LM1 Xia Hang.

Collection
Horological Machines
Series
Horological Machine N°10 Bulldog
Common name
HM10 Bulldog, HM10
Movement
Developed in-house by MB&F
Winding
Manual
Power reserve
45 hours
Components
301/34 jewels
Functions
Hours on left dome (rotating in 12 hours); minutes on right dome (60 minutes), power reserve indicator
Add. features
Bespoke flying 14 mm balance wheel with four traditional regulating screws floating above the domed dials; power reserve indicated by opening and closing of the jaws (end of power reserve = closed jaws); Super-LumiNova on hour/minute markers
Case
Grade 5 titanium or 18K red gold
Dimensions
54 × 45 × 24 mm
Case components
51
Water resistance
30 m/3 ATM
Strap
Calf leather
Buckle
Red gold and titanium folding buckle or Velcro system with titanium buckle
Introduction
2020
Years in production
2020 – in production
Quantity produced
(at the end of 2020) 37 + 3 prototypes

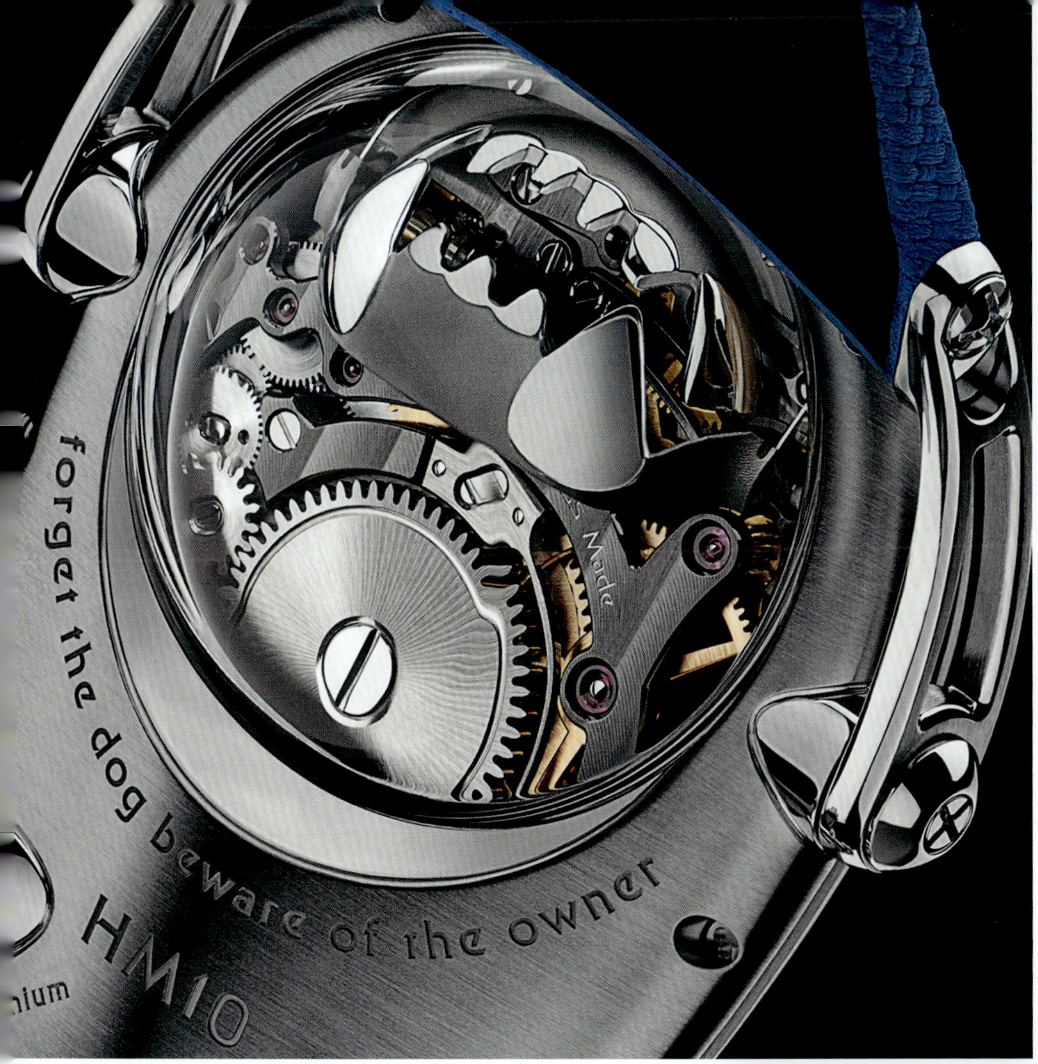
forget the dog beware of the owner
HM10

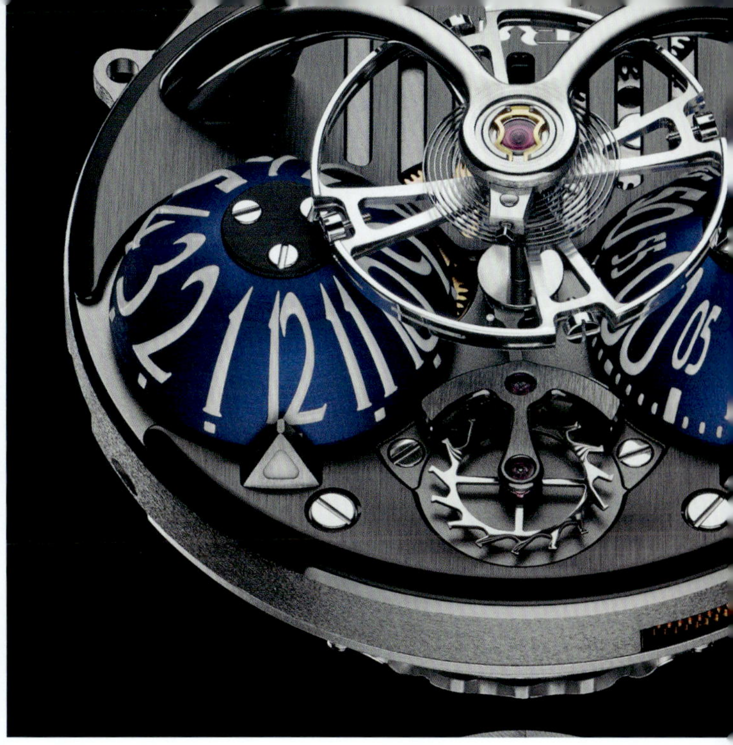

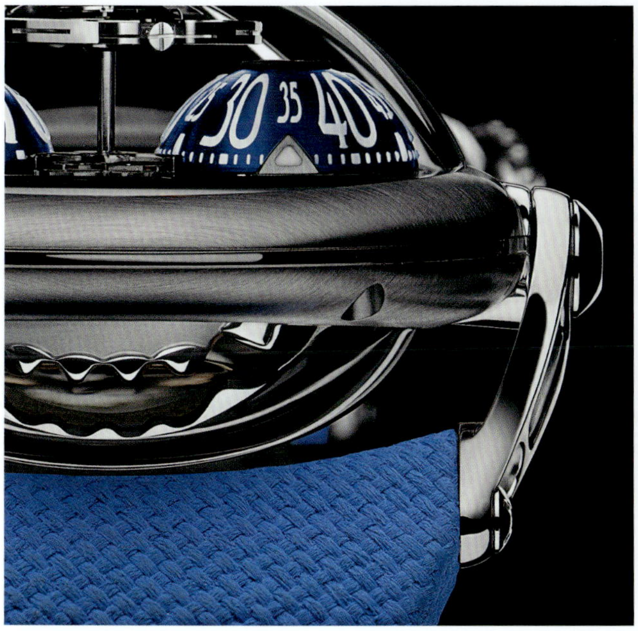

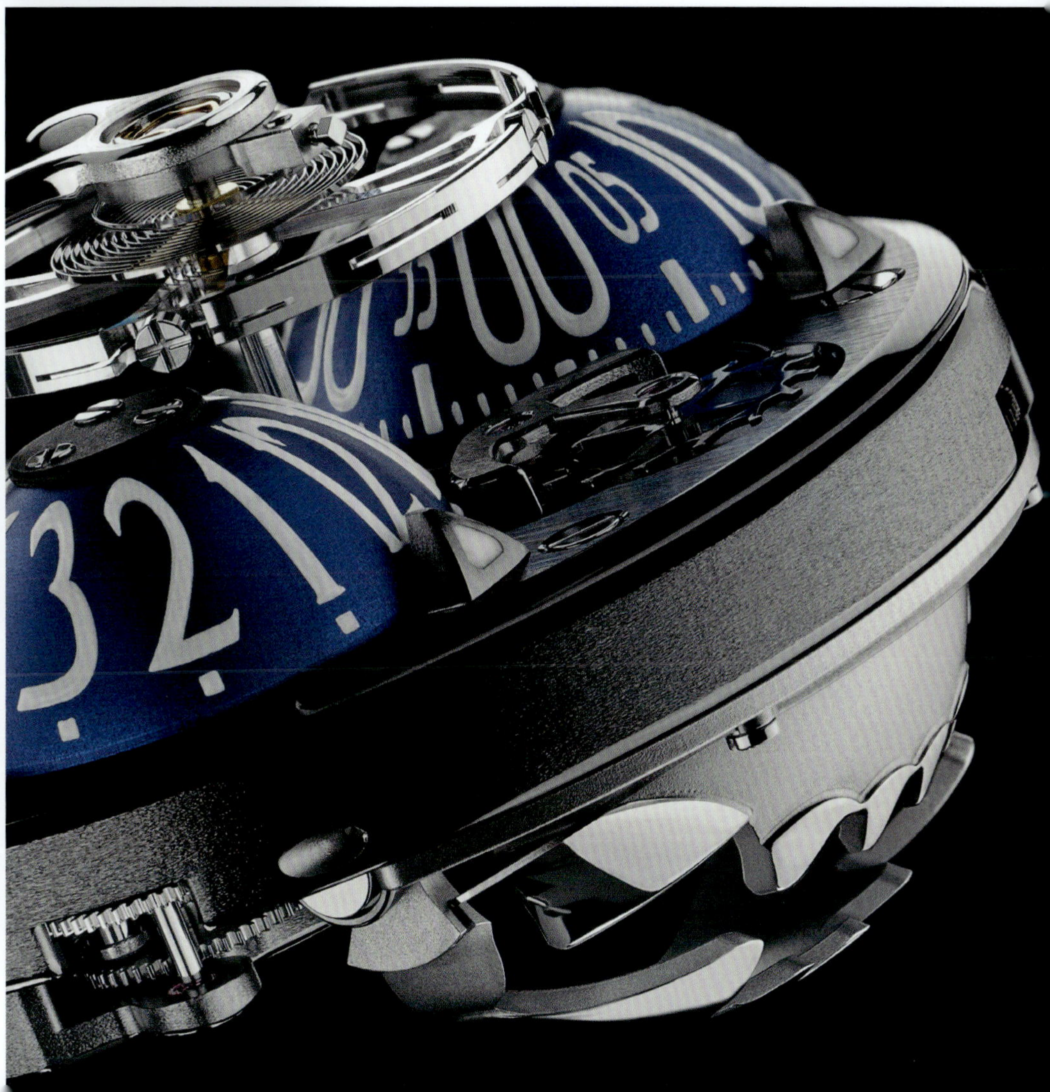

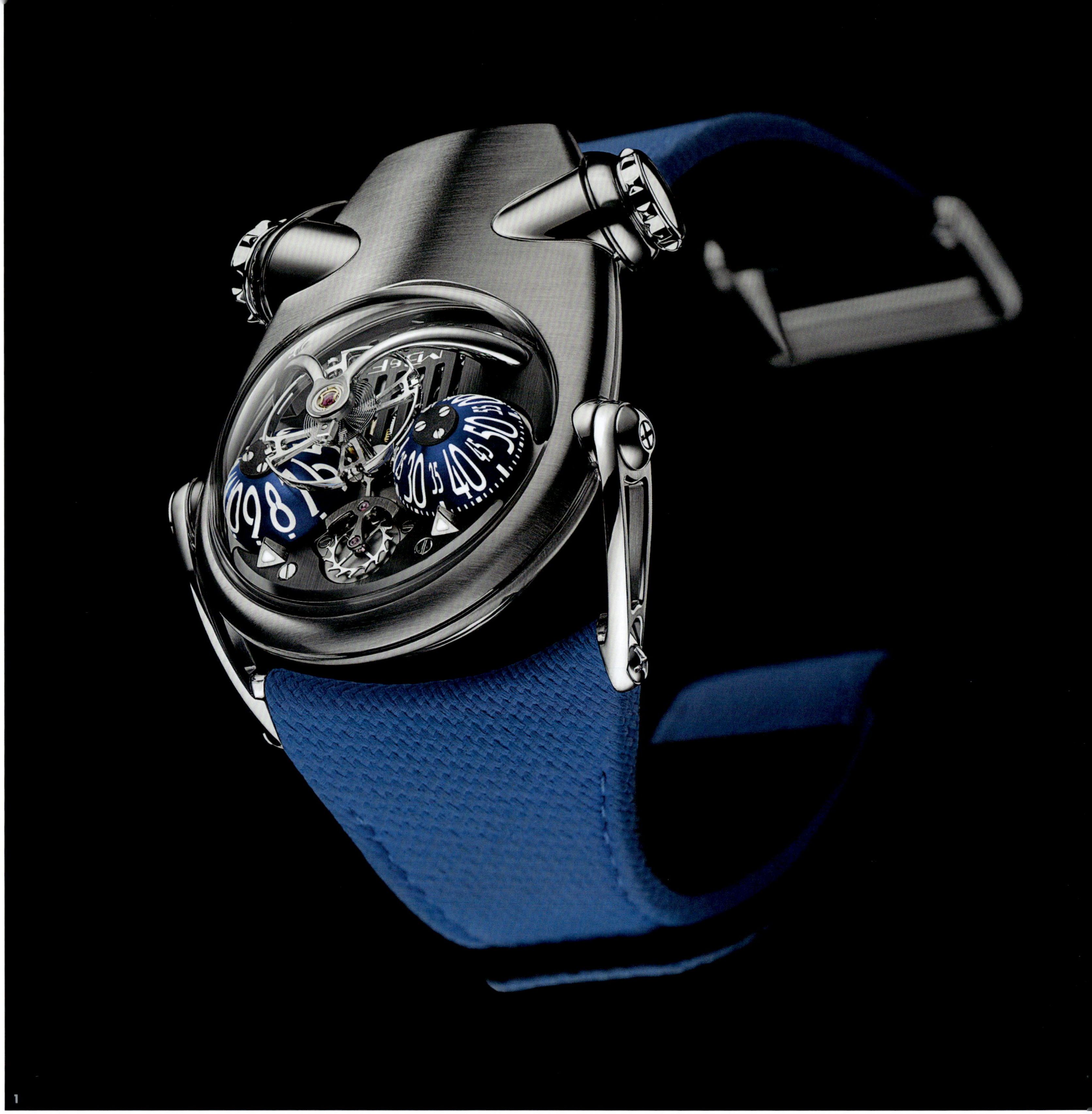

1

1 Horological Machine N°10 Ti
Reference: 100.TL.BL
Material: grade 5 titanium
Limitation: –
Introduction: 2020
Years in production:
2020 – in production
Quantity produced
(at the end of 2020):
24 + 2 prototypes

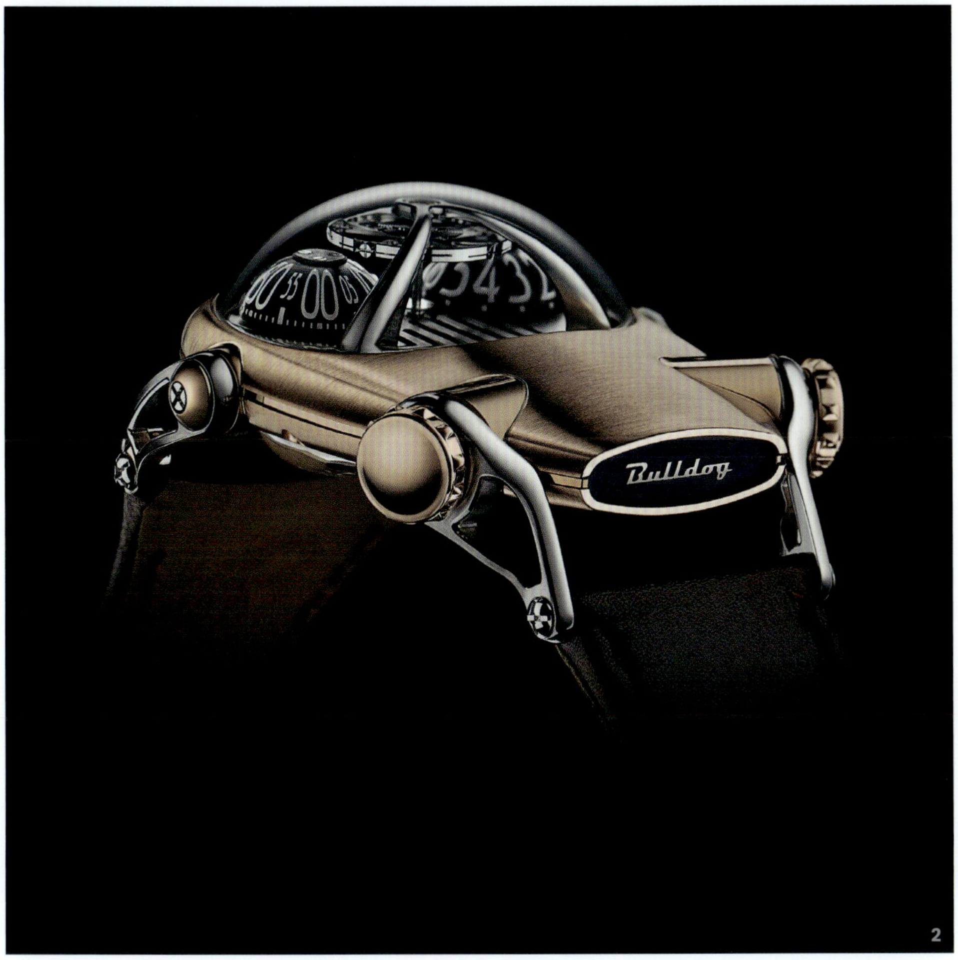

2 Horological Machine N°10 RG
Reference: 100.RL.B
Material: 18K red gold and
grade 5 titanium
Limitation: –
Introduction: 2020
Years in production:
2020 – in production
Quantity produced
(at the end of 2020):
13 + 1 prototype

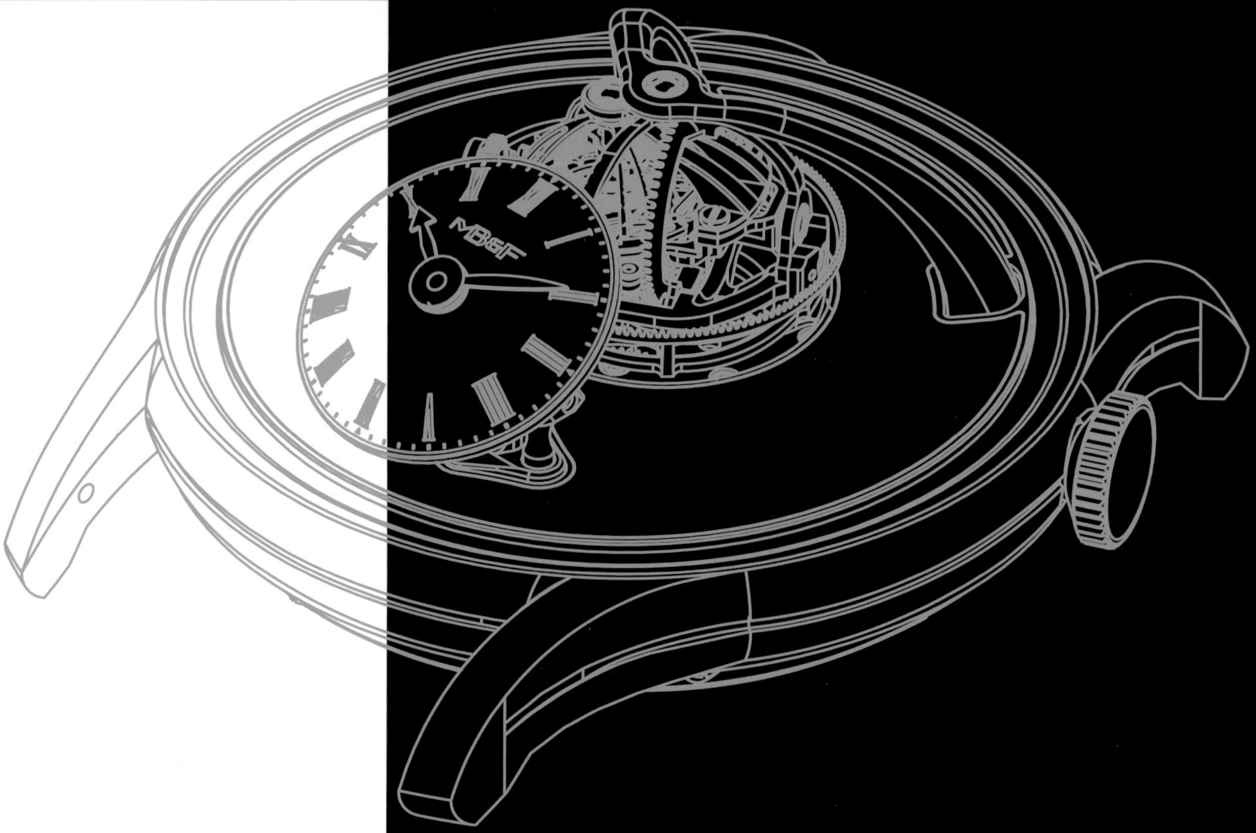

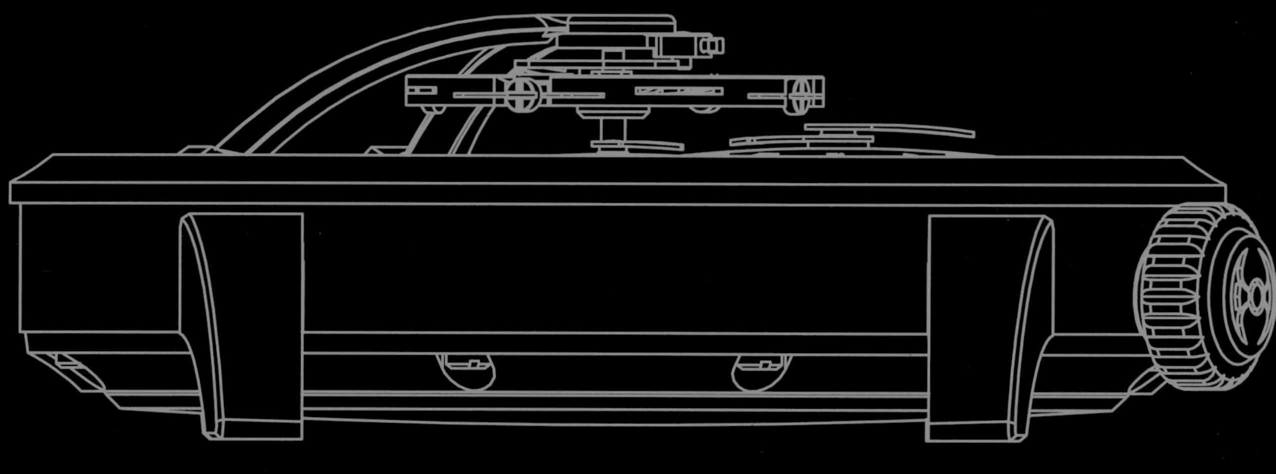

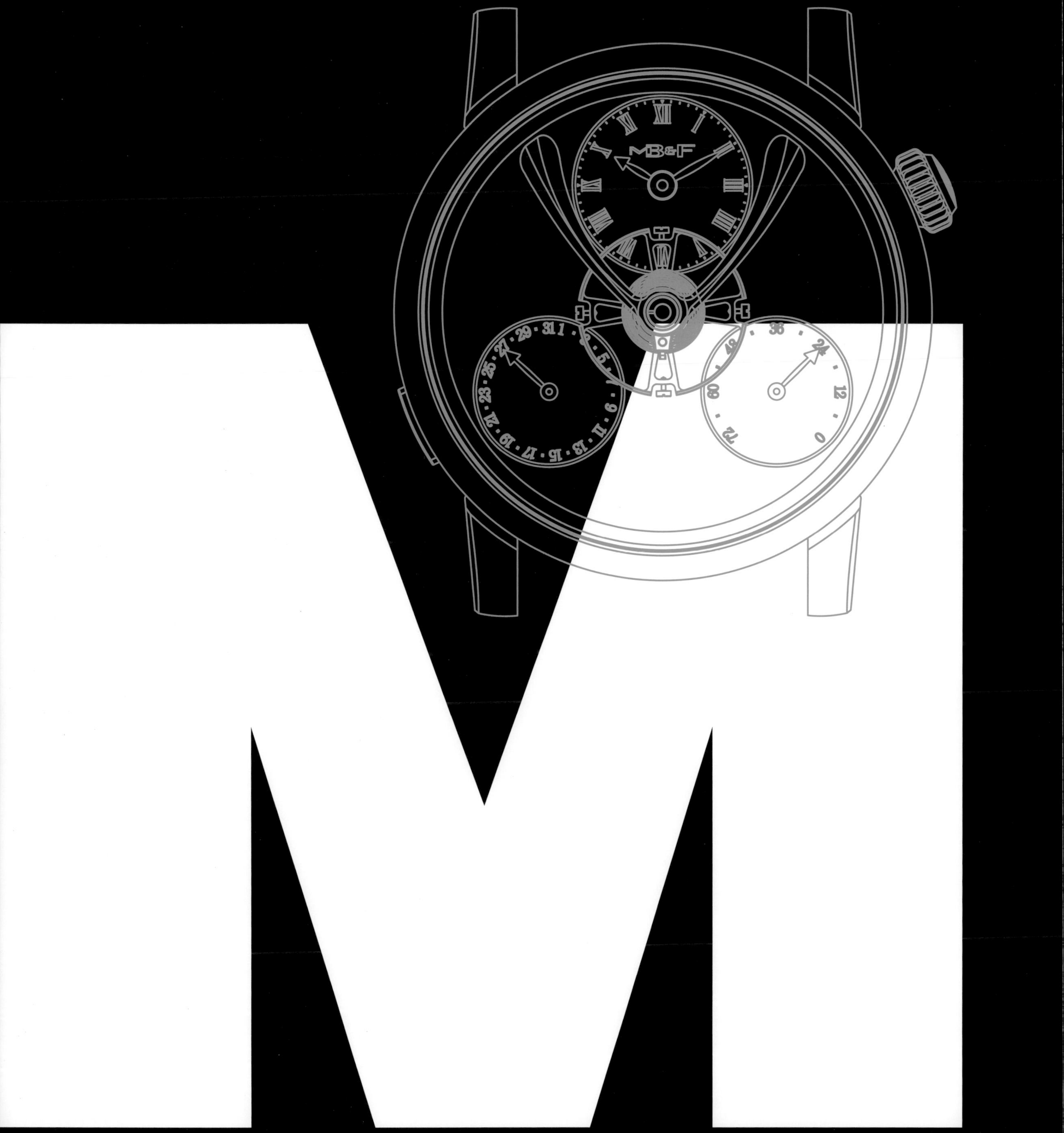
M
MB&F

LM

LEGACY MACHINES

The Legacy Machine Collection is the second collection from MB&F, launched in 2011 with the release of Legacy Machine N°1. To that point, six years from the start, MB&F had devoted its creative energies to a succession of ever-more audacious Horological Machines — three-dimensional and kinetic representations of horological art inspired by automobiles, spaceships and animal forms. But then, LM1 was the brand's first timepiece to be fitted in a round case, and that in its way made it quite as rebellious as the Horological Machines that came before it.

Legacy Machine N°1 has since been joined by a considerable number of siblings: LM2 (2013), LM Perpetual (2015), LM Split Escapement (2017), LM FlyingT (2019), LM Thunderdome (2019), LM Perpetual EVO (2020) and LMX (2021). It nevertheless remains perhaps the collection's most recognized timepiece, with reinterpretations by Chinese artist Xia Hang (2014) and later by designer Alain Silberstein (2016). The spirit of LM1 was furthermore reprised by the LM101 (2014) and LM101 MB&F × H. Moser (2020), whose numbering convention signifies a return to the primary design elements of LM1. Production of the LM1 total 436, with dials in a variety of colours and a range of case metals. In terms of numbers and status, its position is hard to deny.

The design element most closely associated with the Legacy Machine family is probably the flying balance wheel, which appeared first in LM1 and was subsequently featured in almost every timepiece in the collection. In LM1, the balance wheel is oversized to mimic and recall the proportions of a 19th century pocket watch, a touchstone of the entire Legacy Machine collection, drawing inspiration from the architecture (Gustave Eiffel, among others) and literature (Jules Verne, especially) of that era. Aesthetics and finishes reflect these period influences, with an expert contribution from Kari Voutilainen (LM1, LM2, LM101 and LM Thunderdome). Maximilian Büsser once asked himself: what watch would he design if he had been born 100 years earlier, in 1867 instead of 1967? The answer to that question is the LM1.

Further developments of Legacy Machines have focused on the development and mastery of ingenious new technical features, such as MB&F's first in-house movement (LM101); also complications such as the split escapement and perpetual calendar (LM Perpetual); and the triple-axis tourbillon (LM Thunderdome).

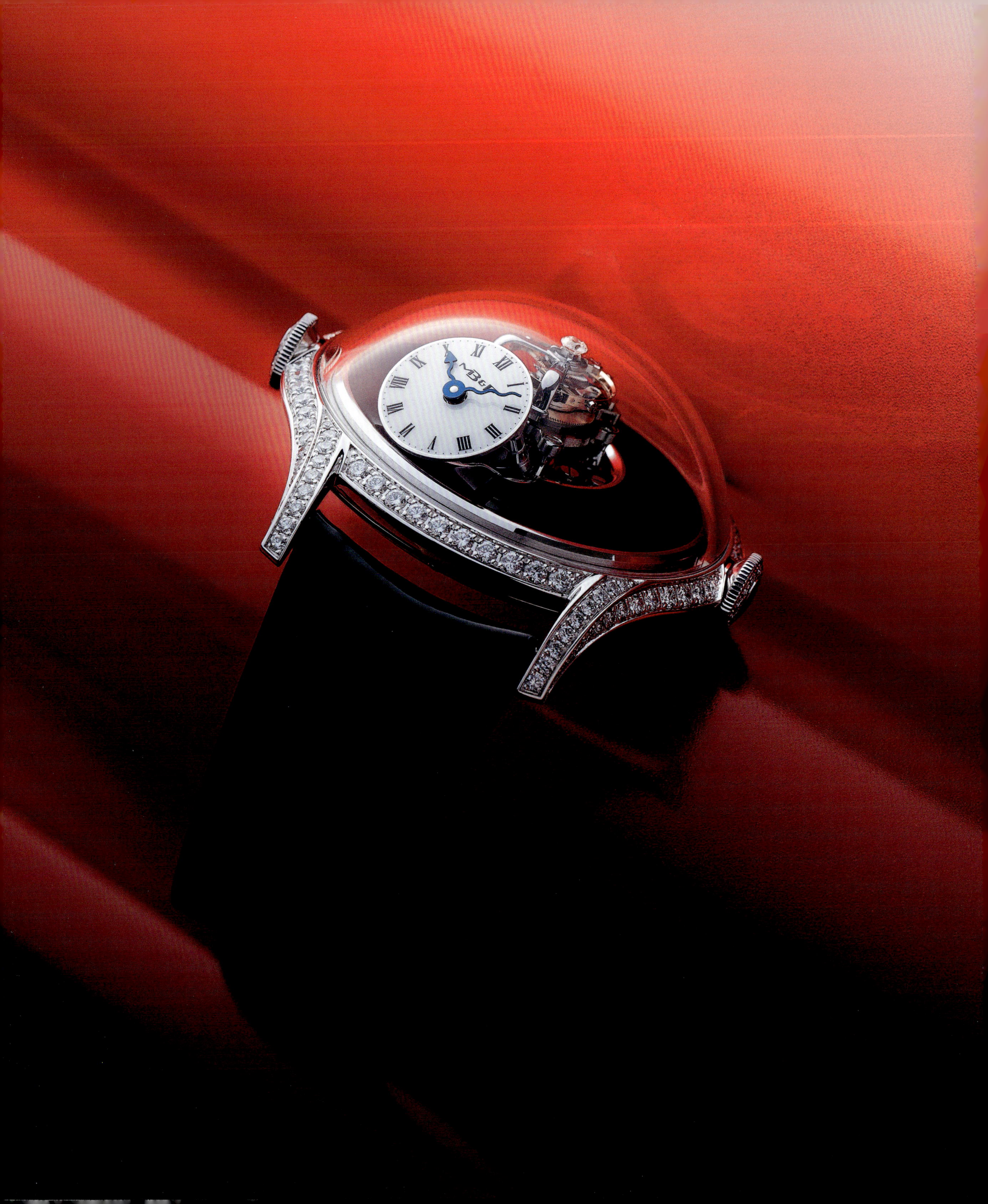

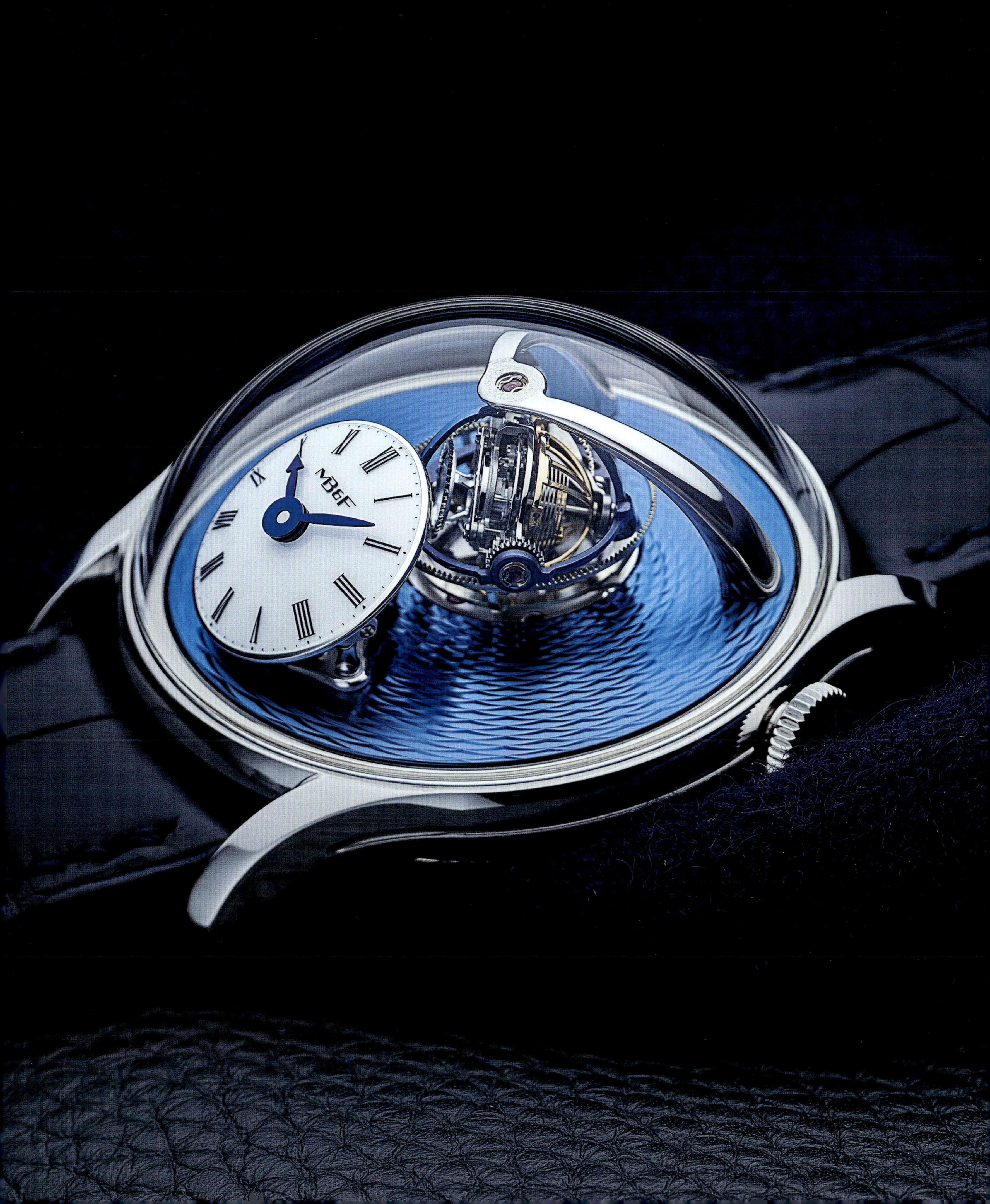
MB&F

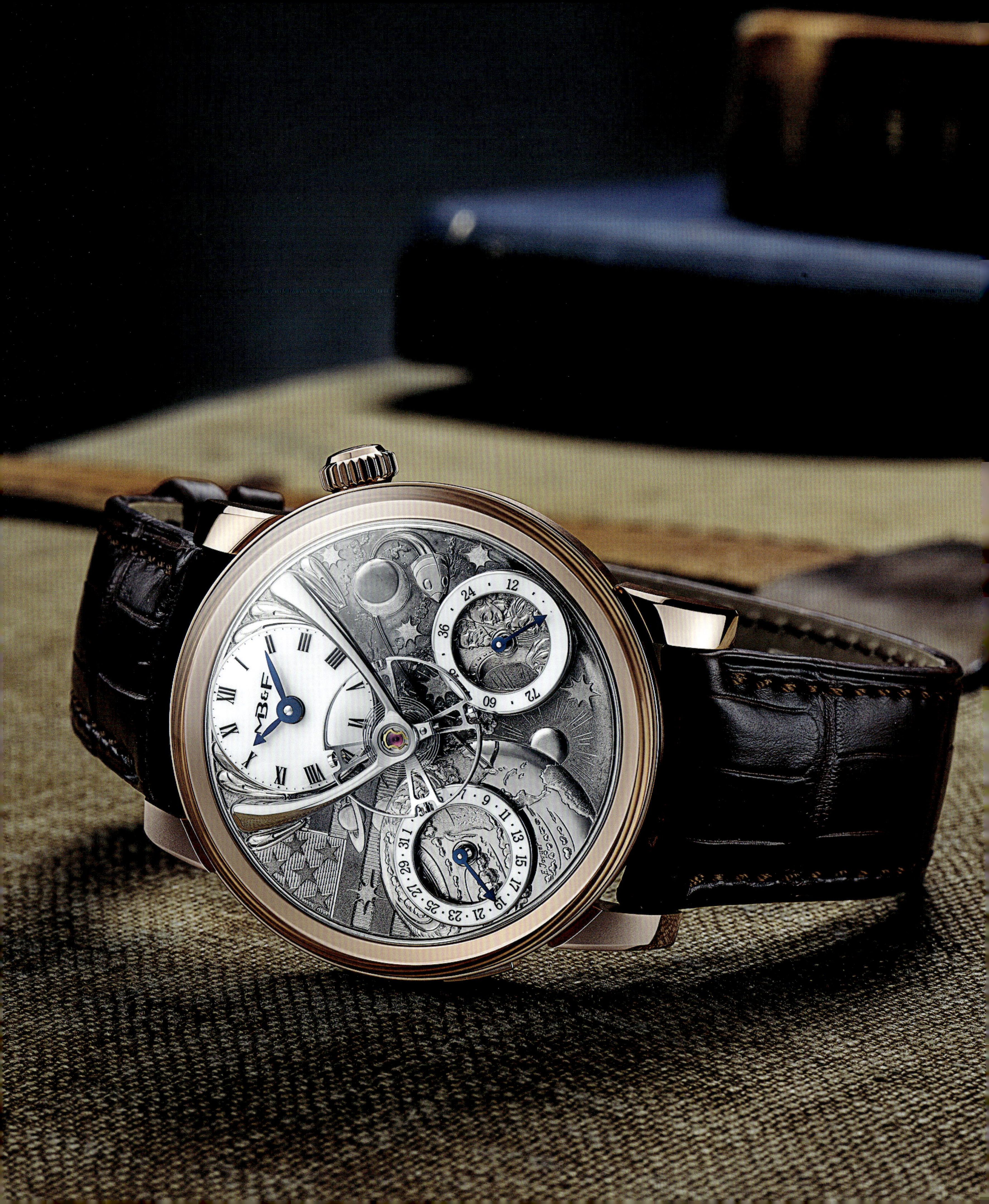
MB&F

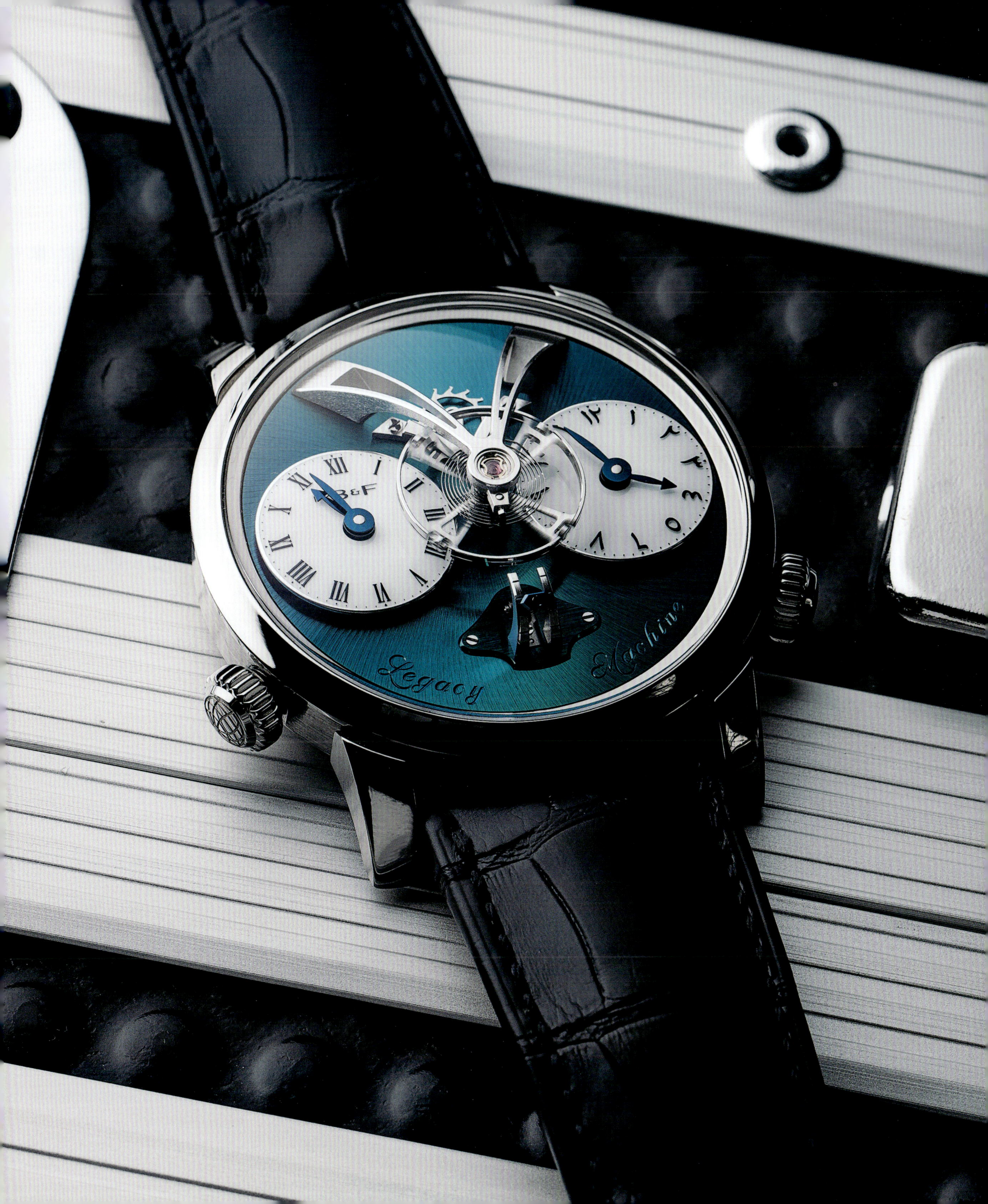

MB&F
Legacy
Machine

LM1

Maximilian Büsser says the creation of the Legacy Machine N°1 began with a rhetorical question: "What watch would I design if I had been born 100 years earlier, in 1867 instead of 1967?" His answer was the LM1, the first in the Legacy Machine series. Launched in 2011, MB&F's sixth year, the Legacy Machines would become the group's second collection, following on from the Horological Machines.

LM1 takes its inspiration from 19th century science fiction, such as the novels by Jules Verne, blended with the traditions of high watchmaking. Three novel features make this a landmark moment in watchmaking.

Perhaps the most evidently unique feature of the LM1 is its signature flying balance wheel, suspended from an arched bridge. Two completely independent time displays are the second unique distinction, one to each side of the balance wheel. The third innovation is a vertical power reserve for a quick and clear reading of barrel wind. Ironically perhaps, these entirely original features are then presented in a traditional round case, something never before employed by MB&F.

LEGACY MACHINE Nº1

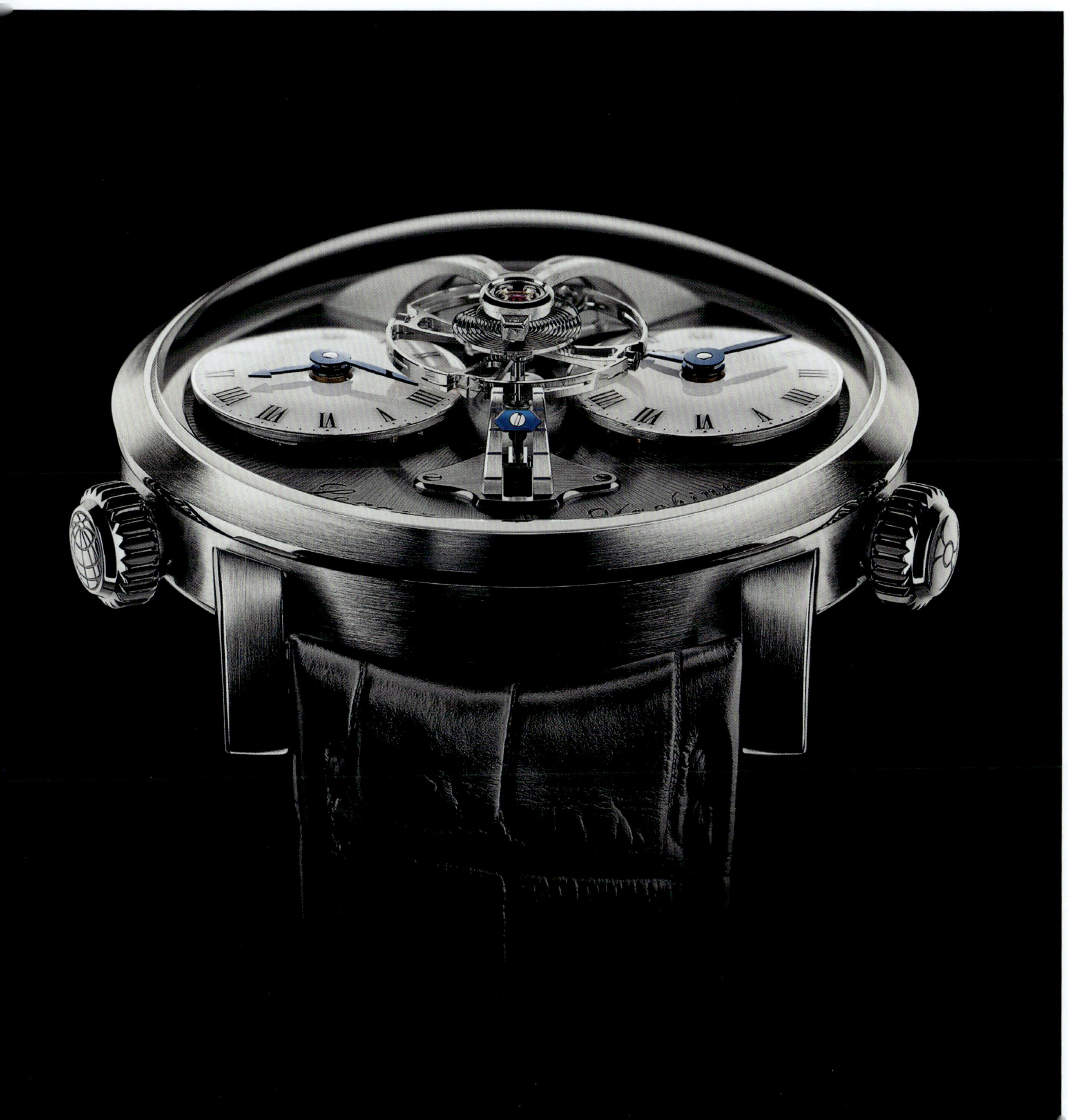

LM1 CALIBRE

The three-dimensional movement of the Legacy Machine N°1 was developed by Jean-Francois Mojon (awarded the Best Watchmaker Prize at the 2010 Grand Prix d'Horlogerie de Genève). Aesthetic design is the work of famed independent watchmaker Kari Voutilainen. True to the spirit of the LM1, the movement is a hybrid of tradition and innovation.

The iconic feature of the LM1 is once again the heart of the movement — an oversized 14 mm "flying" balance wheel — culminating above 279 components and 23 jewels, that hangs suspended from an arched bridge above the LM1's twin dials. The large size of the balance wheel gives the LM1 something of the character of a pocket watch – a novel feature that also evokes the past.

The two dials, showing hours and minutes, run independently to serve two distinct time zones. Below them is the innovative vertical power reserve indicator, made possible by the height of the flying balance wheel and the domed crystal that protects it.

Finishing throughout reflects the 19th century style from which the LM1 draws its inspiration. Bridges are rhodium-plated decorated with Geneva stripes, and the italicized signature of the movement's makers again evokes the LM1's historical roots.

Collection
Legacy Machines
Series
Legacy Machine N°1
Common name
LM1
Movement
Developed for MB&F by Jean-François Mojon/Chronode and Kari Voutilainen
Winding
Manual
Power reserve
45 hours
Components
279/23 jewels

Functions
Hours and minutes, independent dual time zones, vertical power reserve
Add. features
Flying balance wheel
Case
18K red gold or white gold, platinum 950, grade 5 titanium or stainless steel
Dimensions
44 × 16 mm/42.5 × 17 mm for Alain Silberstein editions

Case components
65
Strap
Alligator
Water resistance
30 m/3 ATM
Buckle
Pin or folding buckle, titanium, 18K red or white gold, stainless steel or platinum matching the case
Introduction
2011
Years in production
2011–2017

Quantity produced
429 + 5 prototypes + 2 artist's proofs (LM1 Xia Hang)
Comments
The Legacy Machine N°1 Final Edition bears the redesigned flying balance wheel bridge from the LM101 and the following Legacy Machine series

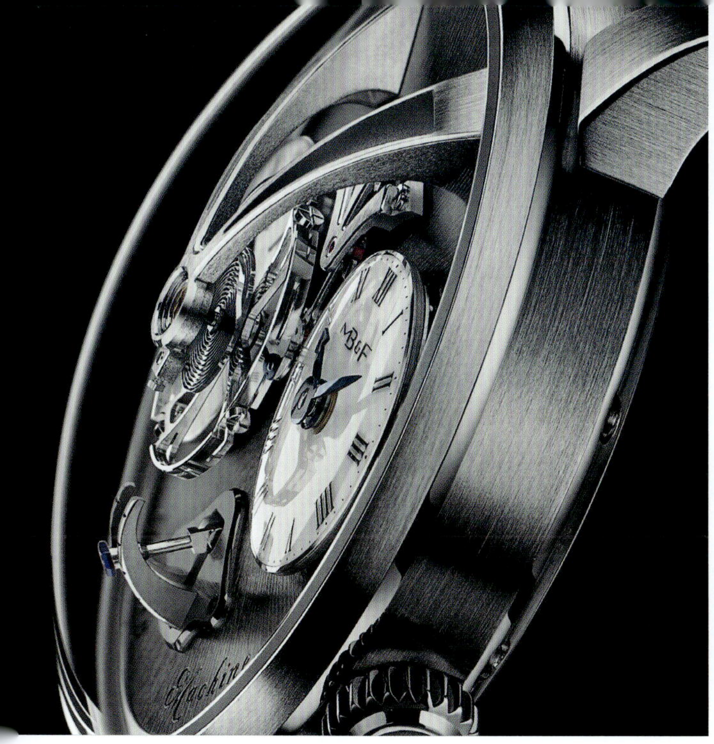

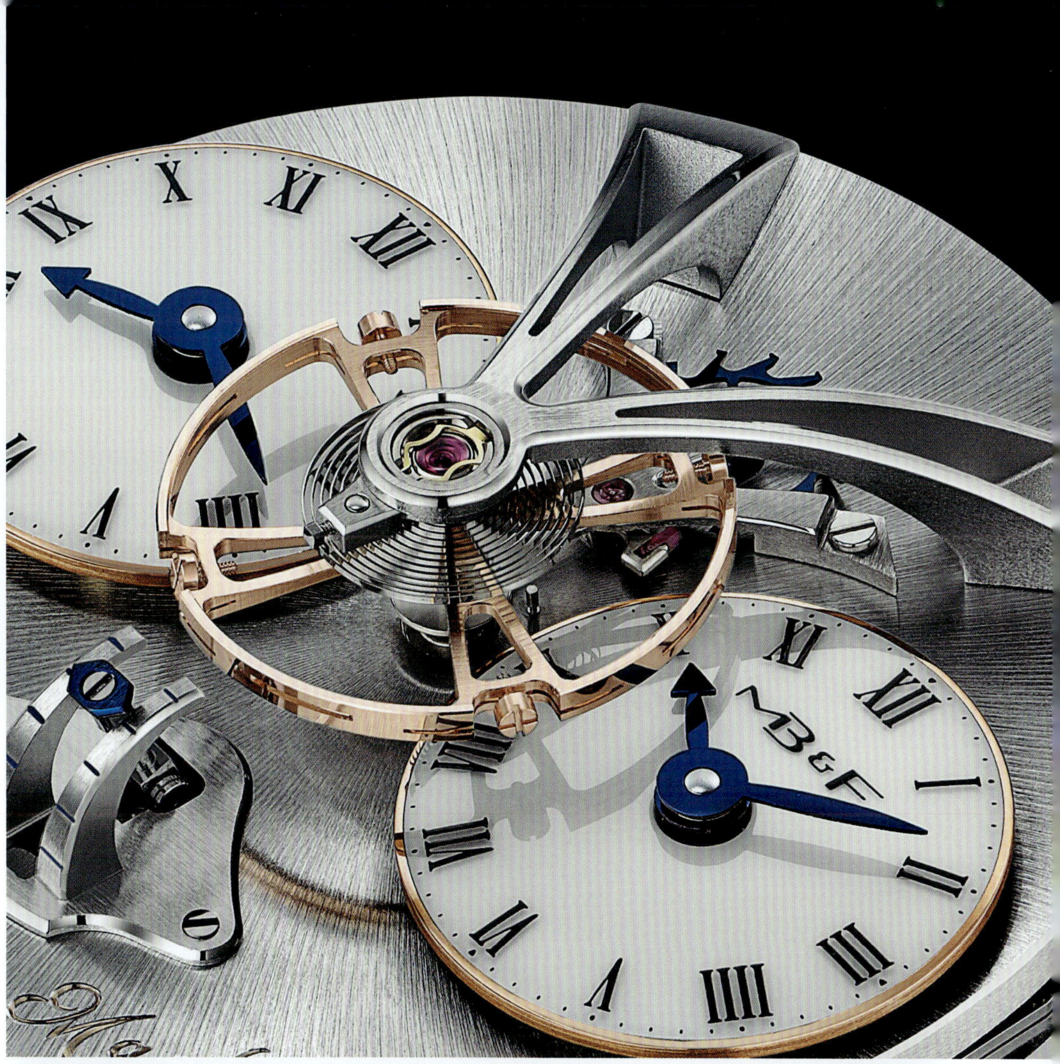
MB&F

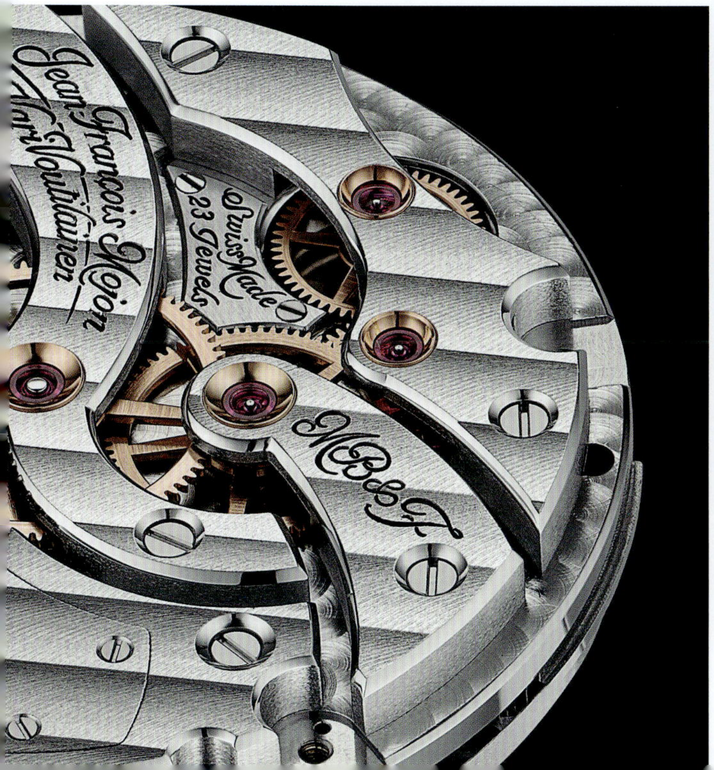
Jean-François Mojon
Kari Voutilainen
Swiss Made
23 Jewels
MB&F

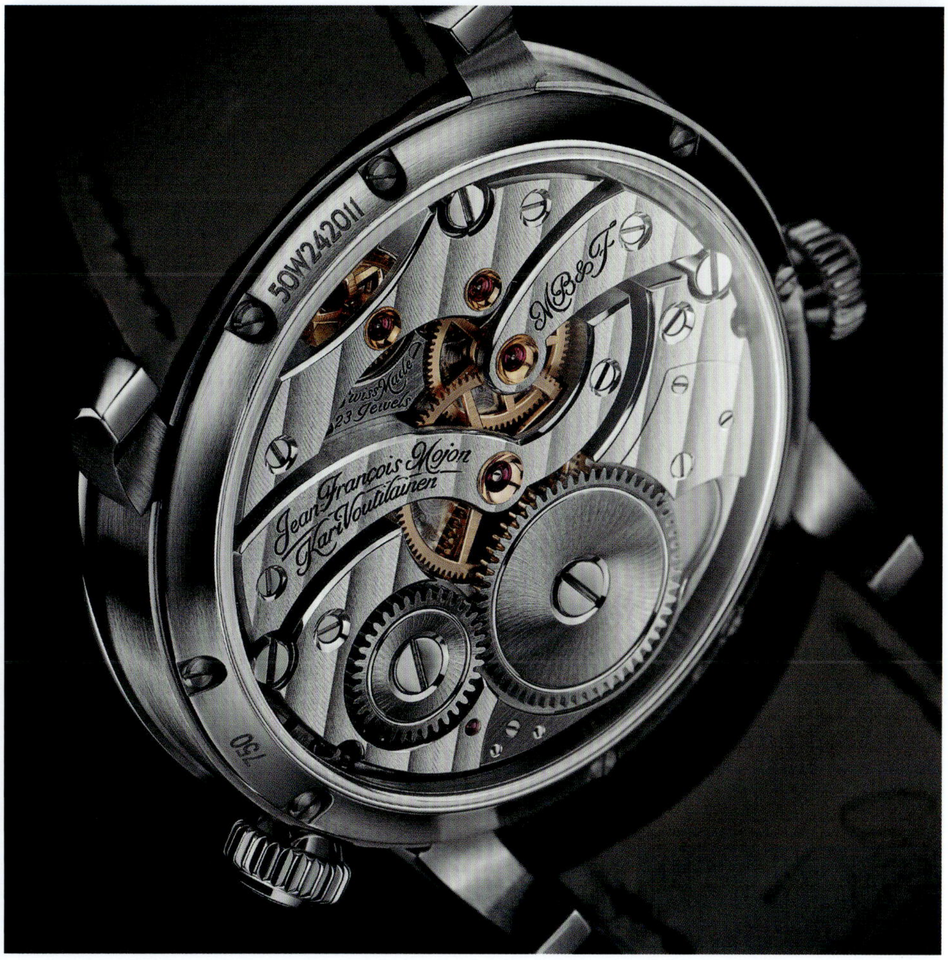
50W242011
MB&F
Swiss Made
23 Jewels
Jean-François Mojon
Kari Voutilainen
750

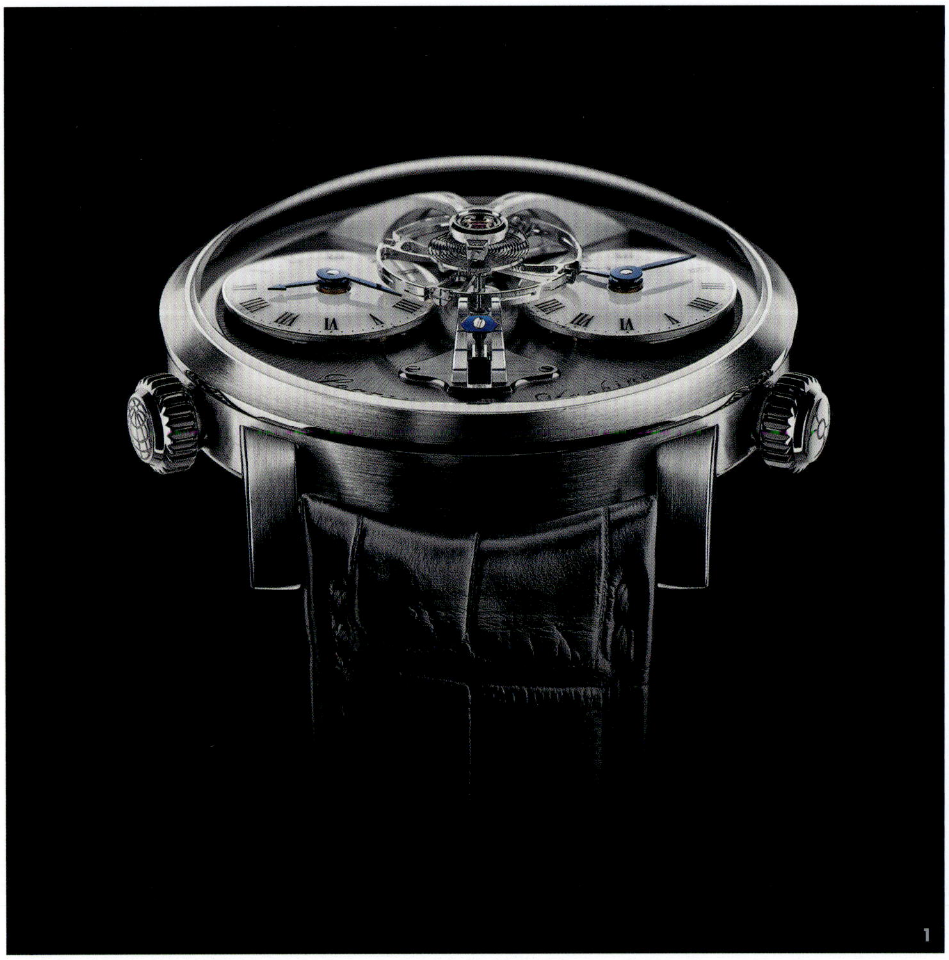

1

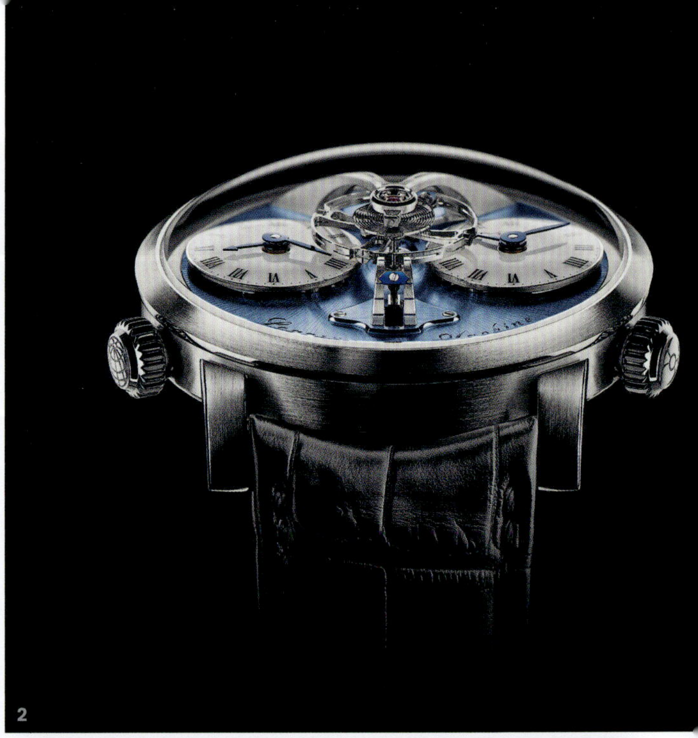

2

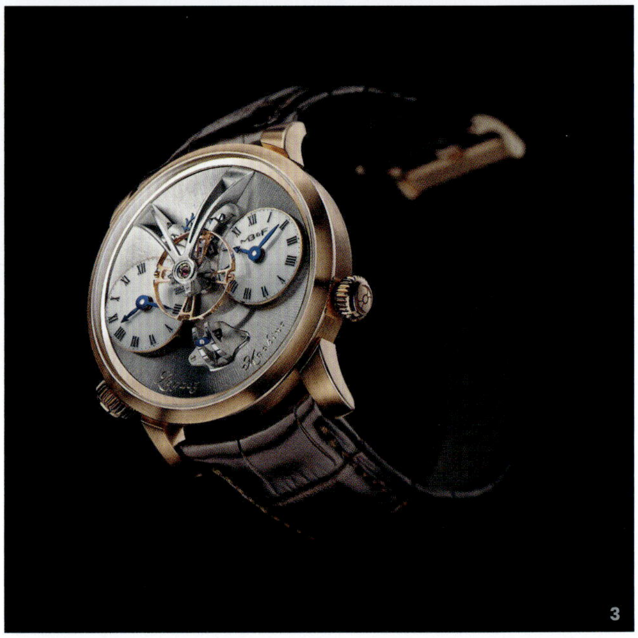

3

1 Legacy Machine N°1 WG
Reference: 01.WL.W
Material: 18K white gold
Limitation: –
Introduction: 2011
Years in production:
2011–2017
Quantity produced: 139

2 Legacy Machine N°1 PT
Reference: 01.PL.W
Material: platinum 950
Limitation: 33
Introduction: 2014
Years in production:
2014–2016
Quantity produced:
32 + 1 prototype

3 Legacy Machine N°1 RG
Reference: 01.RL.W
Material: 18K red gold
Limitation: –
Introduction: 2011
Years in production:
2011–2017
Quantity produced: 166

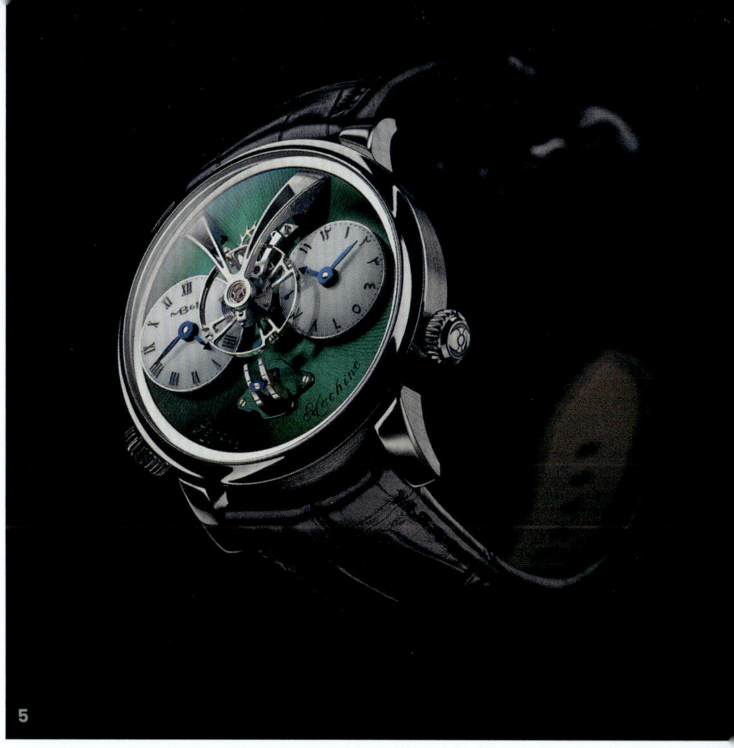

5

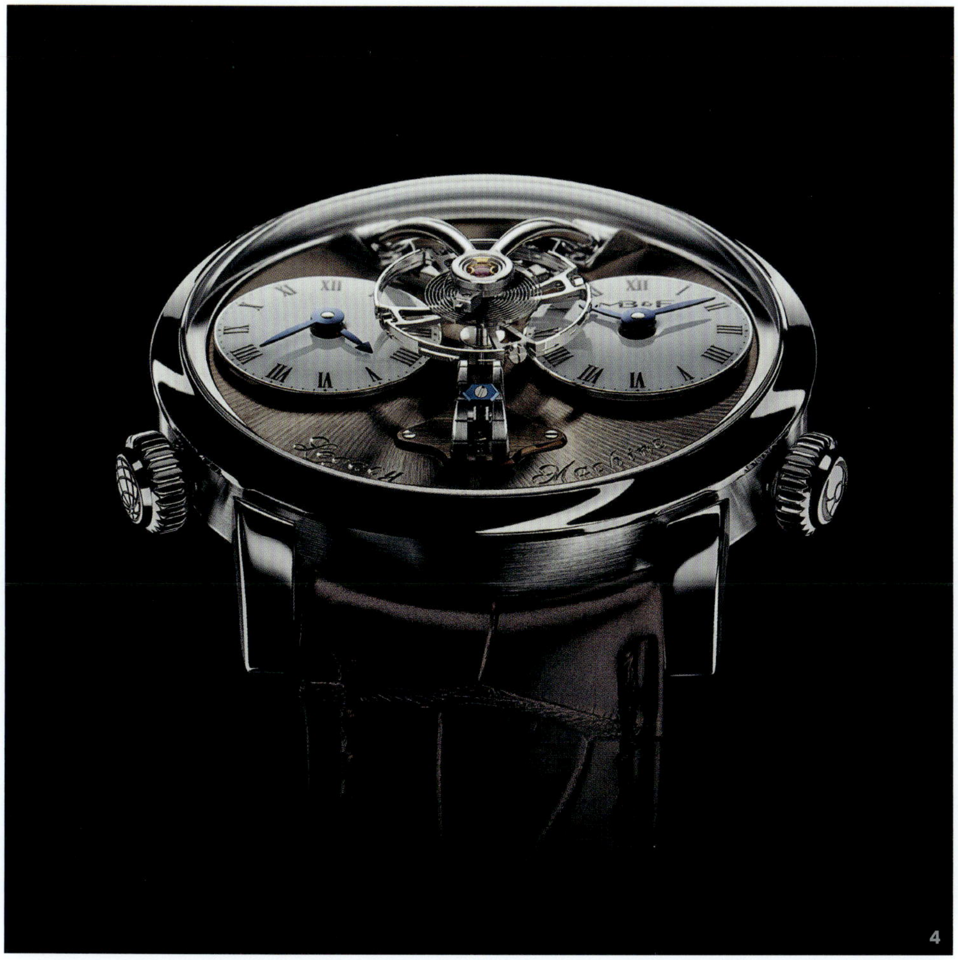

4

4 Legacy Machine N°1
Final Edition
Reference: 01.SL.C
Material: stainless steel
Limitation: 18
Introduction: 2017
Year in production: 2017
Quantity produced:
18 + 1 prototype

5 Legacy Machine N°1
M.A.D. Dubai
Reference: 01.TL.W
Material: grade 5 titanium
Limitation: 13
Introduction: 2016
Years in production: 2016
Quantity produced:
13 + 1 prototype

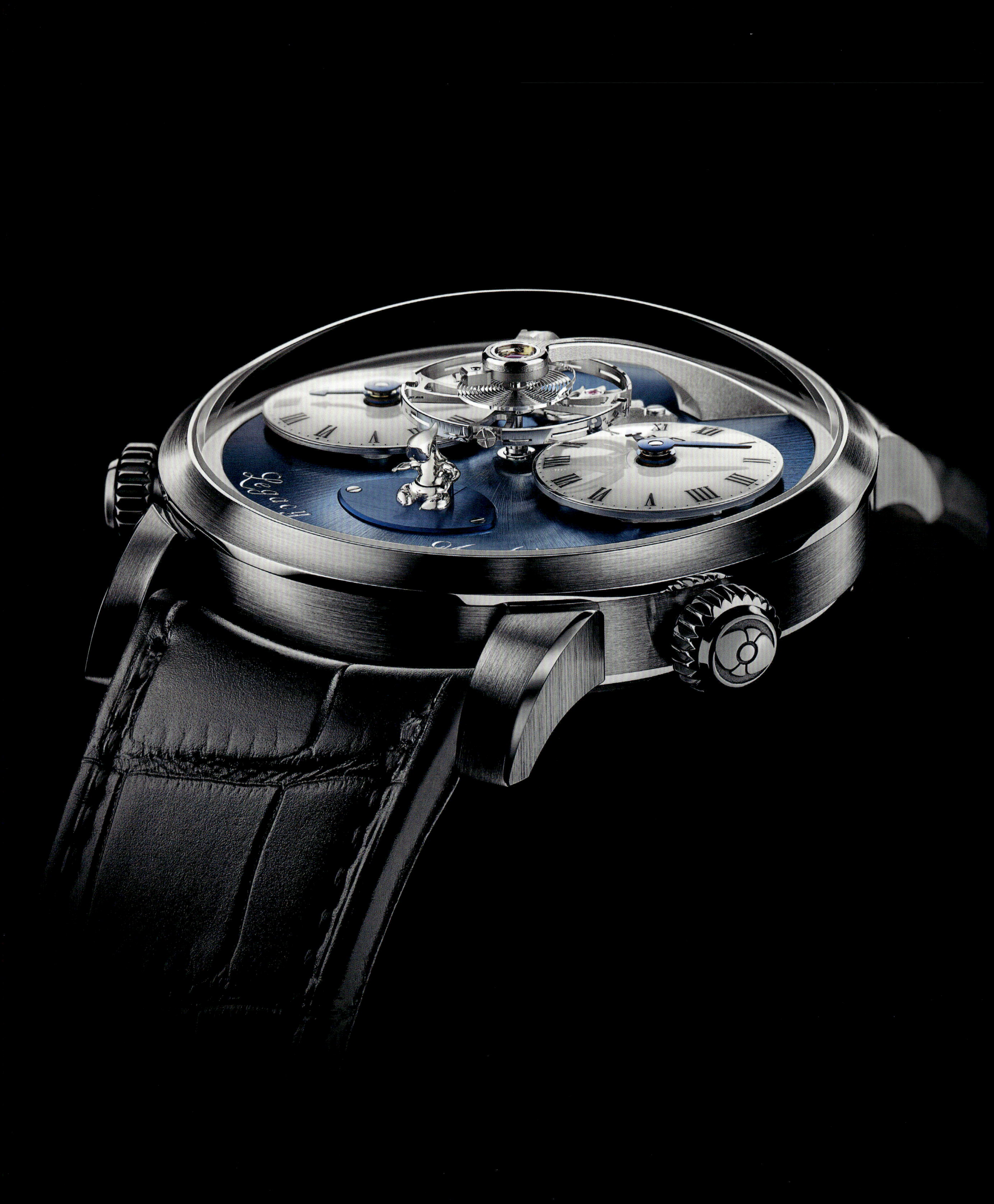

LM1
MB&F × XIA HANG

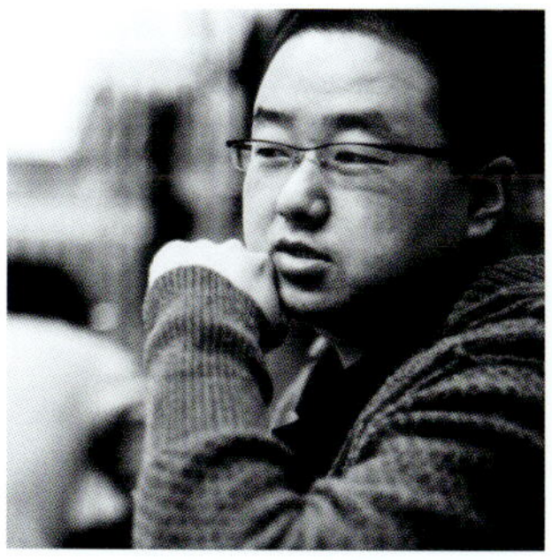

XIA HANG

Born in 1978, in Shenyang, China, Hang is a graduate of the Lu Fun Academy of Fine Arts and China's Central Academy of Fine Arts. He is currently based in Beijing. Known for his sculptures, installations, and public works, his work explores themes relating to cognition and human nature expressed in industrial media, polished steel in particular.

His most recognizable motif is the "comma man", developed in the period 2005 –2008. The artist sees the figure as a little "chubby boy", a comma-shaped, expressive character most often represented in mirror-polished steel.

Hang invites playful interaction between his art and its viewers. His 2008 Beijing exhibition was titled "Please don't touch", with the word "don't" crossed out, to encourage a physical dialogue between audience and art. It is his view that sculptures can be toys, and toys can be sculptures.

Hang brought a lighthearted spirit to his 2014 collaboration on the Legacy Machine N°1. One of his signature "Comma Men", entitled "Mr. Up and Mr. Down", takes the place of the original vertical power reserve, replacing a purely mechanical measure of barrel wind with a more playful, more human expression of physical and mental fatigue.

When the power reserve of the LM1 is at its full 45 hours, the little polished aluminum "Comma Man" sits upright, with his head in an alert position. As power begins to run down and the watch is due for rewinding, the little metal man appears slouched, his head slumping downwards, as if ready for a nap.

This charming effect reflects the good-humored playfulness of the artist, not to mention an inquisitive sense of humour that is very much in the spirit of MB&F.

This timepiece was produced as a limited edition of twelve pieces in rose gold, with a further twelve pieces in white gold. Each comes with two large-scale sculptures of "Mr. Up and Mr. Down," each measuring 6 inches tall and signed by the artist.

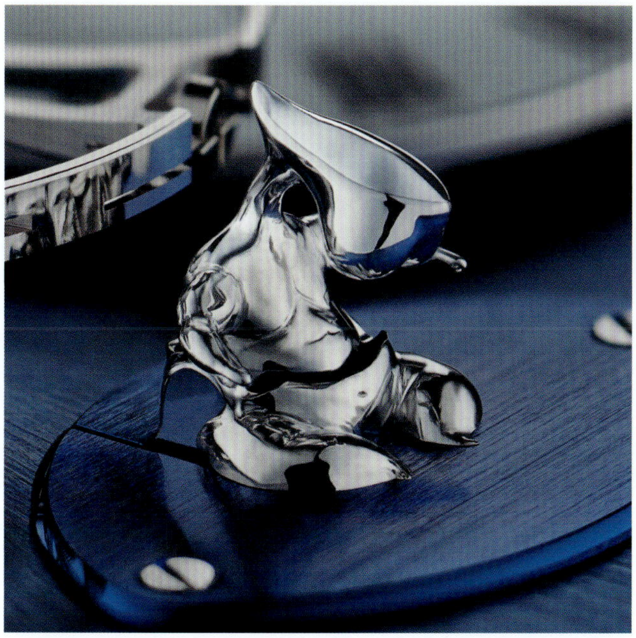

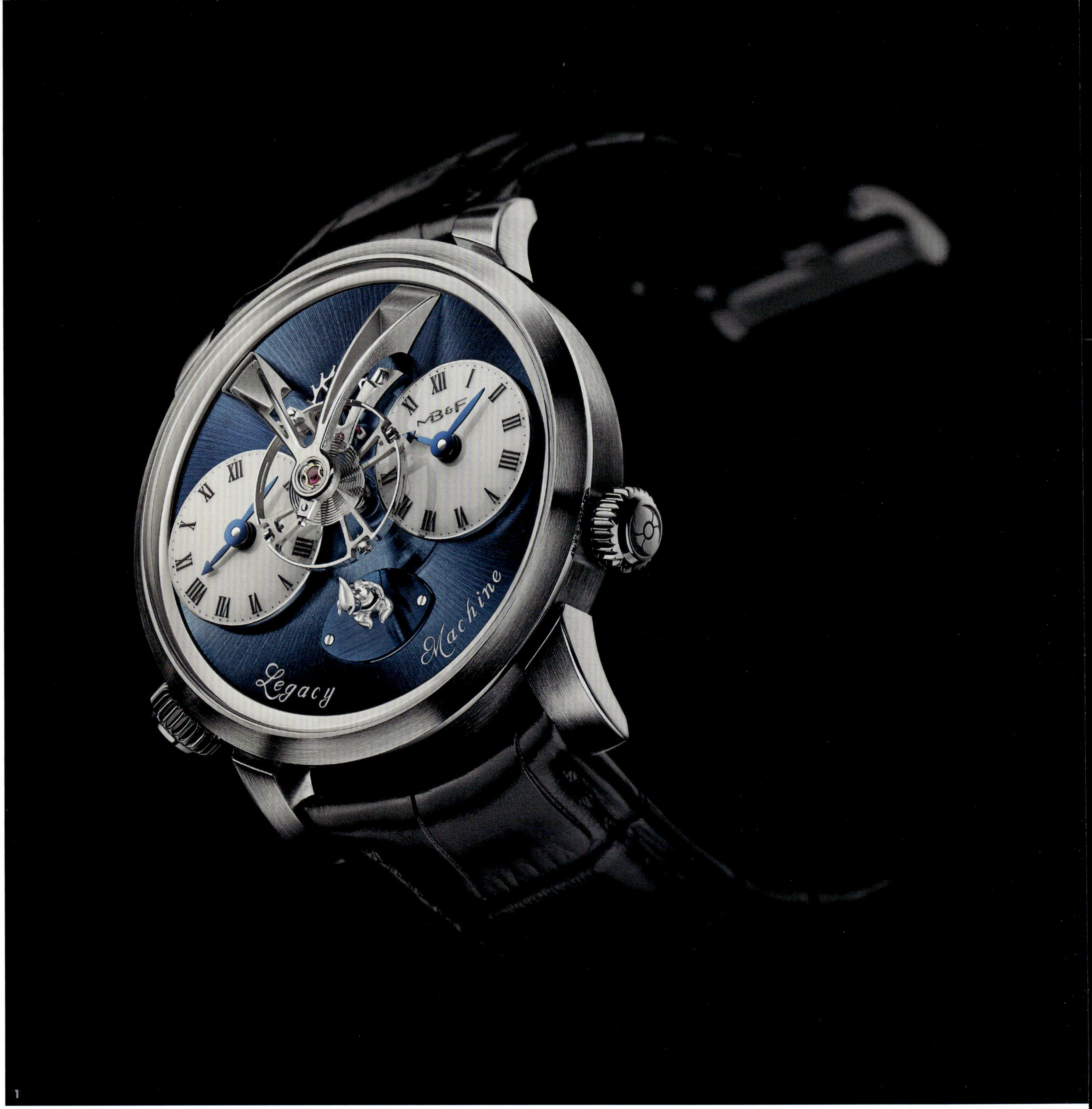

1

1 Legacy Machine N°1
Xia Hang WG
Reference: 52.WL.W
Material: 18K white gold
Limitation: 12
Introduction: 2014
Year in production: 2014
Quantity produced:
12 + 1 artist's proof

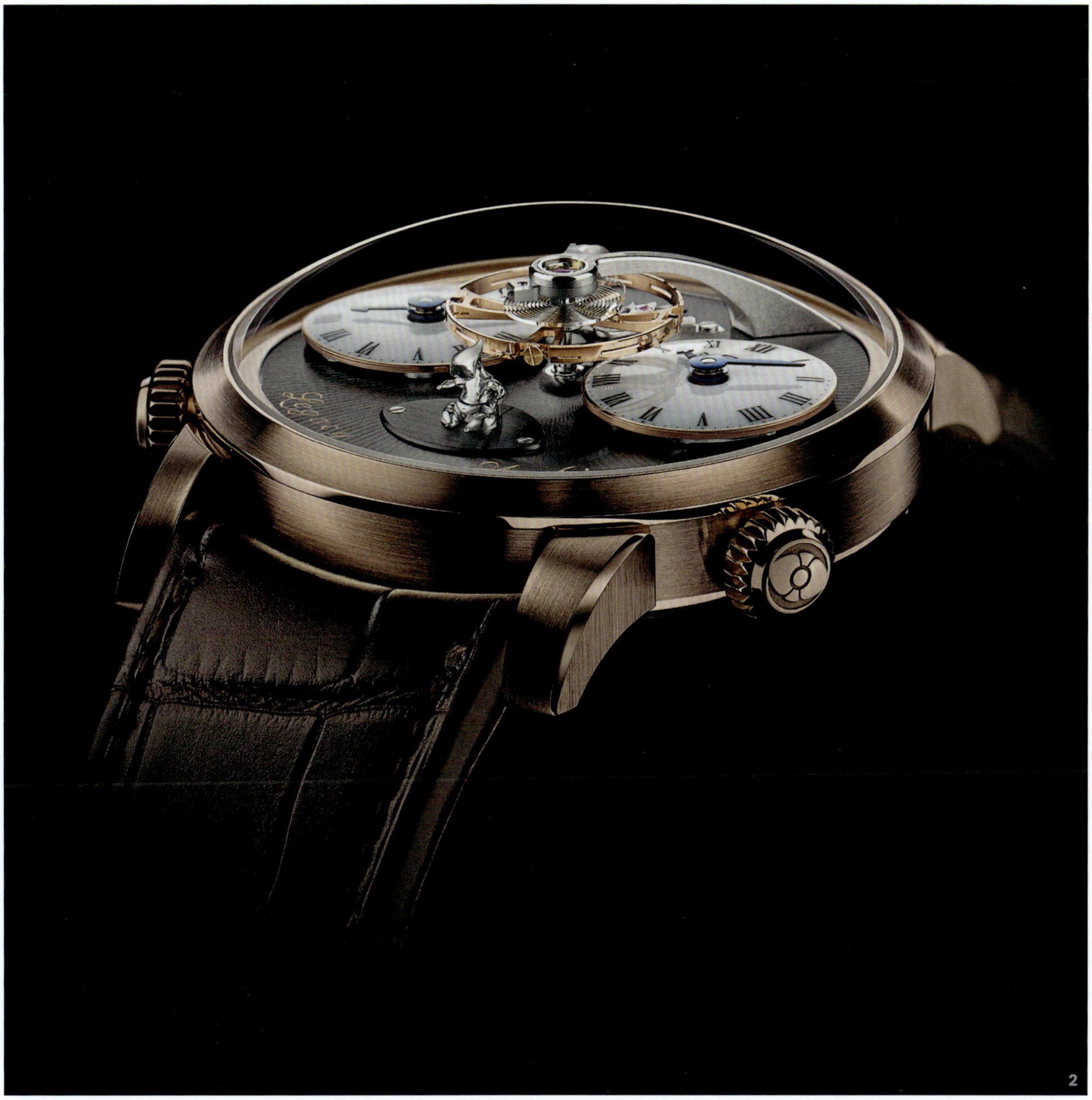

2 Legacy Machine N°1
Xia Hang RG
Reference: 52.RL.W
Material: 18K red gold
Limitation: 12
Introduction: 2014
Year in production: 2014
Quantity produced:
12 + 1 artist's proof

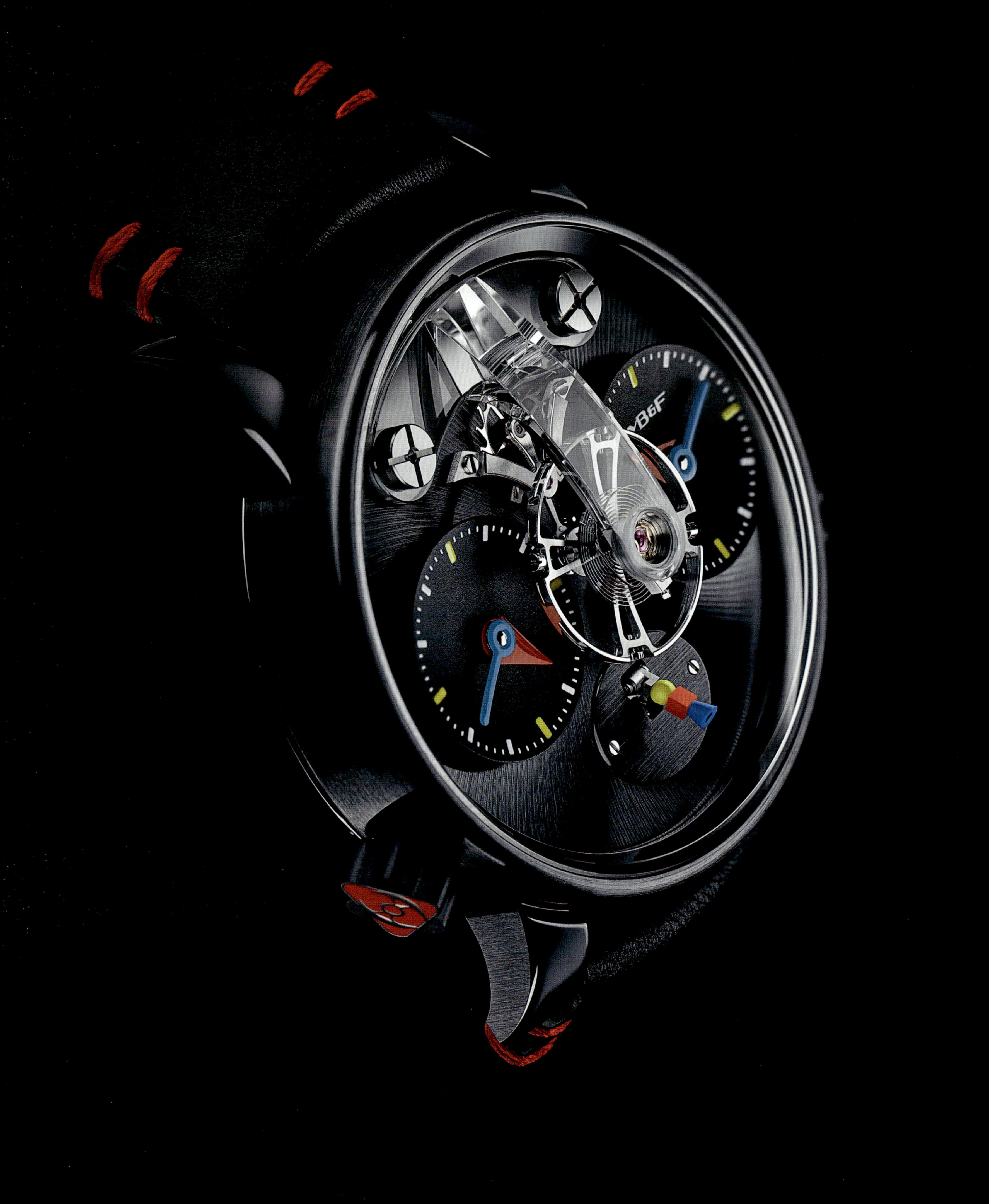
MB&F

LM1 MB&F × ALAIN SILBERSTEIN

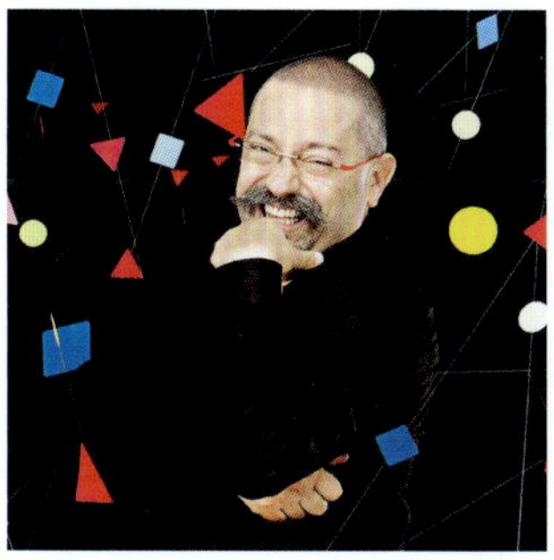

ALAIN SILBERSTEIN

Alain Silberstein was born in Paris, France, in 1950. A trained interior architect and model maker, he worked in Paris as an interior designer before moving to Besançon, the heart of French watchmaking, in 1979. In 1990, Silberstein created his own brand, Alain Silberstein Créations, which quickly developed a reputation for its bright, playful use of primary colours and geometric shapes, taking inspiration from the German Bauhaus movement. Silberstein was the first watchmaker to design a watch with a sapphire case, a material that he would employ again in his 2016 collaboration with MB&F when he reinterpreted the Legacy Machine N°1 with his signature Silberstein flair.

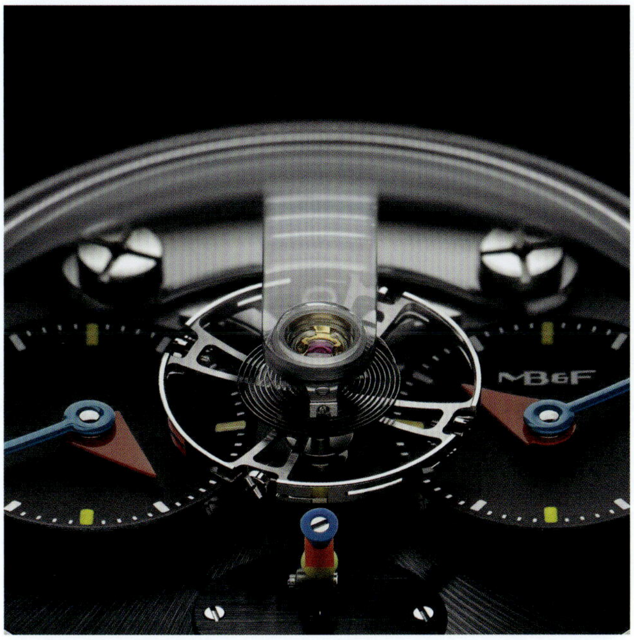

Sapphire crystal is the hallmark element in Silberstein's interpretation of the LM1, his second collaboration with MB&F. A single, clear bridge of sapphire replaces the twin arched, steel bridges of the original, so all is clearly visible and the balance wheel takes centre stage.

The original convex subsidiary dials of the LM1 are replaced by concave versions with matching hands. Silberstein sees this as a symbolic expression of "eternal time" as it is broken down into smaller increments by the beating oscillations of the balance spring — the subsidiary dials serve as receptacles that capture Silberstein's "eternal time."

Silberstein's signature use of primary colours on the subsidiary dials is not simply aesthetic. Glossy red hour hands against the matte black subsidiary dials serve to draw the attention of the eye, making the hour quickly and clearly legible. High watchmaking is not generally known for its playfulness or sense of humour; but we see it here, in the natural partnership of Silberstein and MB&F who together bring a touch of whimsy to the world of luxury.

This limited edition is represented by twelve pieces in red gold, twelve pieces in titanium, and another twelve in black PVD-treated titanium.

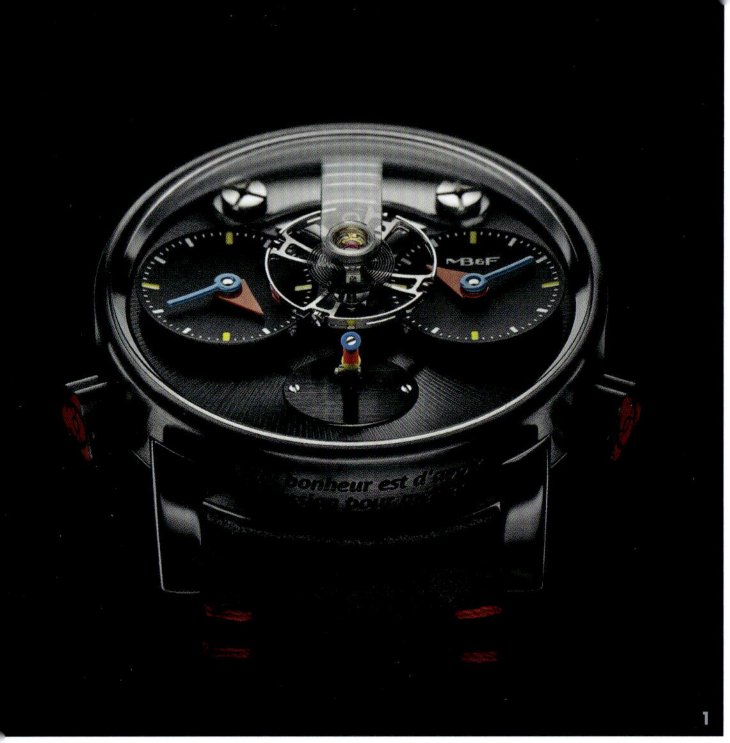

1

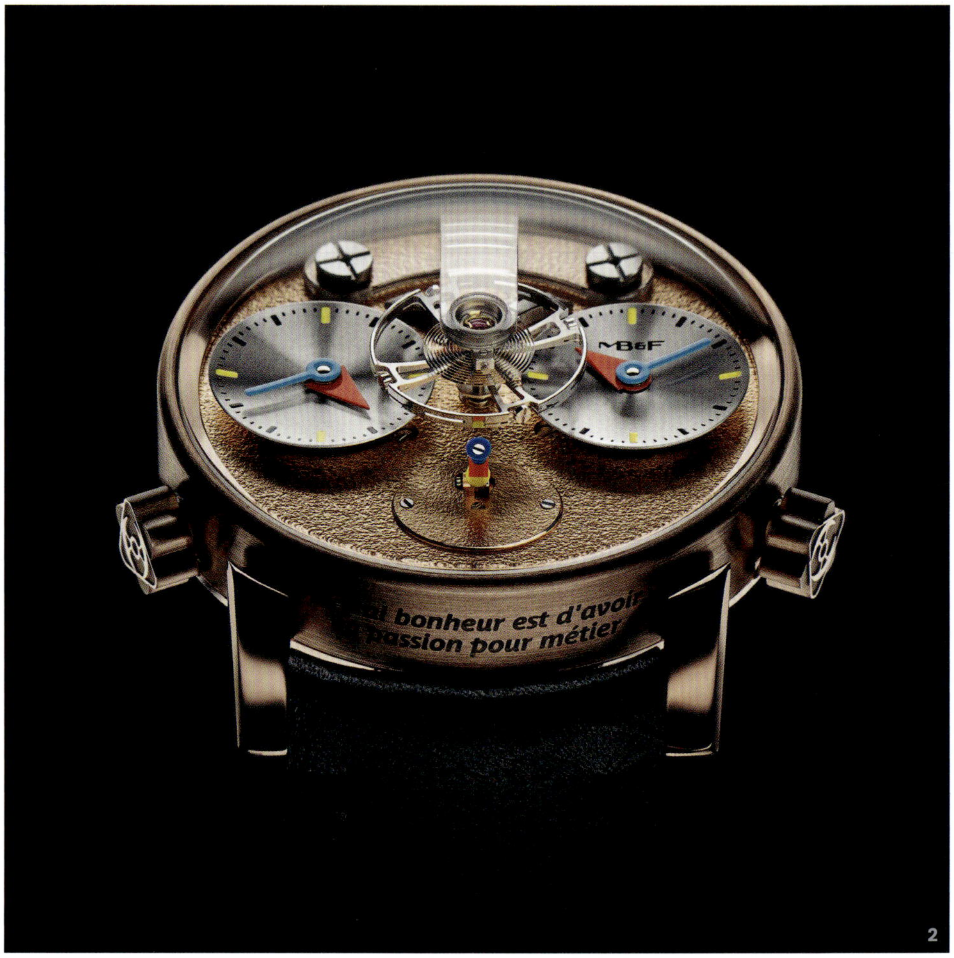

2

1 Legacy Machine N°1
Silberstein Black
Reference: 53.BTL.B
Material: black PVD grade 5 titanium
Limitation: 12
Introduction: 2016
Year in production: 2016
Quantity produced:
12 + 1 prototype

2 Legacy Machine N°1
Silberstein RG
Reference: 53.RL.FS
Material: 18K red gold
Limitation: 12
Introduction: 2016
Year in production: 2016
Quantity produced:
12 + 1 prototype

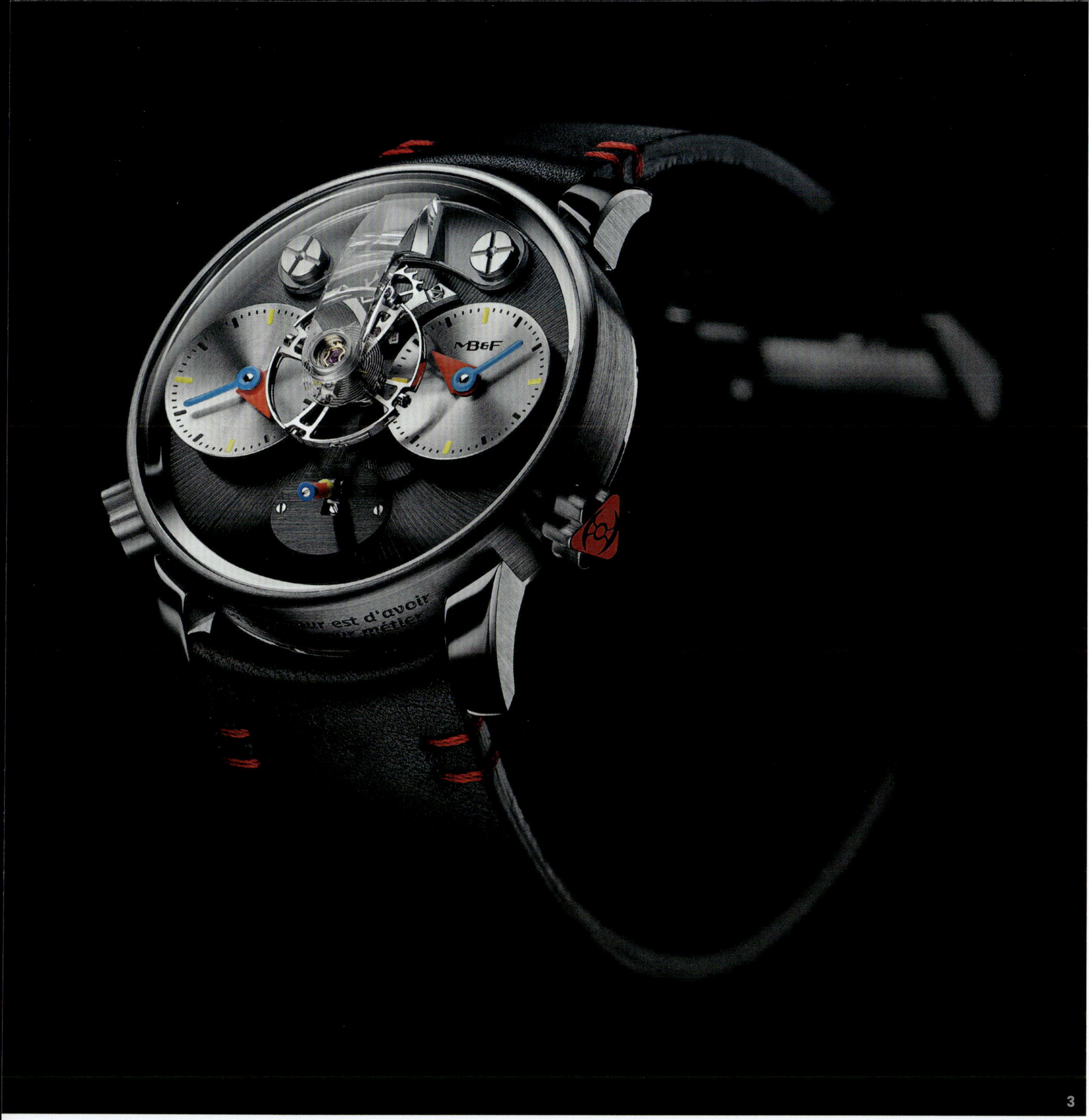

3

3 Legacy Machine N°1
Silberstein Ti
Reference: 53.TL.S
Material: grade 5 titanium
Limitation: 12
Introduction: 2016
Year in production: 2016
Quantity produced:
12 + 1 prototype

“While we are all big fans of MB&F’s heroic Horological Machines, Max’s Legacy Machines caught us by surprise. But that’s the thing about Max – he is one of the smartest and savviest people I know and is tapped into the horological zeitgeist in a way that few others are. The Legacy Machines demonstrate a scholarly understanding of classical watchmaking that is incredibly impressive. They are, to me, some of the most sublime and collectable watches created in the past decade.”

Michael Tay
The Hour Glass

“I feel that the most interesting aspect about watches is the people behind them. MB&F don’t simply credit those involved, they build teams where everyone feels an essential part of the project. Their “no-limit” approach to creating incredible watches has gone way, way beyond traditional Haute Horlogerie.”

Ian Skellern
Co-founder, Quill & Pad

LM 2

The Legacy Machine N°2, or LM2, is the direct descendant to the LM1. Introduced in 2013, it takes its inspiration from the complications of master watchmakers Abraham-Louis Breguet (1747-1823), Ferdinand Berthoud (1727-1807) and Antide Janvier (1751-1835), and their work on one of watchmaking's most intriguing complications – the dual regulator.

Like the LM1, this was an evolution, more than a revolution, with the same focus on a flying balance wheel suspended high over the dial and protected by a sapphire dome. This time however, instead of a single flying balance wheel, there are two.

Sitting proud of the dial is a raised differential wheel at 6 o'clock that serves three functions: transferring power to the dual regulators; receiving the individual timing rates for each balance wheel; and transmitting the average rate of the two regulators to the gear train, which then displays the time. This is the dual regulator system: a rare and technically difficult complication rarely attempted by any previous watchmaker. The result is a beautiful and complex high-precision timepiece.

LEGACY MACHINE N°2

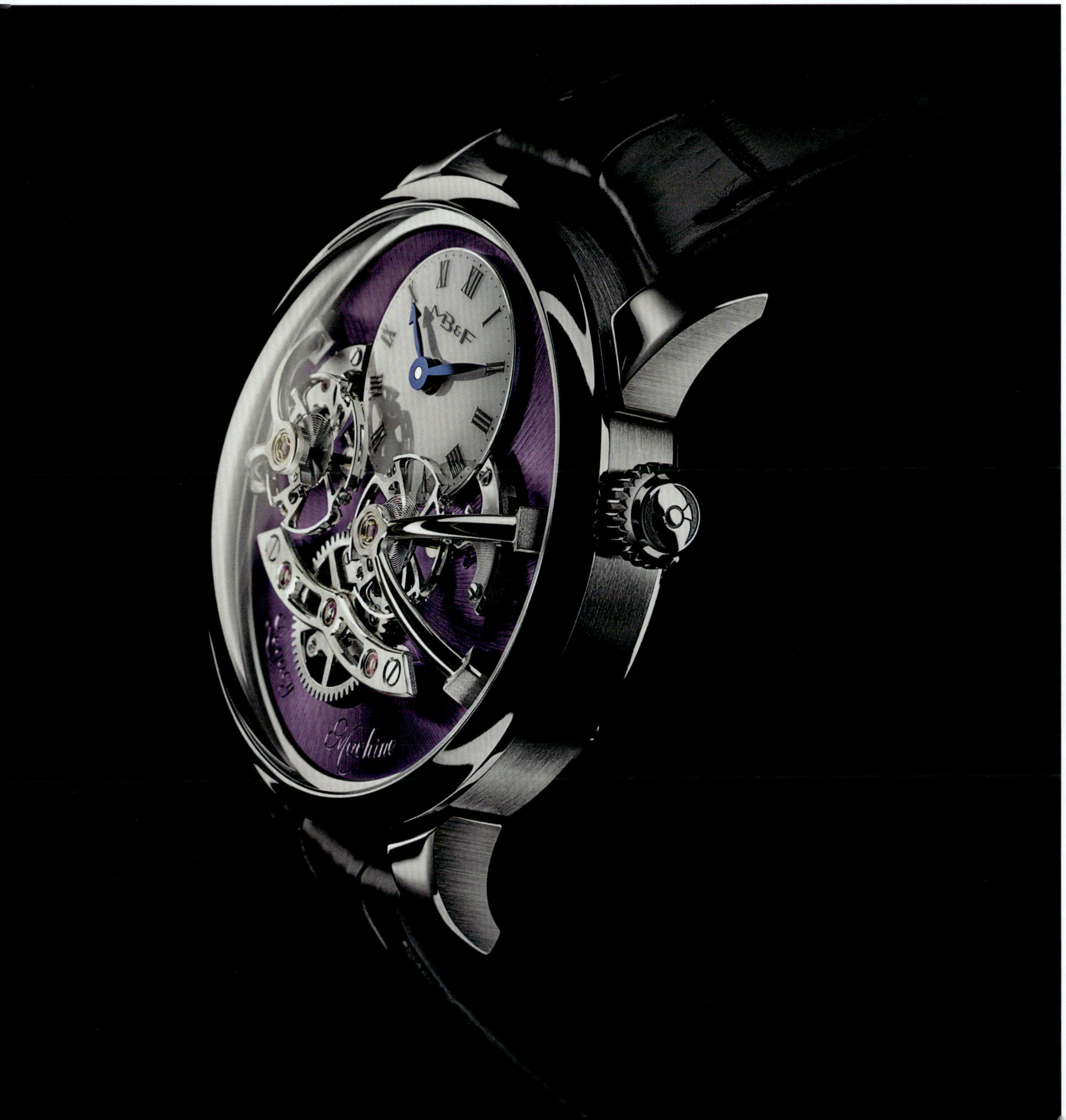

LM2 CALIBRE

Like the LM1, the movement of the LM2 was developed by Jean-Francois Mojon: a three-dimensional movement, with 44 jewels and 241 components, including the twin, bespoke 11 mm suspended balance wheels at centre stage. The LM2 is housed in the same round case as the LM1, evoking the shape and scale of a 19th century pocket watch, with a diameter of 44mm and a thickness of 20 mm.
Kari Voutilainen endowed the movement with an aesthetic style consistent with high-quality traditional timepieces of the 19th century, endorsed by superlative hand-finishing.
The LM2 is both fantastical and traditional: an evolution of the LM1 that celebrates Haute Horlogerie, stripping away certain elements of the LM1 (gone are the vertical power reserve and the independent time displays) to showcase its most complicated structure — the dual regulator.

Collection
Legacy Machines
Series
Legacy Machine N°2
Common name
LM2
Movement
Developed for MB&F by Jean-François Mojon/Chronode and Kari Voutilainen
Winding
Manual
Power reserve
45 hours
Components
241/44 jewels
Functions
Hours and minutes, planetary differential
Add. features
Two flying balance wheels; planetary differential comprising 3 gears and 5 pinions
Case
18K red or white gold, platinum 950, or grade 5 titanium
Dimensions
44 mm × 20 mm; redesigned edition in titanium: 44 mm × 19 mm
Case components
45 or 41 for redesigned edition in titanium
Water resistance
30 m/3 ATM
Strap
Alligator
Buckle
Pin or folding buckle, titanium, 18K red or white gold or platinum matching the case
Introduction
2013
Years in production
2013–2019
Quantity produced
120 + 3 prototypes
Comments
The Legacy Machine N°2 RG Blue, WG Purple and Ti bear the redesigned flying balance wheel bridges from the LM101 and the following Legacy Machine series. The titanium version is an evolution with a redesigned case.

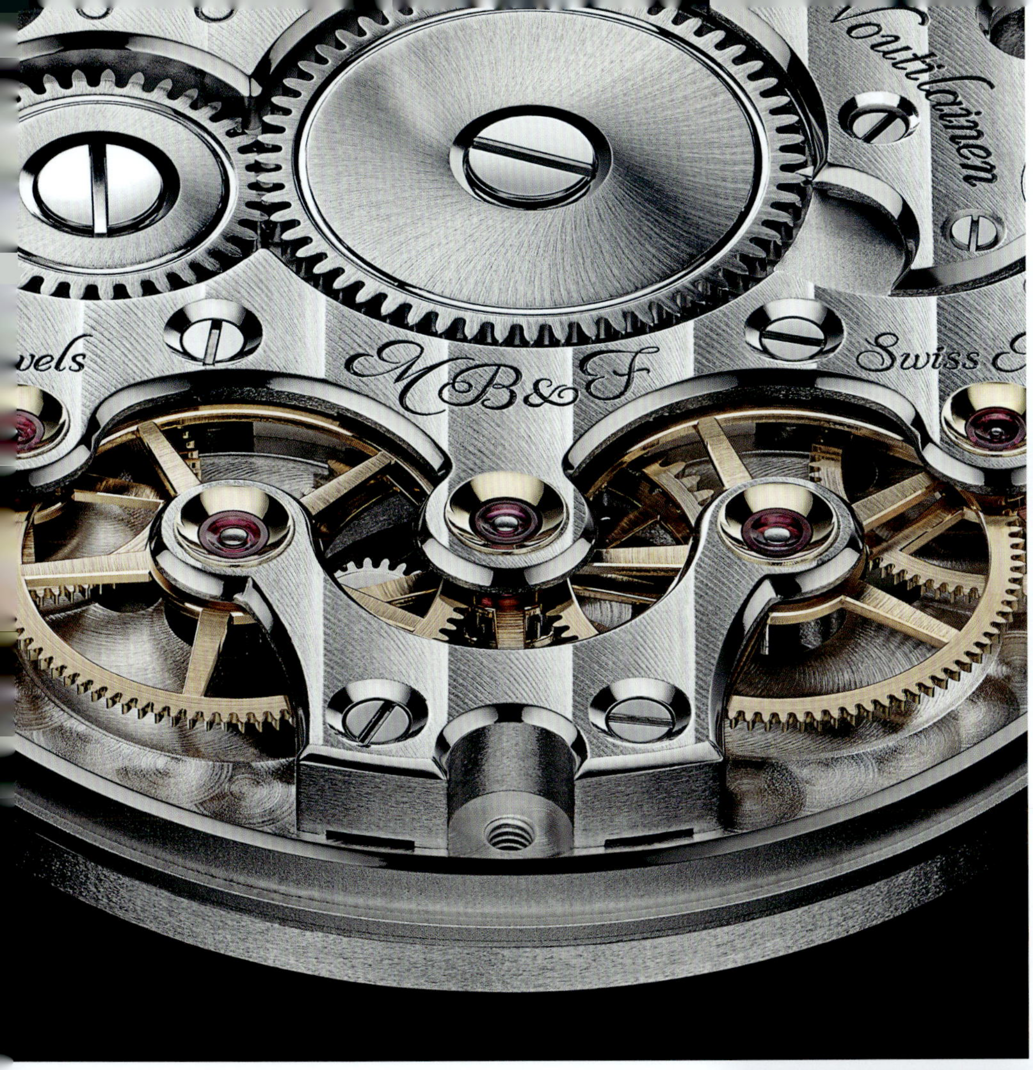
Voutilainen
MB&F
Swiss
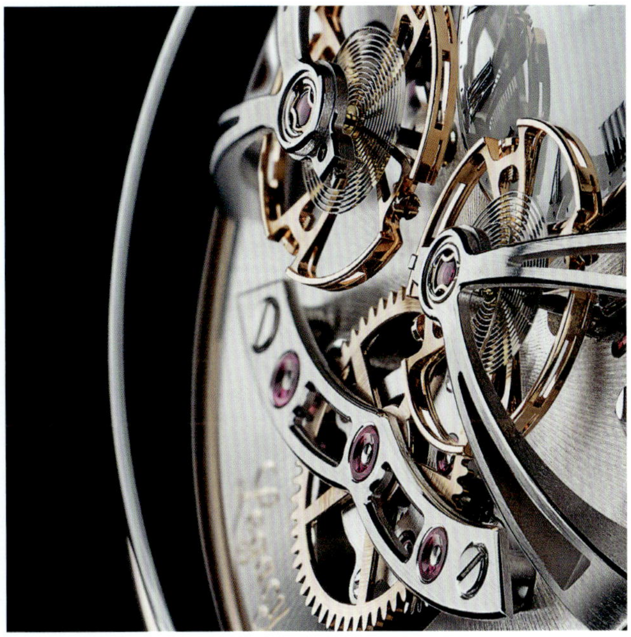
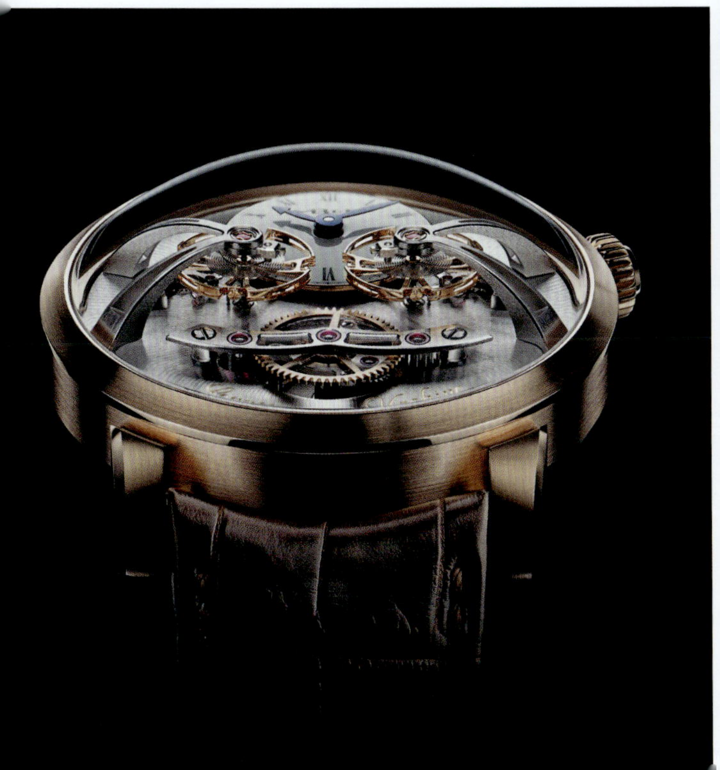
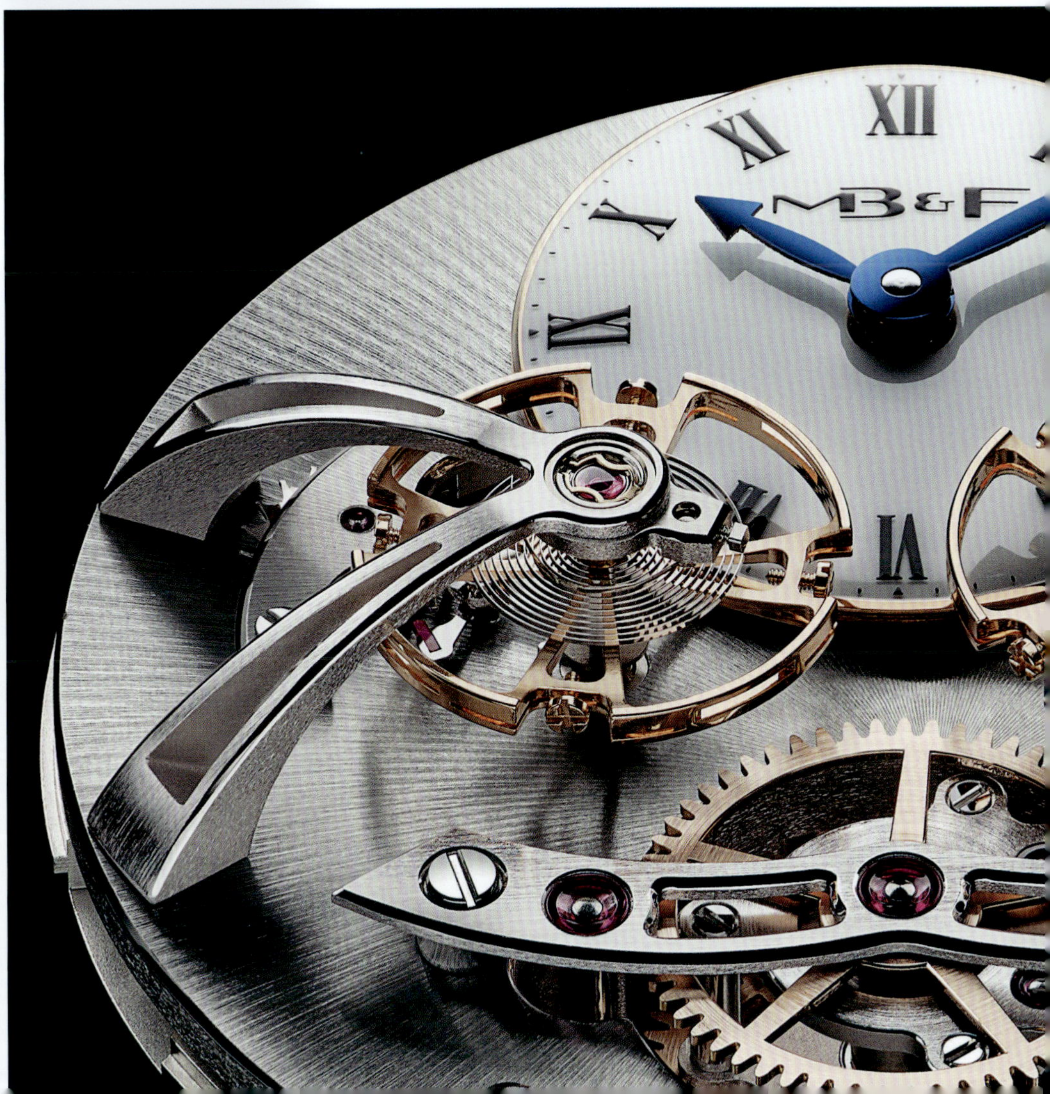
MB&F

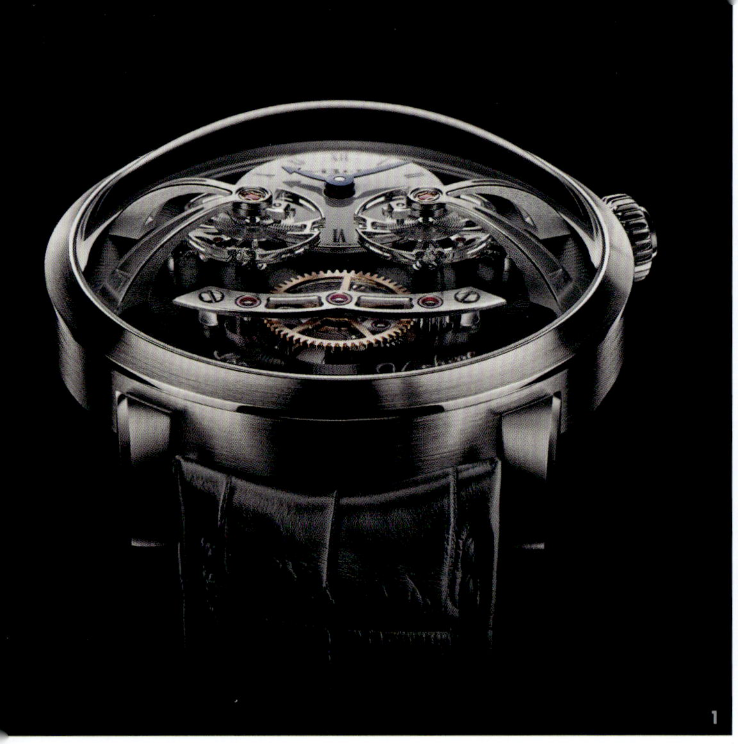

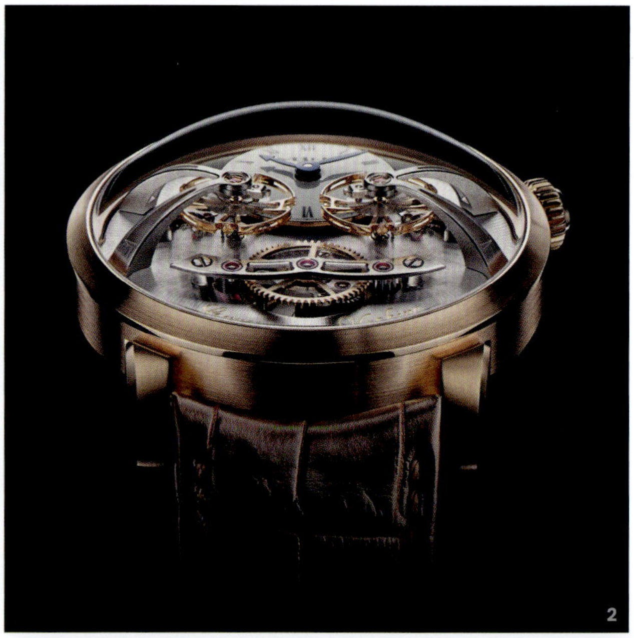

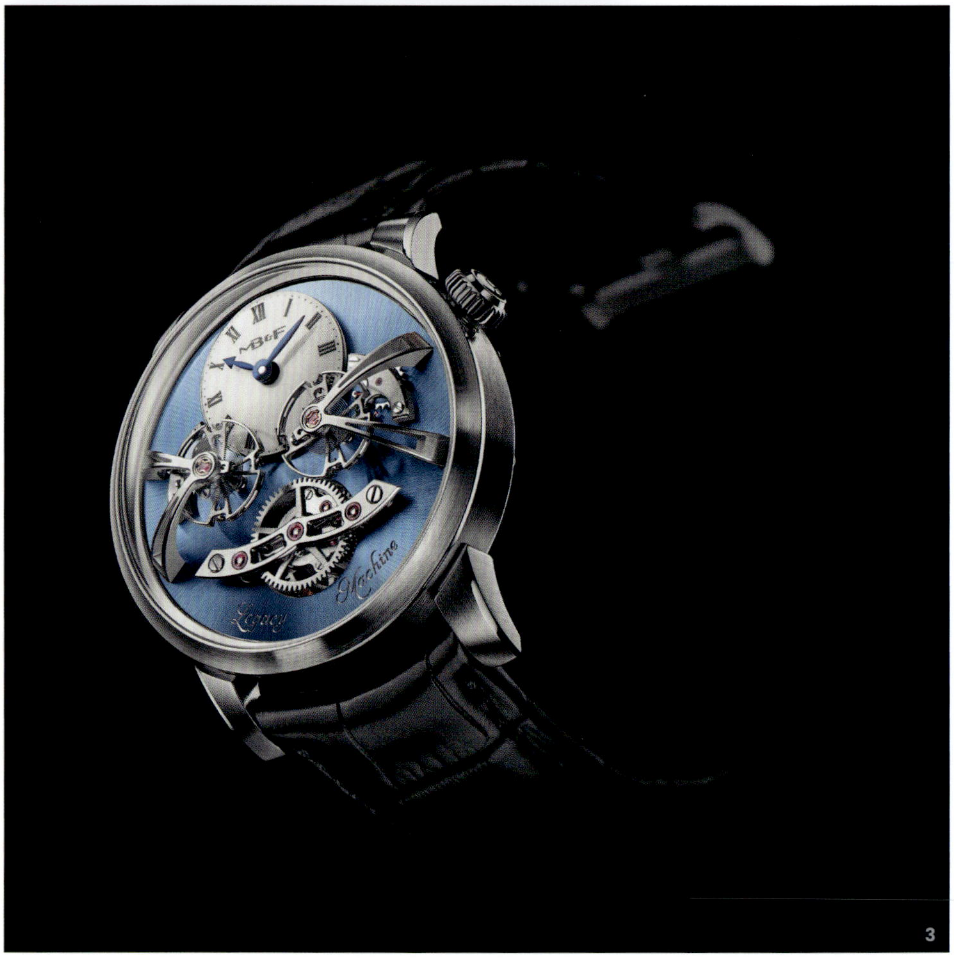

1 Legacy Machine N°2 WG
Reference: 02.WL.W
Material: 18K white gold
Limitation: –
Introduction: 2013
Years in production:
2013–2015
Quantity produced:
29 + 1 prototype

2 Legacy Machine N°2 RG
Reference: 02.RL.W
Material: 18K red gold
Limitation: –
Introduction: 2013
Years in production:
2013–2015
Quantity produced: 31

3 Legacy Machine N°2 PT
Reference: 02.PL.W
Material: platinum 950
Limitation: 18
Introduction: 2013
Years in production:
2013–2014
Quantity produced:
18 + 1 prototype

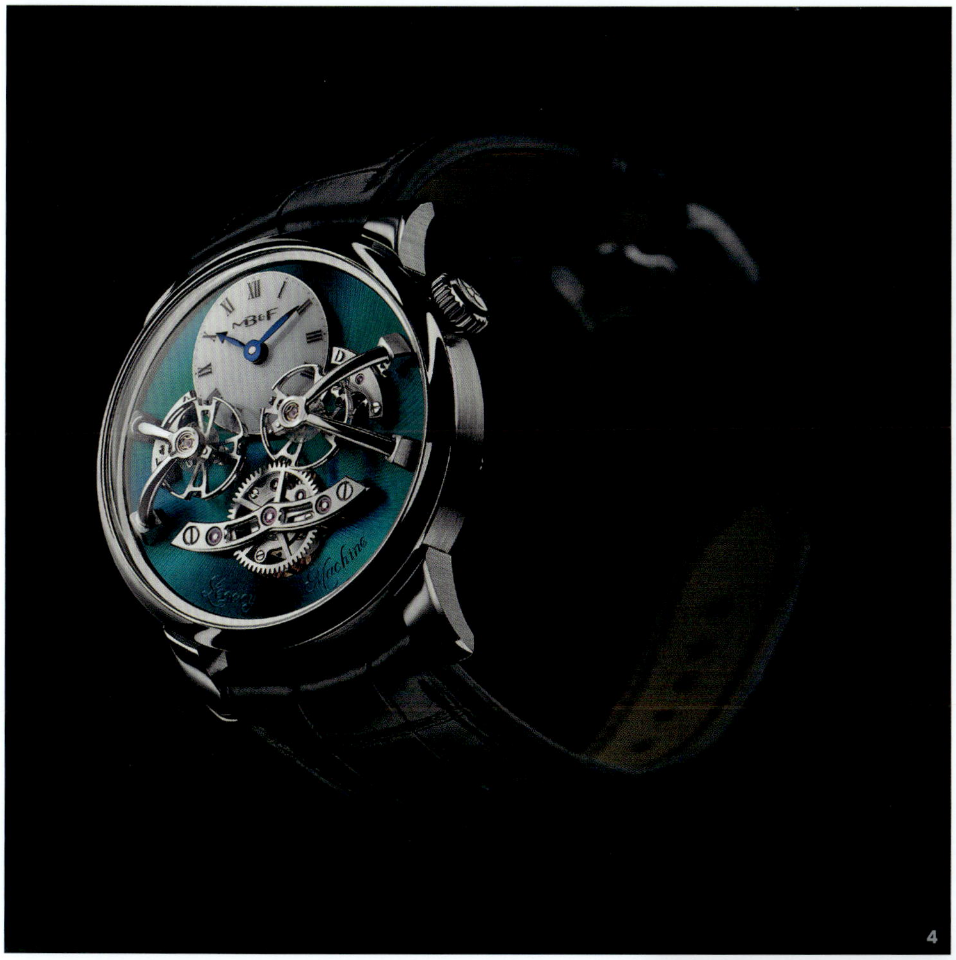

4

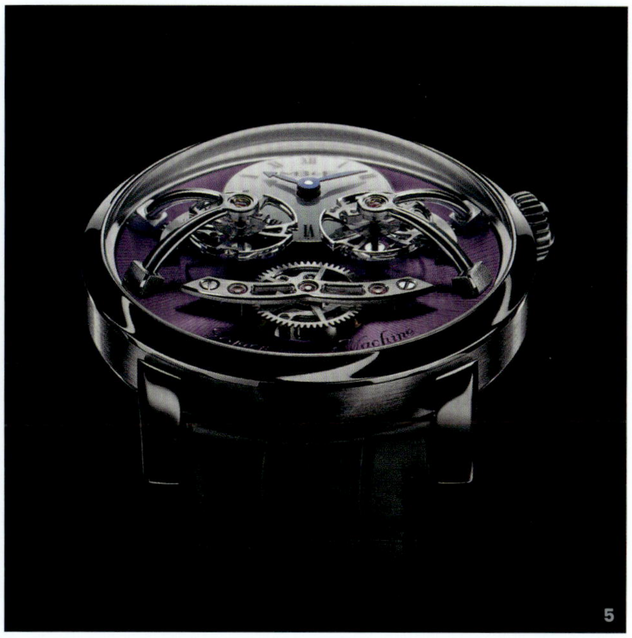

5

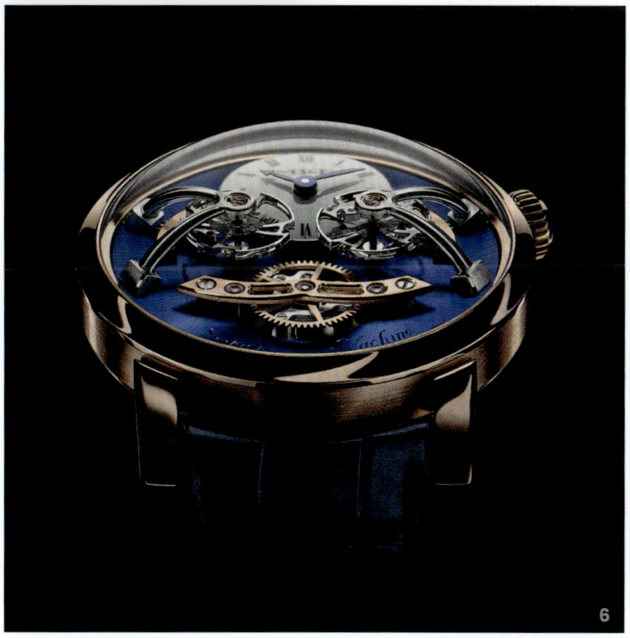

6

4 Legacy Machine N°2 Ti
Reference: 02.TL.G
Material: grade 5 titanium
Limitation: 18
Introduction: 2017
Year in production: 2017
Quantity produced:
18 + 1 prototype

5 Legacy Machine N°2
WG Purple
Reference: 02.WL.P
Material: 18K white gold with
purple dial
Limitation: 12
Introduction: 2018
Year in production: 2018
Quantity produced: 12

6 Legacy Machine N°2 RG Blue
Reference: 02.RL.BL
Material: 18K red gold
with blue dial
Limitation: 12
Introduction: 2019
Year in production: 2019
Quantity produced: 12

LM 101

The Legacy Machine 101, or LM101, is the third evolution of the Legacy Machine concept. Released in 2014, this was the first MB&F machine to feature a movement entirely conceived in-house, an important milestone for MB&F in terms of growth and maturity. As suggested by its "101" name, it focuses on the fundamentals of a mechanical wristwatch.

The 101 is also one of the most compact members of the Legacy Machine family, with a case measuring just 40mm — a full 4mm smaller than the LM1 and LM2. Reprising the dial featuring just a single flying balance wheel, this looks at first glance to be the younger, smaller brother of the LM1. Where there were two independent time displays, the LM101 features just one. Where there was once a vertical power-reserve indicator, the 101 features a small, flat lacquered subdial.

At the centre of the dial and hovering above it is the flying balance wheel – the same 14mm size as the LM1 balance wheel, but appearing even more dominant due to the smaller scale of the LM101. A Legacy Machine, pared down to its purist form.

LEGACY MACHINE 101

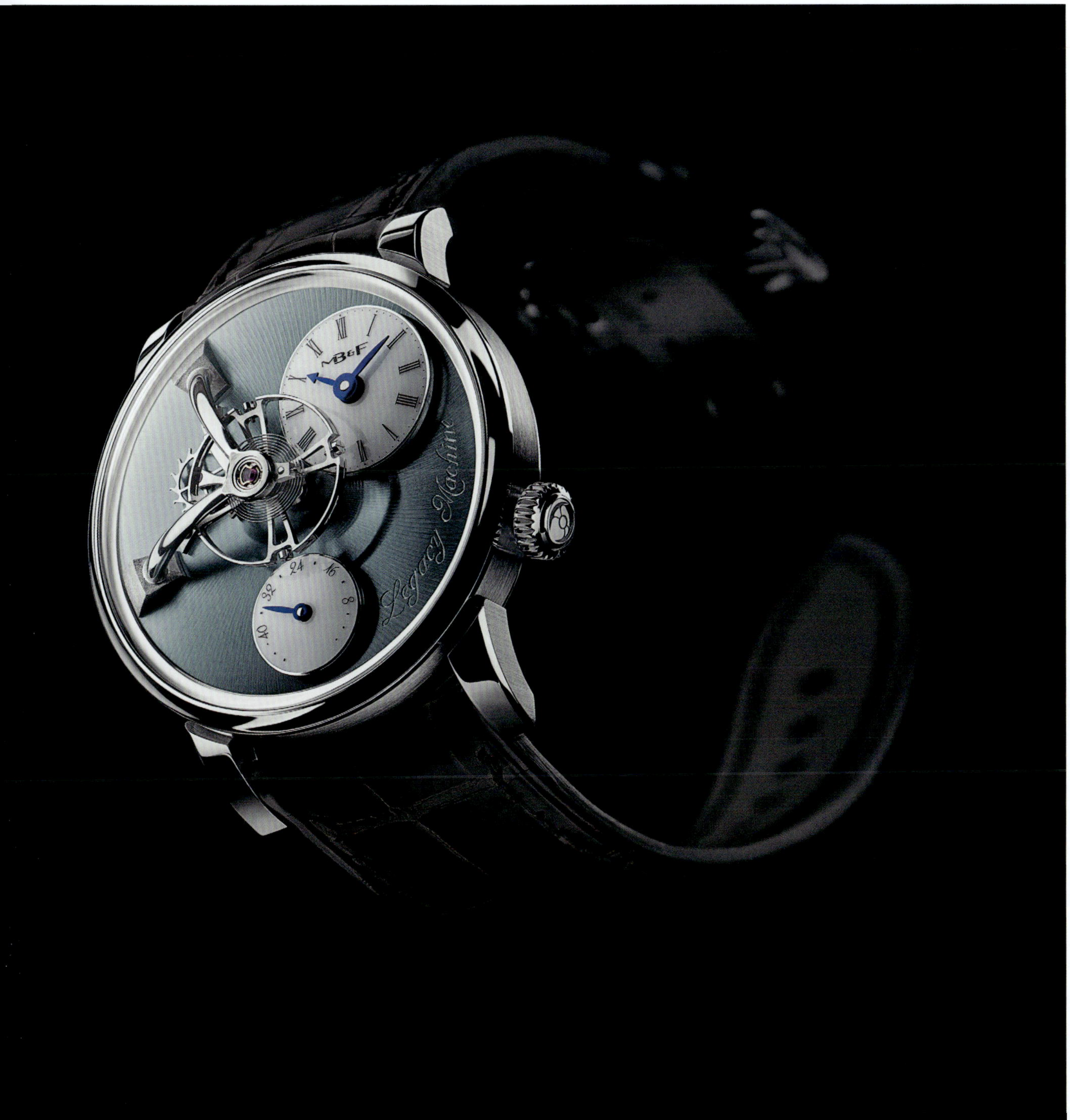

LM 101 CALIBRE

Previous Legacy Machines featured movements developed by Jean-Francois Mojon. The LM101 calibre was the first movement conceived and developed entirely in-house by MB&F. With its 23 jewels, 229 components and 45 hours of power reserve, the LM101 Calibre was an important milestone in the Legacy Machine story, which had begun just three years earlier.

The aesthetic of the LM101 calibre — defined by Kari Voutilainen, whose signature is on the movement – remains consistent with its predecessors, viewable through the sapphire case back.

Collection
Legacy Machines
Series
Legacy Machine 101
Common name
LM101
Movement
Developed in-house by MB&F with Kari Voutilainen
Winding
Manual
Power reserve
45 hours
Components
229/23 jewels

Functions
Hours, minutes and power reserve indicator
Add. features
Flying balance wheel
Case
18K white, red or yellow gold, platinum 950, palladium 950, stainless steel
Dimensions
40 × 16 mm
Case Components
35

Water resistance
30 m/3 ATM
Strap
Alligator
Buckle
Pin or folding buckle, 18K white or red gold, platinum, titanium or stainless steel matching the case
Introduction
2014
Years in production
2014 – in production
Quantity produced
(at the end of 2020) 262 + 11 prototypes

Comments
The LM101 MB&F × H. Moser editions feature a double hairspring, thinner bezel and no subdials for the time and power reserve

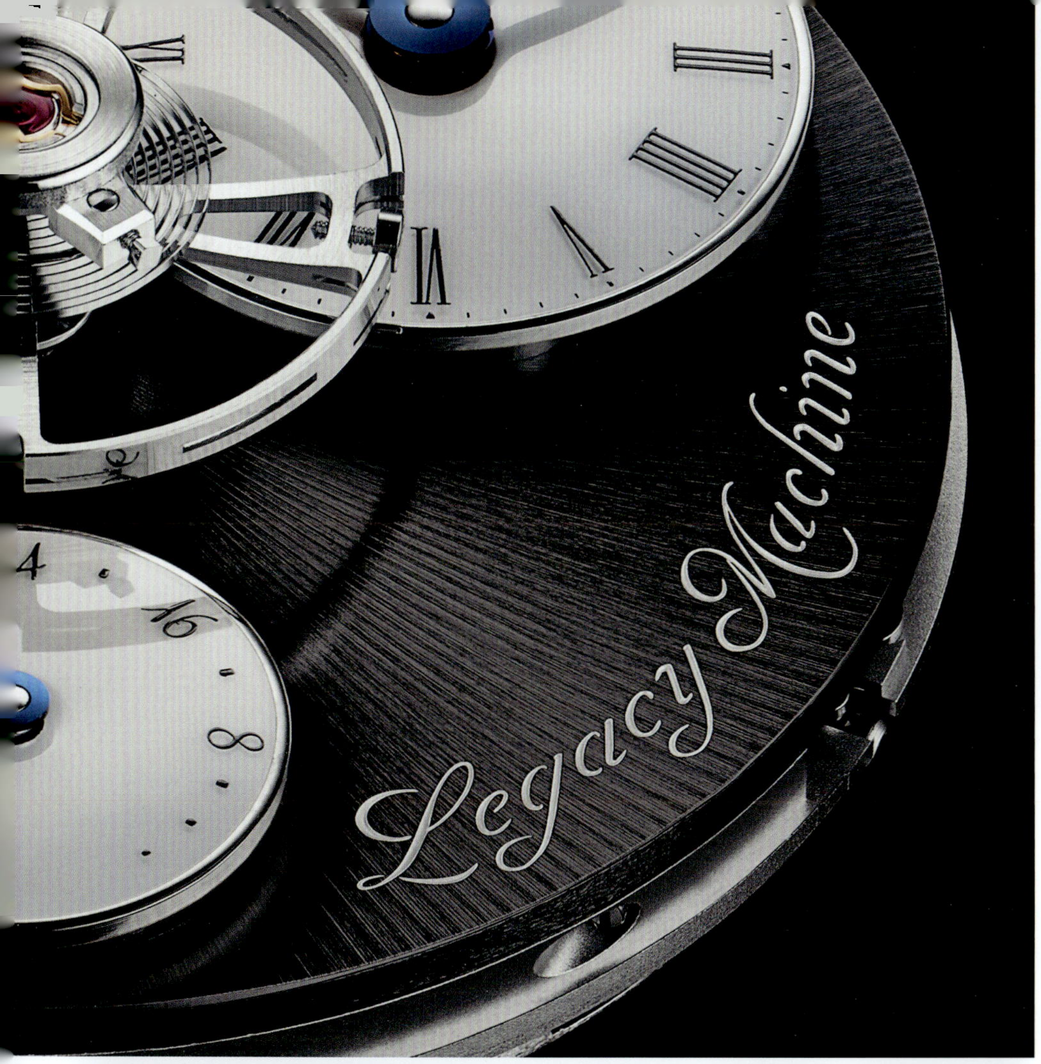
Legacy Machine

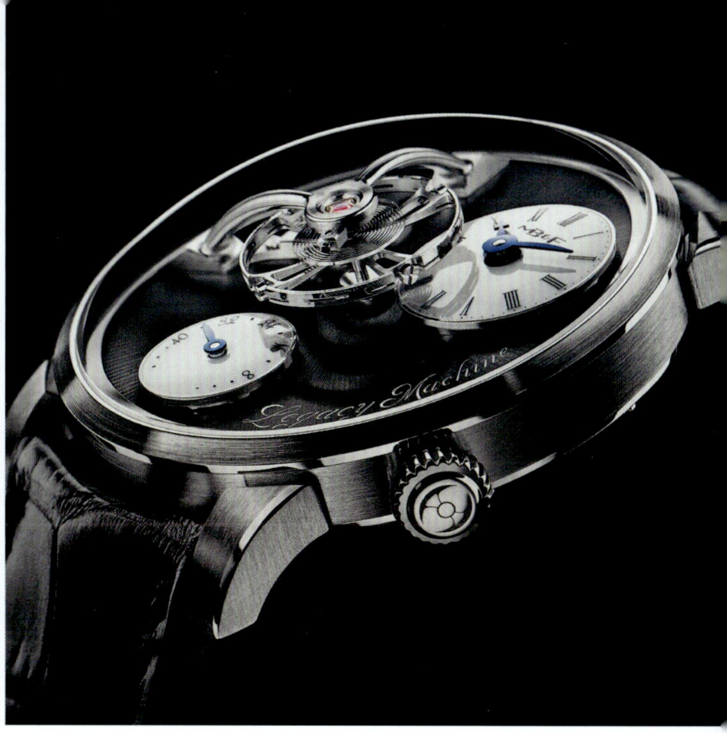
Legacy Machine

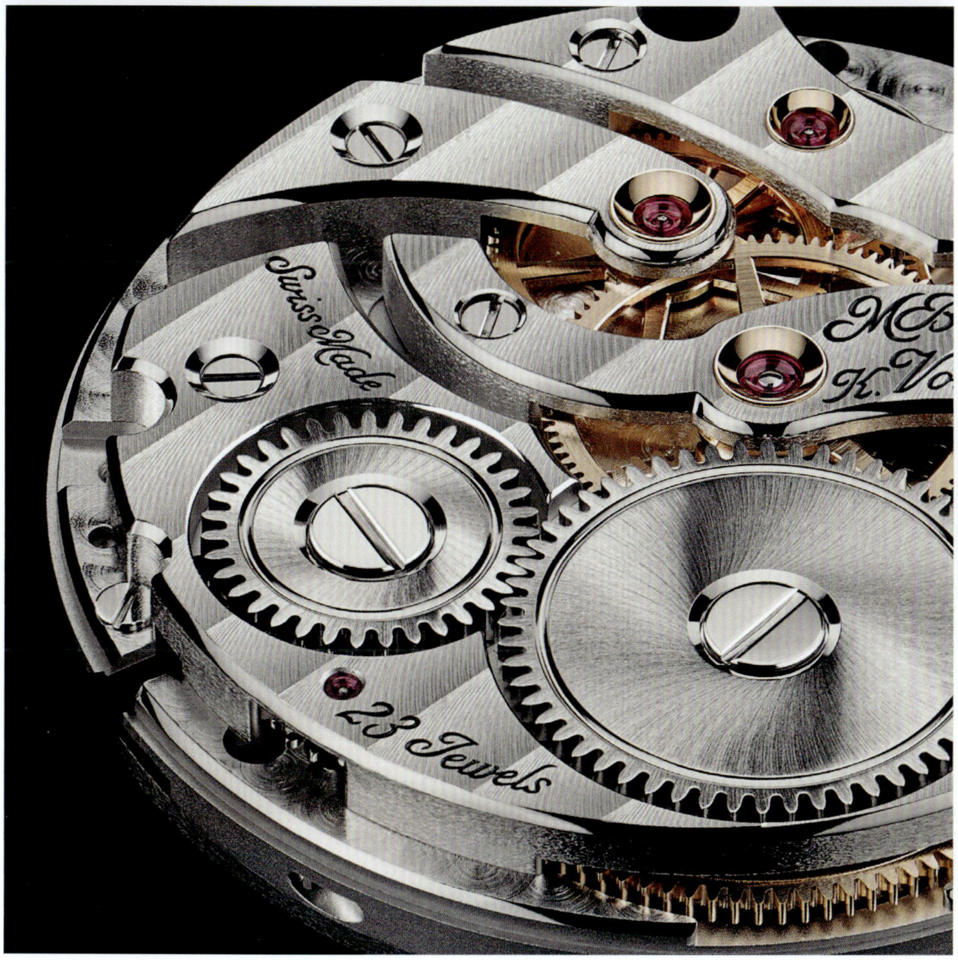
Swiss Made
23 Jewels

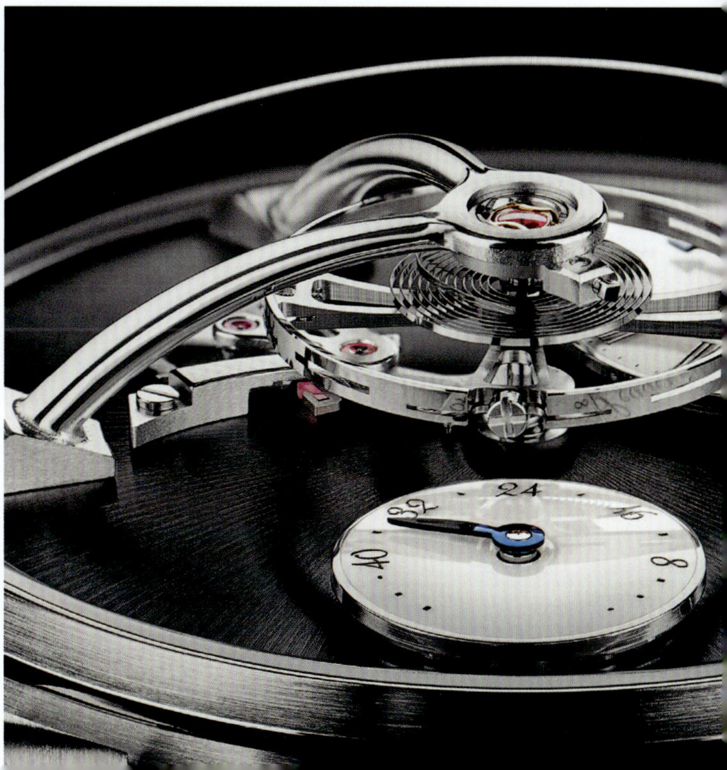

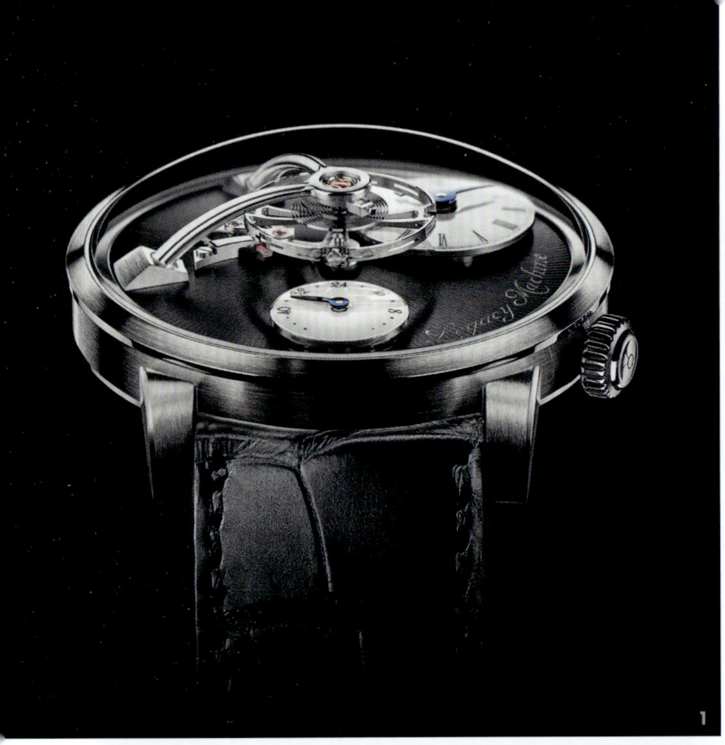

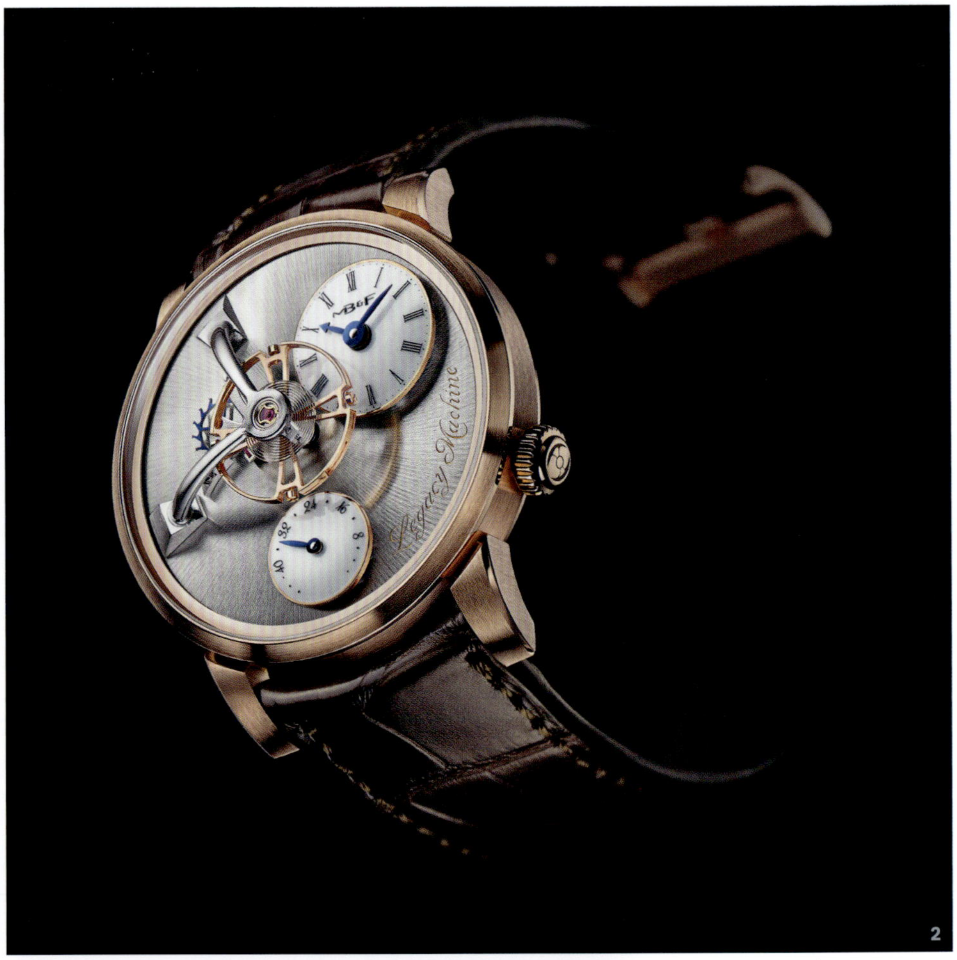

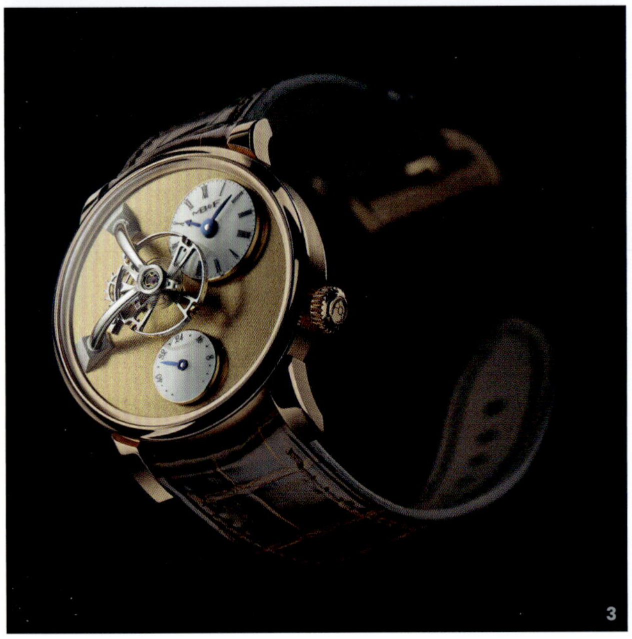

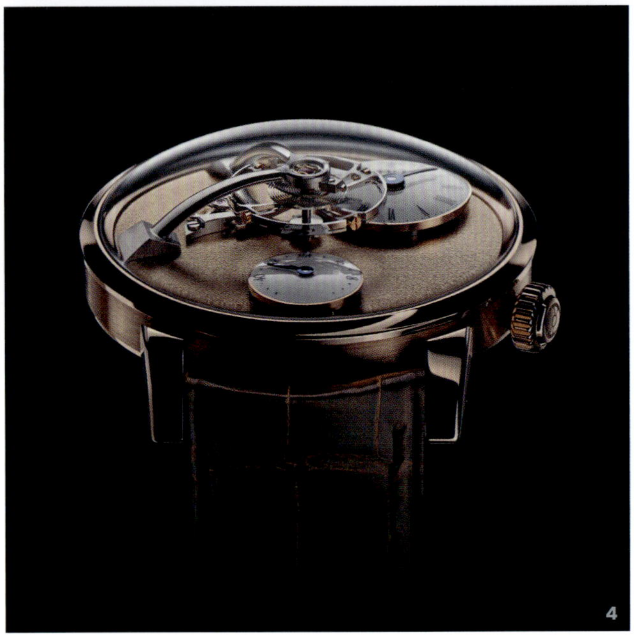

1 LM101 WG
Reference: 51.WL.W
Material: 18K white gold
Limitation: –
Introduction: 2014
Years in production:
2014 – in production
Quantity produced
(at the end of 2020):
46 + 1 prototype

2 LM101 RG
Reference: 51.RL.W
Material: 18K red gold
Limitation: –
Introduction: 2014
Years in production:
2014-2019
Quantity produced:
53 + 1 prototype

3 LM101 Frost YG
Reference: 51.YL.FW
Material: 18K yellow gold
Limitation: 18
Introduction: 2015
Year in production: 2015
Quantity produced:
18 + 1 prototype

4 LM101 Frost RG
Reference: 51.RL.FW
Material: 18K red gold
Limitation: 33
Introduction: 2015
Years in production:
2015-2016
Quantity produced:
24 + 1 prototype

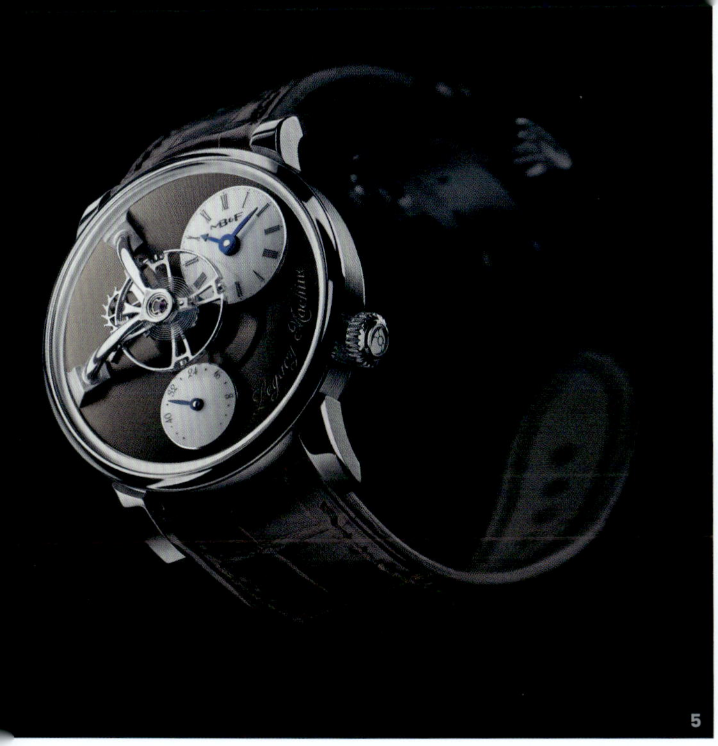

5

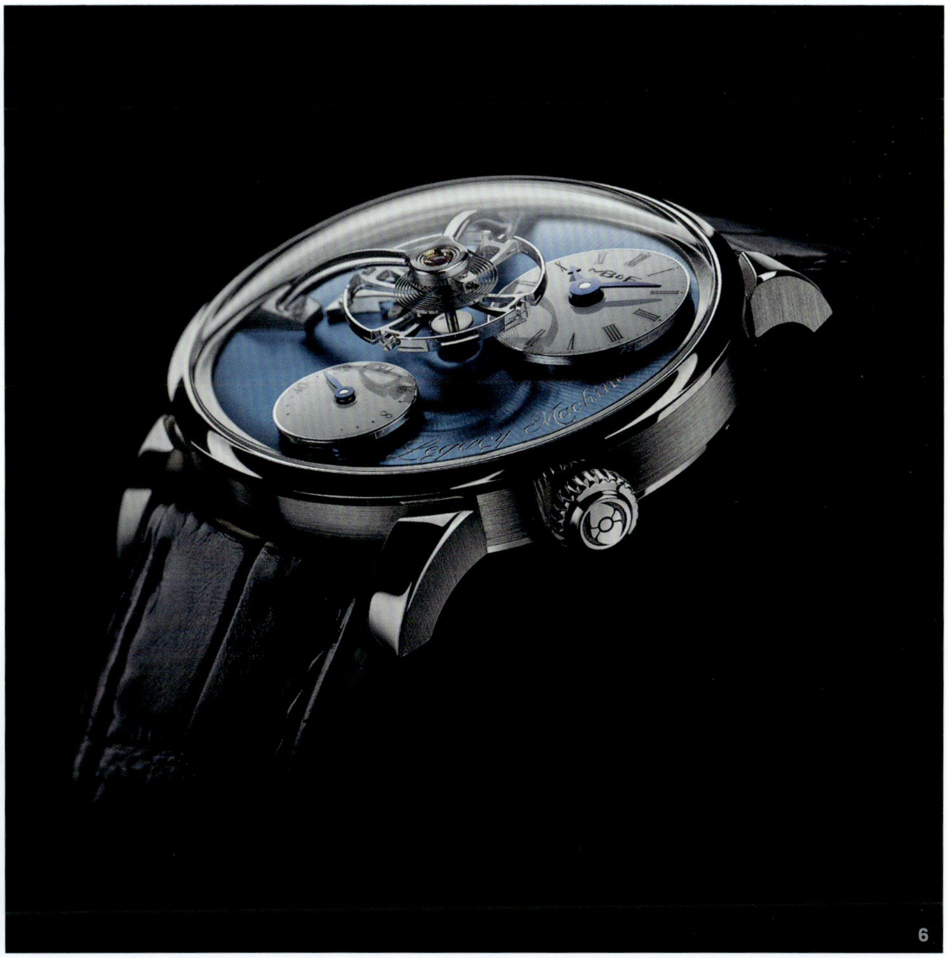

6

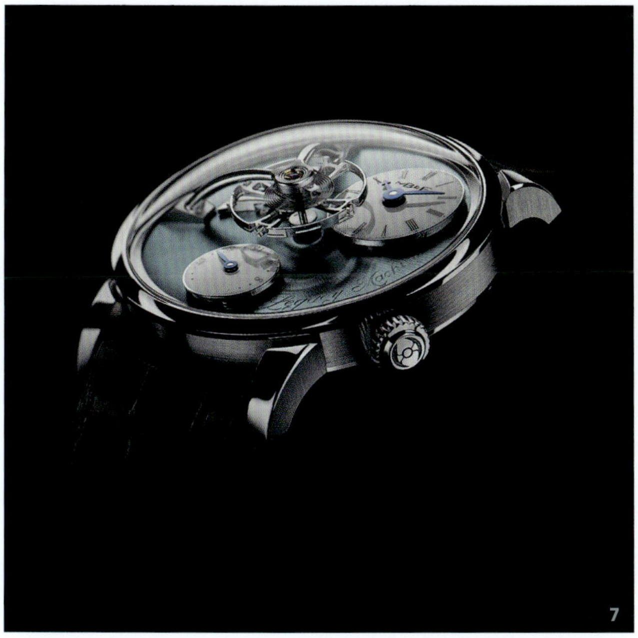

7

5 LM101 Steel Hodinkee
Reference: 51.SL.W
Material: stainless steel 316
Limitation: 10
Introduction: 2015
Year in production: 2015
Quantity produced:
10 + 1 prototype

6 LM101 PT
Reference: 51.PL.W
Material: platinum 950
Limitation: 33
Introduction: 2016
Years in production:
2016–2017
Quantity produced:
33 + 1 prototype

7 LM101 Palladium
Reference: 51.PLD.B
Material: palladium 950
Limitation: 18
Introduction: 2019
Year in production: 2019
Quantity produced:
18 + 1 prototype

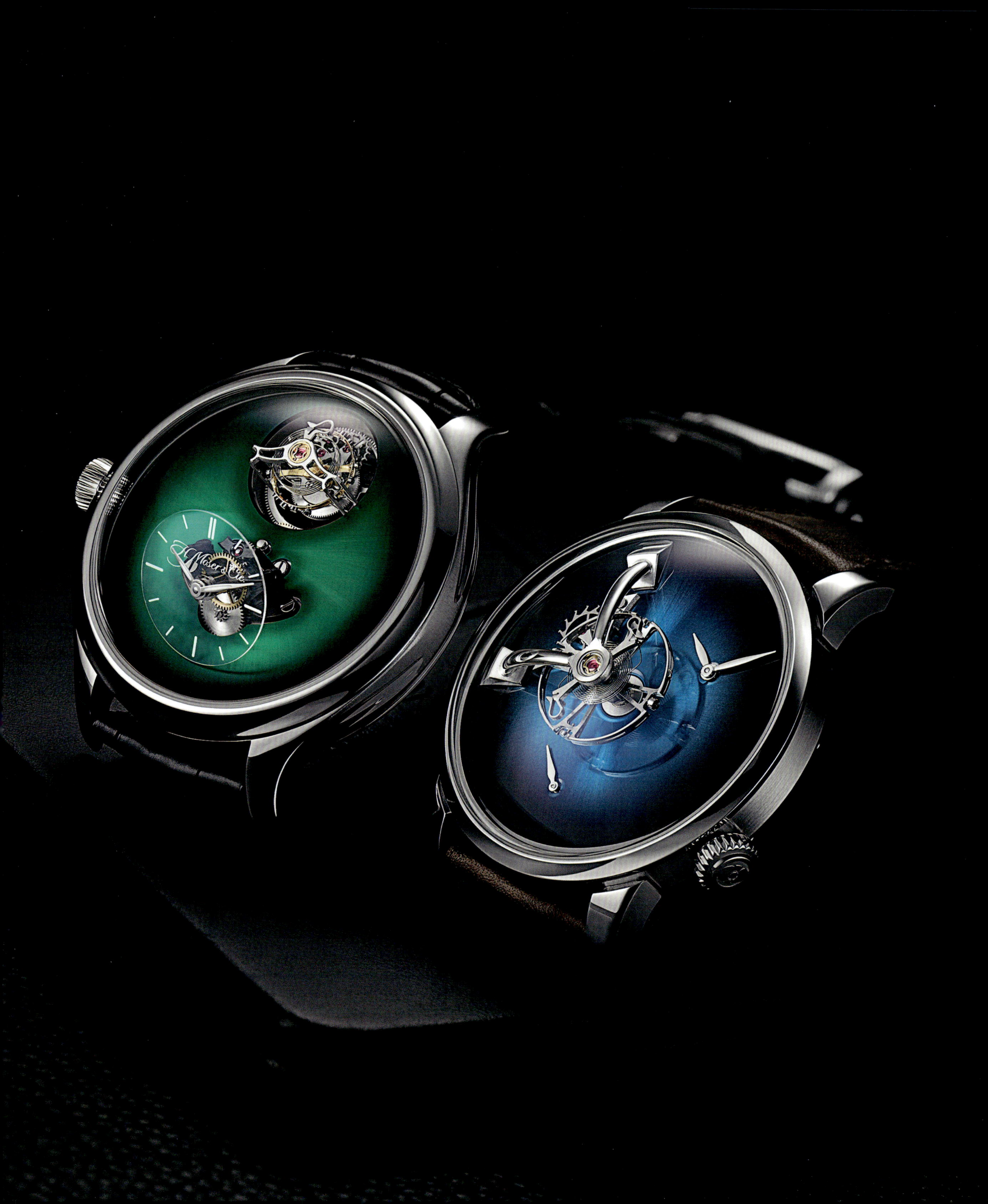
H. Moser & Cie.

LM101
MB&F× H. MOSER

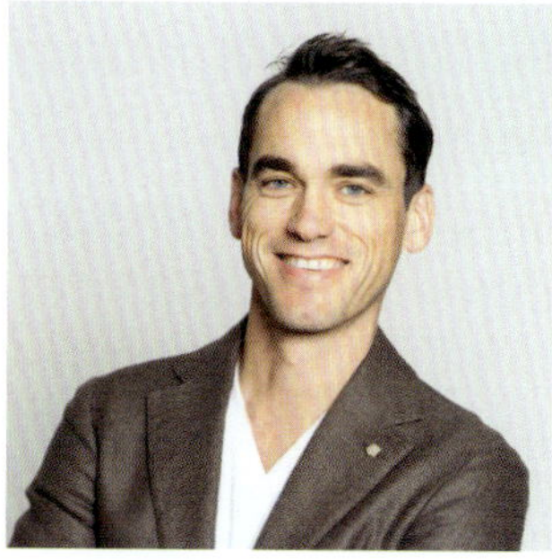

Édouard Meylan,
CEO of H. Moser & Cie

H. MOSER & CIE.

H. Moser & Cie. was founded by Swiss watchmaker Heinrich Moser in St. Petersburg, Russia in 1828. After his death, the company was sold and eventually forced out of watch production by the "quartz crisis." H. Moser & Cie. was officially relaunched in 2005, establishing a new watch factory in Schaffhausen where it quickly gained a reputation for its independent spirit and willingness to break the rules.

The LM101 MB&F × H. Moser represents a natural, perhaps even inevitable, collaboration between two, independent Swiss houses with a like-minded flair for the unexpected. All the more so considering that H. Moser & Cie.'s sister company, Precision Engineering AG, had been supplying MB&F's balance springs for more than ten years.

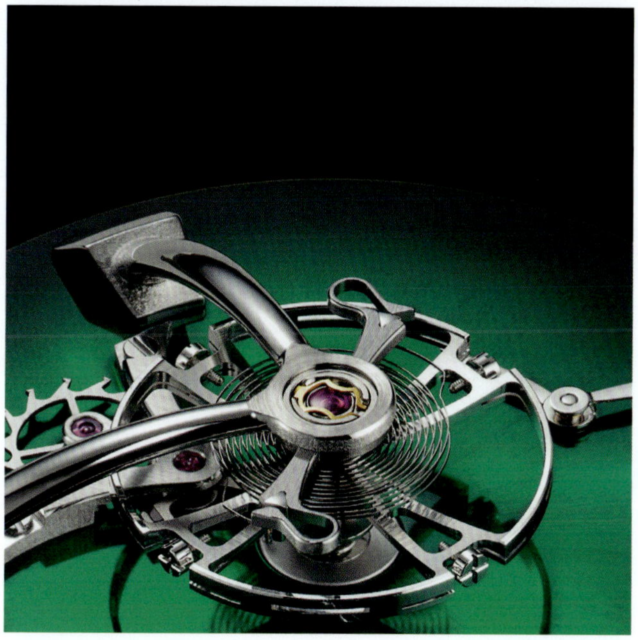

To celebrate their joint 15-year anniversaries, the two brands came up with the idea for a collaborative project. MB&F would style the eventual Endeavour Cylindrical Tourbillon H. Moser × MB&F, and H. Moser & Cie. would reinterpret the LM101.

The LM101 MB&F × H. Moser borrows two key elements from the Moser side: its vibrant, jewel-tone fumé dials with sunray-brushed finish; and a signature double balance-spring from Precision Engineering AG.

The signature fumé dial is probably the most recognizable Moser element on the new LM101, available here in four colours: Funky Blue, Cosmic Green, Red or an Aqua Blue edition. Hovering above the dial is the double balance-spring, which improves isochronism and precision.

Gone are the white lacquer subsidiary dials for time and power reserve. Gone too is the MB&F logo or any other markings. The result is a stripped-down look that lets the colour of the fumé dial really shine through – Moser's flair for colour working to reinforce the minimalist spirit of the LM101.

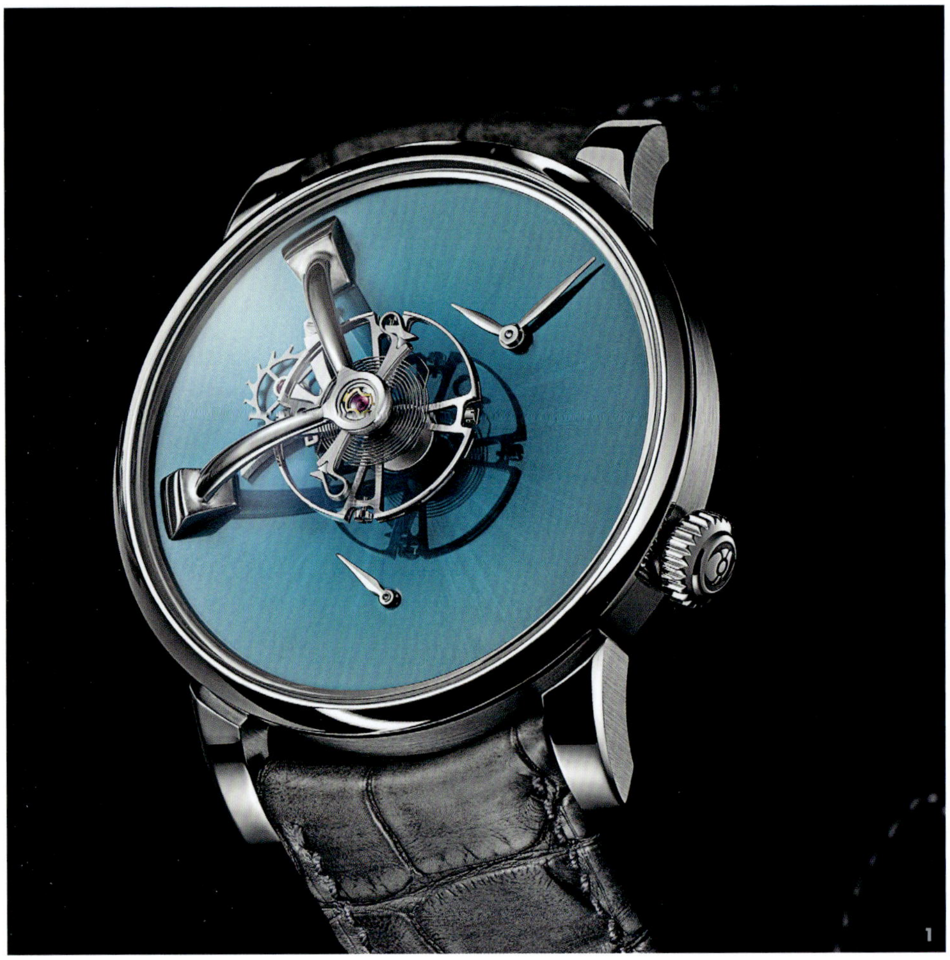

1

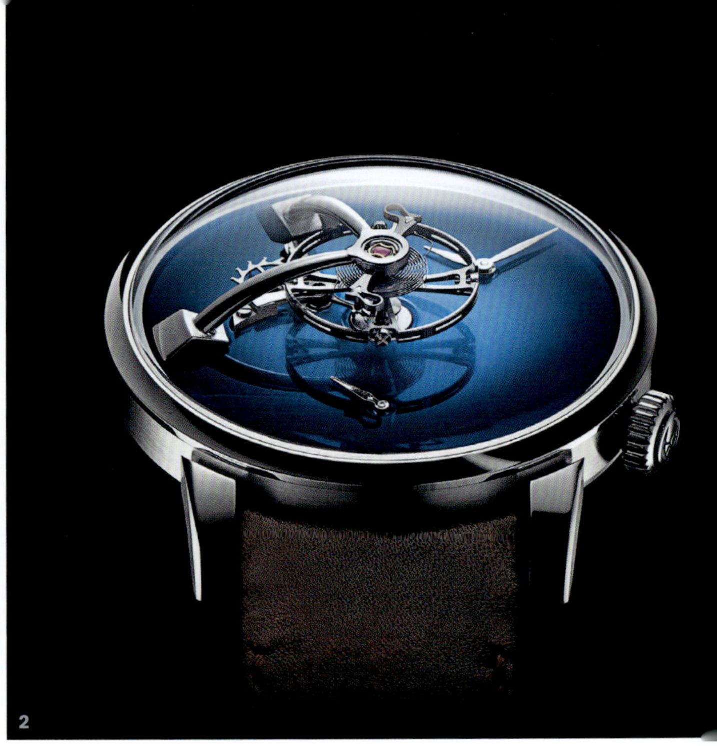

2

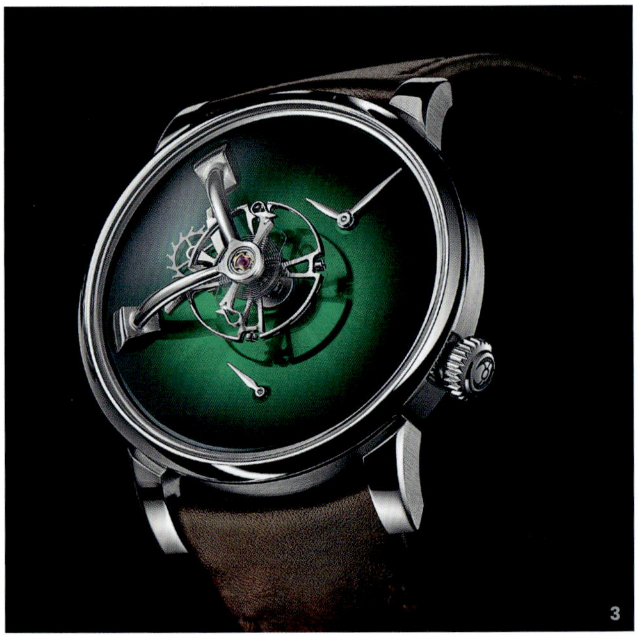

3

1 LM101 MB&F × H. Moser
Aqua Blue Seddiqi Edition
Reference: 51.SL.MS
Material: stainless steel
Limitation: 15
Introduction: 2020
Year in production: 2020
Quantity produced:
15 + 1 prototype

2 LM101 MB&F × H. Moser
Funky Blue
Reference: 51.SL.MB
Material: stainless steel
Limitation: 15
Introduction date: 2020
Year in production: 2020
Quantity produced:
15 + 1 prototype

3 LM101 MB&F × H. Moser
Cosmic Green
Reference: 51.SL.MG
Material: stainless steel
Limitation: 15
Introduction: 2020
Year in production: 2020
Quantity produced:
15 + 1 prototype

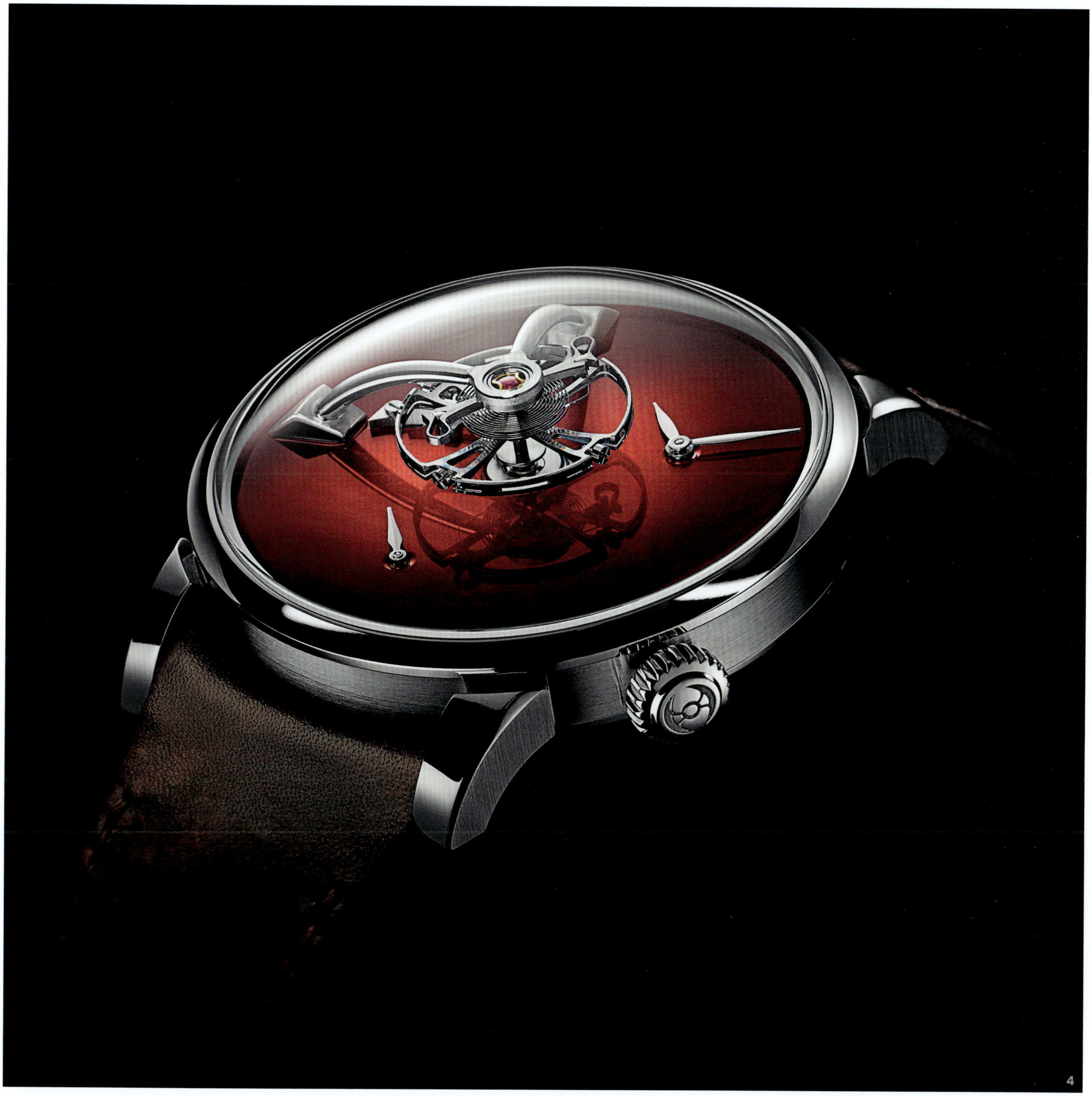

4 LM101 MB&F × H. Moser Red
Reference: 51.SL.MR
Material: stainless steel
Limitation: 15
Introduction: 2020
Year in production: 2020
Quantity produced:
15 + 1 prototype

LM PERPETUAL

The Legacy Machine Perpetual, released in 2015, would be one of the most technically ambitious machines in the Legacy Machine family, introducing MB&F's own perpetual calendar and featuring a visually stunning in-house movement. Conceived by Northern Irish watchmaker (and Oxford-educated theologian) Stephen McDonnell, MB&F would design a perpetual calendar that was more reliable and less prone to mechanical failure than traditional perpetual calendar complications.

The focus would rest entirely on this one highly technical complication – which is perhaps a first for a Legacy Machine. The flying balance wheel is joined by four registers for time, day of the week, month, and day of the month; two indicators with retrograde hands indicate the power reserve and leap year. This ensemble hovers over an intricate dial, skeletonised here to showcase the beauty of the movement's 581 highly complicated components. The result is a striking 3-dimensional effect rarely seen before in watchmaking.

LEGACY MACHINE PERPETUAL

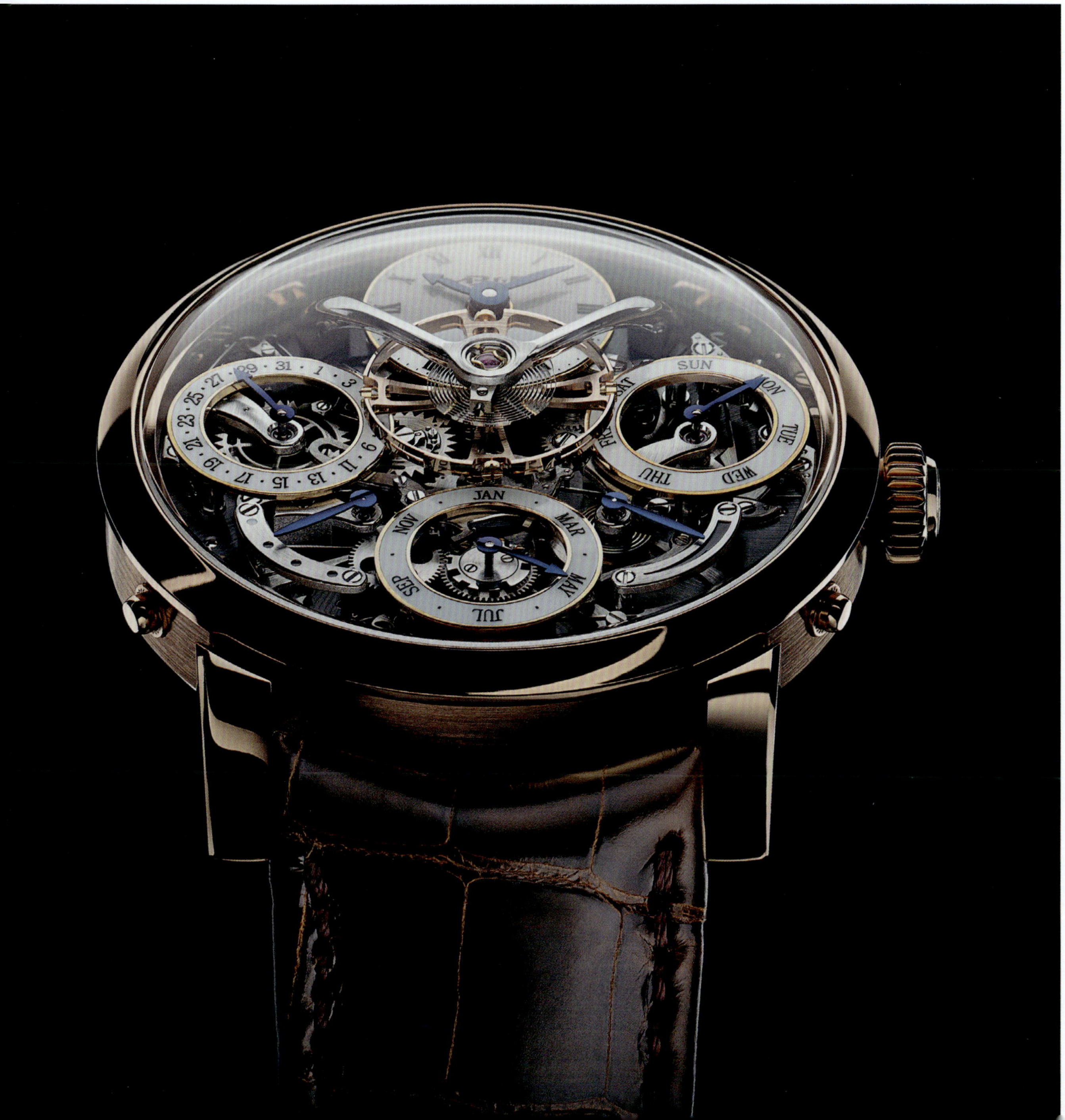

LM PERPETUAL CALIBRE

In a traditional perpetual calendar complication, the default month is set to 31 days. At the end of any month with fewer than 31 days, the complication skips the redundant days at midnight. The MB&F Legacy Machine Perpetual upends this convention by setting the default month to 28 days, and then adding extra days as required. This ensures a reliable reading of day and date, helped by a safety system that blocks the correctors from being used when the calendar is switching over, so preventing risk of damage to the mechanism.

Another feature of this novel design is that the sundials appear to float above the movement without any visible attachment. They in fact rest on non-visible posts – which is not a workable option in a traditional perpetual calendar mechanism that depends for its action on a system of levers, or "grand levier", that would have been blocked by the posts used here.

Instead, the McDonnell perpetual calendar mechanism employs a system of stacked gears, providing for the split escapement that is essential to create the "floating" effect of the subsidiary dials. The result is a striking and unusual design

Collection
Legacy Machines
Series
Legacy Machine Perpetual
Common name
LM Perpetual, LMP
Movement
Developed for MB&F by Stephen McDonnell
Winding
Manual
Power reserve
72 hours
Components
581/41 jewels
Functions
Hours, minutes, day, retrograde date, month, retrograde leap year and power reserve indicators
Add. features
Fully integrated perpetual calendar featuring dial-side complication and mechanical processor system architecture with inbuilt safety mechanism; bespoke 14mm balance wheel
Case
18K red, white gold, or yellow gold, platinum 950 or grade 5 titanium
Dimensions
44 × 17.5 mm
Case Components
69
Water resistance
30 m/3 ATM
Strap
Alligator
Buckle
Folding buckle, 18K red, white or yellow gold, platinum or titanium matching the case
Introduction
2015
Years in production
2015 – in production
Quantity produced
(at the end of 2020): 202 + 6 prototypes

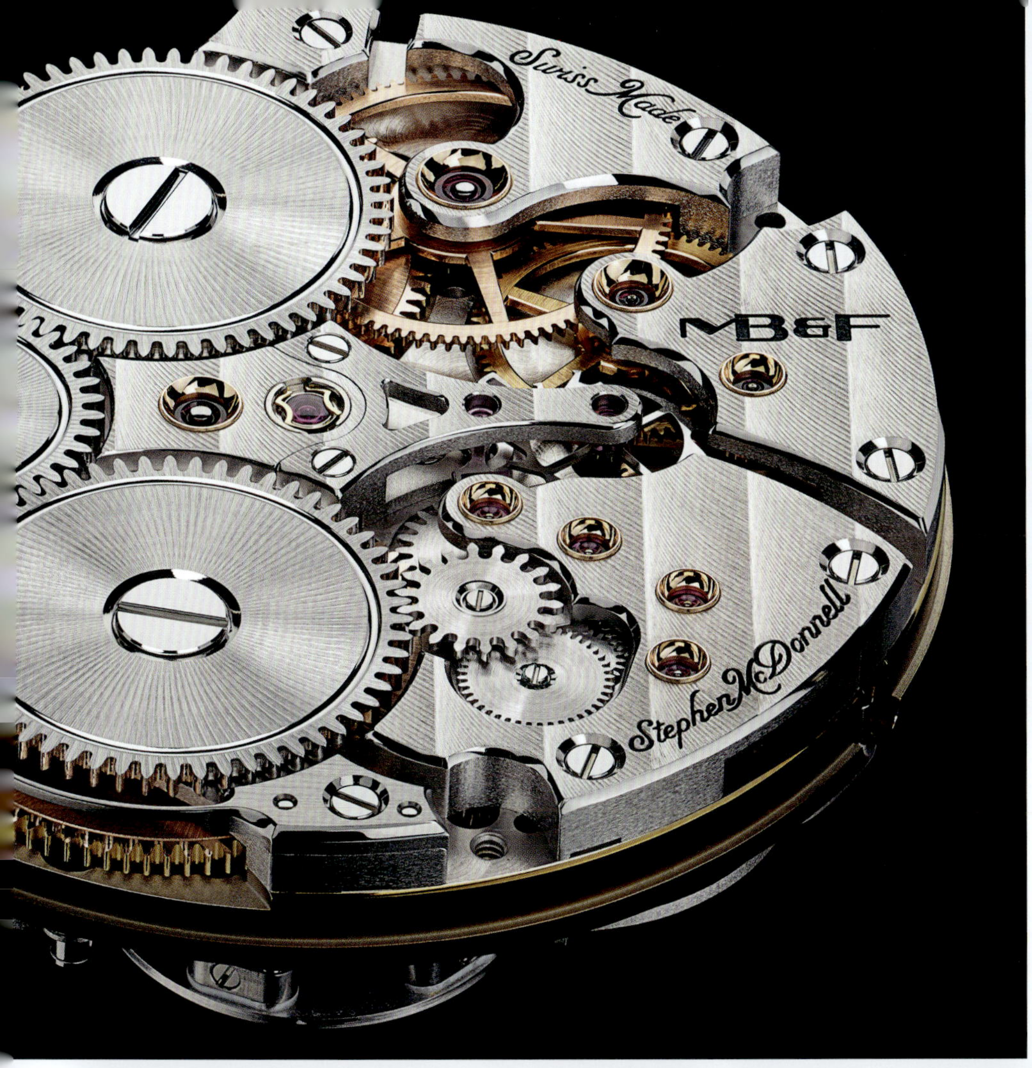
Swiss Made
MB&F
Stephen McDonnell

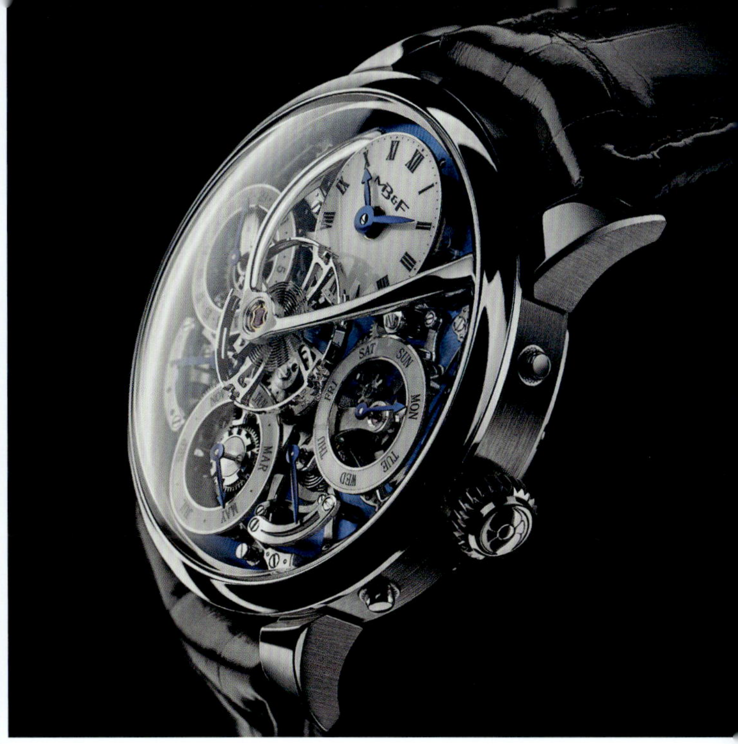

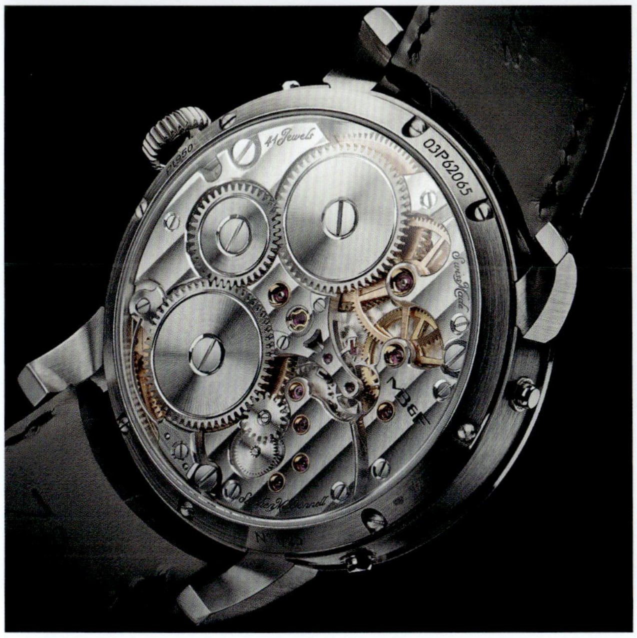
41 Jewels
03P62065

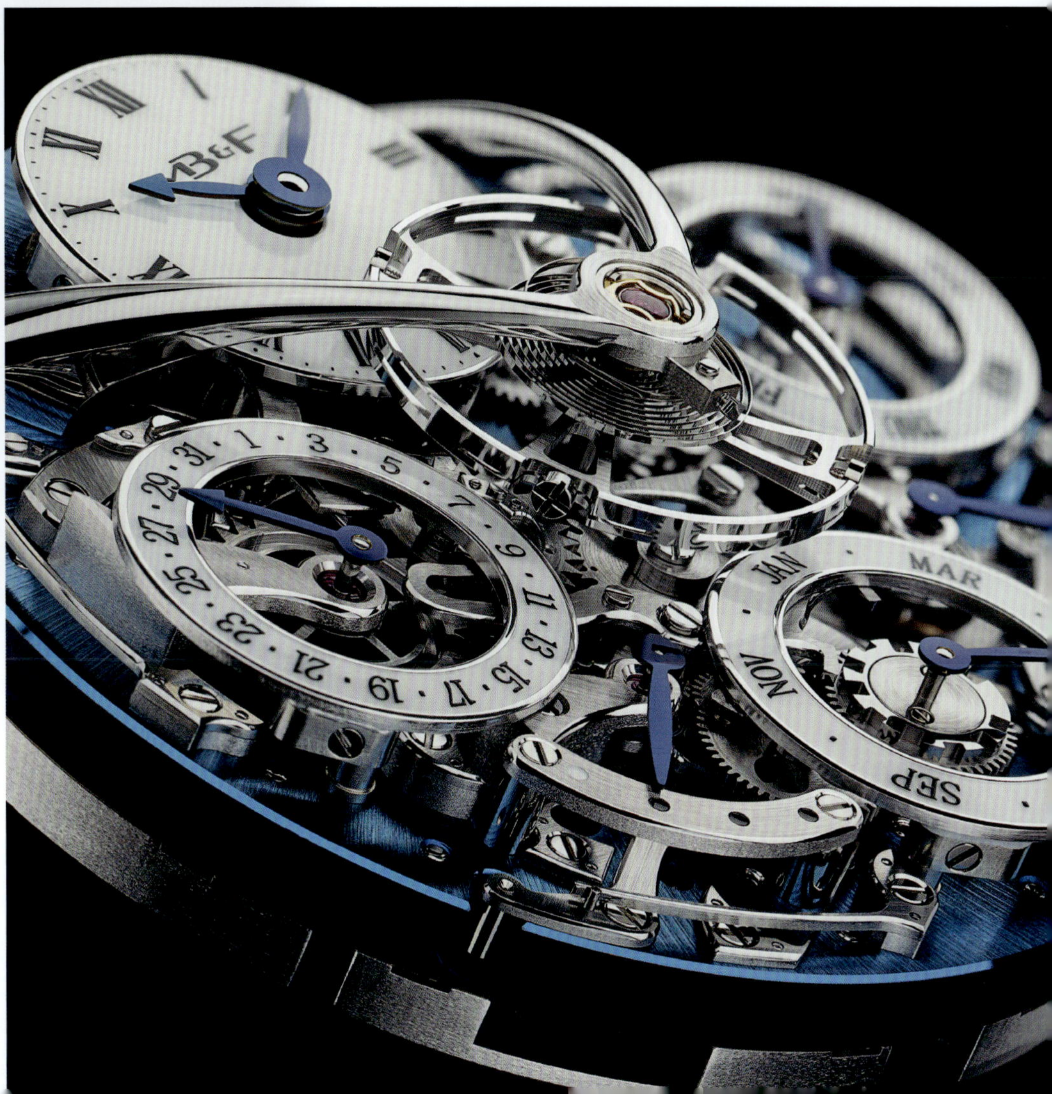
MB&F
JAN
MAR
NOV
SEP

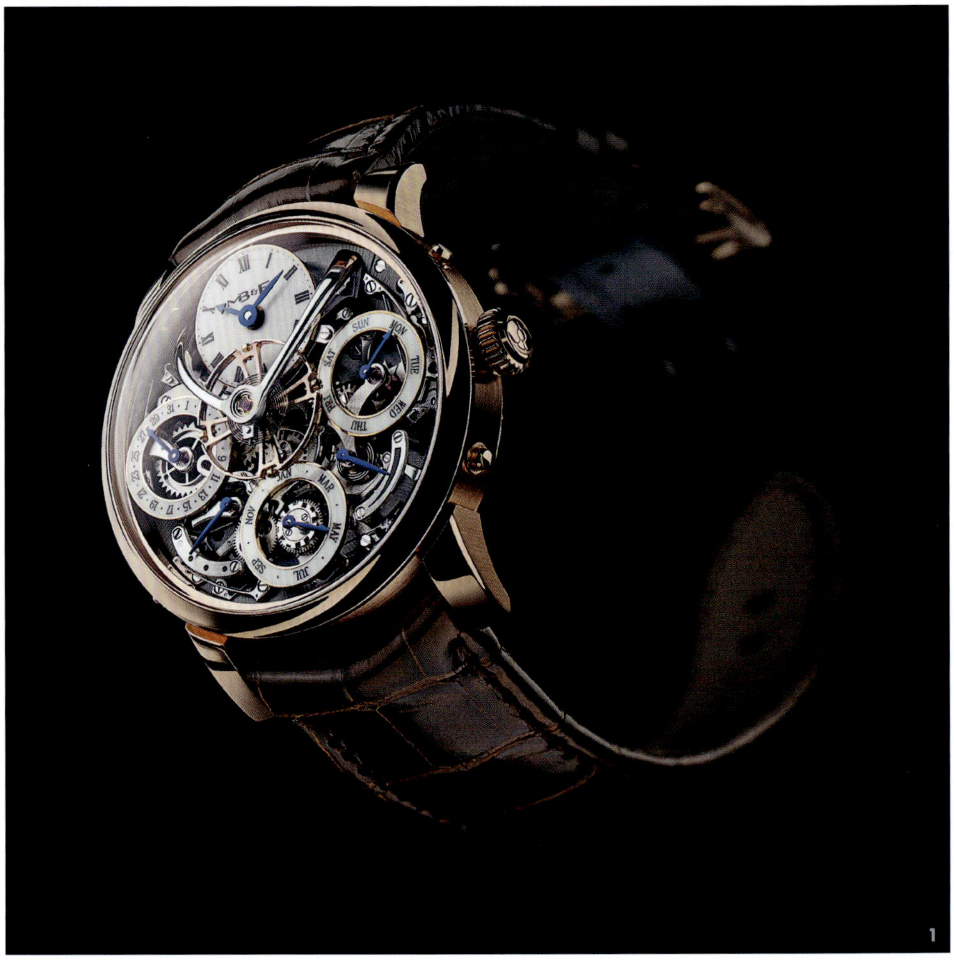

1 Legacy Machine
Perpetual RG
Reference: 03.RL.W
Material: 18K red gold
Limitation: 25
Introduction: 2015
Years in production:
2015–2016
Quantity produced:
25 + 1 prototype

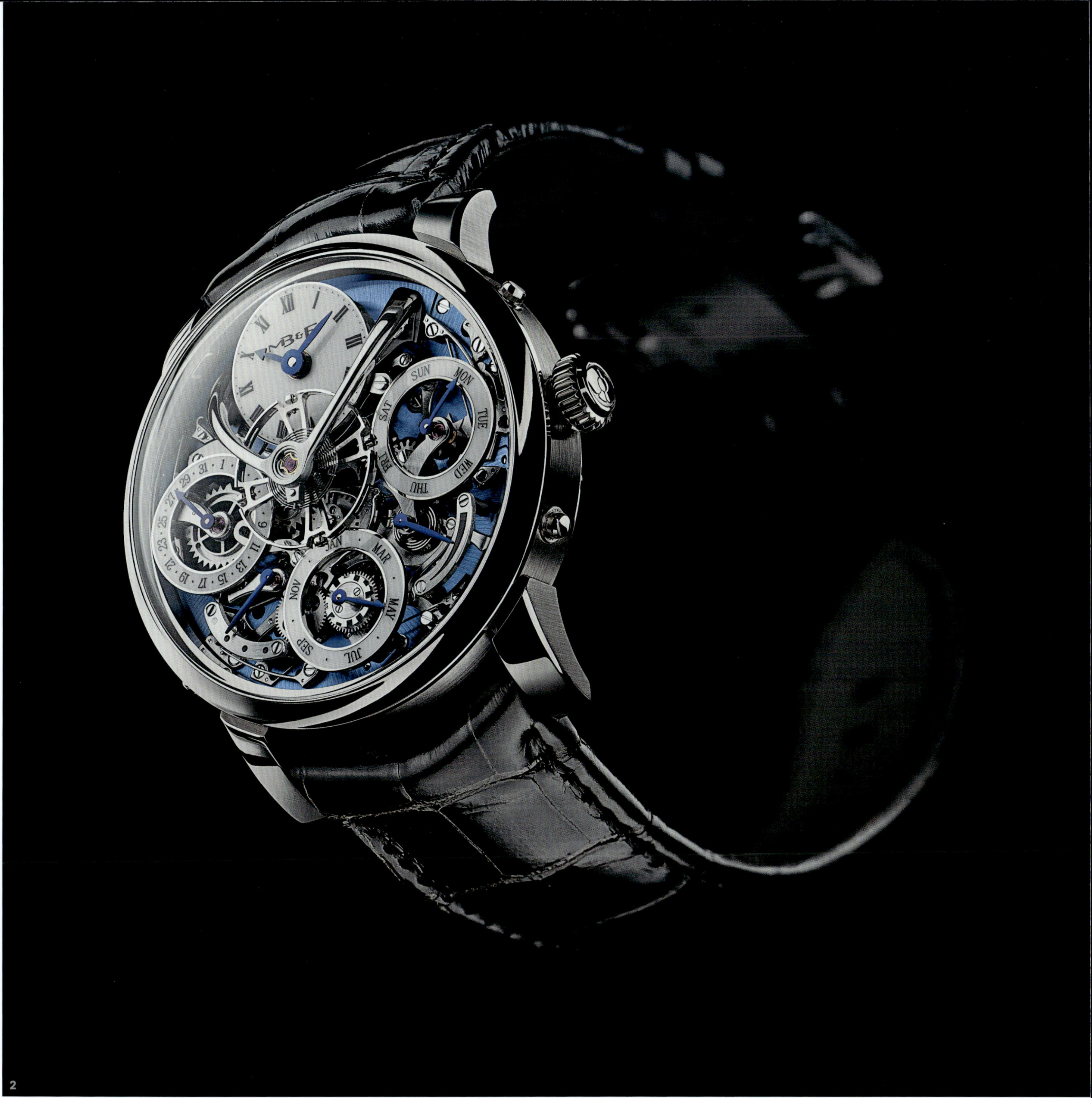

2

2 **Legacy Machine Perpetual PT**
Reference: 03.PL.W
Material: Platinum 950
Limitation: 25
Introduction: 2015
Years in production: 2015–2016
Quantity produced: 25 + 1 prototype

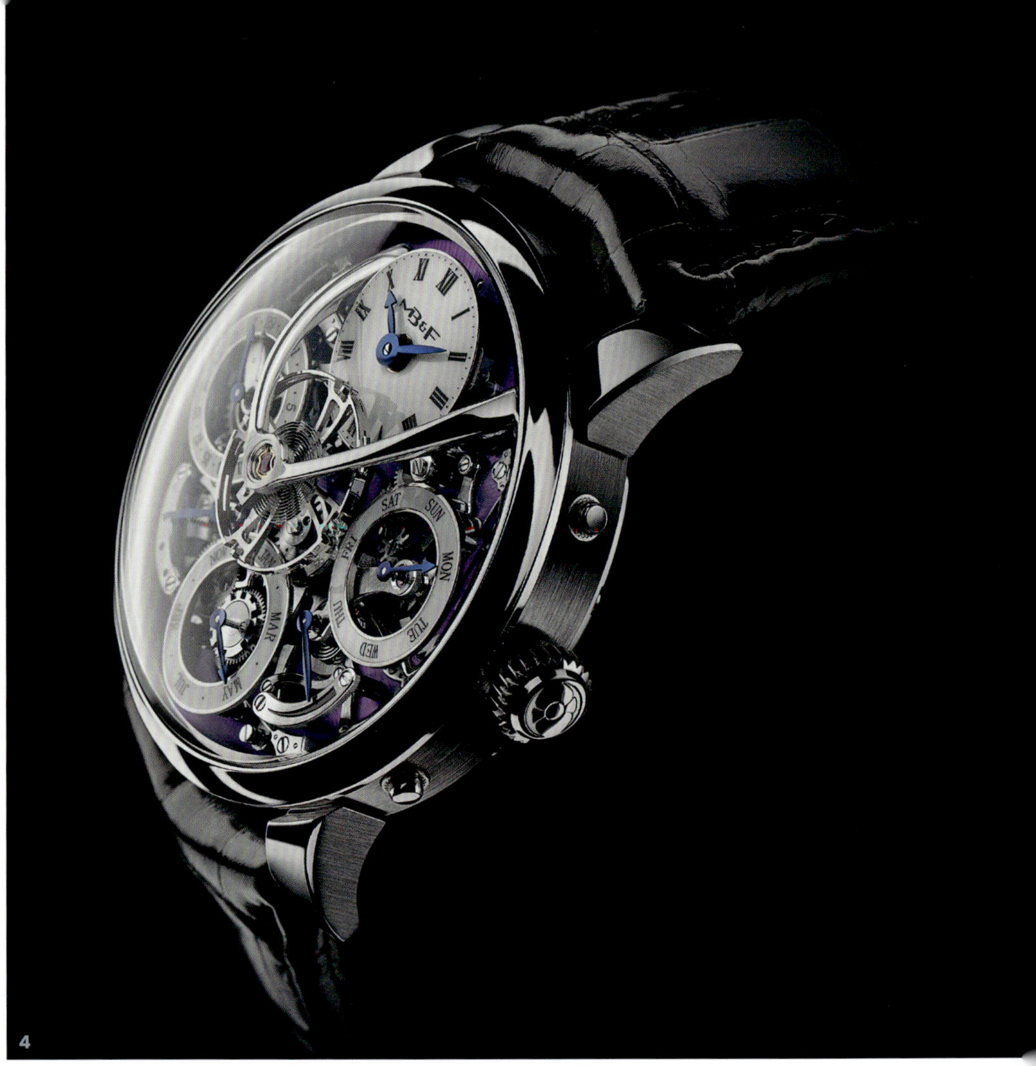

4

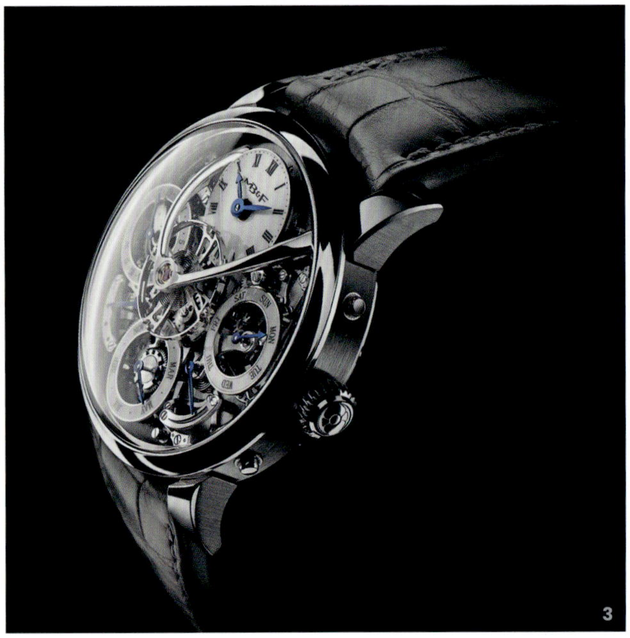

3

3 Legacy Machine Perpetual WG
Reference: 03.WL.B
Material: 18K white gold
Limitation: –
Introduction: 2016
Years in production: 2016 – in production
Quantity produced: 56

4 Legacy Machine Perpetual WG Purple
Reference: 03.WL.P
Material: 18K white gold
Limitation: 25
Introduction: 2016
Years in production: 2016–2017
Quantity produced: 25 + 1 prototype

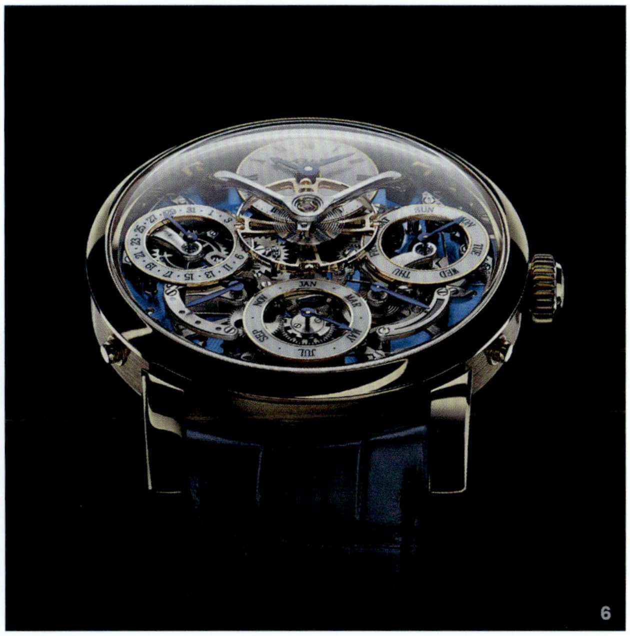

6

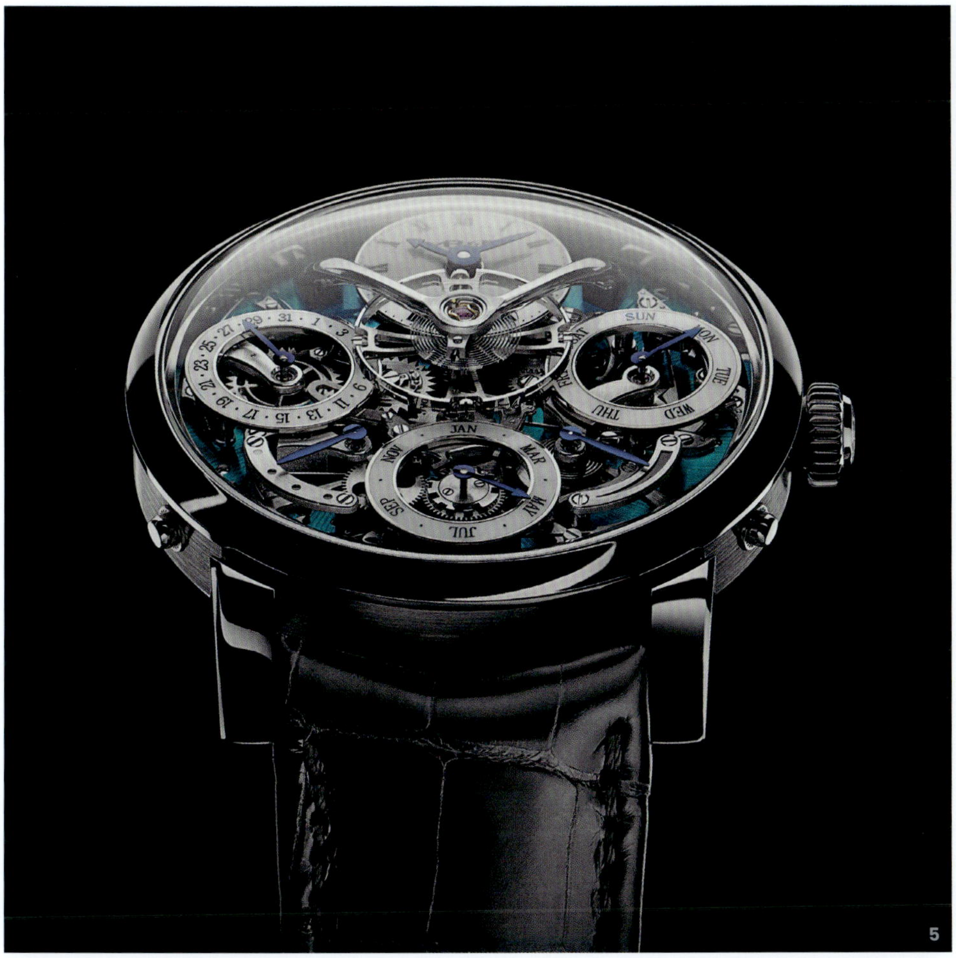

5

5 Legacy Machine
Perpetual Ti
Reference: 03.TL.G
Material: grade 5 titanium
Limitation: 50
Introduction: 2018
Years in production:
2018–2019
Quantity produced:
50 + 2 prototypes

6 Legacy Machine
Perpetual YG
Reference: 03.YL.BL
Material: 18K yellow gold
Limitation: 25
Introduction: 2020
Years in production:
2020 – in production
Quantity produced (at the end
of 2020): 21 + 1 prototype

LM
PERPETUAL EVO

The Legacy Machine Perpetual EVO was released in 2020 and, as the name implies, represents an evolution of 2015's Legacy Machine Perpetual. The perpetual calendar complication, a novel design by Northern Irish watchmaker Stephen McDonnell, was the first such complication to appear in the Legacy Machine series and is reprised here with the same pioneering spirit.
The very essence of the Perpetual EVO lies in an active lifestyle. The wearer is an explorer, a navigator, an individual with energy and direction. It is this same spirit of adventure that prompted the design changes from "Perpetual" to "Perpetual EVO".
The watchcase is made of zirconium, a rare and potentially explosive metal when in powder form. Lighter than steel and more durable than titanium, zirconium also happens to be hypoallergenic and anti-microbial, making it the perfect watch for an active lifestyle. Underscoring its athletic character is an integrated rubber strap (a first for MB&F) along with 80m of water resistance. Inside the Perpetual EVO, between case and movement, a novel stainless-steel dampener called the FlexRing helps protect the perpetual calendar from exterior shock. To improve dial legibility, the redesign also features a no-bezel construction, with the sapphire crystal fused directly to the zirconium case by means of a sophisticated thermal bonding system. Double-sprung oblong actuators meanwhile replace the previously circular pushers used to adjust the perpetual calendar, making adjustment easier while also improving the feel of the watch.

LEGACY MACHINE PERPETUAL EVO

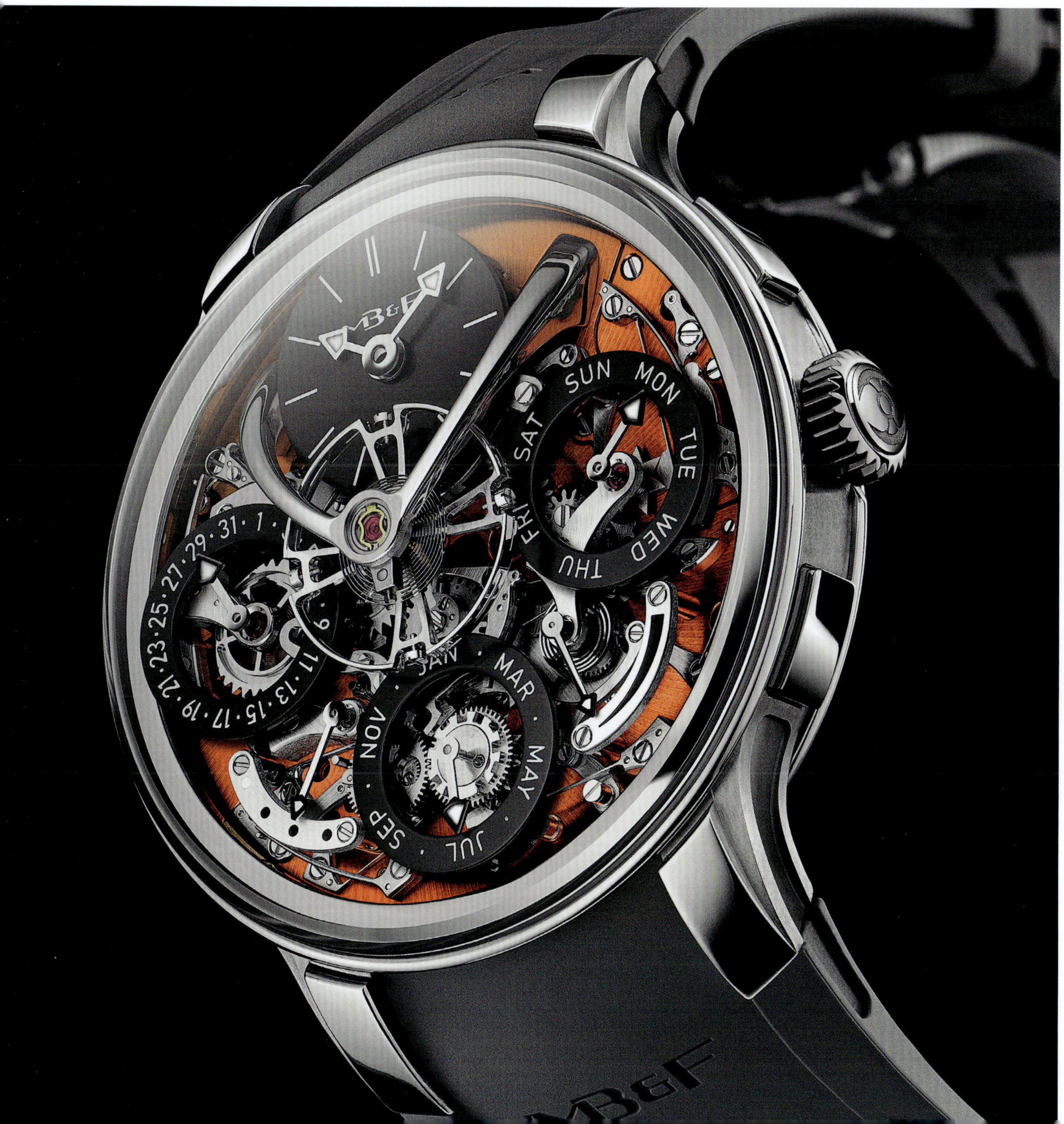

LM PERPETUAL EVO CALIBRE

First presented in 2015, the perpetual calendar complication in MB&F's Legacy Machine Perpetual was a revelation that upended the horological world's traditional understanding of that complication.
Instead of using a 31-day month as the default, Stephen McDonnell's perpetual calendar utilizes a 28-day month and adds extra days as required. It also features a new safety system that blocks the use of the correctors while the date is in process of switching over to prevent any damage to the movement.
With McDonnell's novel creation, the subsidiary dials rest on non-visible posts that make the dials appear to float above the movement without visible attachments. A traditional perpetual calendar mechanism depends for its action on a system of levers that would have been blocked by the posts used here. Instead, the McDonnell perpetual calendar mechanism employs a system of stacked gears, essential to create the "floating" effect of the subsidiary dials. The result is a striking and unusual design.

Collection
Legacy Machines
Series
Legacy Machine Perpetual EVO
Common name
LM Perpetual EVO, LMP EVO
Movement
Developed for MB&F by Stephen McDonnell
Winding
Manual
Power reserve
72 hours
Components
581/41 jewels

Functions
Hours, minutes, day, retrograde date, month, retrograde leap year and power reserve indicators
Add. features
Fully integrated perpetual calendar featuring dial-side complication and mechanical processor system architecture with inbuilt safety mechanism; bespoke 14 mm balance wheel

Case
Zirconium
Dimensions
44 × 17.5 mm
Case components
70
Water resistance
80m/8 ATM
Strap
Rubber, moulded in aircraft-grade Fluorocarbon FKM 85 Shores A elastomer
Buckle
Folding buckle in titanium

Introduction
2020
Years in production
2020 – 2021
Quantity produced
(at the end of 2021): 45 + 2 prototypes

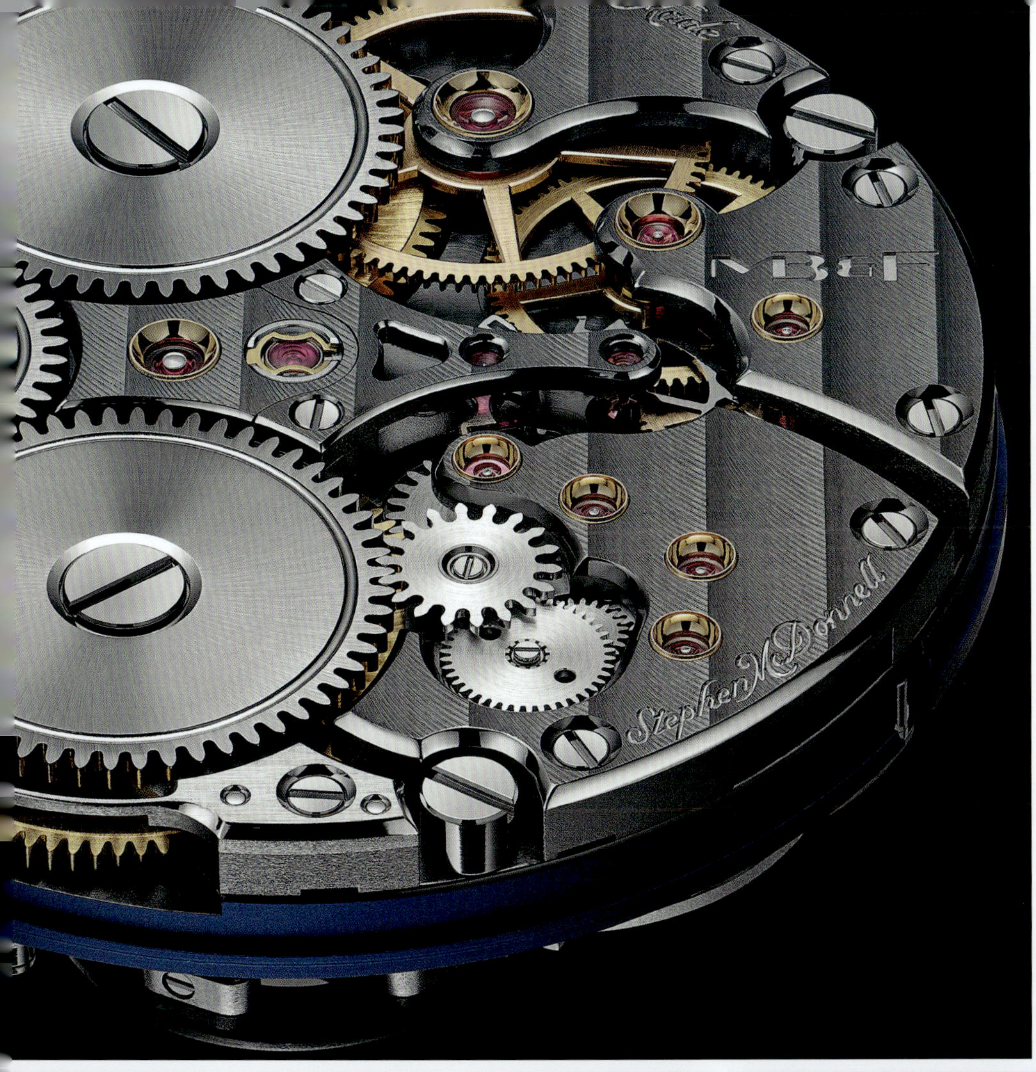
MB&F
StephenMcDonnell
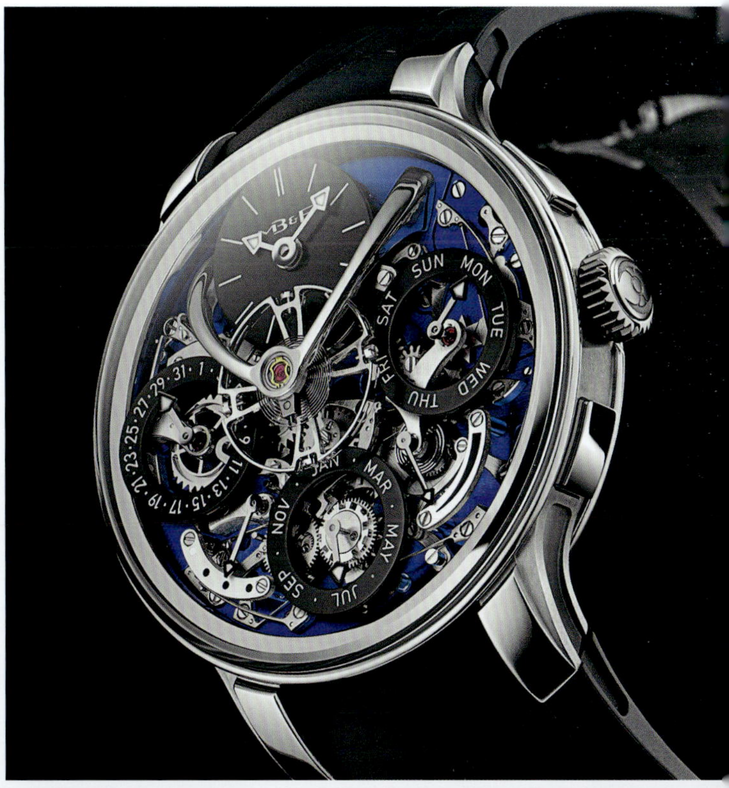
MB&F
SUN MON TUE WED THU FRI SAT
JAN MAR MAY JUL SEP NOV
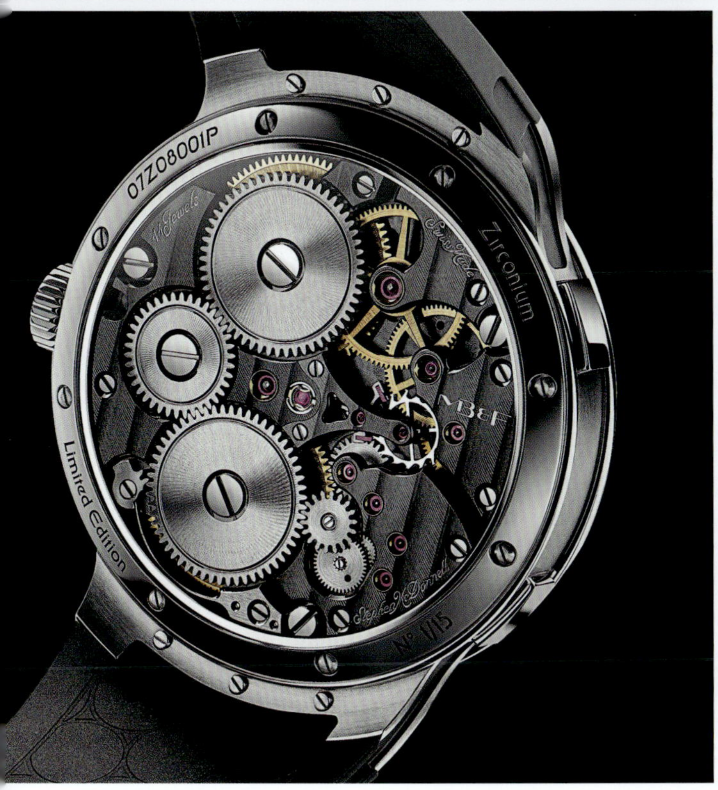
07Z0800IP
Zirconium
Limited Edition
MB&F
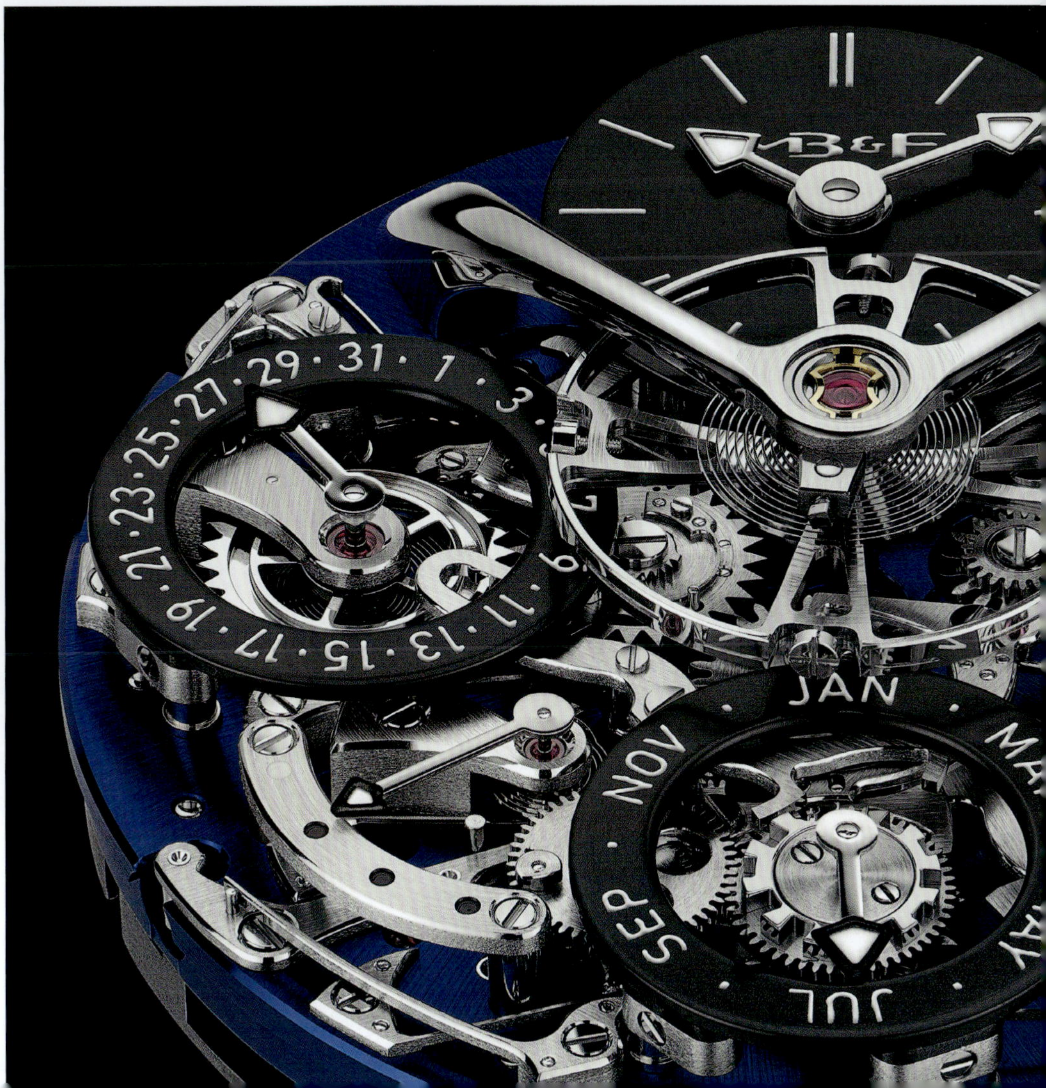
MB&F
25 · 27 · 29 · 31 · 1 · 3
21 · 23 ·
19 · 17 · 15 · 13 · 11
JAN · MAY · JUL · SEP · NOV

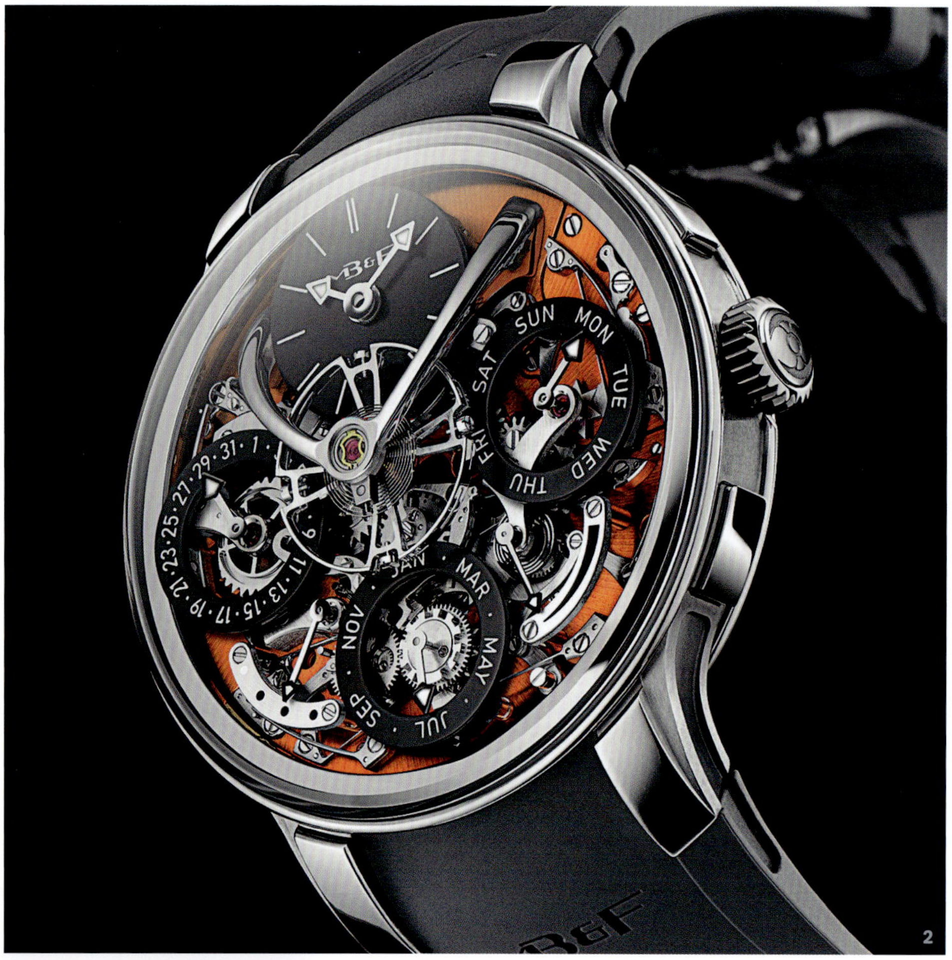

2

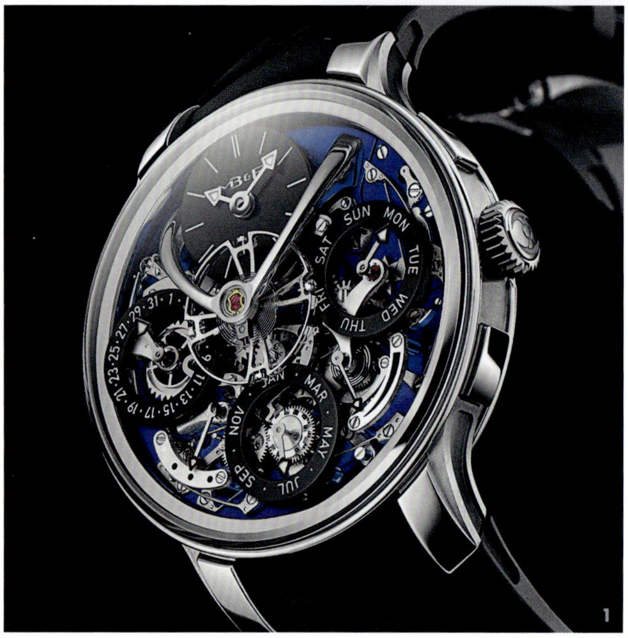

1

1 Legacy Machine Perpetual EVO Blue
Reference: 07.ZR.BU
Material: zirconium with blue CVD dial
Limitation: 15
Introduction: 2020
Years in production: 2020–2021
Quantity produced (at the end of 2021): 15 + 1 prototype

2 Legacy Machine Perpetual EVO Orange
Reference: 07.ZR.OR
Material: zirconium with orange CVD dial
Limitation: 15
Introduction: 2020
Years in production: 2020–2021
Quantity produced (at the end of 2021): 15 + 1 prototype

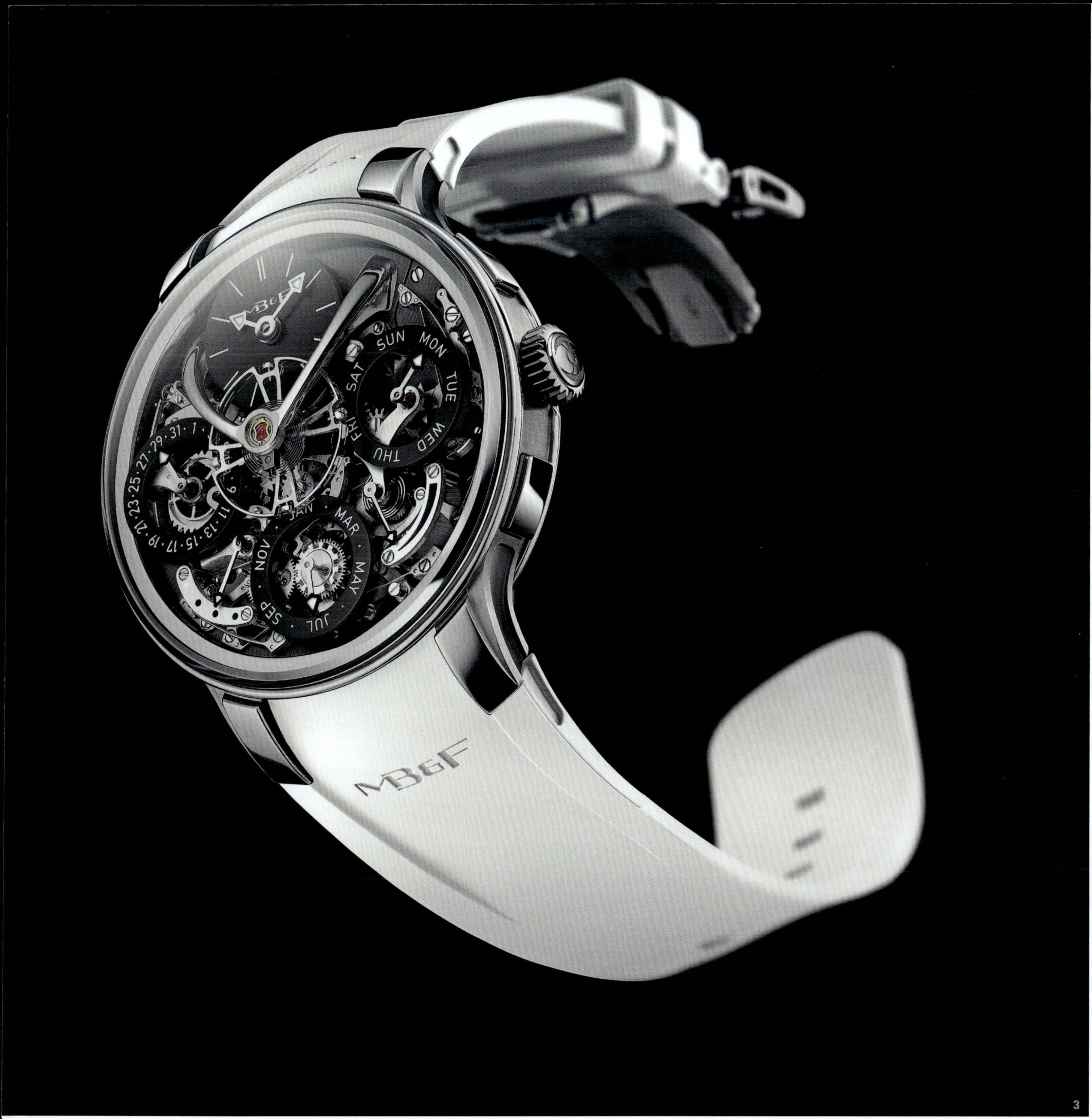

3 Legacy Machine Perpetual EVO Black
Reference: 07.ZR.BL
Material: zirconium with black PVD dial
Limitation: 15
Introduction: 2020
Years in production: 2020–2021
Quantity produced (at the end of 2021): 15

“When I proposed the idea of developing a completely new perpetual calendar mechanism, who else but MB&F would have been insane enough to allow me to attempt this ambitious challenge, and to give me their complete support? Because, at MB&F, the ethos is to seek out and embrace technical innovation – not to play safe. Only in this way can truly exceptional watches be created, and therein lies all my interest and motivation.”

Stephen McDonnell
Independent watchmaker

“Max Busser is the single most influential individual in independent watchmaking. In the initial chapter of his career he was the first to shine a spotlight on the independents with his Opus project. In the second he created one of the most creatively significant independent watchmaking brands of all time. MB&F has simultaneously unveiled the most thrilling avant-gardist acts of horological sculpturalism with his Horological Machines but also given us one of the most sublime visions for Neo-classical watchmaking with his Legacy Machines. On top of this Max has always remained the kind, honest and humble individual that we all know and love.”

Wei Koh
Founder, The Rake and Revolution magazines

LM SE

First launched in 2017, the Legacy Machine Split Escapement gives pride of place to the famed flying balance wheel of the Legacy Machine family. It hovers above the dial, hallmark of the Legacy Machine collection. Any sign of energy intake or drive is however hidden from view. These essential parts are concealed on the opposite side of the dial, almost 12mm below the suspended balance wheel. Hence the name, "Split Escapement."

This is not the first Legacy Machine with a split escapement — that honour goes to the Legacy Machine Perpetual. But this one strips away other design elements to present the split escapement in all its glory. Free of any visible support, the effect of the flying wheel is close to magical — a clean, legible dial, with an almost unobstructed view of the three subsidiary dials showing time, date, and power reserve.

LEGACY MACHINE SPLIT ESCAPEMENT

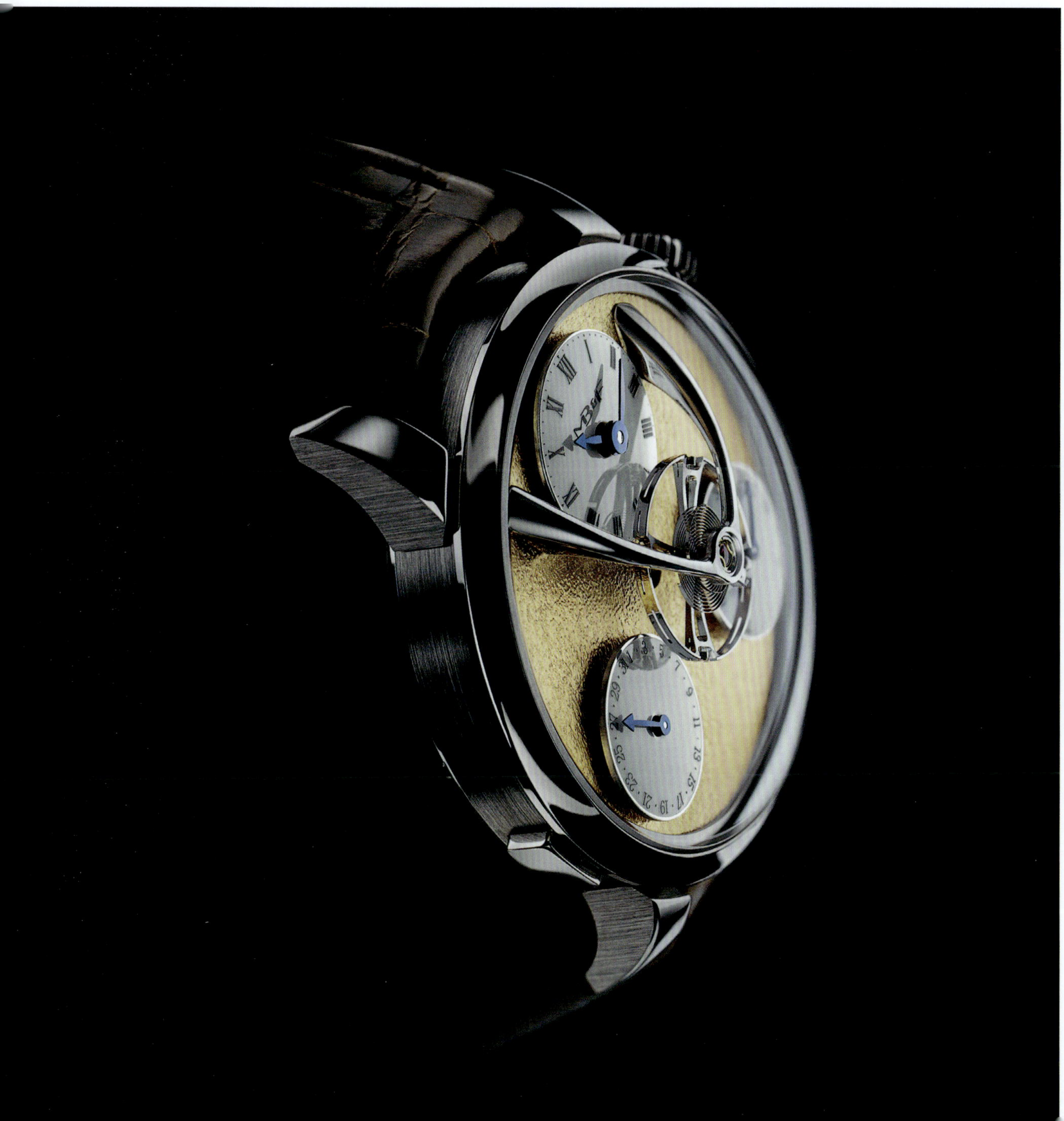

LM SPLIT ESCAPEMENT CALIBRE

The movement of the LMSE was developed for MB&F by Northern Irish watchmaker Stephen McDonnell and features an unusually long balance arbour of 11.78mm in length that posed unique challenges. In traditional watchmaking, the balance assembly, anchor and escape wheel (all parts of what is termed an escapement) are placed as close together as possible to limit disruptions from outside influences. The LM SE by comparison deconstructs, or "splits", these components one from another, which could leave the movement vulnerable. To compensate, the engine of the LM SE is cleverly designed to anticipate problems arising from the potential distortion of an elongated axle under continuous torsion, from the inertia of the balance wheel and from the rigidity of the long arbour. The increased mass of the elongated arbour is compensated by twin barrels that allow up to 72 hours of optimal timekeeping.

Collection
Legacy Machines
Series
Legacy Machine Split Escapement
Common name
LMSE
Movement
Developed for MB&F by Stephen McDonnell
Winding
Manual
Power reserve
72 hours
Components
314/35 jewels
Functions
Hours, minutes, date and power reserve indicator
Add. features
Split escapement, flying balance wheel, double mainspring barrels
Case
18K white gold for launch editions, grade 5 titanium or 18K red gold
Dimensions
44 × 17.5mm / 44.5 × 18.2mm for Eddy Jaquet editions
Case components
49 / 50 for Eddy Jaquet editions
Water resistance
30m/3 ATM
Strap
Alligator
Buckle
Folding buckle, 18K white or red gold, titanium matching the case
Introduction
2017
Years in production
2017 – in production
Quantity produced
(at the end of 2020) 102 + 4 prototypes
Comment
The LMSE "Eddy Jaquet" pieces have a thinner bezel

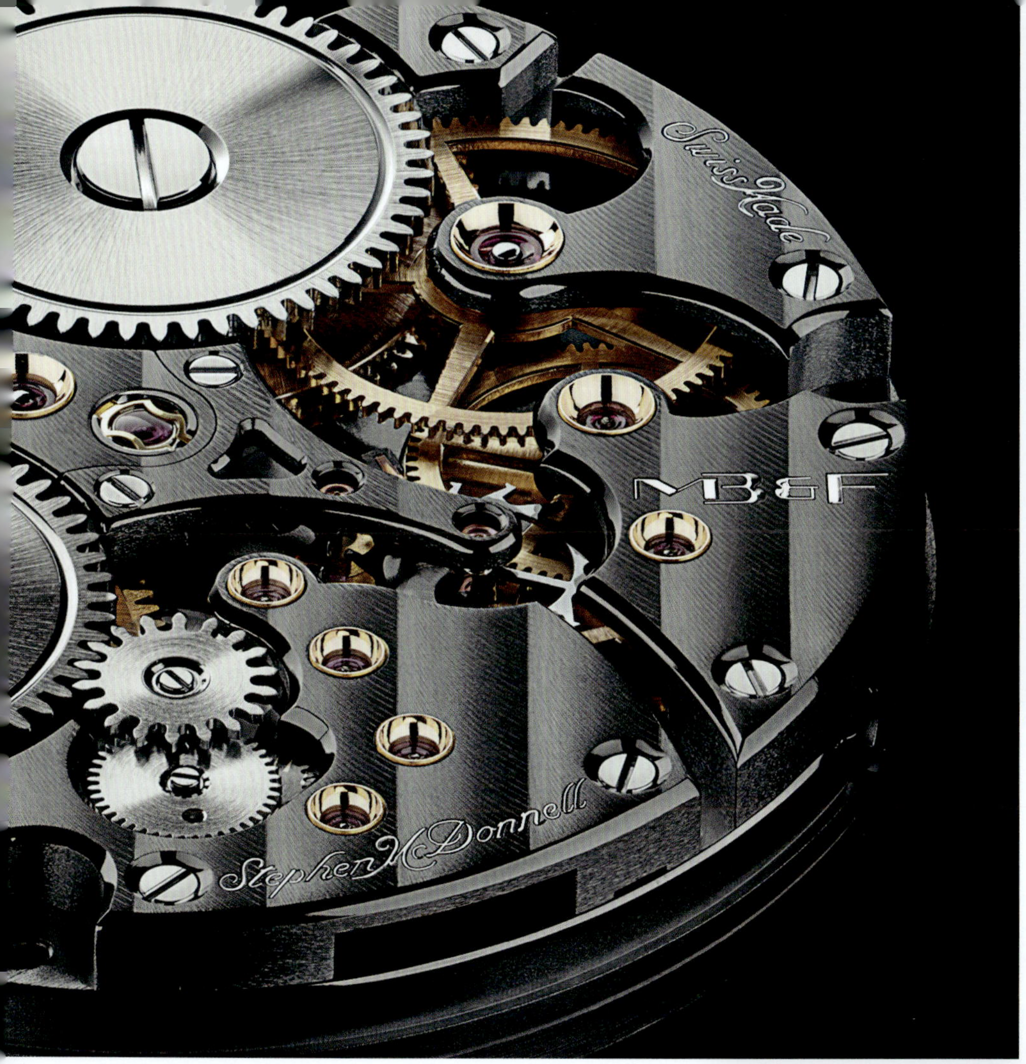
Swiss Made
MB&F
Stephen McDonnell

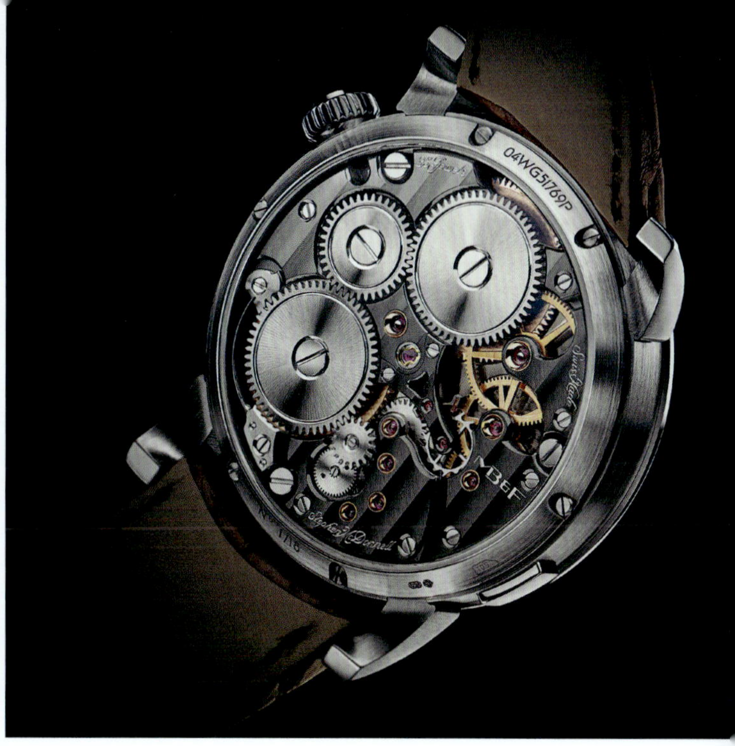
04WG51769P
MB&F

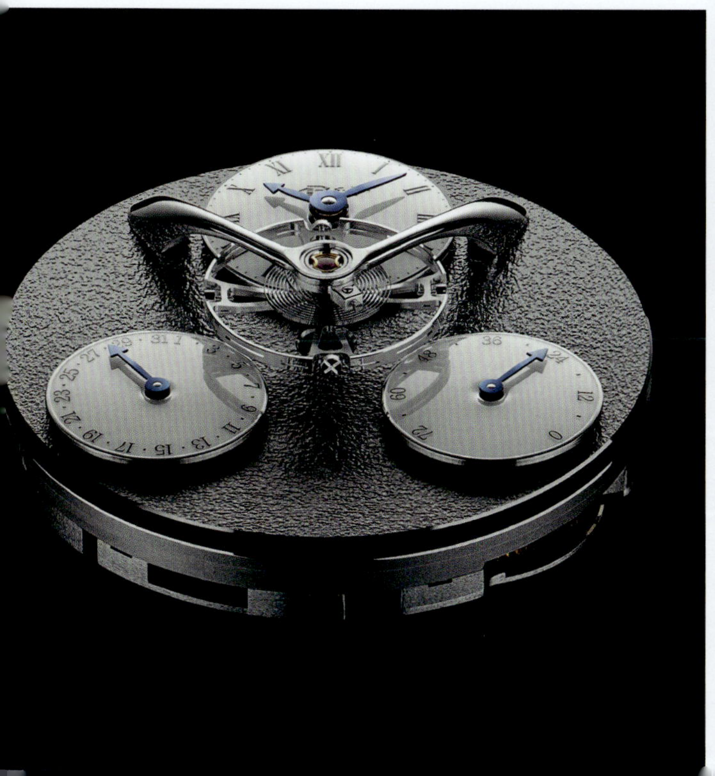

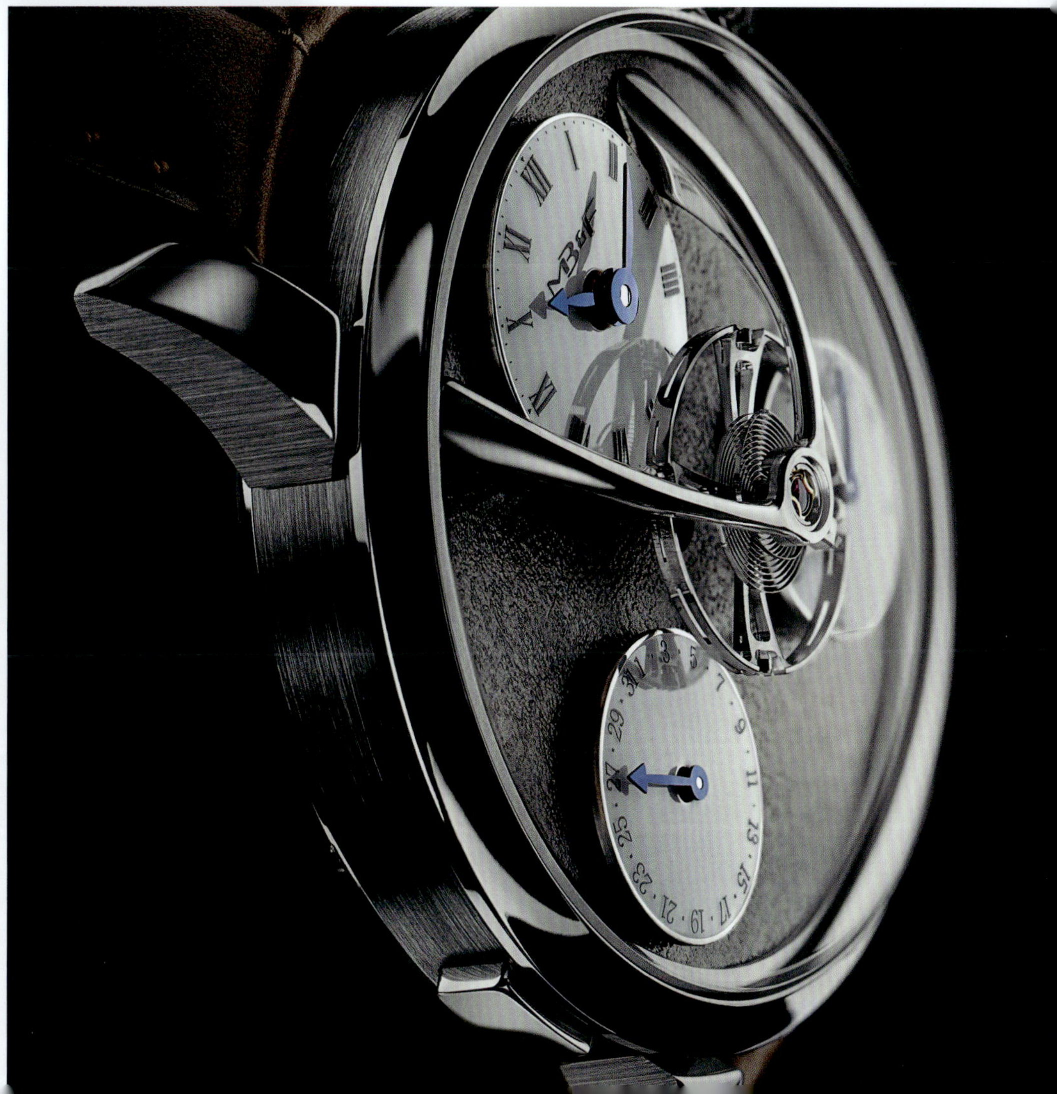
MB&F

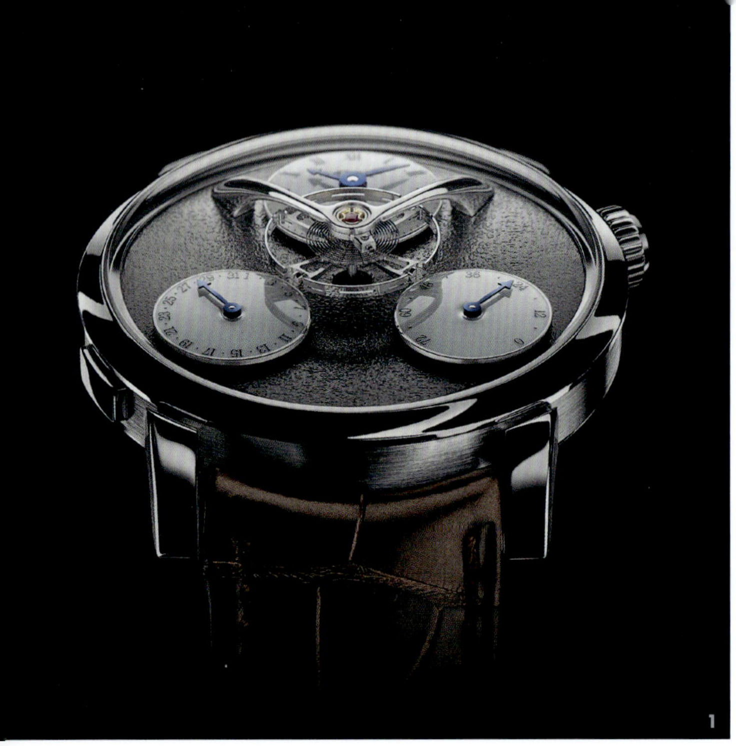

1

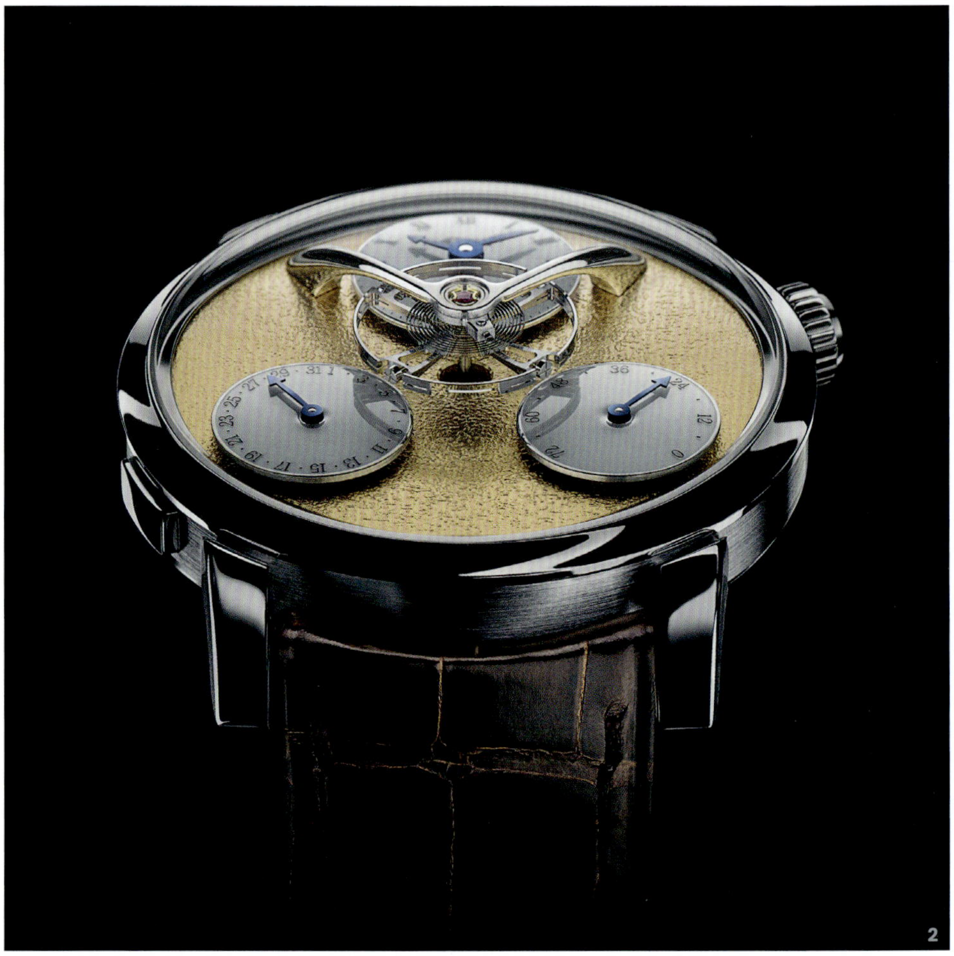

2

1 LMSE WG Grey
Reference: 03.WL.FB
Material: 18K white gold
with ruthenium frosted dial
Limitation: 18
Introduction: 2017
Years in production:
2017-2018
Quantity produced:
18 + 1 prototype

2 LMSE WG Yellow
Reference: 03.WL.FY
Material: 18K white gold
with yellow gold frosted dial
Limitation: 18
Introduction: 2017
Years in production:
2017 – in production
Quantity produced:
14 + 1 prototype

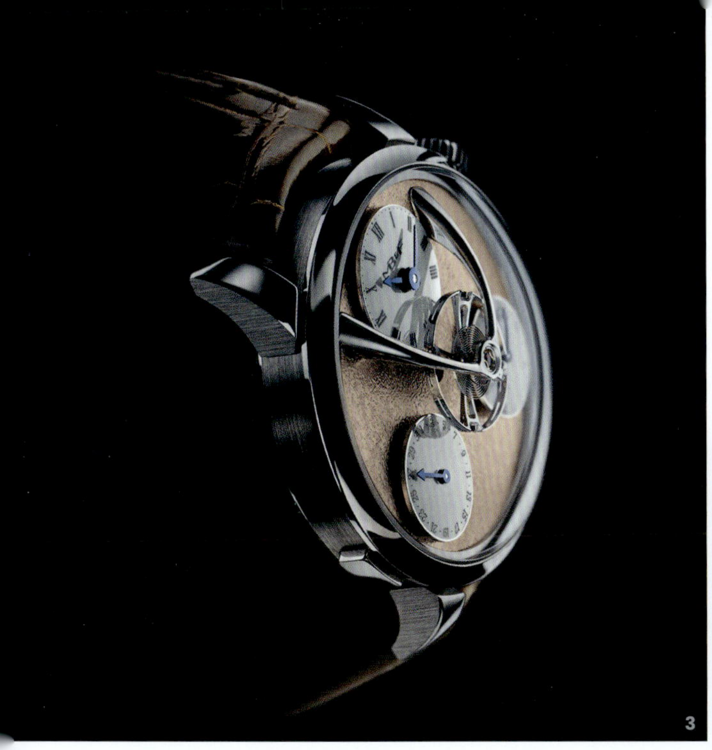

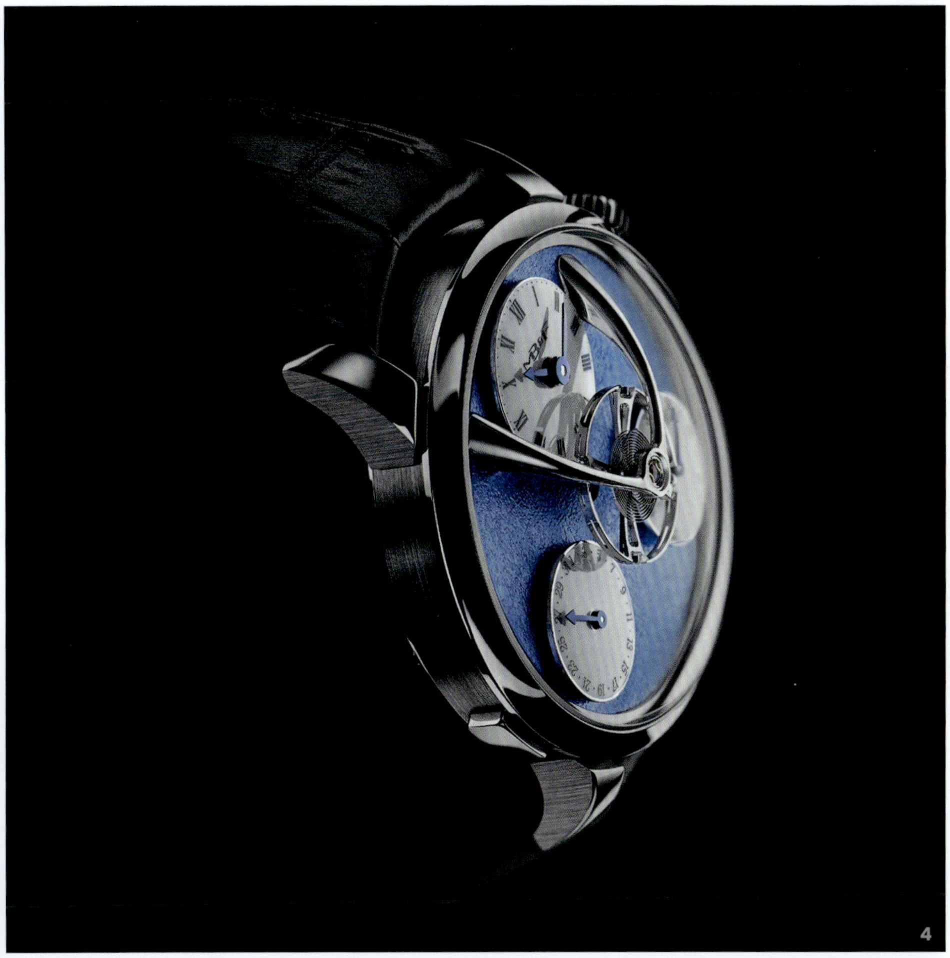

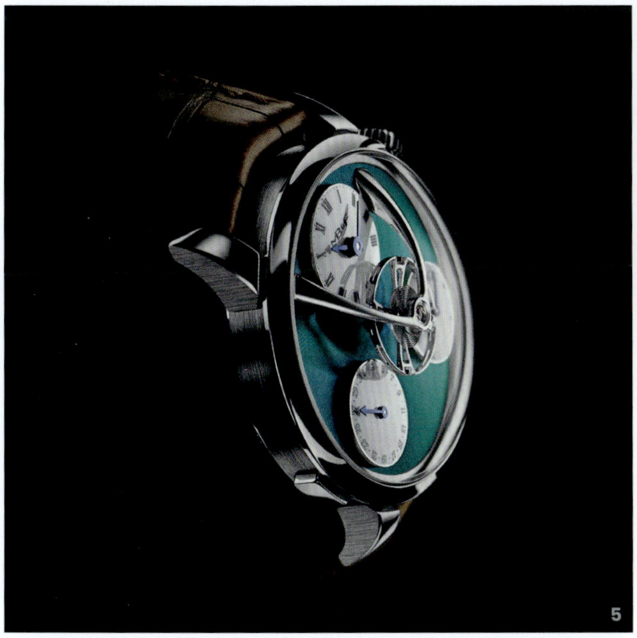

3 LMSE WG Red
Reference: 03.WL.FR
Material: 18K white gold with red gold frosted dial
Limitation: 18
Introduction: 2017
Years in production: 2017-2018
Quantity produced: 18 + 1 prototype

4 LMSE WG Blue
Reference: 03.WL.FBL
Material: 18K white gold with blue frosted dial
Limitation: 18
Introduction: 2017
Years in production: 2017-2018
Quantity produced: 18

5 LMSE Ti Green
Reference: 04.TL.G
Material: grade 5 titanium with green CVD-treated dial
Limitation: 33
Introduction: 2019
Years in production: 2019 – in production
Quantity produced (at the end of 2020): 26 + 1 prototype

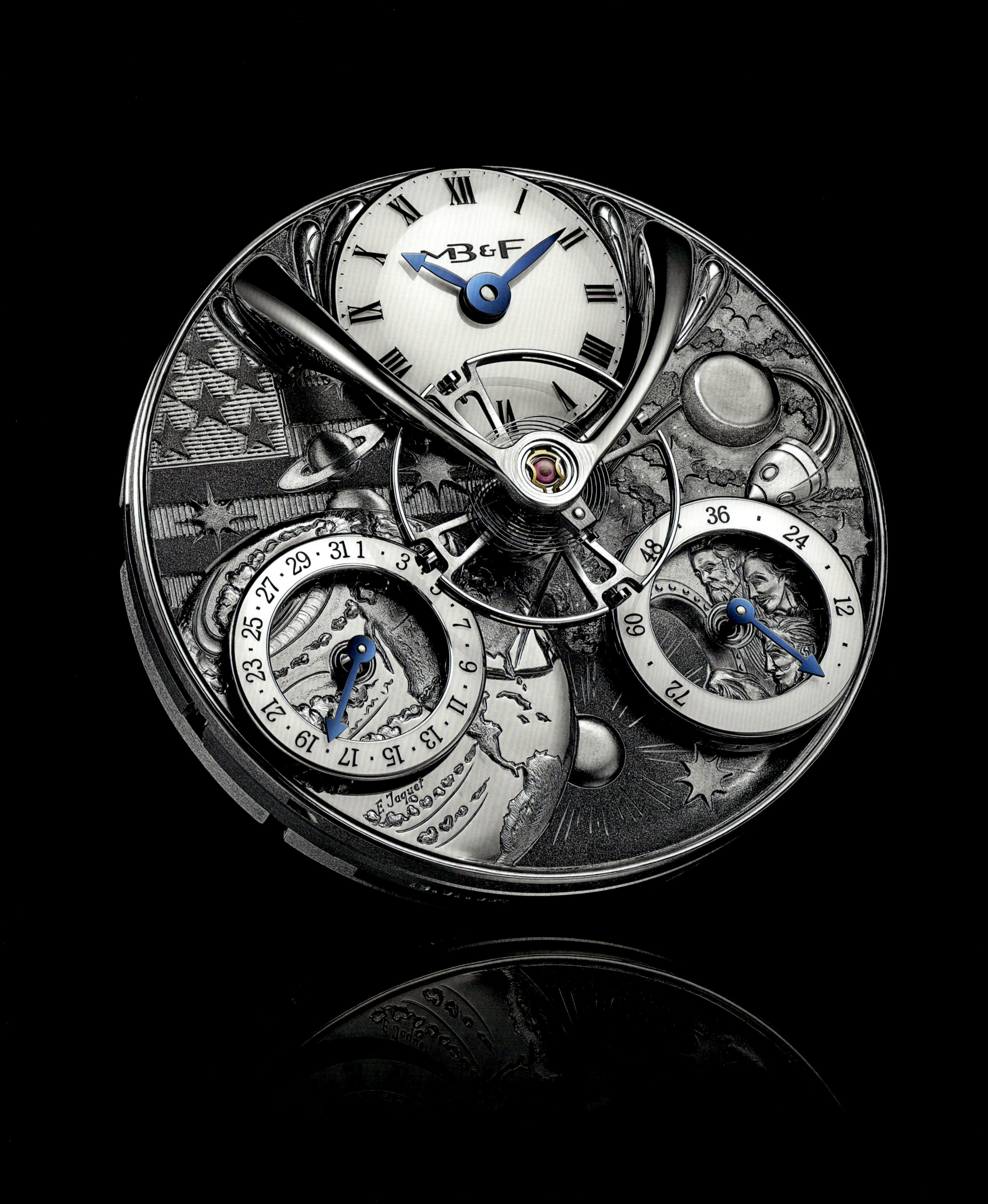
MB&F

LM SE MB&F × EDDY JAQUET

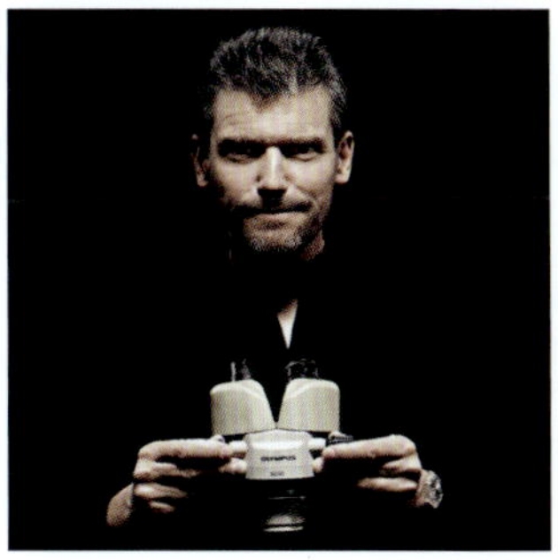

EDDY JAQUET

Eddy Jaquet, artist and master engraver, was born in 1965 in a small village outside Neuchâtel in Switzerland. Having learned his trade at the Ecole d'Arts Appliqués in La Chaux-de-Fonds, the freshly graduated Jaquet launched straight into a career that has continued uninterrupted from 1987 to the present day.

He began working with MB&F in 2011, engraving the names of Kari Voutilainen and Jean-François Mojon on the movement bridge of the first Legacy Machine. Maximilian Büsser would later say that "getting Eddy Jaquet to engrave names on movements is like asking a piano virtuoso to play Für Elise – you couldn't imagine a more modest use of such an amazing gift." Büsser holds Jaquet's work in such high esteem that their eventual collaboration was almost inevitable.

At the heart of the Legacy Machine project ever since the outset are the science fiction stories of 19th century French writer Jules Verne (prolific author of Around the World in Eighty Days, Journey to the Centre of the Earth, Twenty Thousand Leagues Under the Sea and much besides). His fantastical tales were after all a key inspiration behind the original Legacy Machine, which was born out of the question: what watch would Büsser have designed if he had been born in 1867 instead of 1967? In this alternative reality, MB&F imagines itself taking inspiration from Jules Verne, not Star Wars.

For MB&F's project with Jaquet, the LM SE was a logical choice. Its large, expansive dial is almost a natural canvas, allowing Jaquet the space to bring to life some of his favourite scenes from a variety of Verne's novels. Each engraving is an intricate tableau conceived by Jaquet after reading the source material. Scenes range from a battle of sea monsters taken from Journey to the Centre of the Earth to a fallen Russian soldier from Michael Strogoff – altogether a series of eight unique pieces released in 2020.

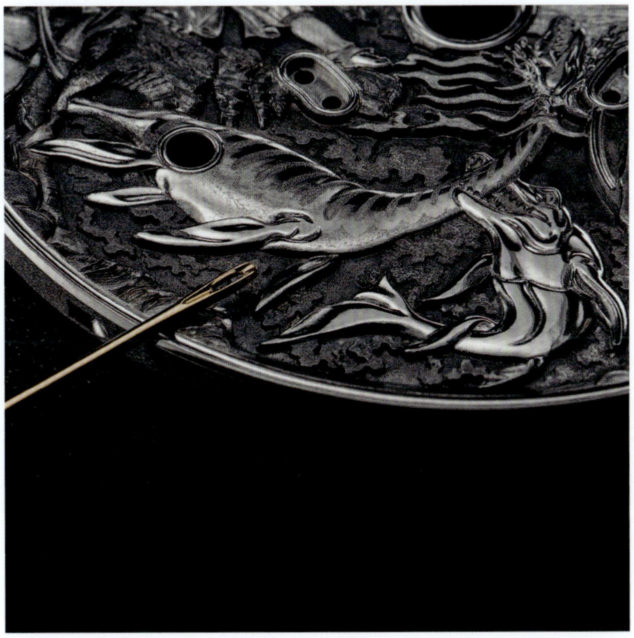

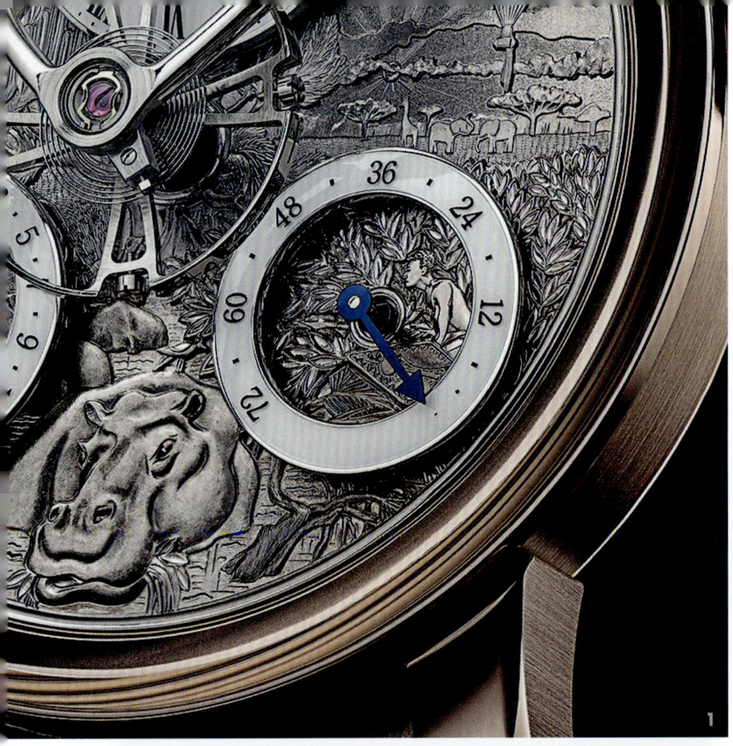

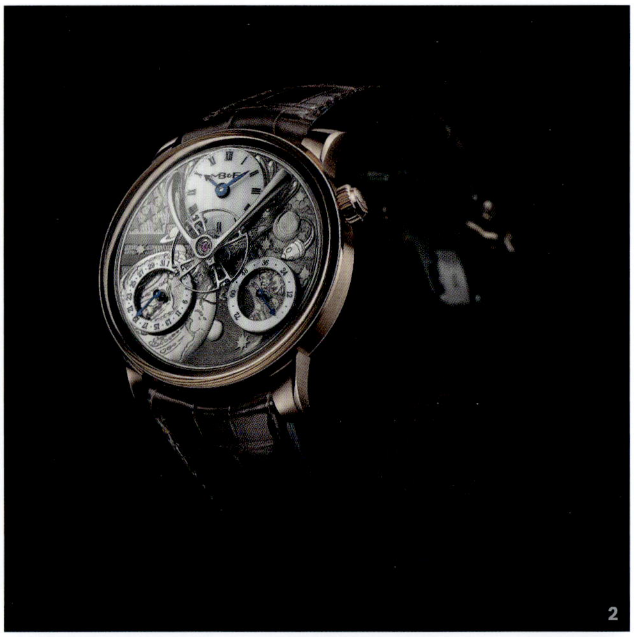

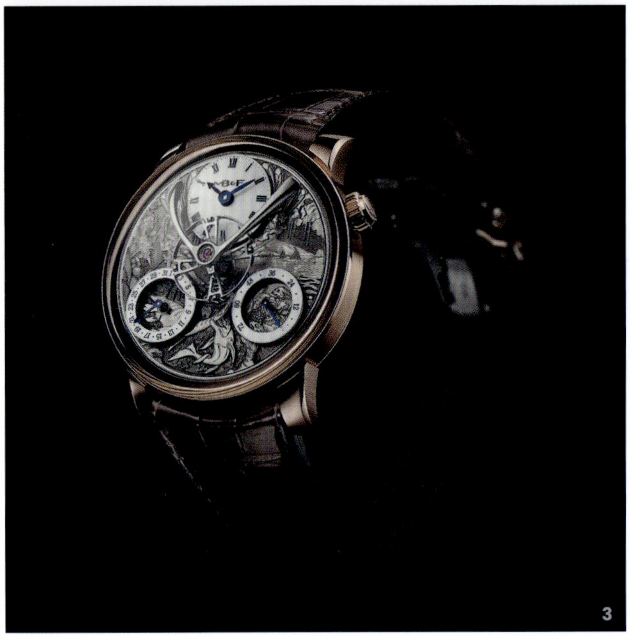

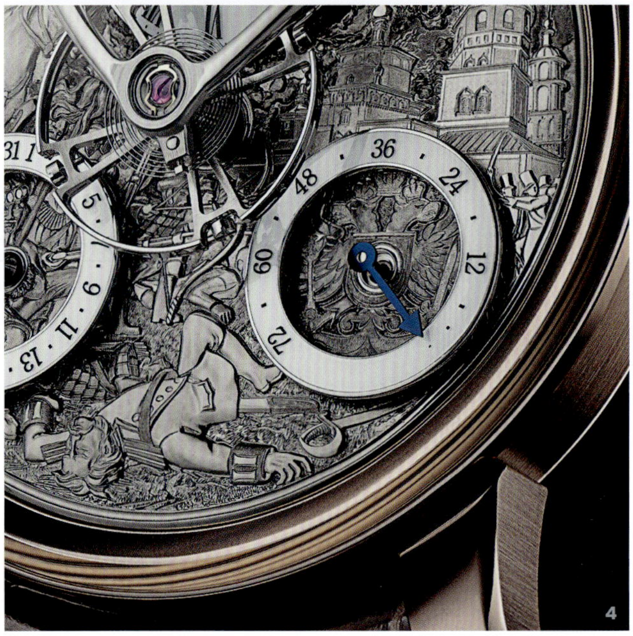

1 Legacy Machine Eddy Jaquet
Five Weeks in a Balloon
Reference: 04.RL.W4
Material: 18K red gold
Limitation: unique piece
Introduction: 2020
Year in production: 2020
Quantity produced: 1

2 Legacy Machine Eddy Jaquet
From the Earth to the Moon
Reference: 04.RL.W2
Material: 18K red gold
Limitation: unique piece
Introduction: 2020
Year in production: 2020
Quantity produced: 1

3 Legacy Machine Eddy Jaquet
Journey to the Centre
of the Earth
Reference: 04.RL.W6
Material: 18K red gold
Limitation: unique piece
Introduction: 2020
Year in production: 2020
Quantity produced: 1

4 Legacy Machine Eddy Jaquet
Michael Strogoff
Reference: 04.RL.W7
Material: 18K red gold
Limitation: unique piece
Introduction: 2020
Year in production: 2020
Quantity produced: 1

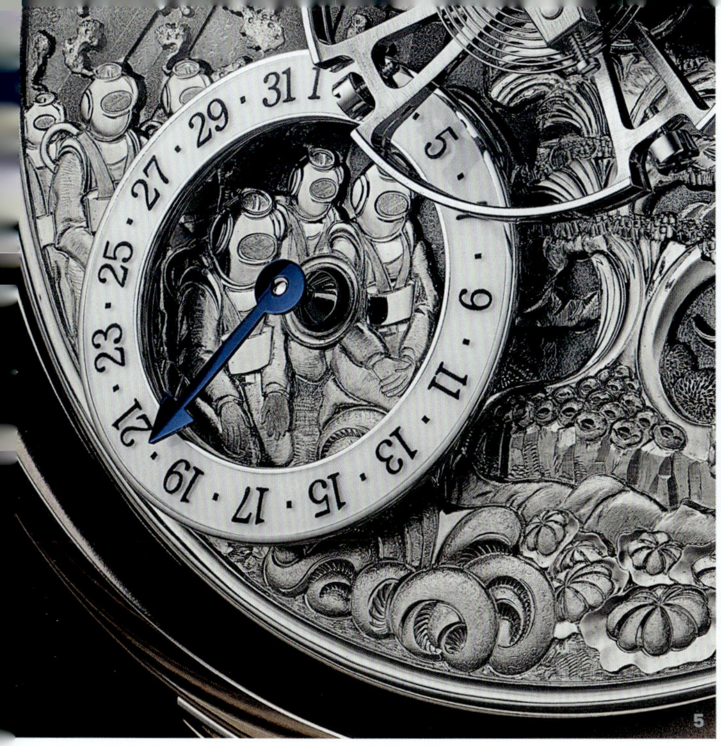
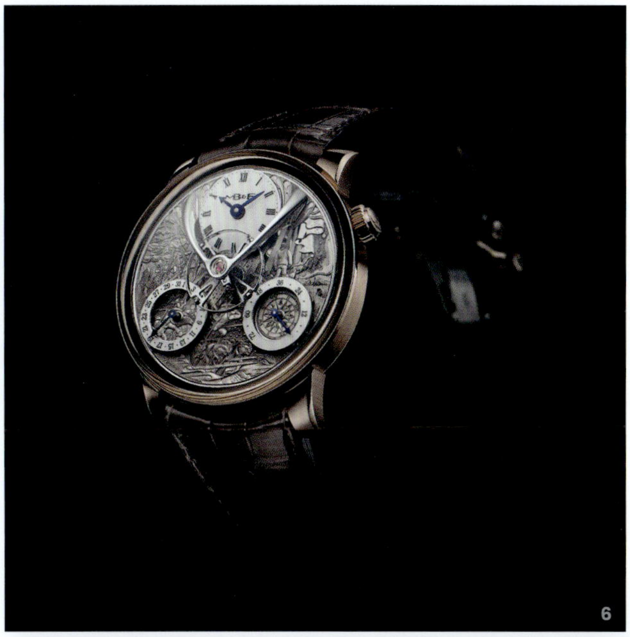
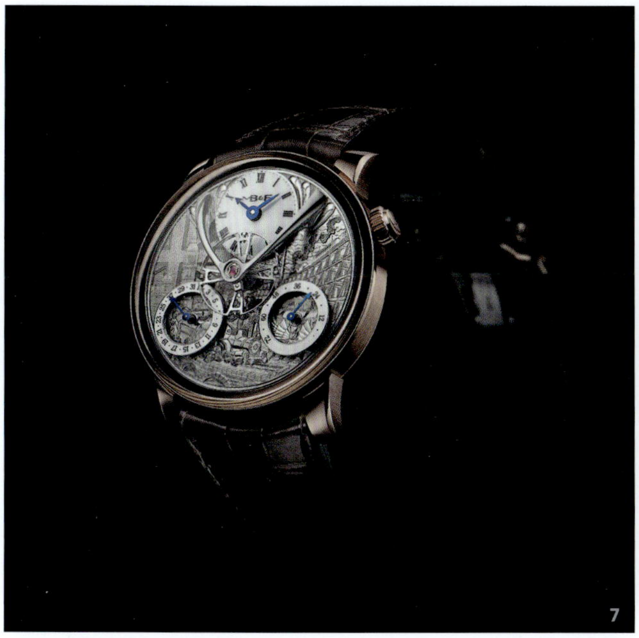
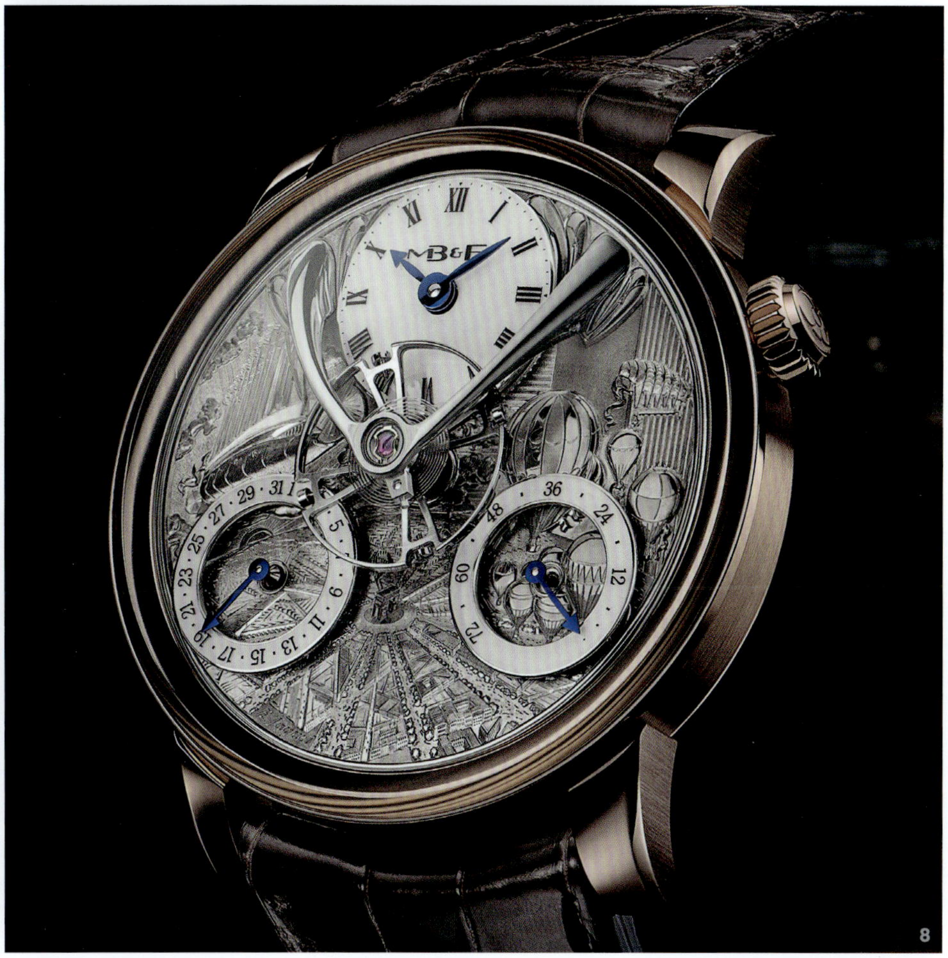

5 Legacy Machine Eddy Jaquet
Twenty Thousand
Leagues under the Sea
Reference: 04.RL.W1
Material: 18K red gold
Limitation: unique piece
Introduction: 2020
Year in production: 2020-2021
Quantity produced: 1

6 Legacy Machine Eddy Jaquet
The Adventures of
Captain Hatteras
Reference: 04.RL.W5
Material: 18K red gold
Limitation: unique piece
Introduction: 2020
Year in production: 2020-2021
Quantity produced: 1

7 Legacy Machine Eddy Jaquet
Around the World
in Eighty Days
Reference: 04.RL.W3
Material: 18K red gold
Limitation: unique piece
Introduction: 2020
Year in production: 2020-2021
Quantity produced: 1

8 Legacy Machine Eddy Jaquet
Robur the Conqueror
Reference: 04.RL.W8
Material: 18K red gold
Limitation: unique piece
Introduction: 2020
Year in production: 2020-2021
Quantity produced: 1

LM
FLYING T

Since the founding of MB&F, all of its creations according to Maximilian Büsser were intended for himself. Which is not to say that all of his creations were intended for men, but none were made expressly for a wider audience including women – until the Legacy Machine FlyingT.
Central to its conception were two qualities he saw in the women close to him, including his mother, wife, and daughters: refined elegance, and energetic dynamism.
Elegant in its lines and its finishes; energetic in its captivating, twisting tourbillon. The Legacy Machine FlyingT was first conceived as a women's timepiece, though it is certainly not exclusive to either gender. Case height and diameter are reduced in the FlyingT, the lugs are slimmer and lines are softer throughout, right down to the serpentine hands of its single subsidiary dial. The dial is located at the 7 o'clock position and inclined at a 50° tilt so that the time can only be read by the wearer. At the top of the tourbillon for which this LM is named is a single, sparkling diamond that draws the eye to the centre of the dial and its source of mechanical power.
Diamonds, baguette or pavé, cover some or all of the surface area depending on the reference. In subsequent red gold and platinum editions, diamonds are traded for elegant guilloché work by Kari Voutilainen. A lapis lazuli version, with diamonds set in white gold was also produced.

LEGACY MACHINE FLYING T

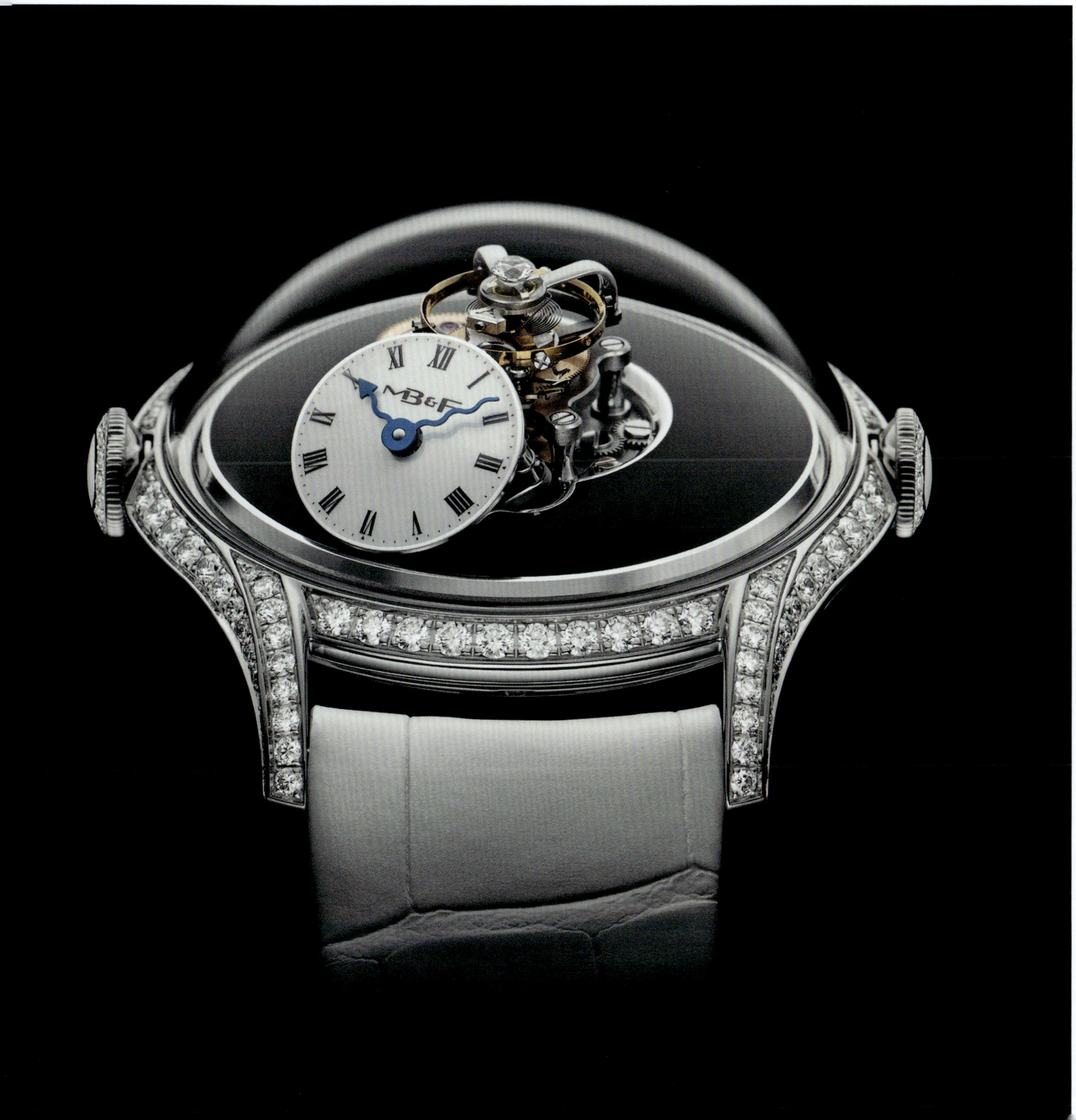

LM FLYING T CALIBRE

Peeking upwards from an asymmetric dial plate is the heart and namesake of the LM FlyingT – a flying tourbillon that stops just short of the full height of its exaggerated crystal dome. This technically difficult complication was achieved only by hiding invisible counterweights within the tourbillon cage to maintain the balance and integrity of the mechanism.
On the reverse, its automatic winding rotor assumes the shape of a three-dimensional red gold sun with brilliant, outstretched rays. The sun stands as a symbol of life-giving and sustenance, recalling the dynamic feminine energy that inspired Büsser to undertake this project; and it quite literally supplies energy to the whole machine, delivering a total power reserve of 100 hours, one of the highest among all MB&F machines.

Collection
Legacy Machines
Series
Legacy Machine FlyingT
Common name
LM FlyingT
Movement
Developed in-house by MB&F
Winding
Automatic
Power reserve
100 hours
Components
280/30 jewels
Functions
Hours, minutes displayed on a 50° vertically tilted dial with two serpentine hands
Add. features
Central flying 60-second tourbillon
Three-dimensional sun winding rotor in 18K red gold
Two crowns: winding on left and time setting on right
Case
18K white gold and diamonds,
18K red gold or platinum 950
Dimensions
38.5 × 20mm
Case components
17
Water resistance
30 m/3 ATM
Strap
Calf or alligator
Buckle
Pin buckle, 18K white or red gold, platinum matching the case
Introduction
2019
Years in production
2019 – in production
Quantity produced
(at the end of 2020)
91 + 6 prototypes

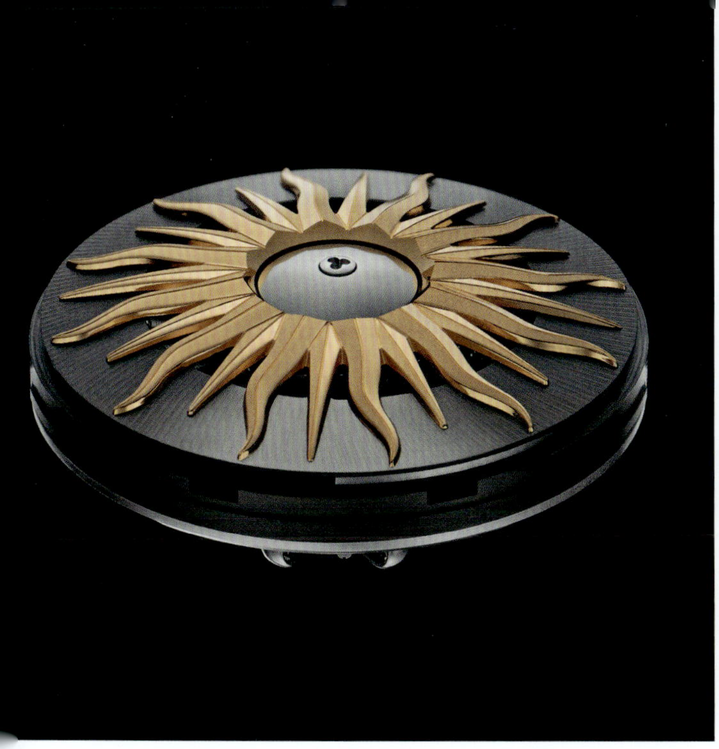

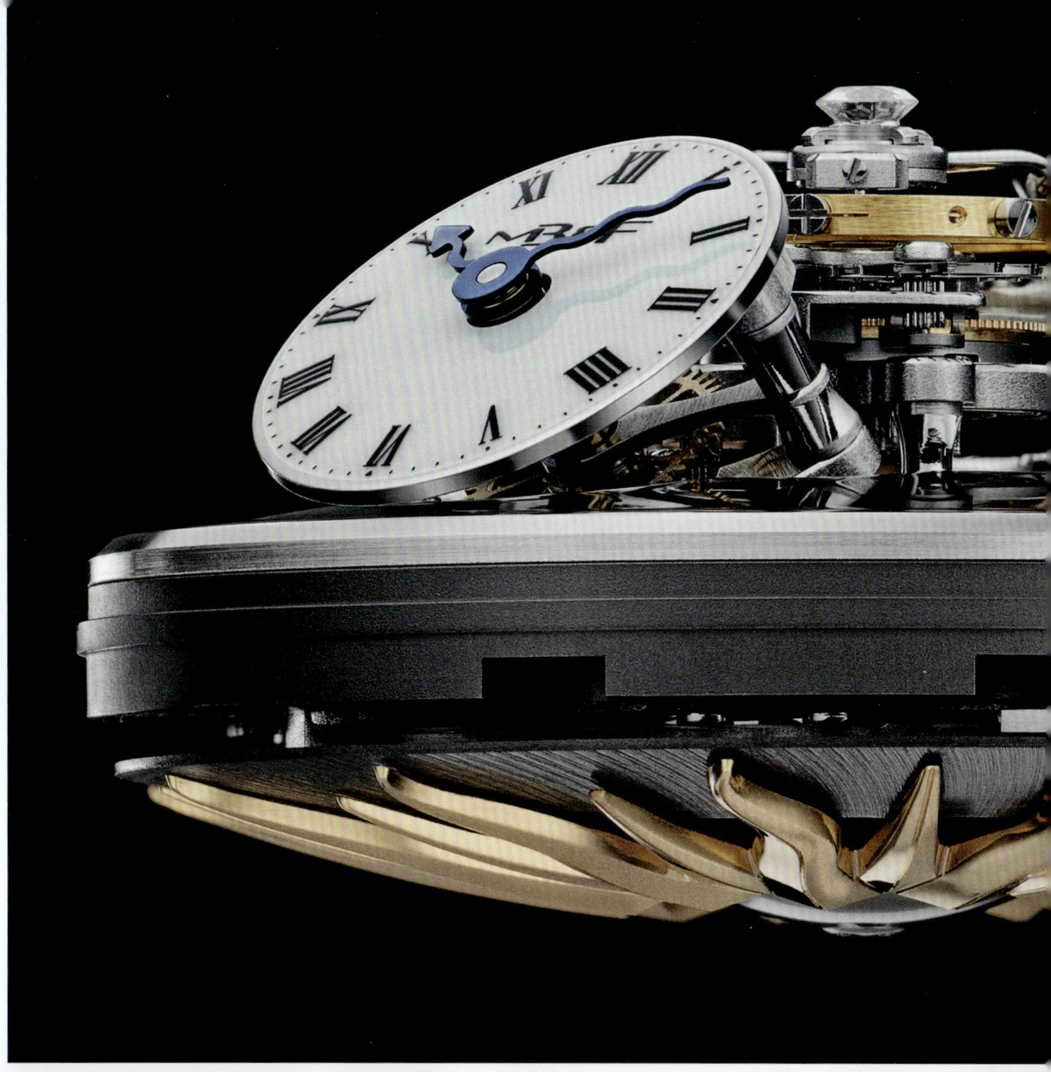
MB&F

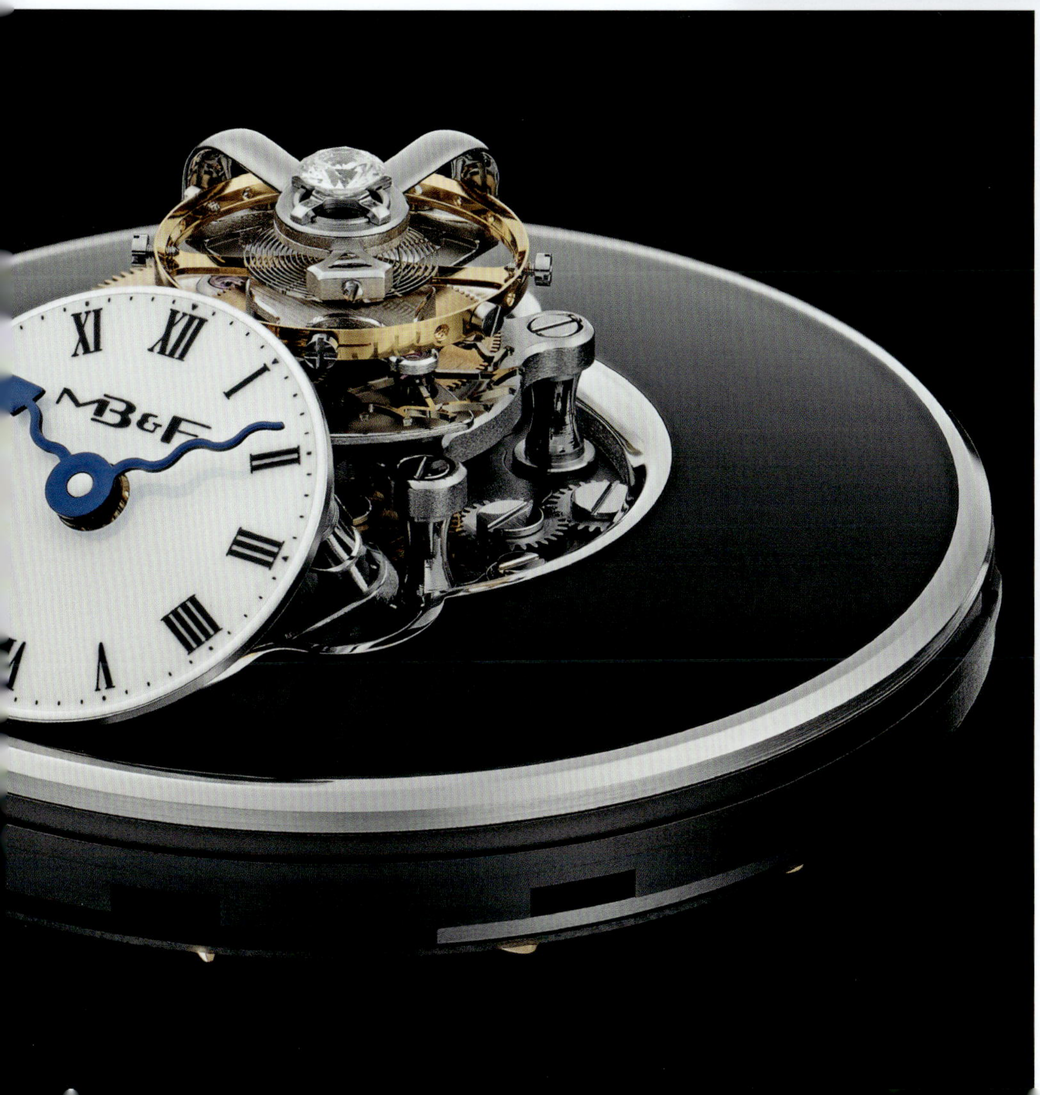
MB&F

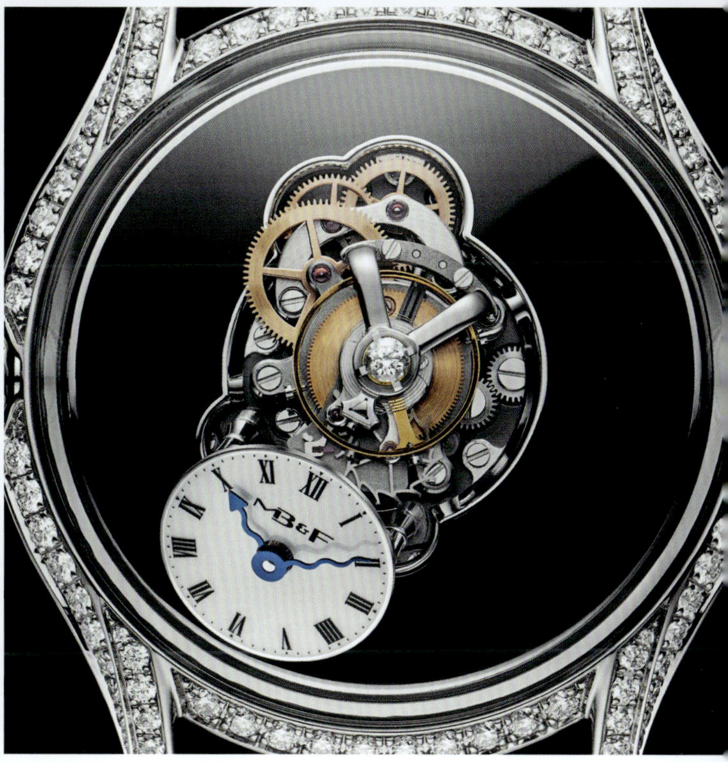
MB&F

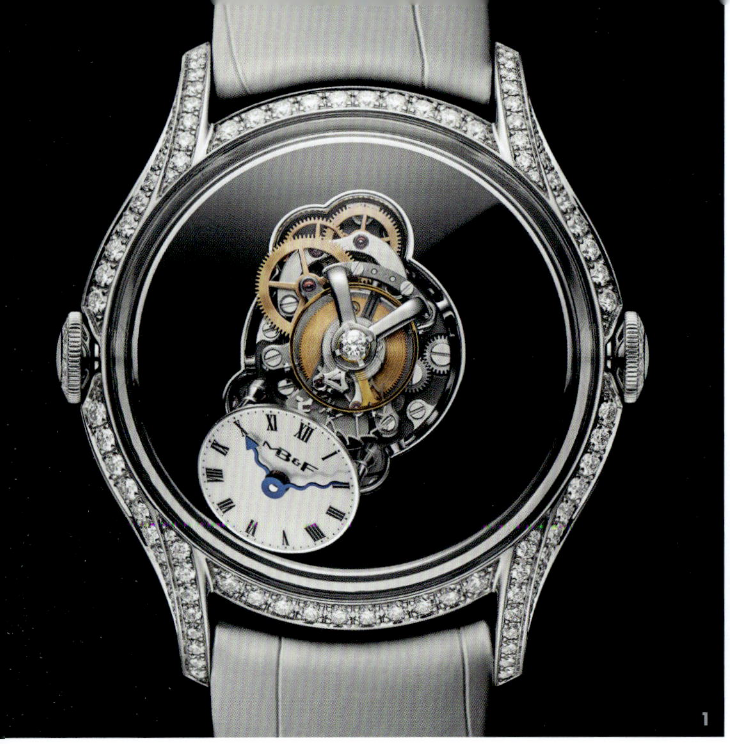

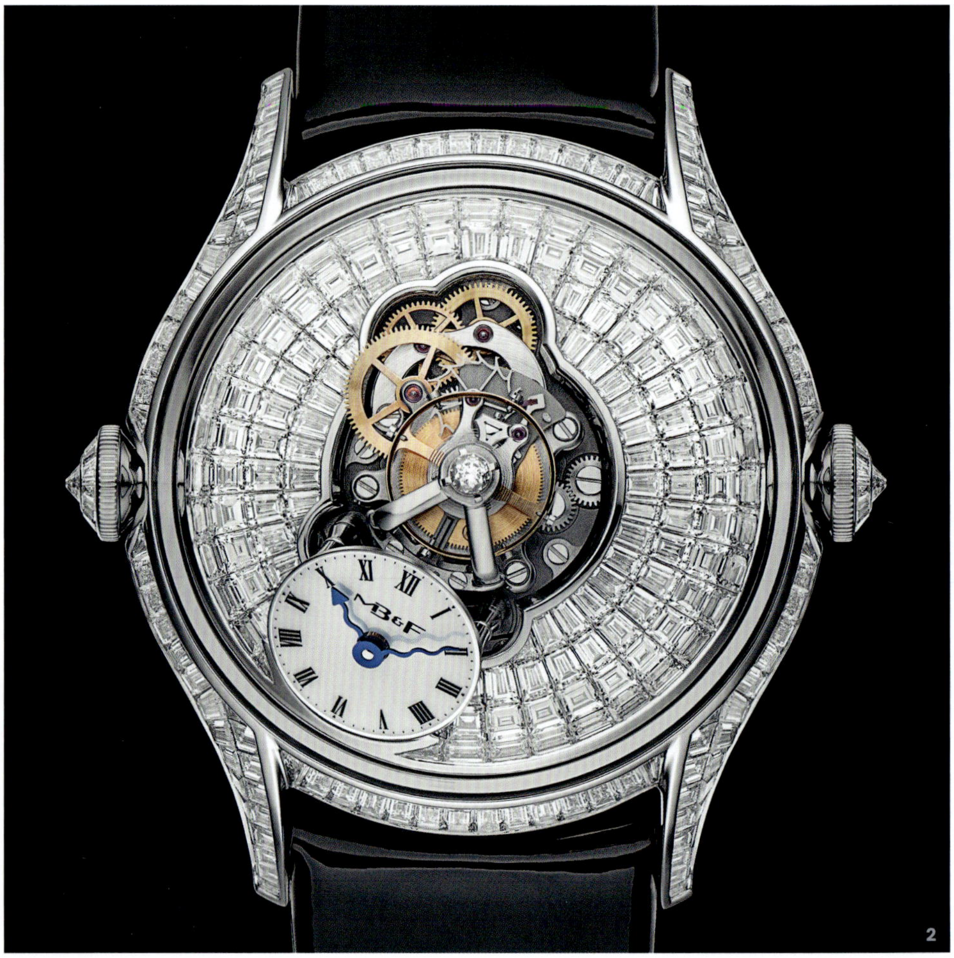

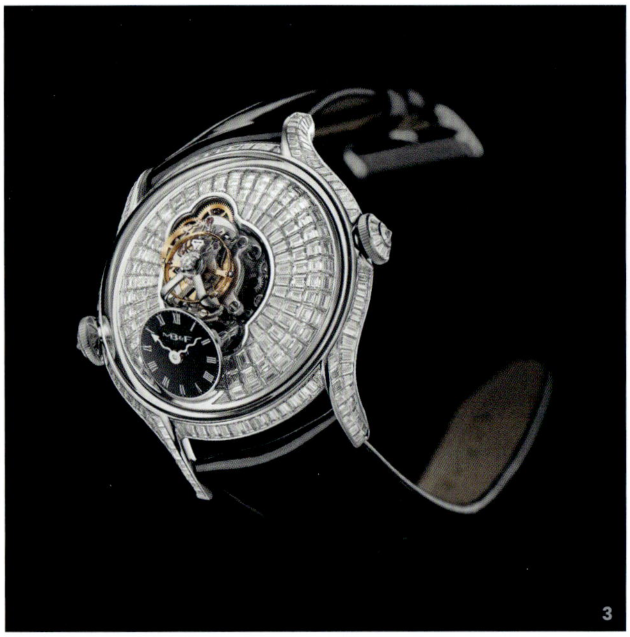

1 Legacy Machine FlyingT
Black Lacquer, white dial
Reference: 05.WSL.B
Material: 18K white gold and diamond-set case, with black lacquer dial plate
Limitation: -
Introduction: 2019
Years in production:
2019 – in production
Quantity produced (at the end of 2020): 31 + 1 prototype

2 Legacy Machine FlyingT
Baguette, white dial
Reference: 05.WSL.B1
Material: 18K white gold, case and dial plate fully set with baguette diamonds
Limitation: -
Introduction date: 2019
Years in production:
2019 – in production
Quantity produced (at the end of 2020): 4 + 1 prototype

3 Legacy Machine FlyingT
Baguette, black dial
Reference: 05.WSL.B2
Material: 18K white gold, case and dial plate fully set with baguette diamonds
Limitation: -
Introduction: 2019
Years in production:
2019 – in production
Quantity produced (at the end of 2020): 1

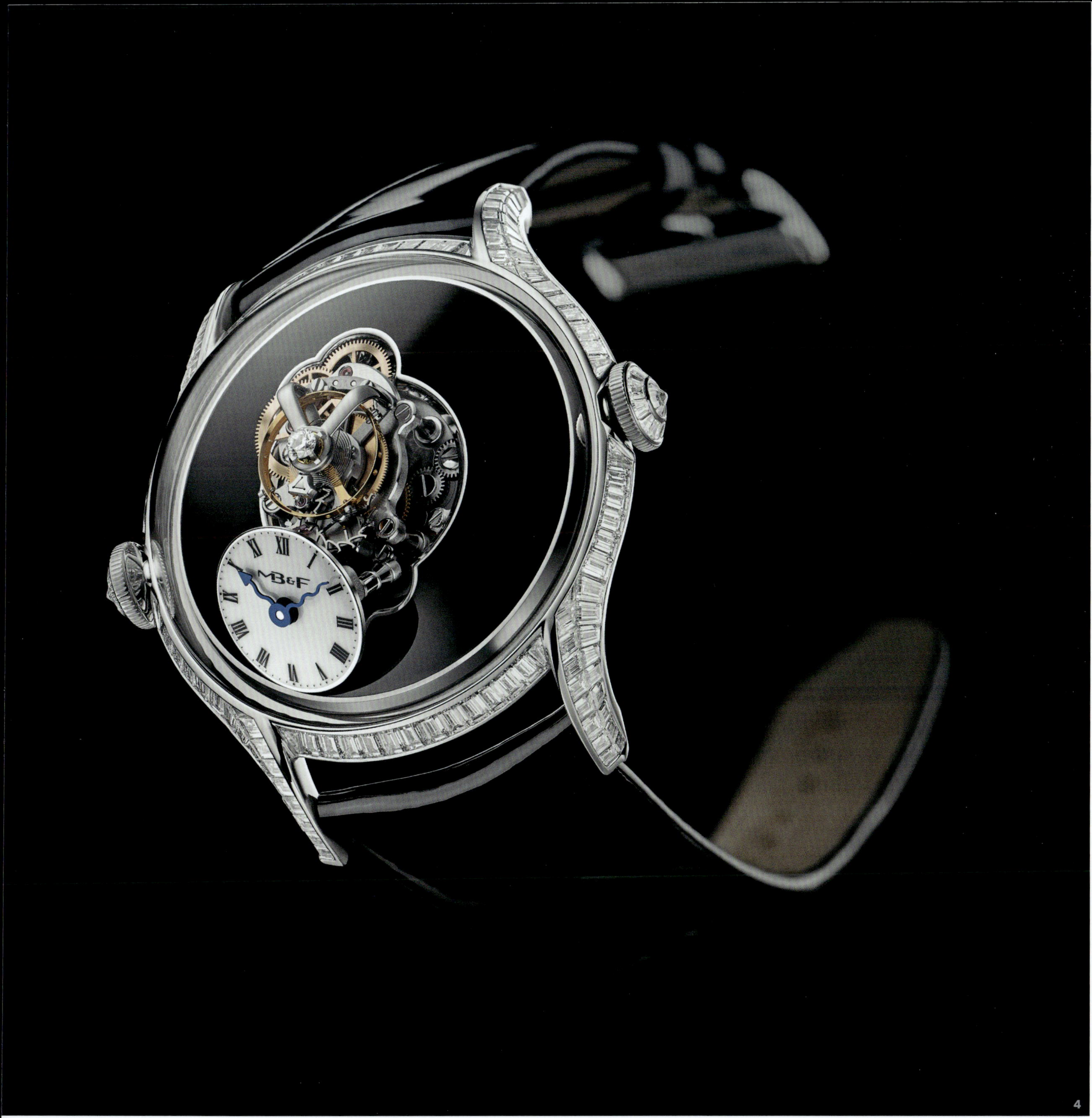

4

4 **Legacy Machine FlyingT Baguette, black lacquered base plate**
Reference: 05.WSL.B3
Material: 18K white gold, case set with baguette diamonds and black lacquered dial plate
Limitation: –
Introduction date: 2019
Years in production: 2019 – in production
Quantity produced (at the end of 2020): 1

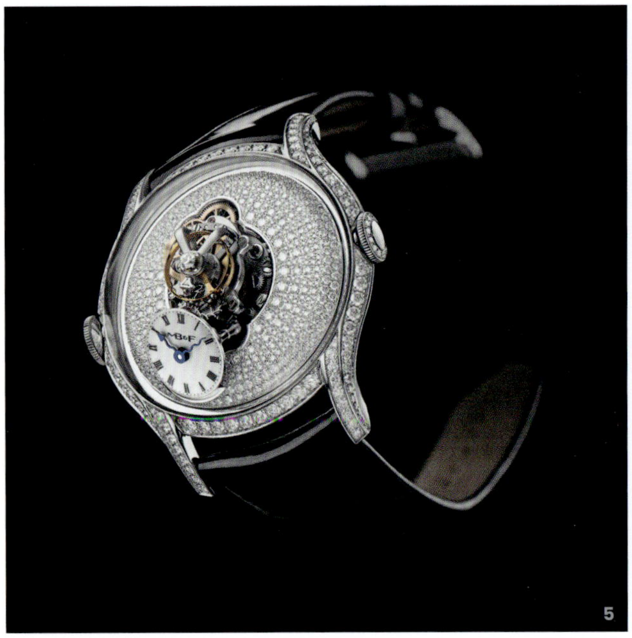

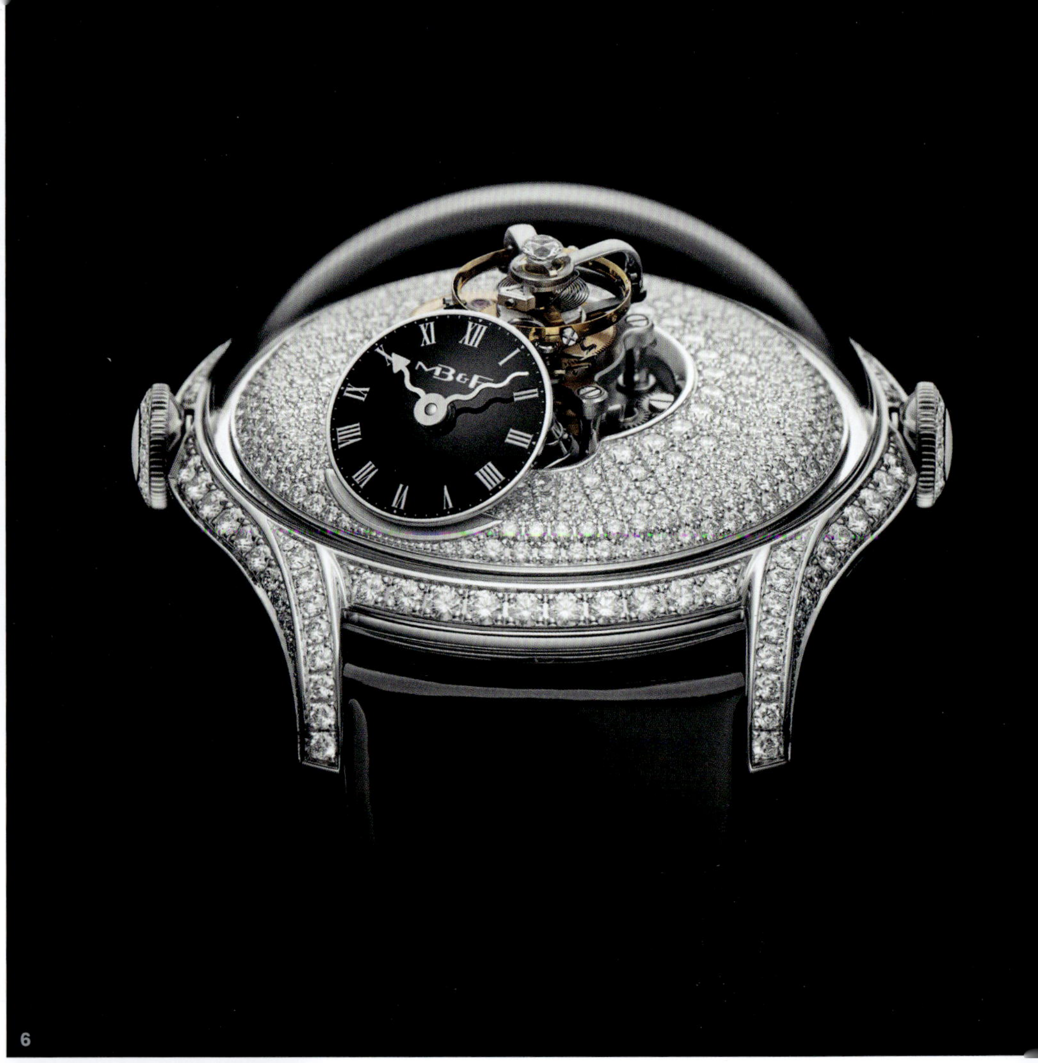

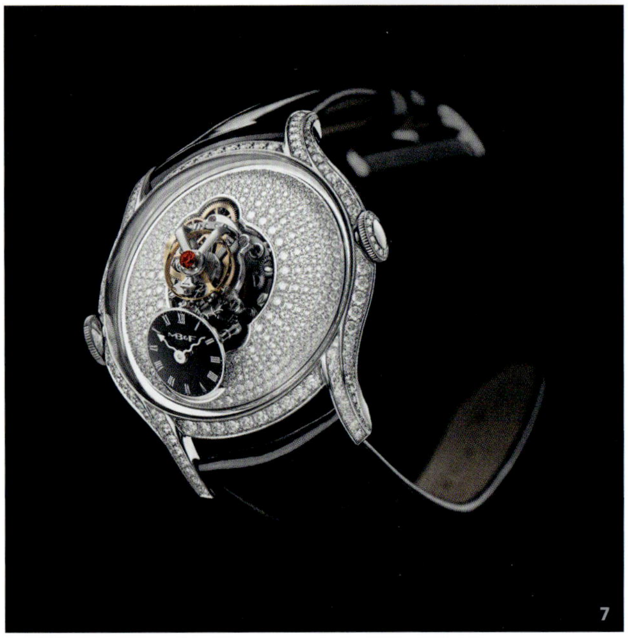

5 Legacy Machine FlyingT Pavée, white dial
Reference: 05.WSL.S1
Material: 18K white gold, case and dial plate fully set with diamonds
Limitation: –
Introduction: 2019
Years in production: 2019 – in production
Quantity produced (at the end of 2020): 12 + 1 prototype

6 Legacy Machine FlyingT Pavée, black dial
Reference: 05.WSL.S2
Material: 18K white gold, case and dial plate fully set with diamonds
Limitation: –
Introduction: 2019
Years in production: 2019 – in production
Quantity produced (at the end of 2020): 17 + 1 prototype

7 Legacy Machine FlyingT Pavée, ruby
Reference: 05.WSL.S3
Material: 18K white gold, case and dial plate fully set with diamonds and red ruby on top of tourbillon cage
Limitation: –
Introduction: 2019
Years in production: 2019 – in production
Quantity produced (at the end of 2020): 1

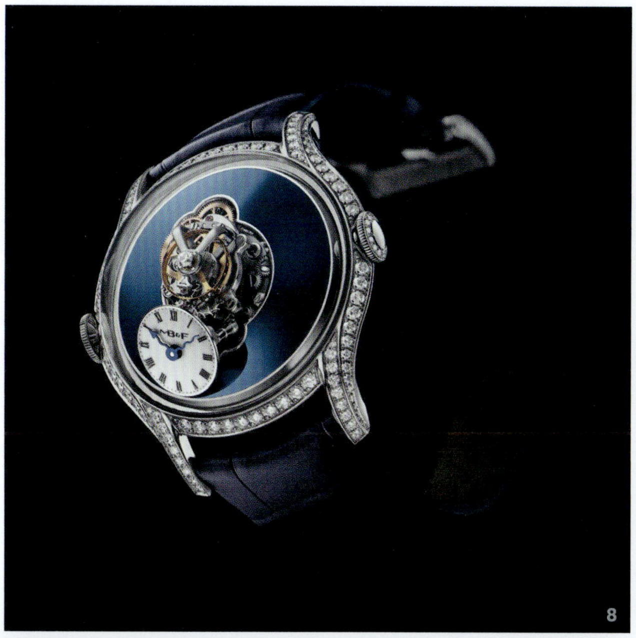

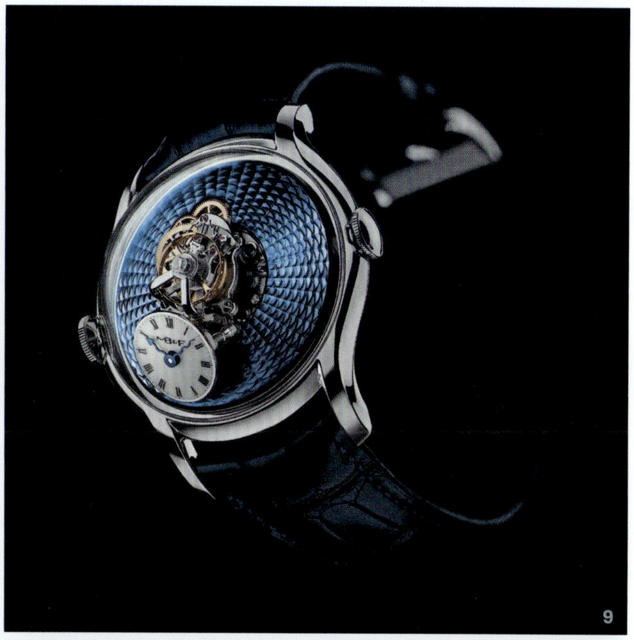

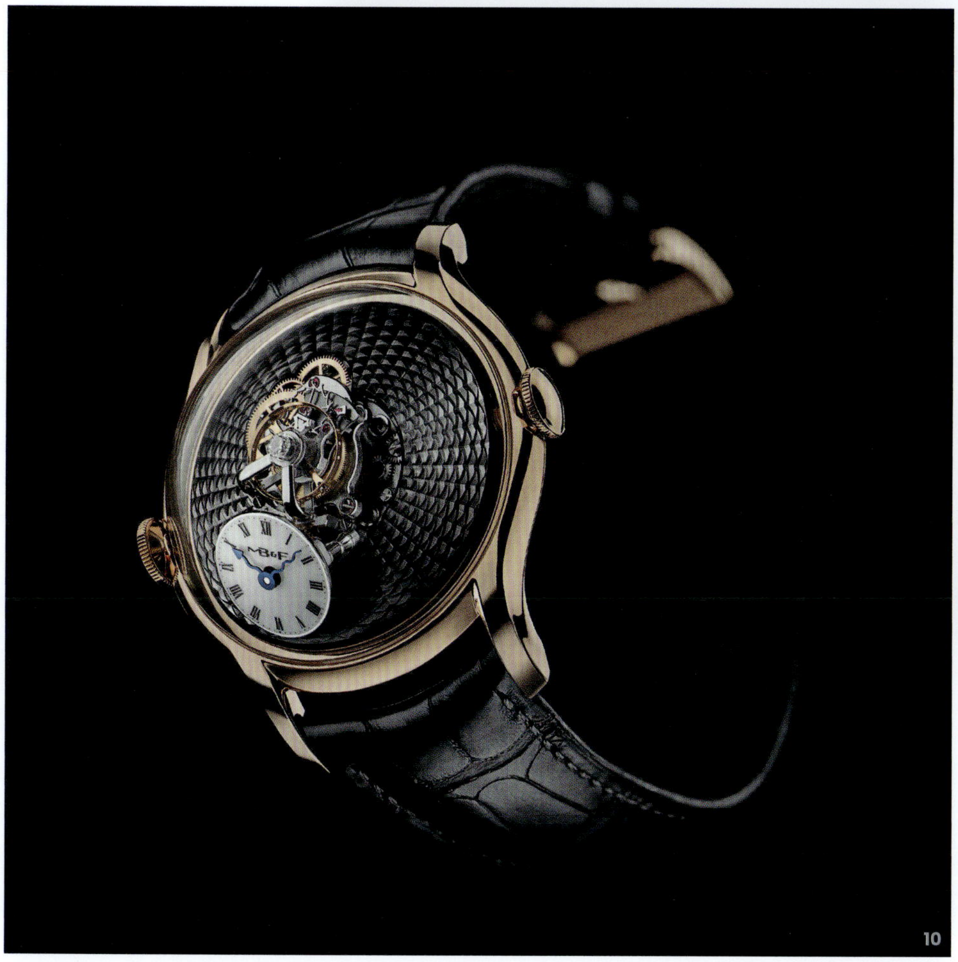

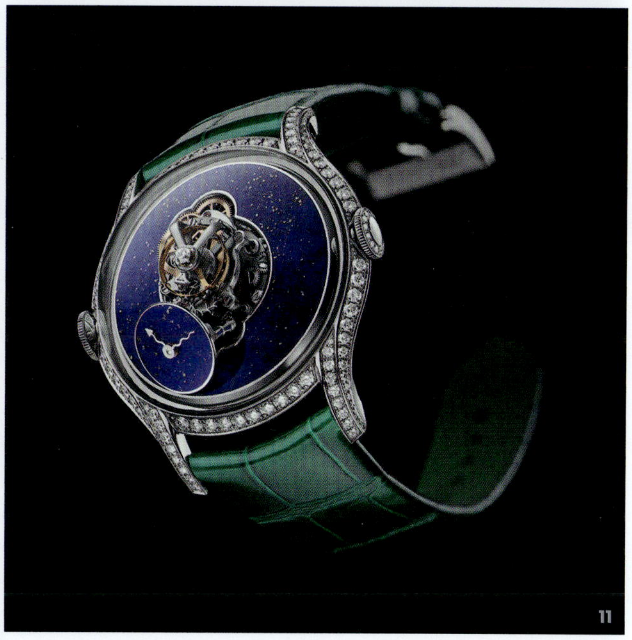

8 **Legacy Machine FlyingT WG Blue**
Reference: 05.WL.BL1
Material: 18K white gold and diamond-set case with blue lacquered dial plate
Limitation: prototype
Introduction: 2019
Year in production: 2019
Quantity produced:
1 prototype

9 **Legacy Machine FlyingT PT**
Reference: 05.PL.BLG
Material: platinum 950
Limitation: 18
Introduction: 2020
Years in production:
2020 – in production
Quantity produced
(at the end of 2020):
9 + 1 prototype

10 **Legacy Machine FlyingT RG**
Reference: 05.RL.BG
Material: 18K red gold
Limitation: 18
Introduction: 2020
Years in production:
2020 – in production
Quantity produced
(at the end of 2020): 10

11 **Legacy Machine FlyingT lapis lazuli**
Reference: 05.WSL.WL
Material: 18K white gold and diamond-set case with lapis lazuli dial and dial plate
Limitation: –
Introduction: 2020
Years in production:
2020 – in production
Quantity produced
(at the end of 2020): 5

LM
THUNDERDOME

Rising from a sea of bright blue guillochage, a spherical assemblage of mechanical wizardry moves and rotates at lightning speed, seeming to turn and churn the reflective surface below it. This little hurricane, encased in a dome of sapphire crystal, is the brainchild of Eric Coudray and Kari Voutilainen, two master watchmakers who had never before joined in collaboration. MB&F's only directive to Coudray was initially to create "the craziest, most cinematic three-axis tourbillon ever."

The maelstrom whipped up by Coudray and Voutilainen is something entirely new, dubbed the MB&F "TriAx." Three axes, each revolving at a different speed, create a visual spectacle never before seen in the world of horology. The end result is as dizzying as it is technically advanced. The innermost structure revolves once every eight seconds, the middle cage revolves every twelve seconds, and the outermost case revolves every twenty seconds.

LEGACY MACHINE THUNDERDOME

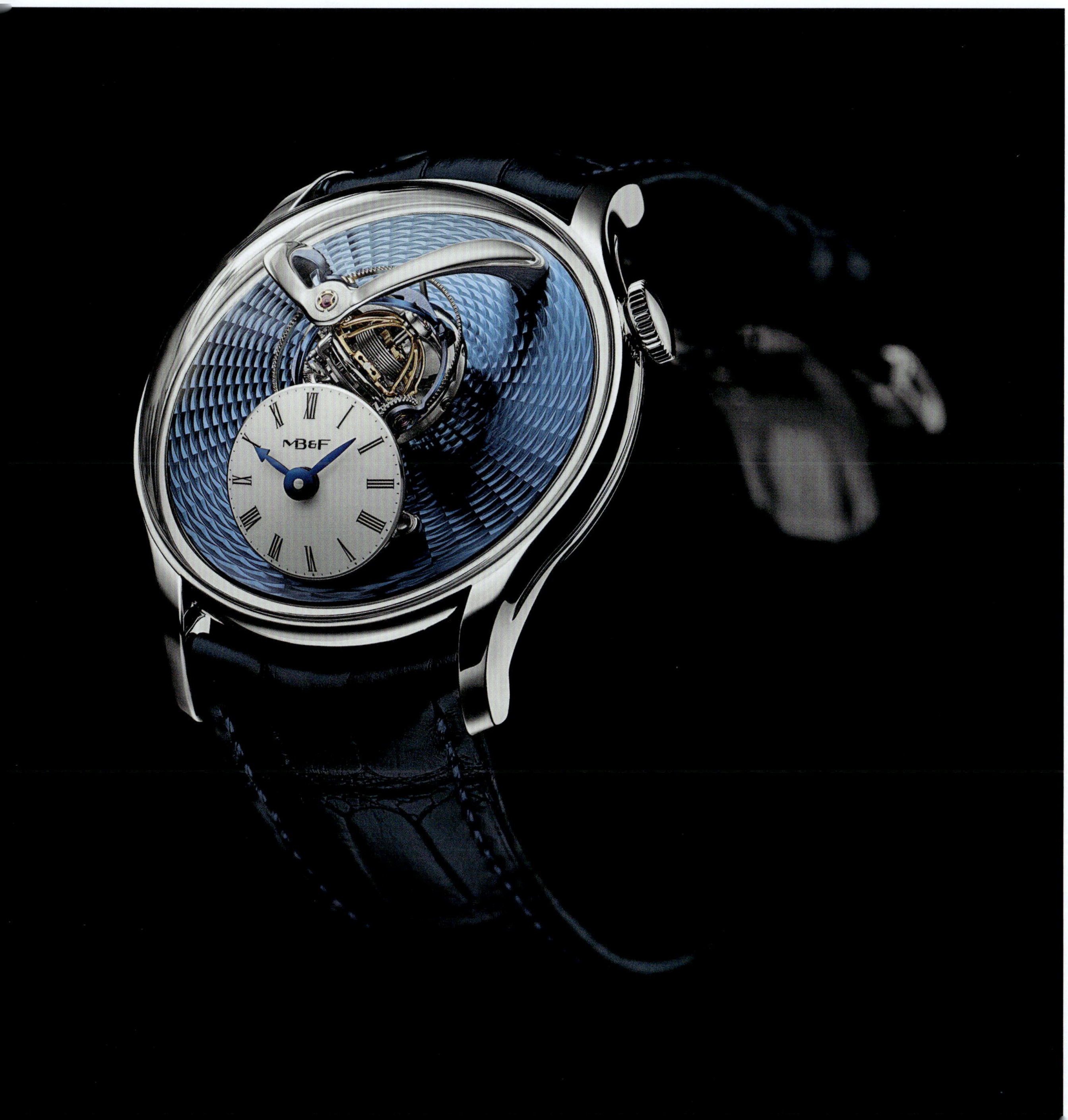

LM THUNDERDOME CALIBRE

The TriAx upends the typical construction of a triple-axis rotating escapement, a system that usually associates one cage with each rotational axis. The TriAx has a three axis, two cage configuration, making the mechanism more visible to an unprecedented degree. A second innovation is the use of a Potter escapement, named after 19th-century watchmaker Albert H. Potter, which features a fixed escape wheel as opposed to a mobile one. The use of the Potter escapement in this case allows for higher rotational speed inside the Thunderdome. A considerable amount of energy is required to achieve this increased speed, particularly given the considerable mass of the TriAx that weighs nearly one gram.

Driving the Thunderdome is a triple-barrel movement with 45 hours of power reserve.

Collection
Legacy Machines
Series
Legacy Machine Thunderdome
Common name
LM Thunderdome
Movement
Developed for MB&F by Eric Coudray and Kari Voutilainen
Winding
Manual
Power reserve
45 hours
Components
413/63 jewels
Functions
Hours, minutes displayed on a 58° vertically tilted dial, power reserve indicator on the back of the movement
Add. Features
Regulating mechanism featuring 3 fast rotation axes revolving at different speeds and on different planes; starting from the centre, respectively 8 seconds, 12 seconds and 20 seconds
Case
Platinum 950 or tantalum
Dimensions
44 × 22.2 mm
Case components
20
Water resistance
30 m/3 ATM
Strap
Alligator
Buckle
Folding buckle, platinum or tantalum matching the case
Introduction
2019
Years in production
2019 – in production
Quantity produced
(at the end of 2020): 30 + 2 prototypes
Comments
The tantalum editions are exclusive to retail partner The Hour Glass

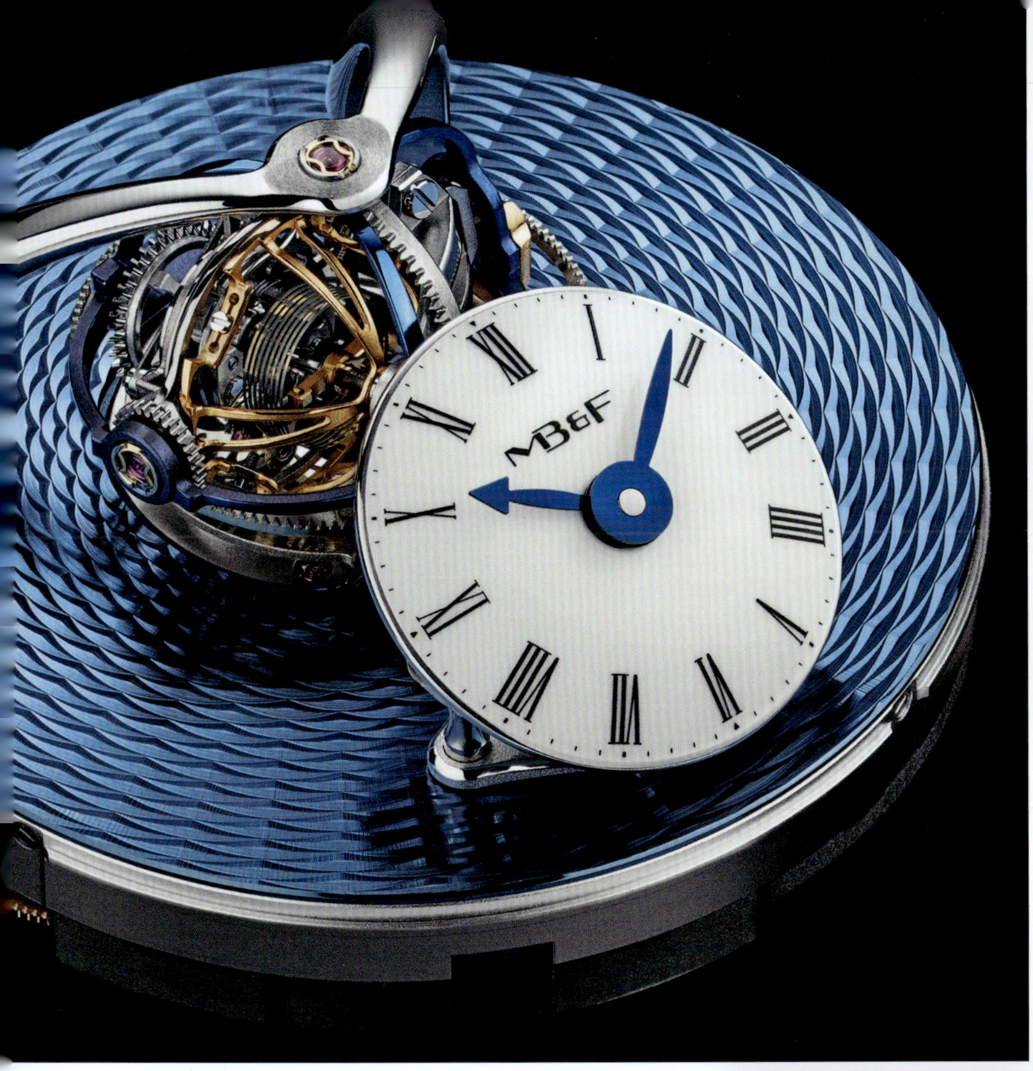
MB&F

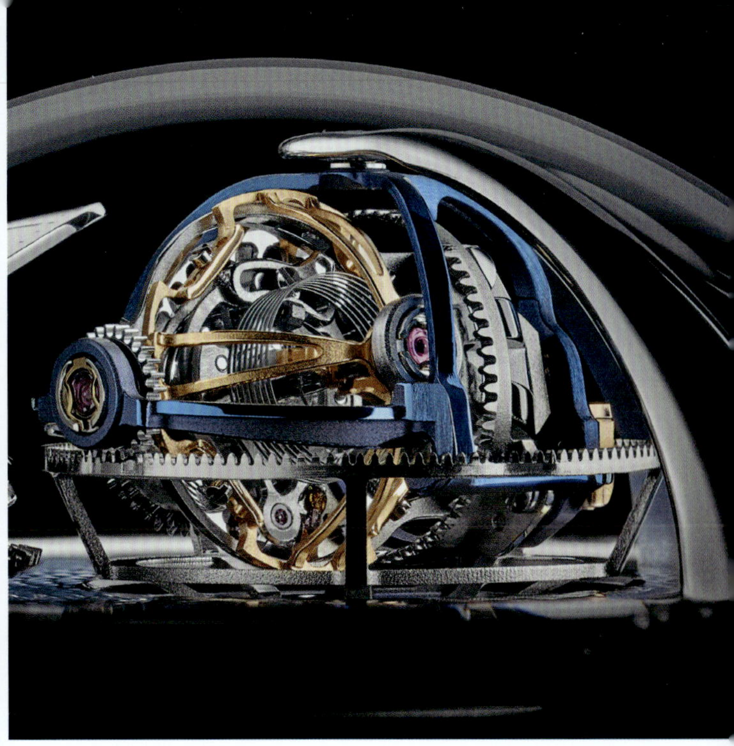

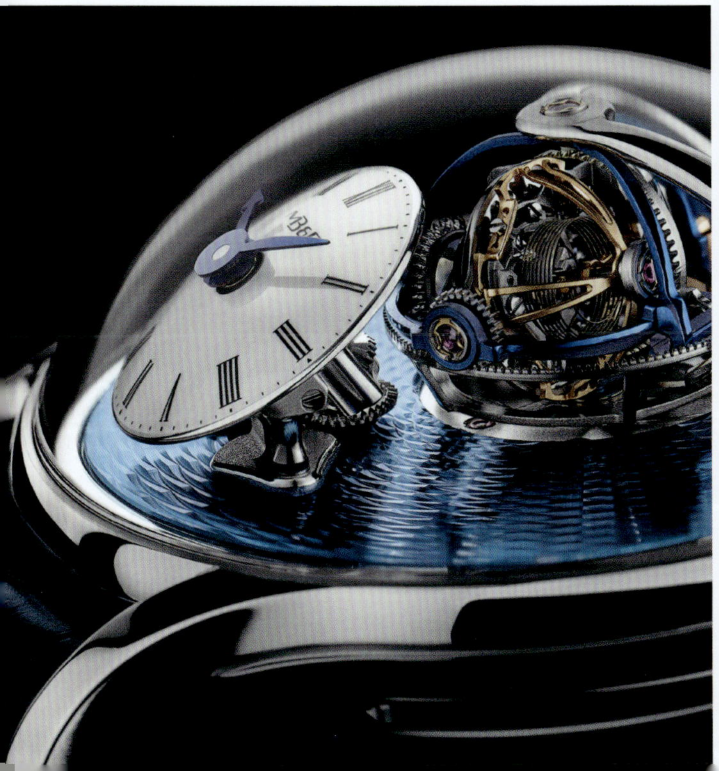

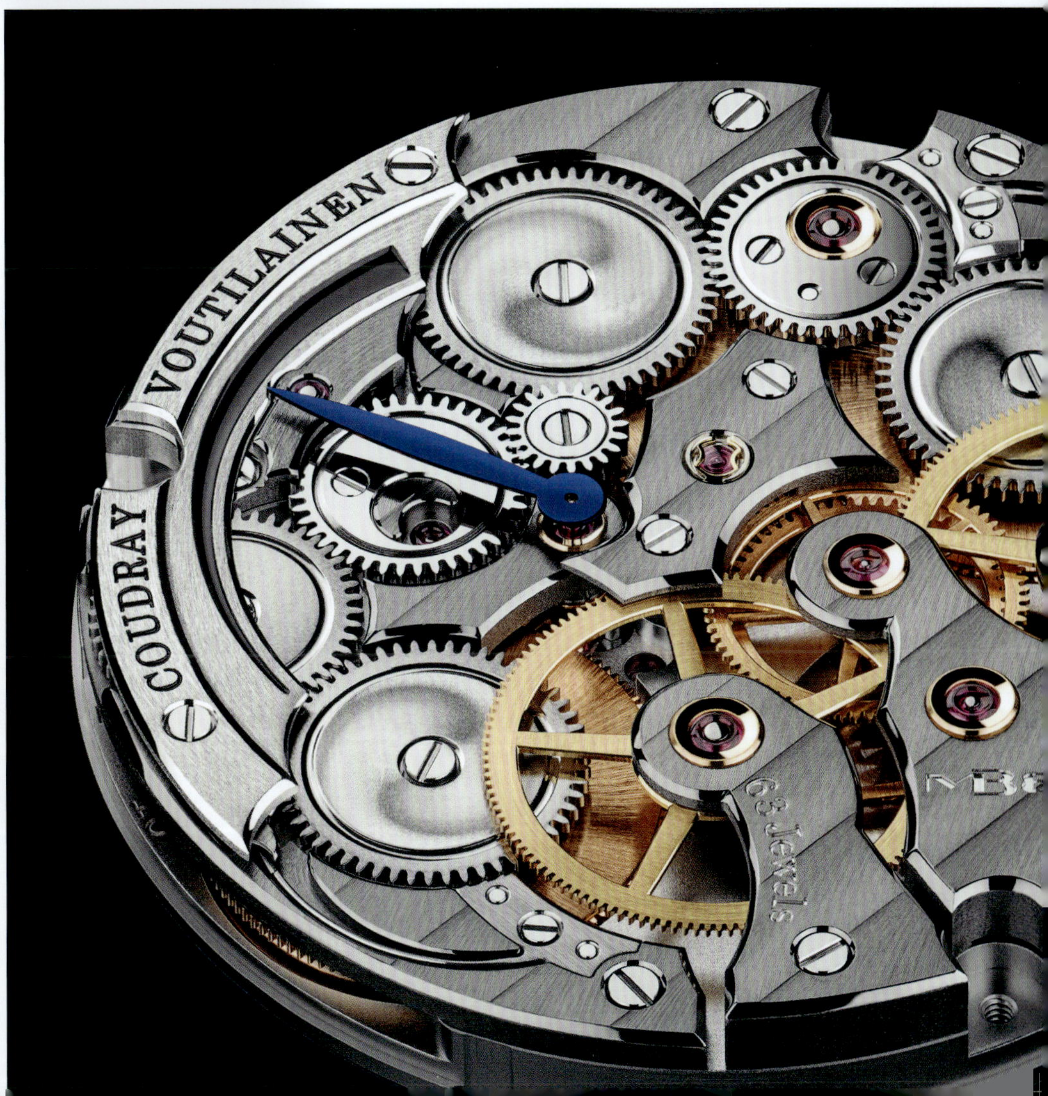
VOUTILAINEN
COUDRAY
63 Jewels

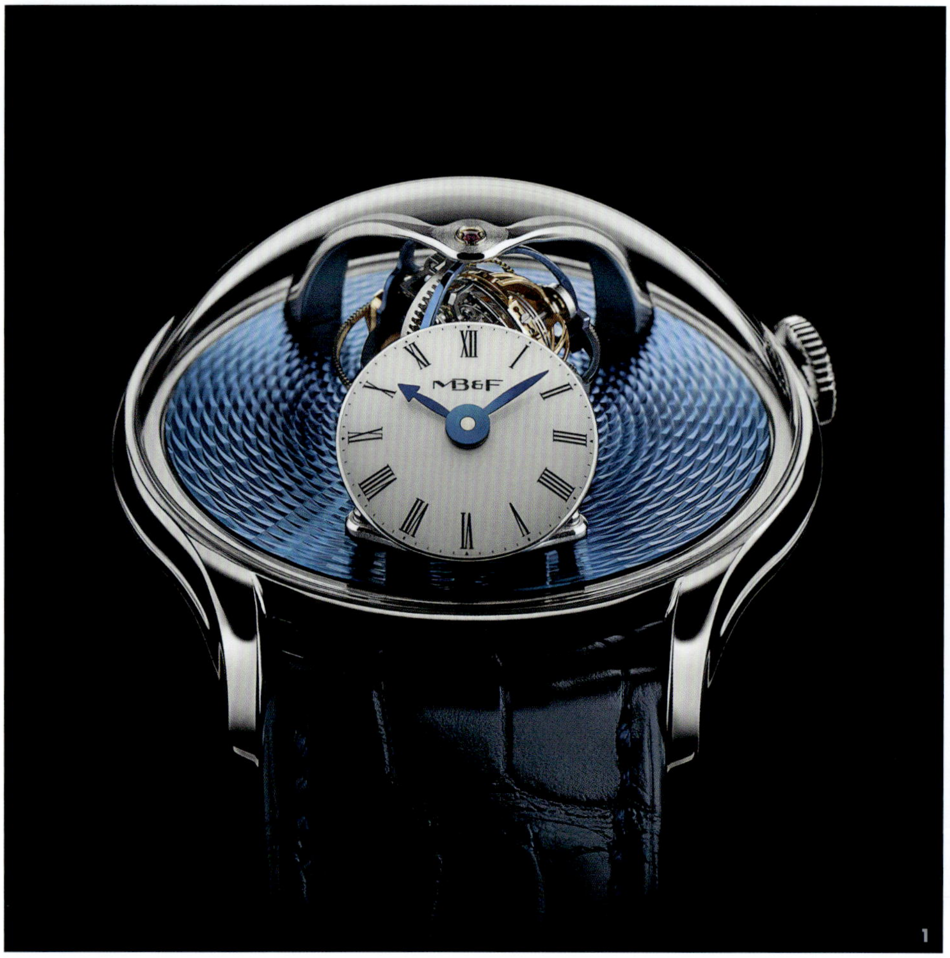

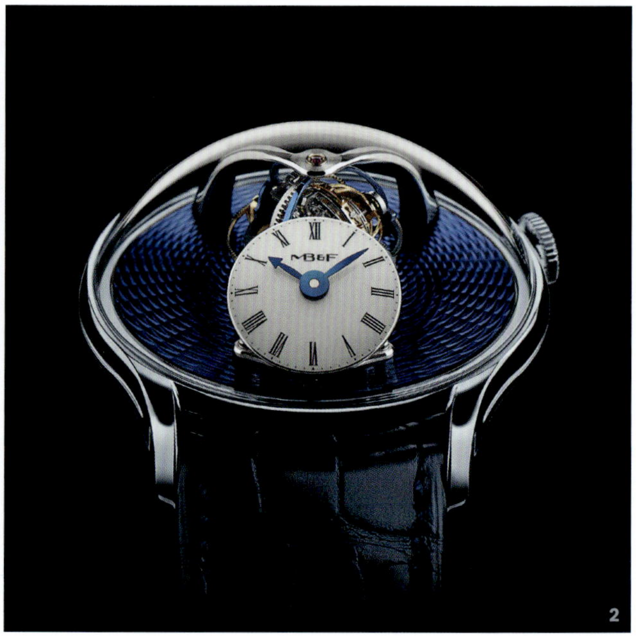

1 Legacy Machine
Thunderdome PT
Reference: 06.PL.B
Material: platinum 950
Limitation: 33
Introduction: 2019
Years in production:
2019 – in production
Quantity produced
(at the end of 2020):
22 + 1 prototype

2 Legacy Machine
Thunderdome
THG Tantalum Blue
Reference: 06.TAL.BLG
Material: tantalum
Limitation: 5
Introduction: 2019
Years in production:
2019 – in production
Quantity produced
(at the end of 2020):
4 + 1 prototype

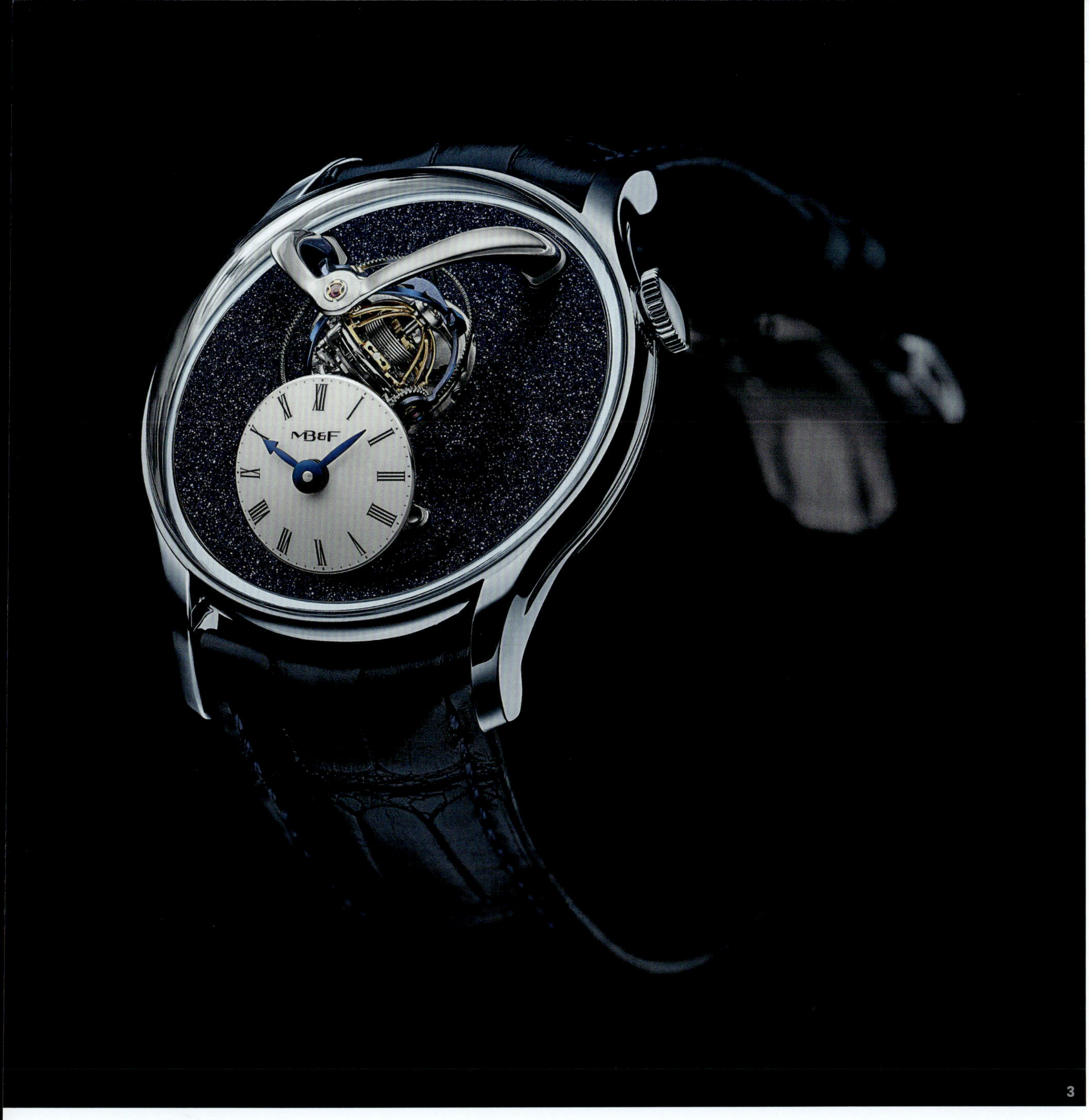

3

3 Legacy Machine
Thunderdome
THG Tantalum Aventurine
Reference: 06.TAL.BLA
Material: tantalum
Limitation: 5
Introduction: 2019
Years in production:
2019 – in production
Quantity produced
(at the end of 2020): 4

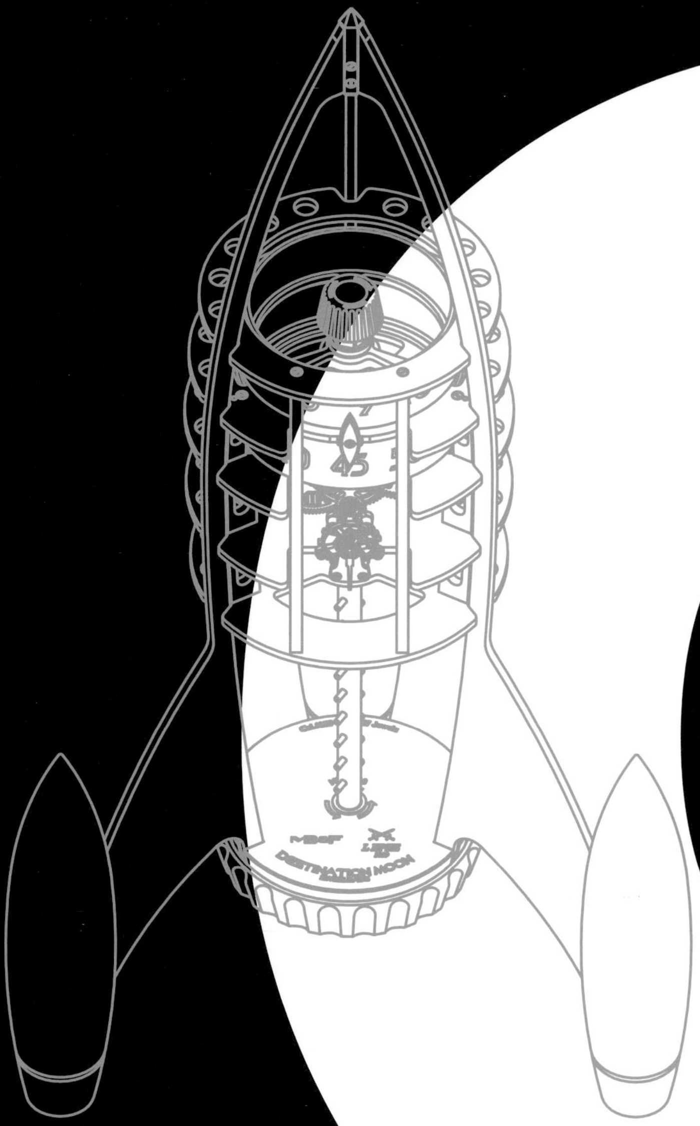

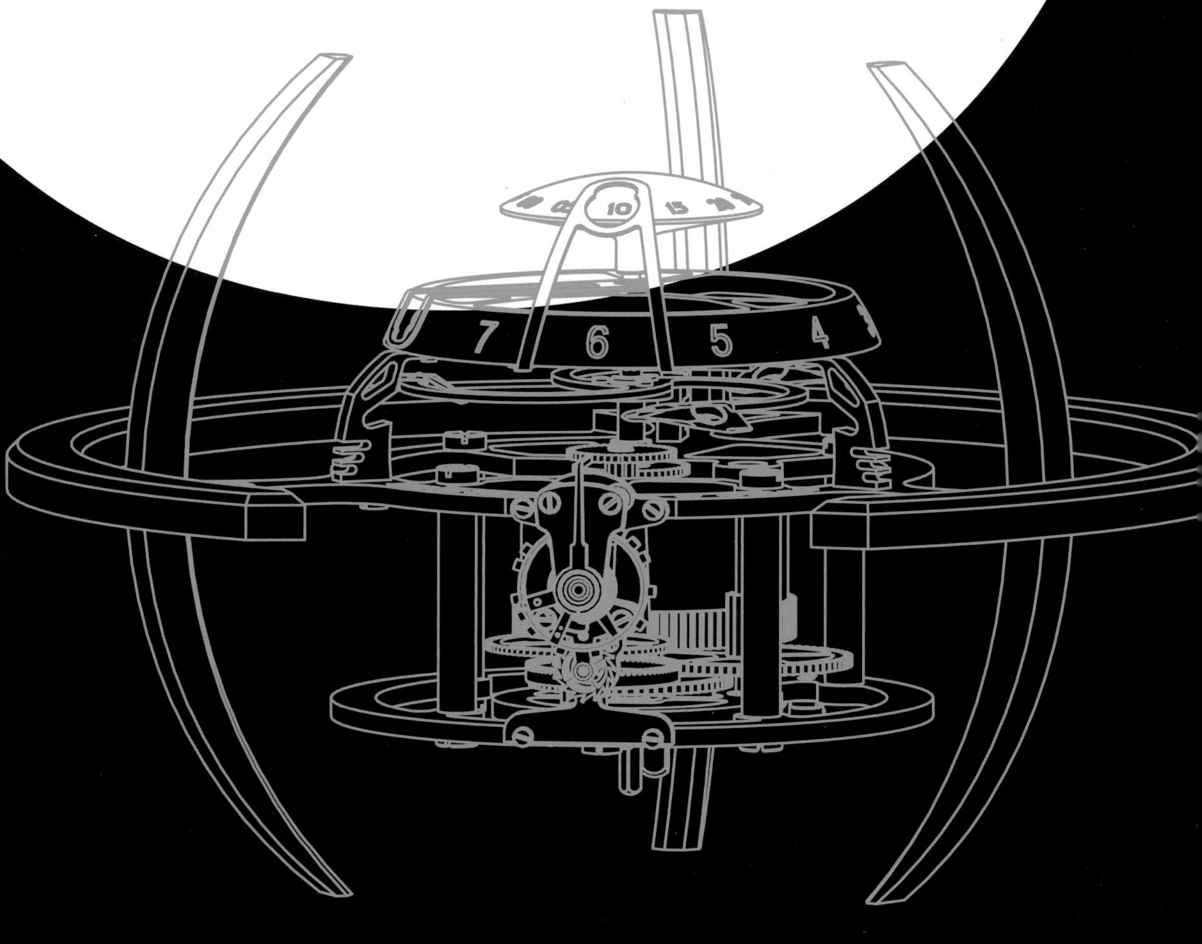

Co-Créations

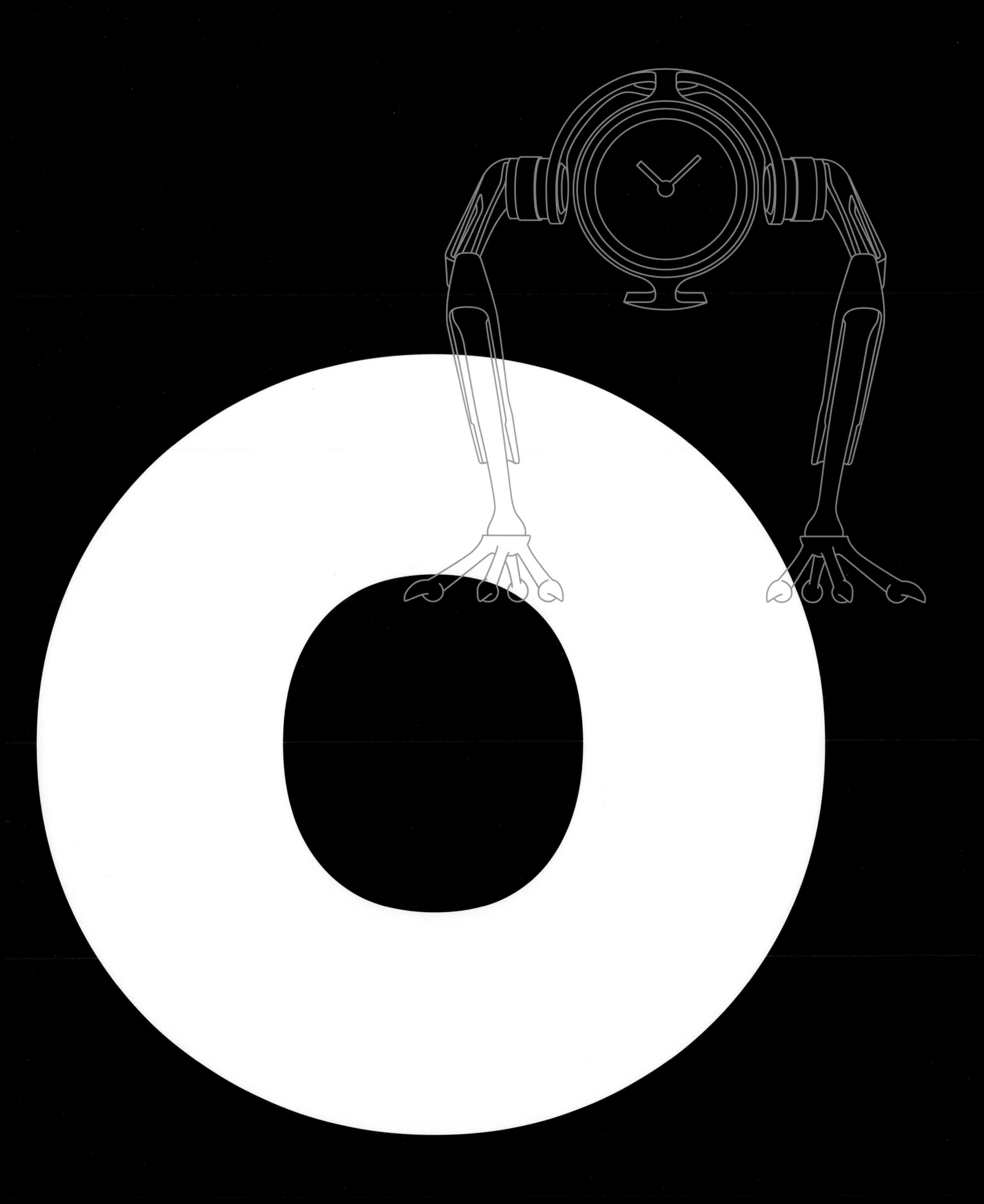
O

CO

CO-CREATIONS

The “F” in MB&F stands for “Friends”: the artists, craftspeople and engineers whose designs and contributions are so much part of MB&F. Co-Creations push this collaborative spirit one step further, forging creative partnerships that merge MB&F concepts with the technical savoir-faire of other independent manufacturers. To date these partners include: L’Epée 1839; Reuge; Caran d’Ache; and Loupe System.

Subjects for Co-Creations are wide-ranging, from table clocks, music boxes and automatons to magnifying lenses and writing instruments. The thematic concepts they apply should be readily familiar to the MB&F enthusiast – air and space travel for instance are common themes. Starfleet Explorer, The Fifth Element, Destination Moon and Starfleet Machine are all fruits of collaboration with L’Epée 1839; collaboration with Reuge produced MusicMachine, MusicMachine 2 and MusicMachine 3; collaboration with Caran d’Ache produced Astrograph; and Loupe System was the partner for Project LpX.

Then there are the zoological and biomorphic shapes: Tripod, T-Rex, Medusa, Octopod and Arachnophobia from L’Epée 1839; and Kelys & Chirp from Reuge. You also have the thematic opposites, in android and robotic forms, like Grant, Balthazar, Sherman and Melchior from collaboration with L’Epée 1839 – all concepts that began as the childhood musings of MB&F founder Maximilian Büsser, given new modes of expression thanks to the participation of MB&F’s partner brands.

The first MB&F Co-Creation was the 2013 MusicMachine with Swiss music box maker Reuge; but it is Swiss clockmaker L’Epée 1839 that has proved the most prolific and industrious partner, with thirteen Co-Creations between 2014 and 2020. Collaborations with Reuge total four, plus one each for Caran d’Ache, the Swiss producer of writing instruments, and Loupe System, a Hong-Kong based producer of magnifying lenses used by watchmakers and watch enthusiasts.

Taken together, what we see in MB&F’s Co-Creations is an amalgam of some of Maximilian Büsser’s most fantastical and treasured dream objects. Who wouldn’t prefer a rocket ship to a pen, a Tyrannosaurus Rex to a clock, or a starfighter to a music box? After all, as the mantra goes: “a creative adult is a child who survived.”

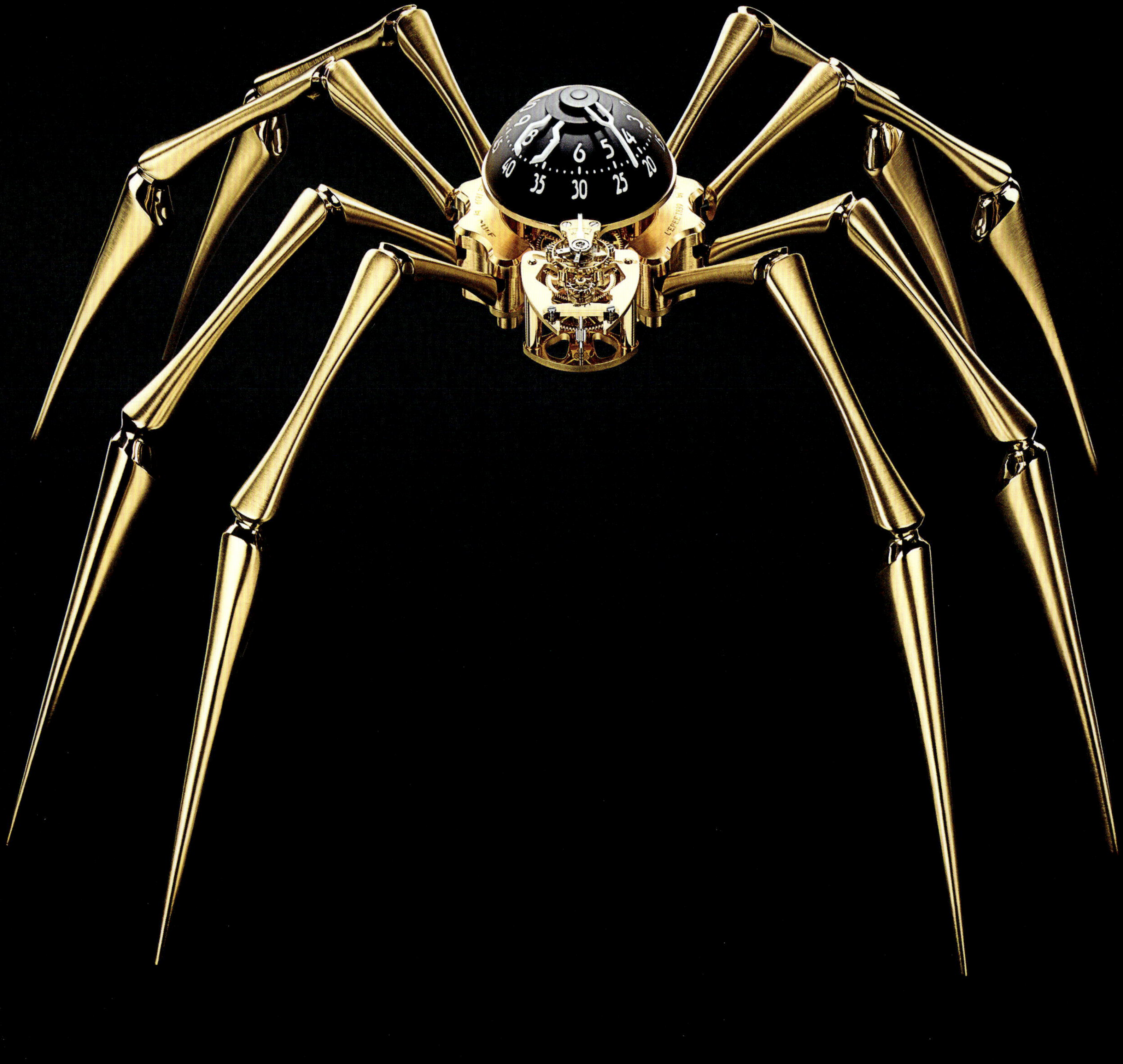
6 5
40 35 30 25 20

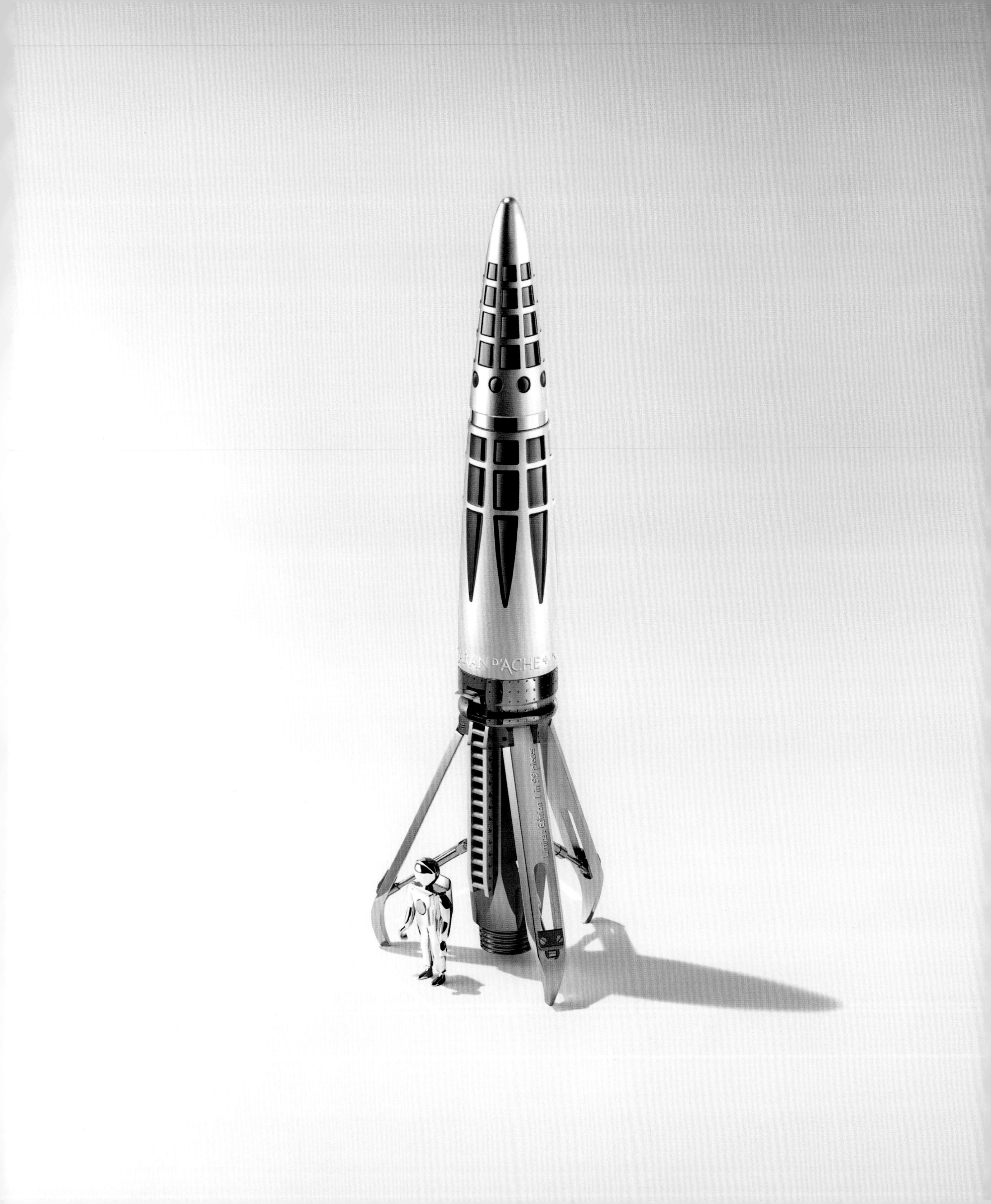
ARAN D'ACHE
Limited Edition 1 in 99 pieces

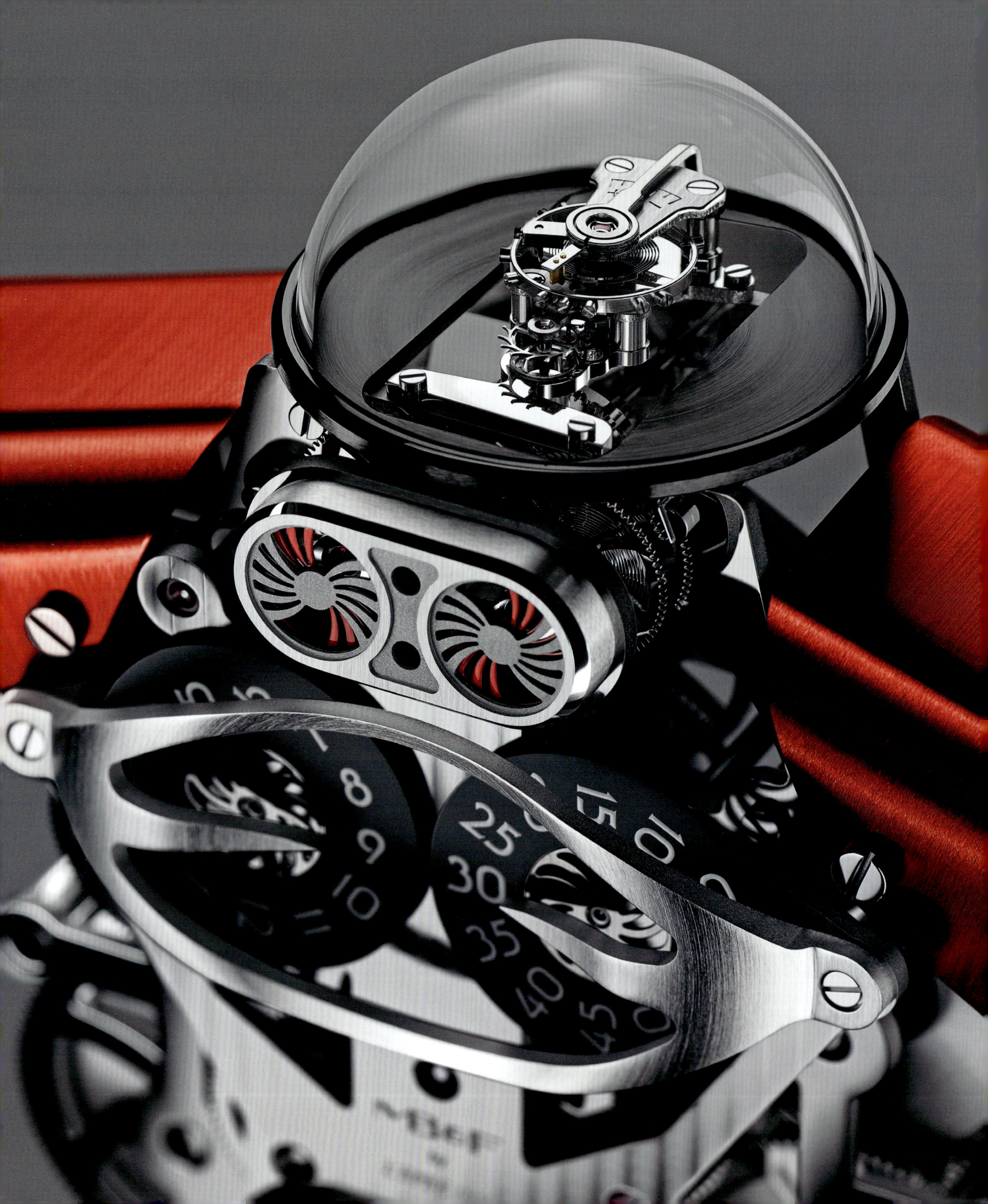
8
9
10
25
15
10
30
35
40
45

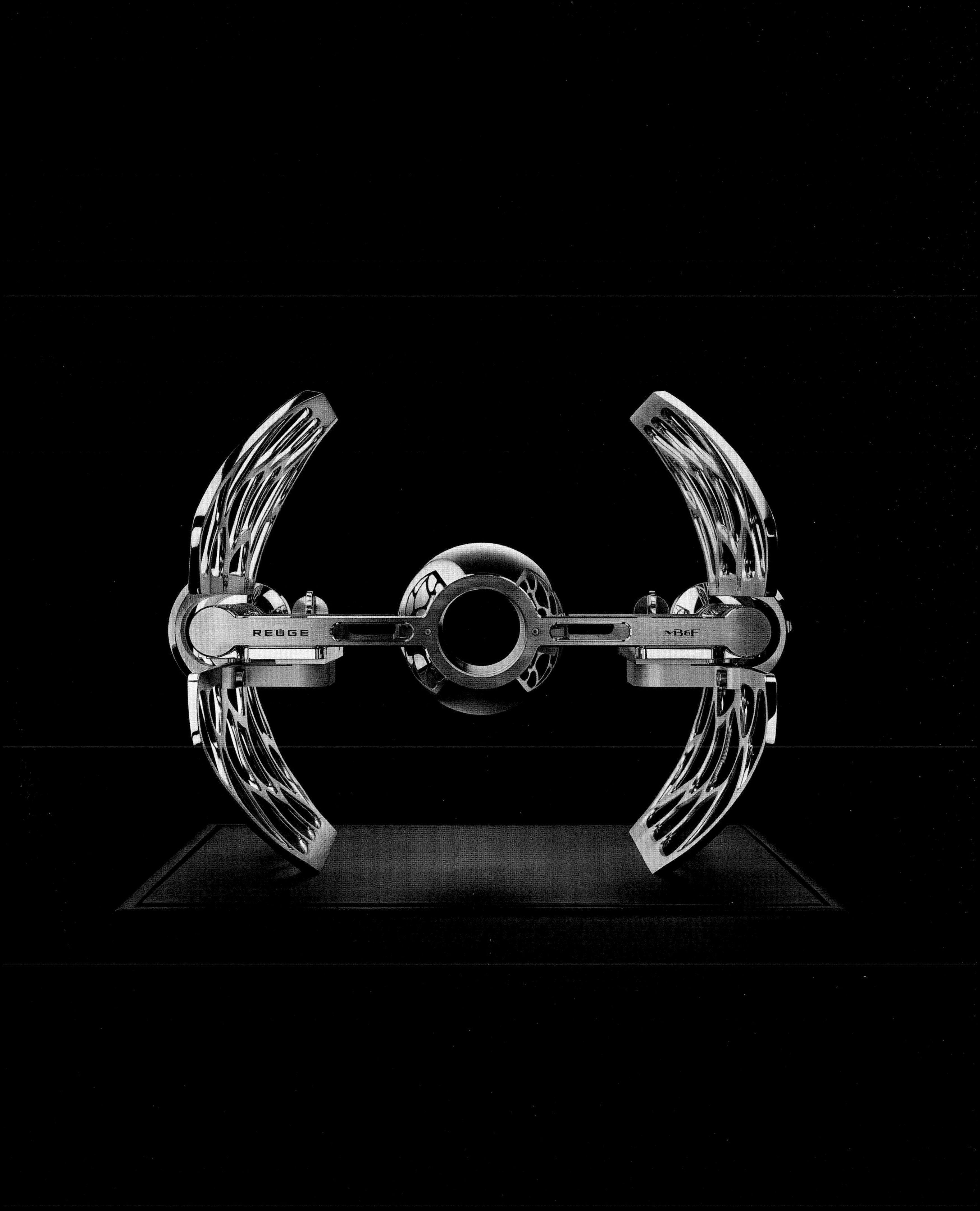
REUGE

L'ÉPÉE 1839

L'Epée 1839 is based in Delémont, Switzerland, and has been the world's leading manufacturer of high-end clocks for more than 180 years. The company was founded by Auguste L'Epée in Besançon, France, initially as a producer of music box and watch components. By the 19th century, L'Epée had developed a solid reputation as the manufacturer of the world's best carriage clocks. In the 1970s, L'Epée famously created the now-iconic wall clocks for the Concorde supersonic aircraft.

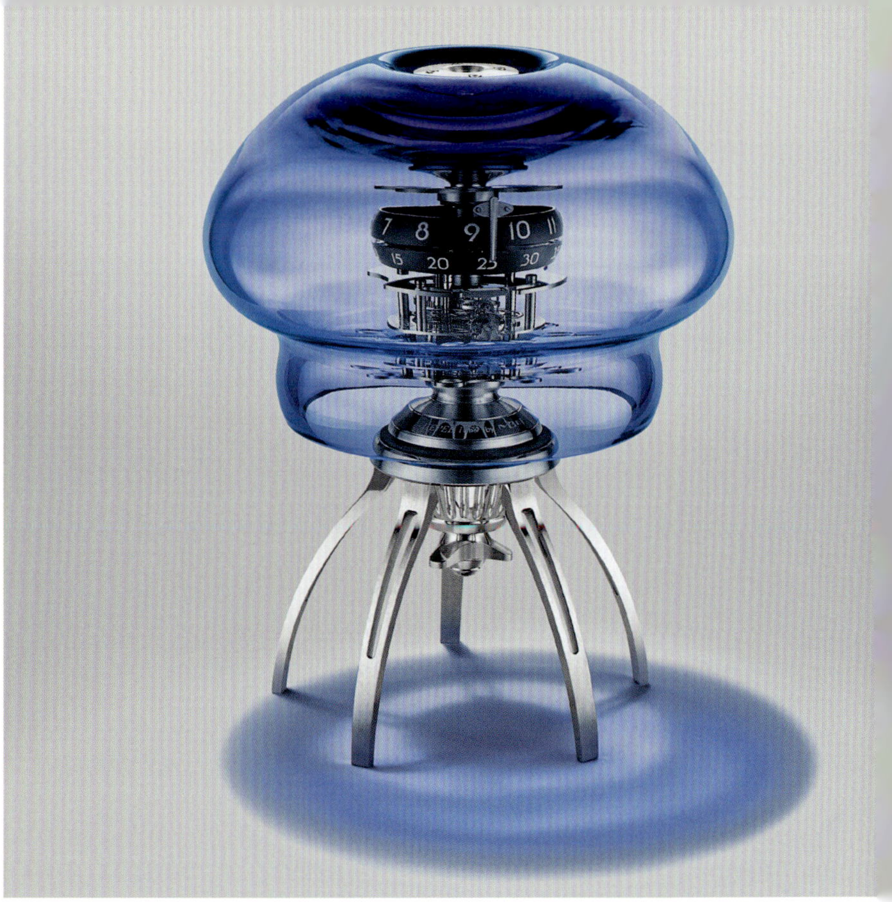

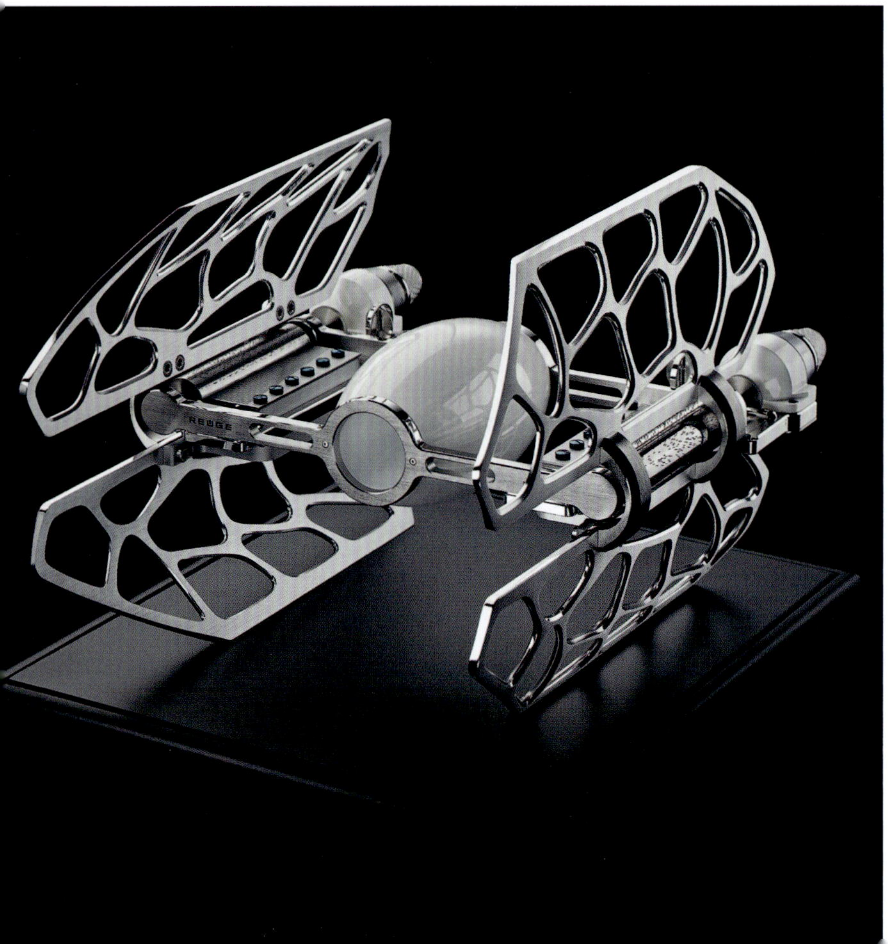

REUGE

Reuge has more than 150 years experience as the world's leading manufacturer of mechanical music boxes. Within a year of the company's foundation by Charles Reuge in 1865, its musical movements could be found in the most unlikely objects, from powder compacts to cigarette lighters. Reuge today produces custom and bespoke music boxes able to replicate any melody.

CARAN D'ACHE

Founded in 1915, Caran d'Ache designs and produces writing instruments and accessories to the highest standards of Swiss manufacturing.

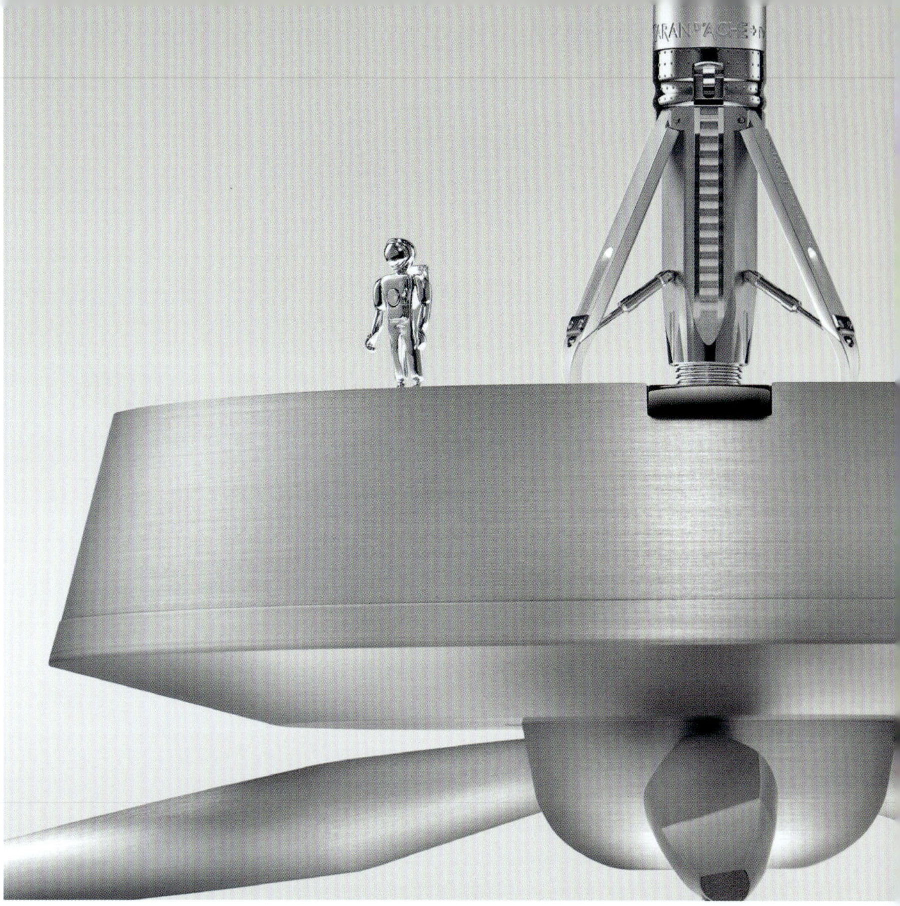

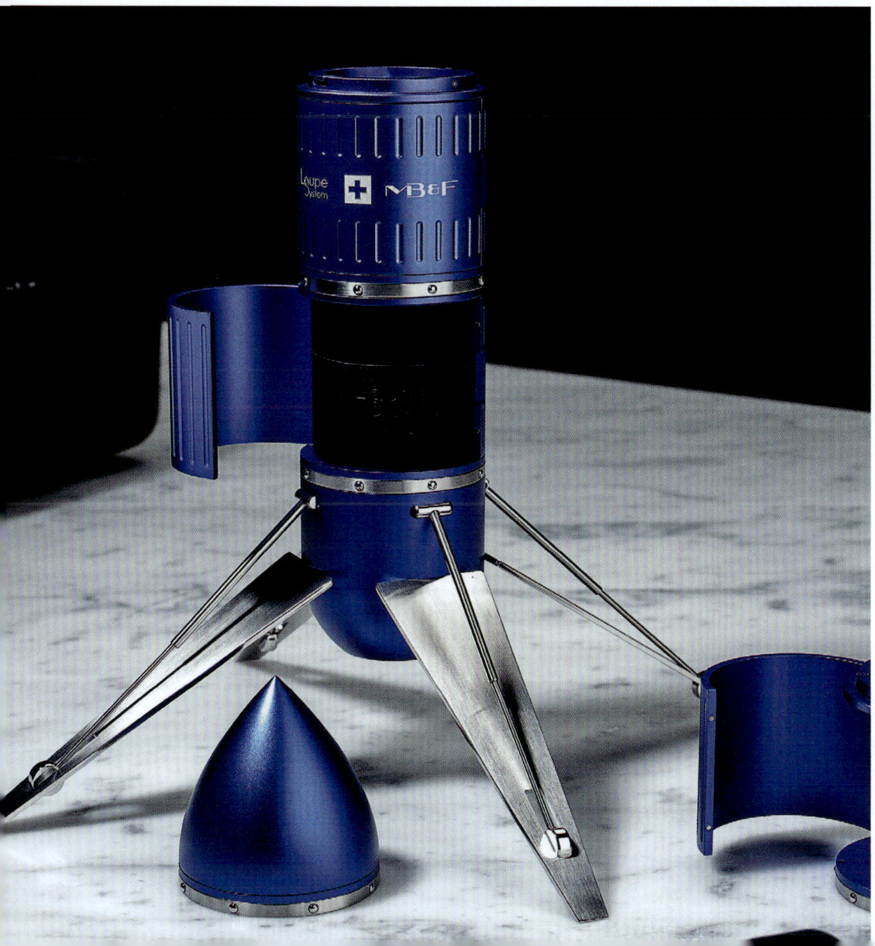

LOUPE SYSTEM

Loupe System, founded in 2012, is the world's premier producer of high-quality portable loupes for watch appreciation. The company founder, disappointed by the poor quality of the loupes commonly used in the watch industry, set out to create his own proprietary loupes. Of superlative quality, the initial pieces sold out within minutes of their launch at the Baselworld trade fair in 2012.

L'ÉPÉE 1839 × MB&F

STARFLEET MACHINE

Inspired by science fiction, Starfleet Machine is almost more spaceship than table clock. Features include hours, minutes, double retrograde seconds and a power reserve indicator, with time displayed on a central black dome. A smaller rotating dome, accompanied by a revolving radar dish, indicates the remaining power reserve by means of bars – when the bars diminish it is time to wind the clock. Starfleet Machine was issued as a 175-piece edition including "light" and "dark" versions, with ruthenium-finished components.The original Starfleet Machine, launched in 2014, was the first clock ever co-created by MB&F and L'Epée 1839; it marked the 175th anniversary of L'Epée and the beginning of a long series of collaborations with MB&F. Inspired by an intergalactic space station, it stood out like a Vulcan at a terrestrial dinner party - even compared to other contemporary table clocks!

Collaboration
L'Epée 1839 × MB&F
Collection
Co-Creations
Series
Starfleet Machine
Editions
Marking L'Epée 175th anniversary: Starfleet Machine "Light" (50.6801/101) and Starfleet Machine "Black Ruthenium" (50.6801/201), limited to 175 pieces; Starfleet Machine "Chocolate" (50.6801/301) Unlimited piece; Starfleet Machine Black Badger: "Purple Reign" (50.6801/402), "Radar Green" (50.6801/401) and "Phantom Blue" (50.6801/400), limited to 18 pieces each
Movement
L'Epée 1839 in-house designed and manufactured
Materials
Palladium-treated brass and stainless steel, luminescent hours and minutes with exclusive lume details for the Black Badger editions
Power Reserve
40 days/manual winding with double-depth square socket key
Dimensions
29 × 21 cm
Functions
Hours and minutes on a central dome; 20-second intervals indicated by double retrograde fly-out cannons emanating from central dome; dome indicator, framed by a hand finished arc, provides an intuitive view of remaining energy as it turns 270° – complemented by a "radar dish" that also revolves 270°
Introduction
2014

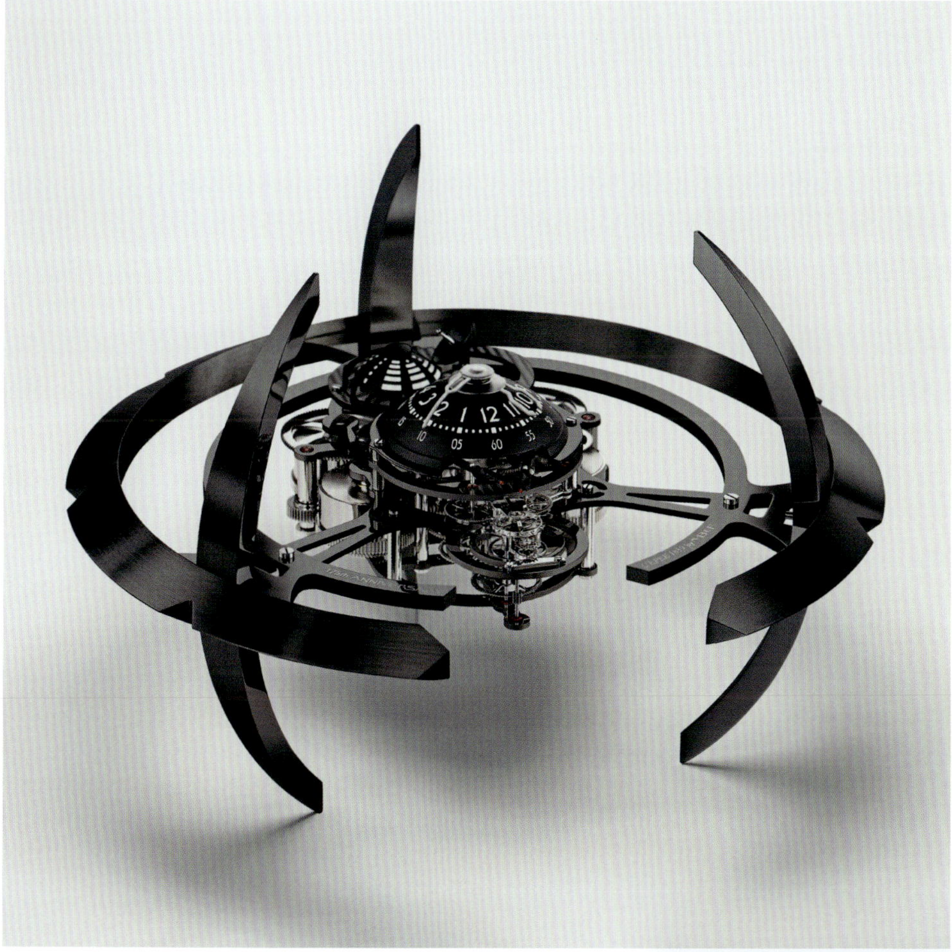

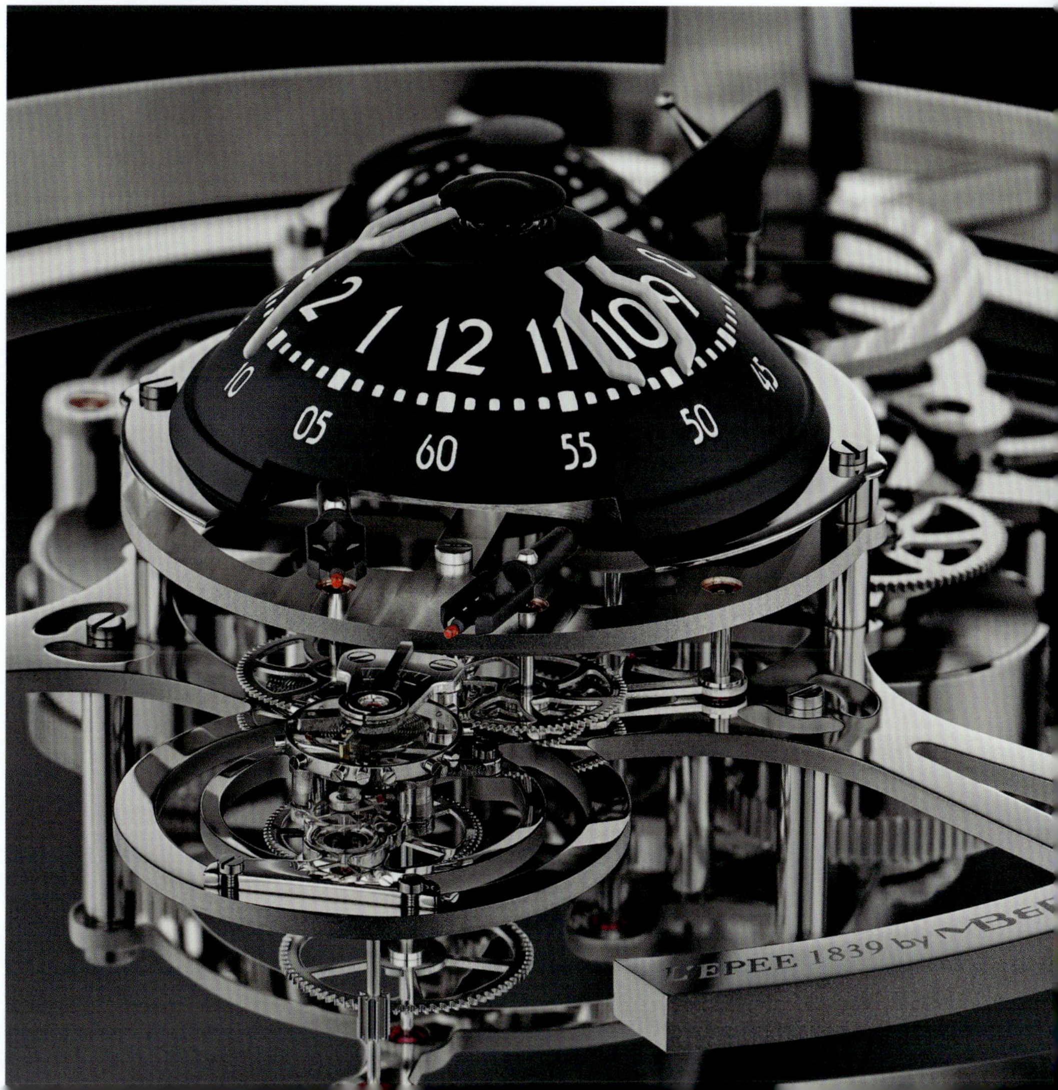
2 1 12 11 10 9
10 05 60 55 50 45
L'EPEE 1839 by

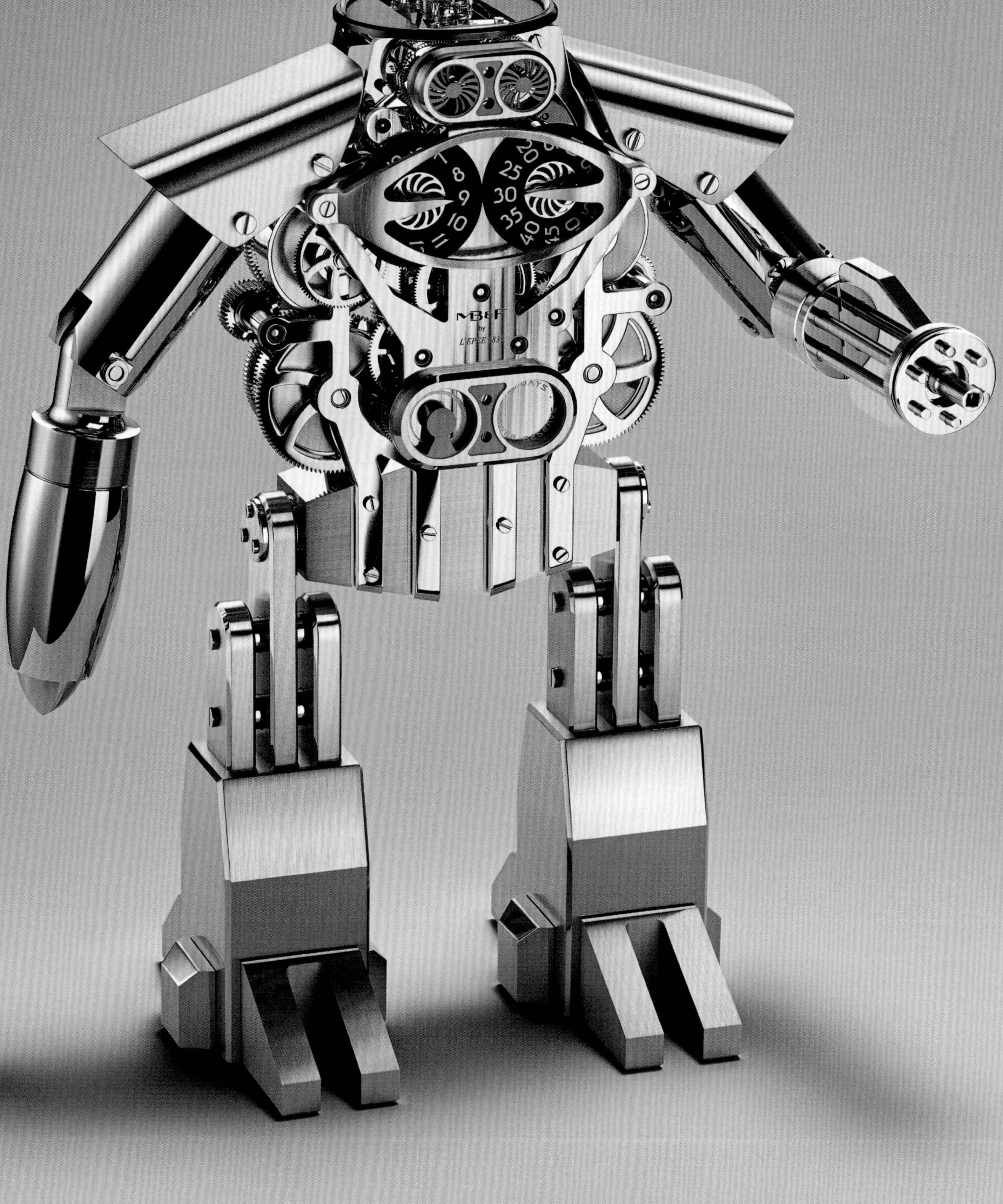
MB&F
by

L'ÉPÉE 1839 × MB&F

MELCHIOR

With his steel and brass armour and articulated arms, the 480-component Melchior table clock resembles the robotic characters in science-fiction series like Star Wars or Lost in Space. The clock features jumping hours, sweeping minutes, double retrograde seconds, and a 40-day power reserve. The arms pivot up and down, and hold a rocket launcher and a detachable gun, which doubles as the machine's winding key. Melchior was released as a 99-piece edition, including a monochromatic "light" version, and a two-tone "dark and light" version with black PVD-treated components. A final 100th piece with red armour was released as a Pièce Unique for the Only Watch charity auction. Presented in 2015 and created with L'Epée 1839, Melchior celebrated MB&F's 10th Anniversary. "A creative adult is a child who survived": Melchior reminds you of your childhood dreams, but this 480-component kinetic and majestic robot is also a carefully crafted mechanical table clock.

Collaboration
L'Epée 1839 × MB&F
Collection
Co-Creations
Series
Melchior
Editions
Monochromatic "Light" (50.6802/101) or "Dark and Light" (50.6802/201), limited to 99 pieces; Unique piece for Only Watch 2015: Melchior "Only Watch" (50.6802/301)
Movement
L'Epée 1839 movement, designed and manufactured in-house; Incabloc shock protection system
Materials
Stainless steel, brass and mineral glass
Power Reserve
40 days/manual winding with double-depth square socket key
Dimensions
30.3 × 21.7 × 11.2 cm
Functions
Jumping hours and sweeping minutes; twin discs forming part of Melchior's breast plate, one disc displaying hours, the other disc minutes; retrograde seconds: flyback discs mark 20-second intervals behind a steel mask; power reserve indicator on Melchior's belly
Introduction
2015

ARACHNO-PHOBIA

Arachnophobia is a Louise Bourgeois-inspired table clock in the form of a spider, with hours and minutes displayed on the creature's abdomen. Attached to the abdomen is the machine's head, incorporating an oscillating balance wheel and even including a set of jaws. Eight articulated legs radiate from the spider's body, each one adjustable so the machine can assume different positions. Arachnophobia was released in a choice of black or 18K yellow gold. Designed by MB&F, Arachnophobia is as extreme as it comes... but the eye-catching three-dimensional sculpture is also an impeccably finished table clock (or wall clock), engineered and manufactured by L'Epée 1839.

Collaboration
L'Epée 1839 × MB&F
Collection
Co-Creations
Series
Arachnophobia
Editions
Arachnophobia "Black" – 76.6000/114 and Arachnophobia "Gold" – 76.6000/11
Limited editions of 500 pieces each
Movement
L'Epée 1839 in-house designed and manufactured; Incablock shock protection system
Materials
Palladium-plated brass, or 18K yellow gold-plated brass
Power reserve
8 days/manual winding with double-depth square socket key
Dimensions/Weight
Movement: 7.5 × 13.4 × 6.3 cm; clock: 20.3 × 40.5 cm
Functions
Hours and minutes: curved hands rotate to indicate hours and minutes on a polished, central dome
Details
An innovative system enables Arachnophobia to be either clipped to a wall or stand on a table, and the legs are articulated.
Introduction
2015

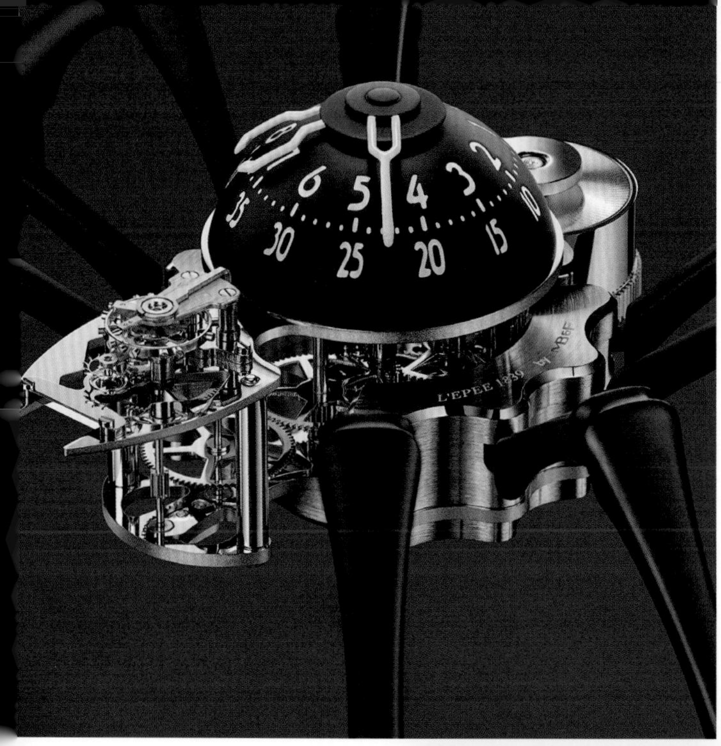
6 5 4 3
30 25 20 15

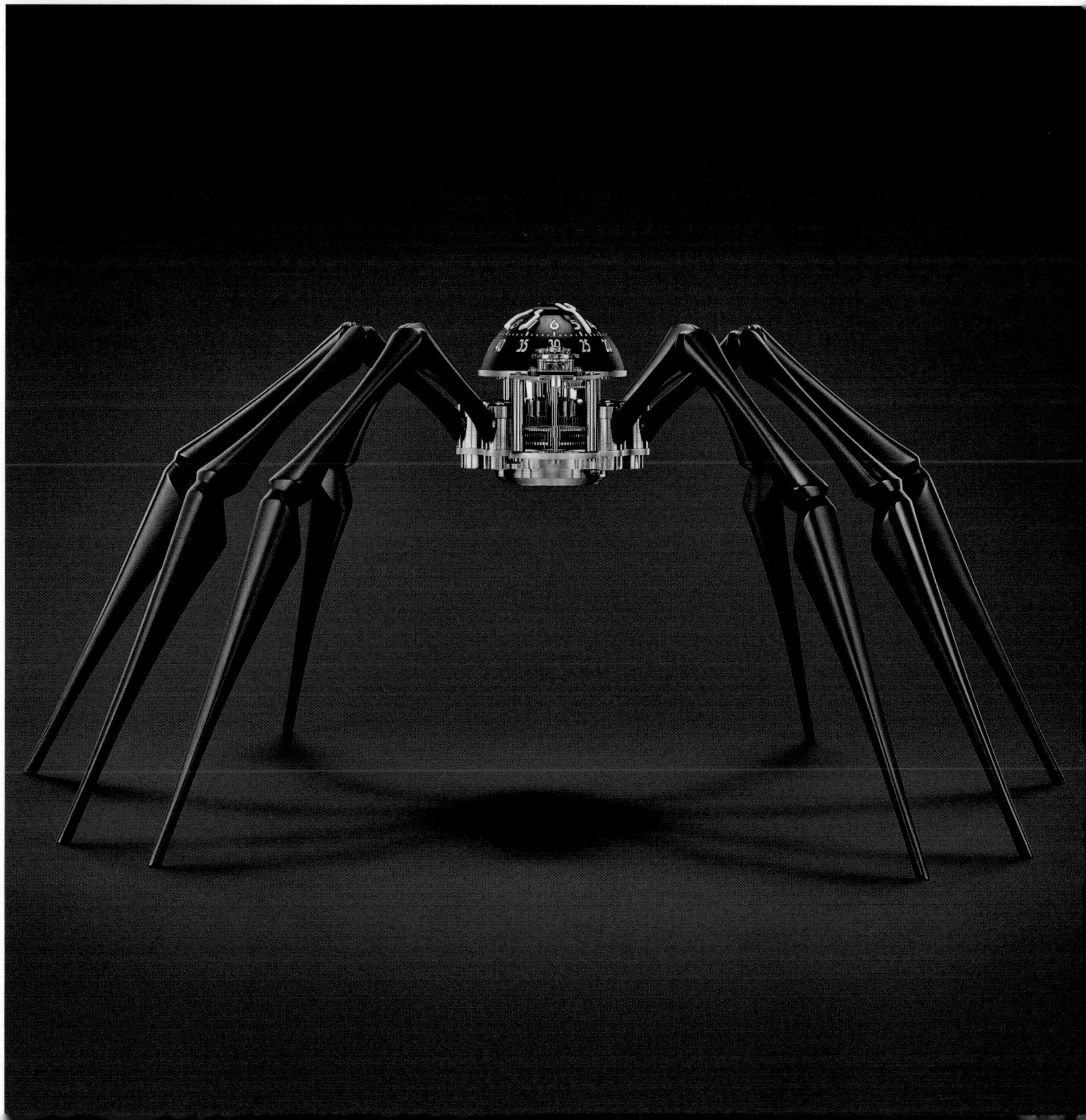

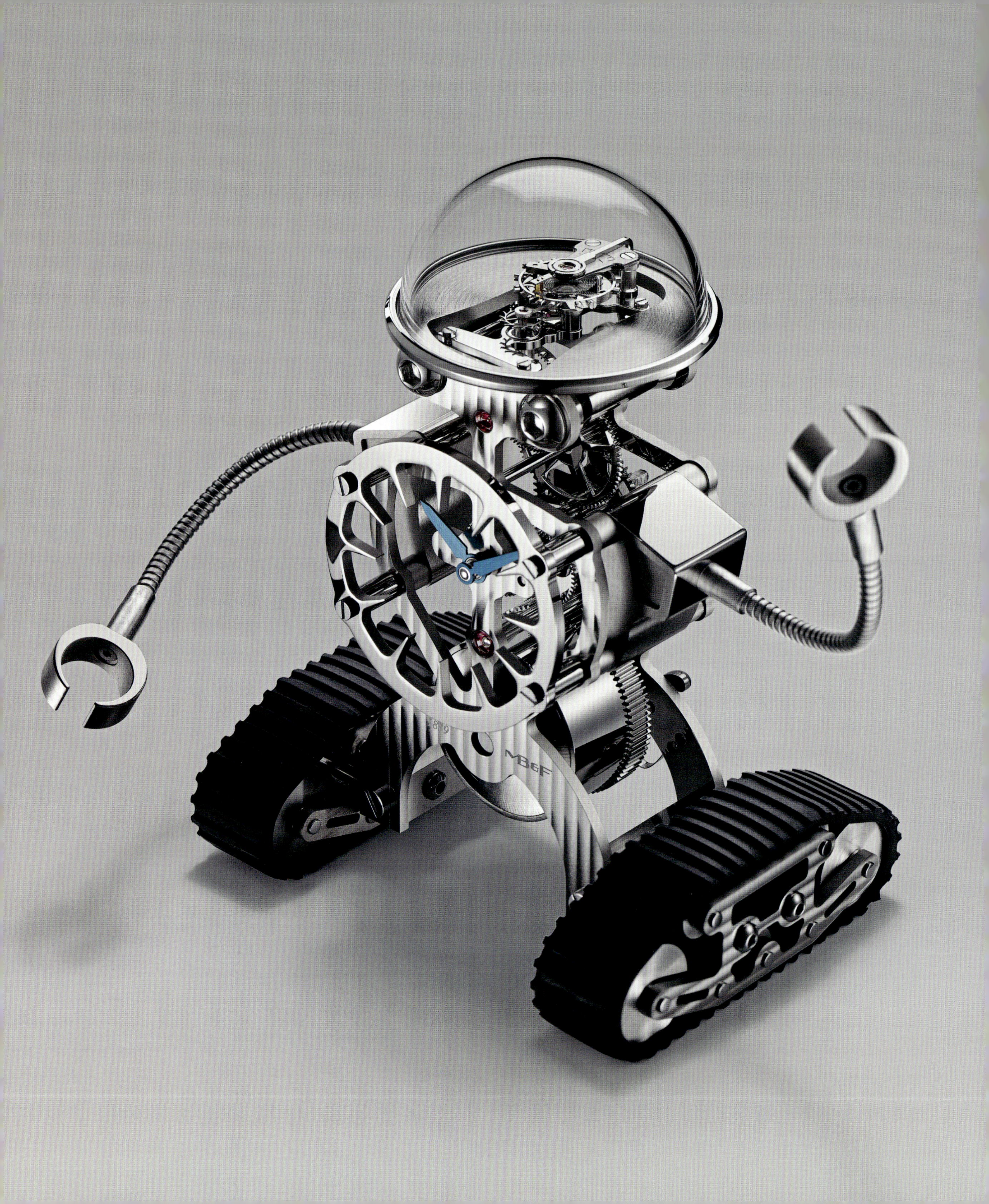

L'ÉPÉE 1839 × MB&F

SHERMAN

Sherman is a table clock in the form of a robot friend, complete with a mechanical brain under a glass dome cranium, articulated arms that can hold anything from a pencil to a winding key, and rubber caterpillar tracks for feet. Hours and minutes are displayed centrally on the robot's torso. Sherman was produced in several limited editions: palladium plating, gold plating, set with diamonds and gold plating...
Sherman doesn't walk, talk, weld cars, or roam Mars. He doesn't try to kill Sarah Connor, help Luke Skywalker, warn Will Robinson, star in feature-length films, or enforce the law. In fact, Sherman really only does two things, but he does both extremely well. Sherman tells the time. And Sherman makes people smile, which is probably the world's most useful and emotionally valuable complication.

Collaboration
L'Epée 1839 × MB&F
Collection
Co-Creations
Series
Sherman
Editions
Sherman "Palladium" (76.6001/101), limited to 200 pieces; Sherman "Gold" (76.6001/041), limited to 200 pieces; Sherman "Gold Diamond set" (76.6001/051), limited to 50 pieces; Sherman "Silver Diamond set" (76.6001/151), limited to 50 pieces; Bad Sherman "Midnight Green" (76.6001/131), limited to 18 pieces; Bad Sherman "Black" (76.6001/121), limited to 18 pieces; Bad Sherman "Silver" (76.6001/141), limited to 18 pieces;
Movement
L'Epée 1839 in-line eight-day movement, in-house designed and manufactured; Incablock shock protection system
Materials
Palladium-plated, gold-plated or gem-set with 477 high-quality VVS diamonds, stainless steel and mineral glass
Power reserve
8 days/manual winding with double-depth square socket key
Dimensions
14.3 × 10.9 × 8 cm
Functions
Hours and minutes
Details
Rubber caterpillar tracks are fully functional
Introduction
2016

L'ÉPÉE 1839 × MB&F

BALTHAZAR

Balthazar is a 618-component table clock that takes the form of a 16-inch-tall robot with two sides – a light side and a dark side. On the light side is a smiling robotic face, but rotate the torso at the hips and you have a pair of glaring ruby eyes set in a skull. On his light side, Balthazar displays "slow" jumping hours and trailing minutes via two discs on the chest, with power reserve indicated on his belly. On the dark side, the time display disappears and is replaced by a moon-phase complication accurate for 122 years. Under a polished glass dome, an animated balance oscillates quietly to let you know the robot is ticking. Weighing in at over 8 kilograms and standing 40 centimetres tall, Balthazar is a sophisticated and imposing high-precision robot clock composed of beautifully finished, micro-engineered components.

Collaboration
L'Epée 1839 × MB&F
Collection
Co-Creations
Series
Balthazar
Editions
Balthazar "Black" (50.6803/201), Balthazar "Blue" (50.6803/401), Balthazar "Nickel" (50.6803/104), Balthazar "Green" (50.6803/301), limited to 50 pieces each; Balthazar "Full Gold" (50.6803/001), limited to 20 pieces; Balthazar "Full Black" (50.6803/211), limited to 20 pieces;
Movement
L'Epée 1839 designed and manufactured in-house; Incablock shock protection system
Materials
Palladium-plated polished brass, stainless steel and mineral glass
Power reserve
35 days/manual winding with double-depth square socket key hidden in the shield
Dimensions
39.4 × 23.8 × 12.4 cm
Functions
"Slow" jumping hours and sweeping minutes; 20-second retrograde second displays in the eyes; double hemisphere moon phase indicator
Details
Balthazar rotates on precision ball bearings with spring click to indicate and hold at resting positions; its centre of gravity is low around the hips to minimise risk of being knocked over
Introduction
2016

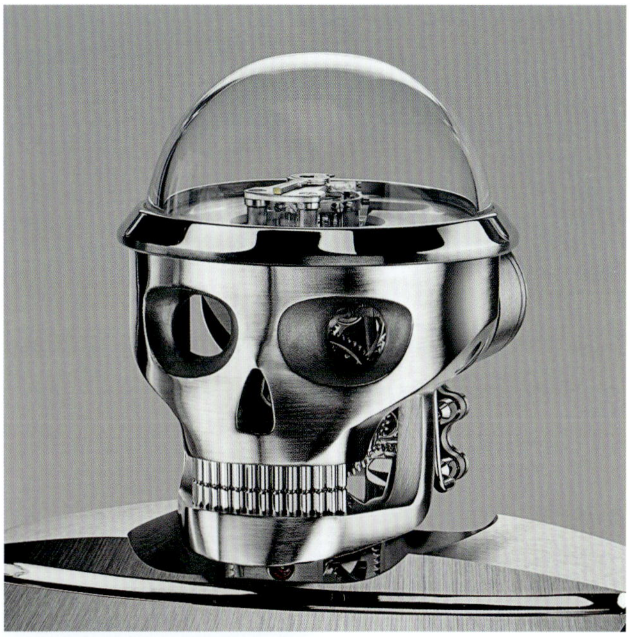

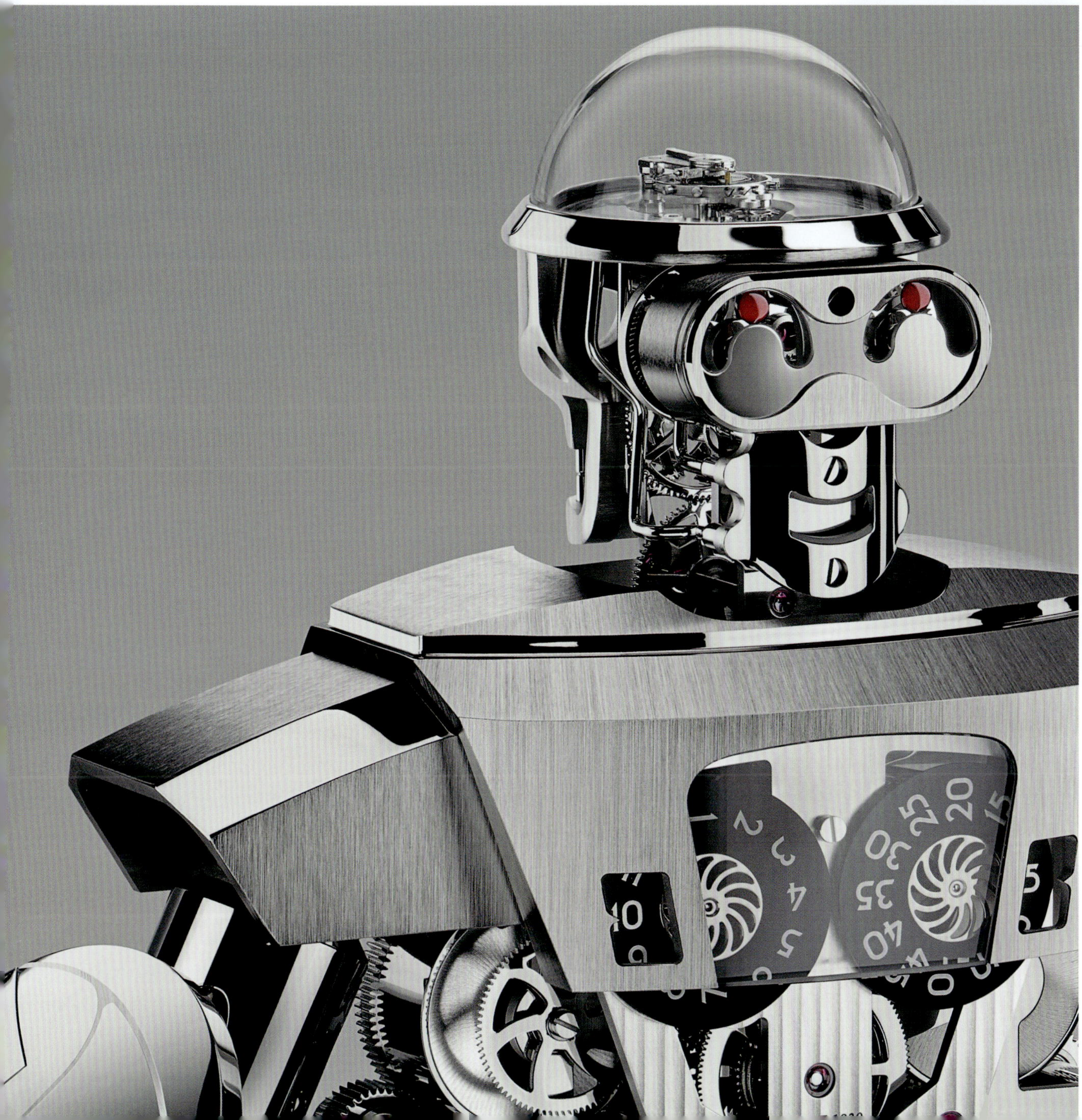

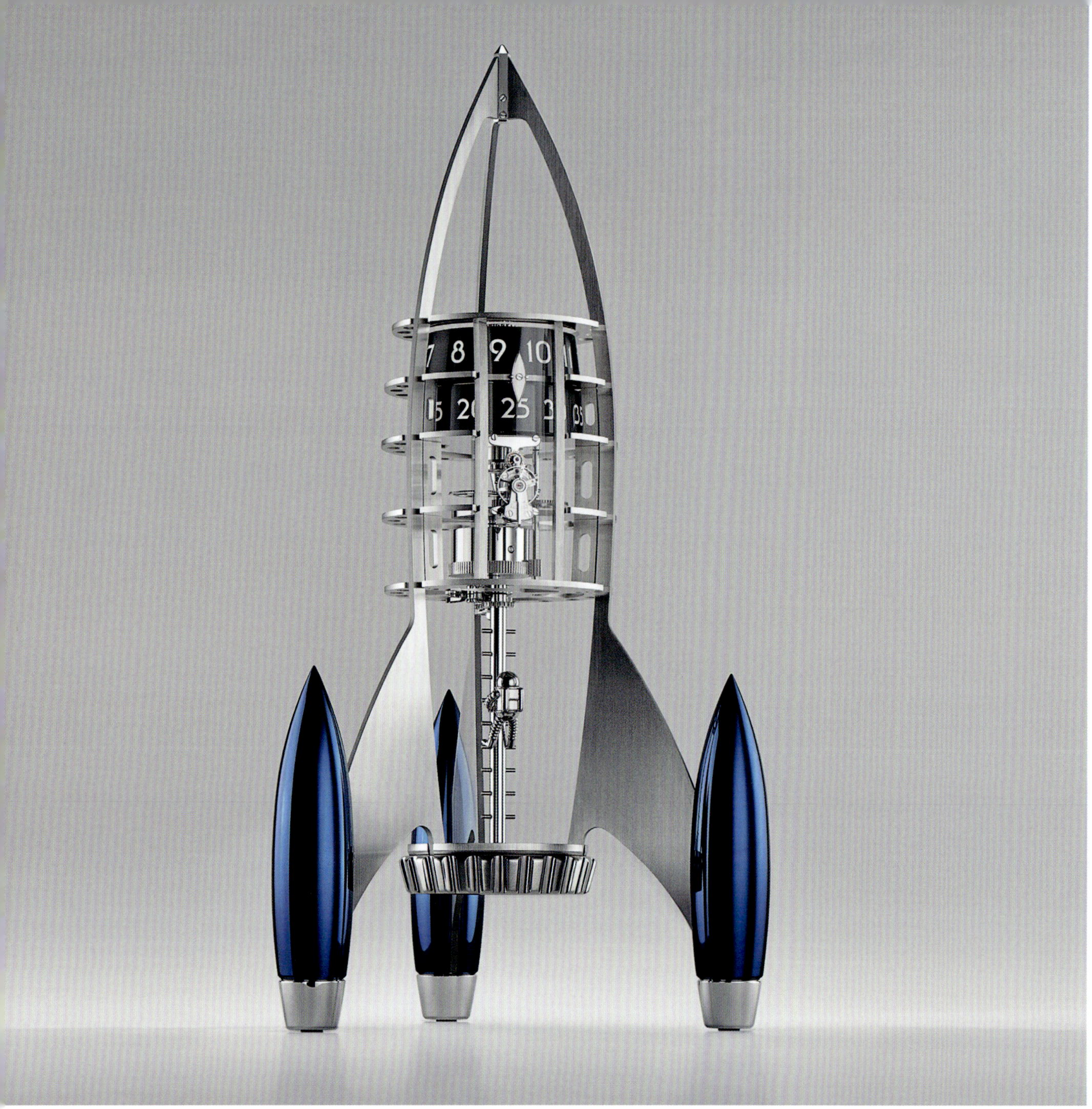

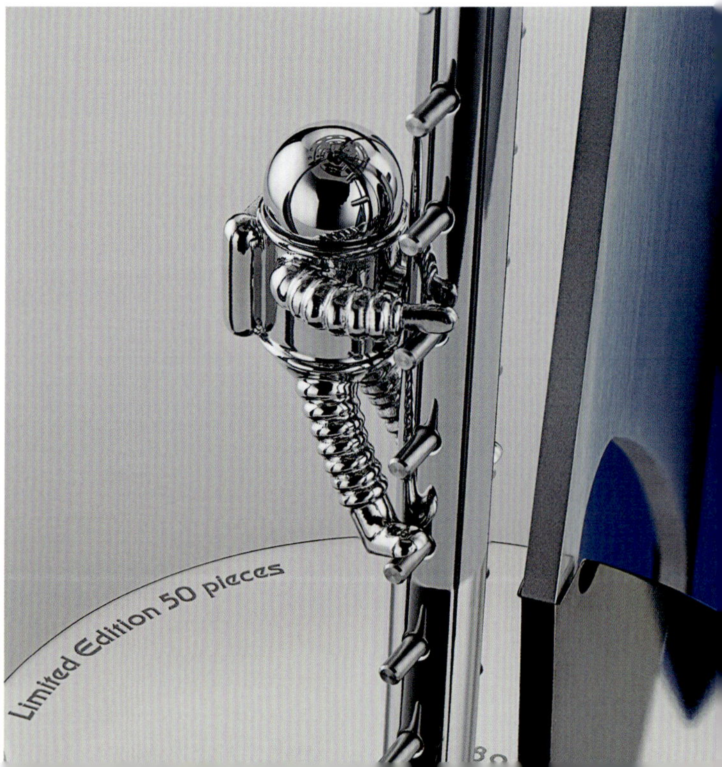
Limited Edition 50 pieces

DESTINATION MOON

Destination Moon is a table clock in the form of a 1960s era science-fiction-inspired space rocket, complete with a tiny stainless steel and silver astronaut. Hours and minutes are indicated by large, stainless steel rotating discs, with power provided by an oversized winding crown at its base. Destination Moon was produced in limited editions of 50 pieces each in black, green, blue and red, and palladium. Conceived by MB&F and built by L'Epée 1839, Destination Moon is the quintessential torpedo-shaped rocket of childhood dreams.

Collaboration
L'Epée 1839 × MB&F
Collection
Co-Creations
Series
Destination Moon
Editions
Destination Moon "Red" (74.6000/154), Destination Moon "Silver" (74.6000/104), Destination Moon "Black" (74.6000/114), Destination Moon "Blue" (74.6000/124), Destination Moon "Green" (74.6000/134), limited to 50 pieces each
Movement
L'Epée 1839 eight-day movement designed and manufactured in-house; Incablock shock protection system
Materials
Palladium-plated brass, nickel-plated stainless steel, PVD coating for the blue, green and black editions, anodized aluminium for the red edition. Solid silver for Neil (astronaut figurine)
Power reserve
8 days/manual winding by propulsion wheel
Dimensions
41.4 × 23.3 cm
Functions
Hour and minute indications stamped on rotating discs, manual winding by rotation of the propulsion wheel
Details
A space-suited figurine is magnetically attached to the ladder
Introduction
2017

OCTOPOD

Octopod is an eight-leg, eight-day table clock inspired by cephalopods. A prominent spherical head in polycarbonate crystal houses the machine's clockwork, which appears to hang unsupported within the sphere. The Octopod's articulated legs can be locked in two positions (standing or extended). The Octopod was produced in limited editions of 50 pieces each, in black PVD, blue PVD or silver. Octopod continues MB&F's exploration of aquatic themes – blending contemporary design with kinetic sculpture and precision horology. The real horological magic and mystery take place in Octopod's completely transparent spherical 'head'. It is gimballed in a similar way to traditional ship chronometers. The pulsating escapement, regulating the clock's precision, is located on its minute hand rather than the more usual position attached to stationary movement plates. Octopod's clockwork is suspended inside its crystalline sphere, so that it appears to be floating in space or water. Like an octopus concealing parts of itself with camouflage, Octopod conceals parts of itself with visual tricks of its own.

Collaboration
L'Epée 1839 × MB&F
Collection
Co-Creations
Series
Octopod
Editions
Octopod "Silver" (11.6000/101), Octopod "Black" (11.6000/201), Octopod "Blue" (11.6000/401), limited to 50 pieces each
Movement
L'Epée 1839 in-house designed and manufactured; Incablock shock protection system
Materials
Stainless steel, nickel, palladium plated brass
Power Reserve
8 days/manual winding with double-depth square socket key
Dimensions
28 × 45 × 22 cm (crouching) or 28 cm (standing)
Functions
Hours, minutes with finely counter-balanced regulator mounted on minute hand
Details
360° rotation of the sphere in both vertical and horizontal planes; 8 articulated legs with release button in each leg, lockable in two positions
Introduction
2017

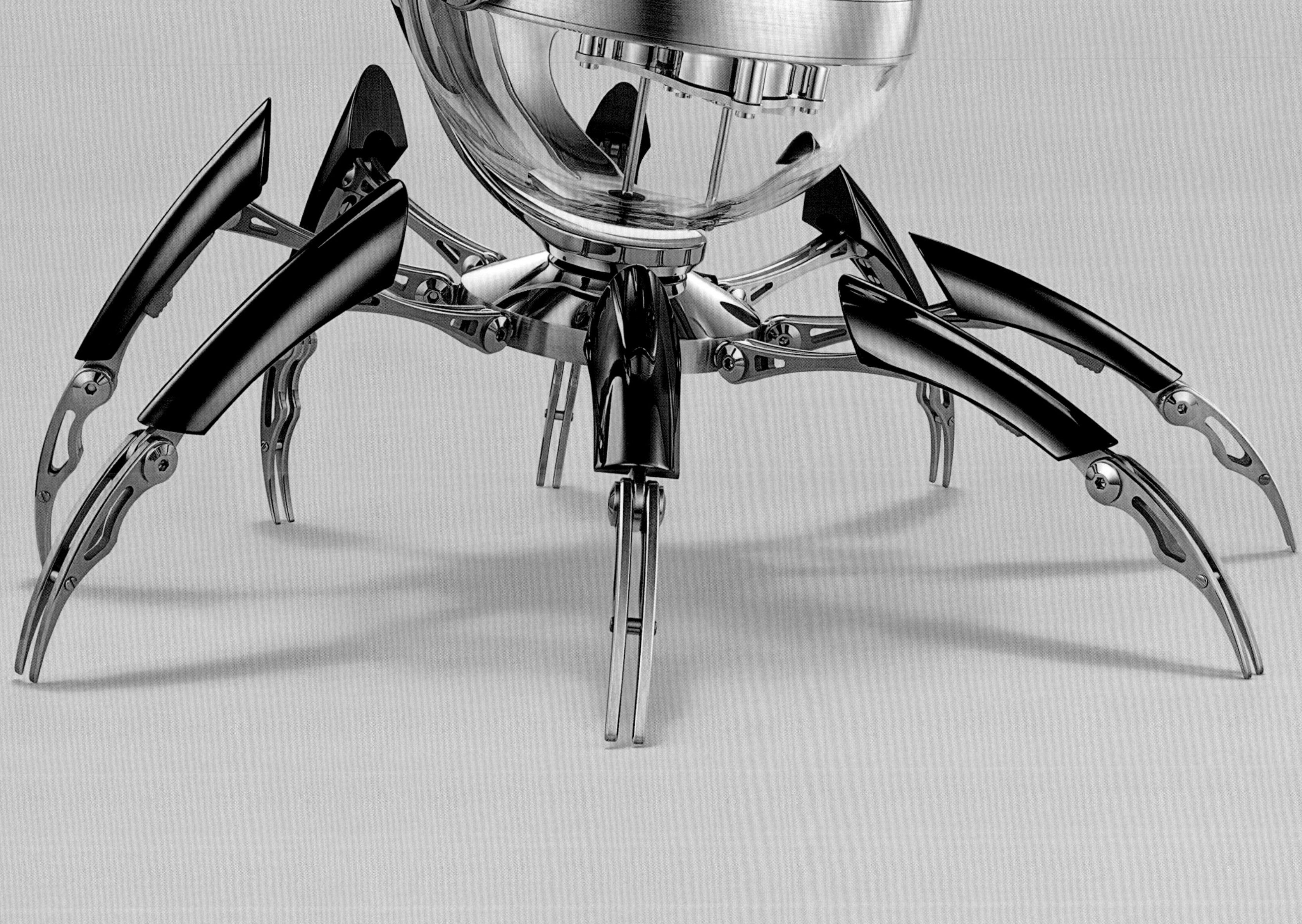

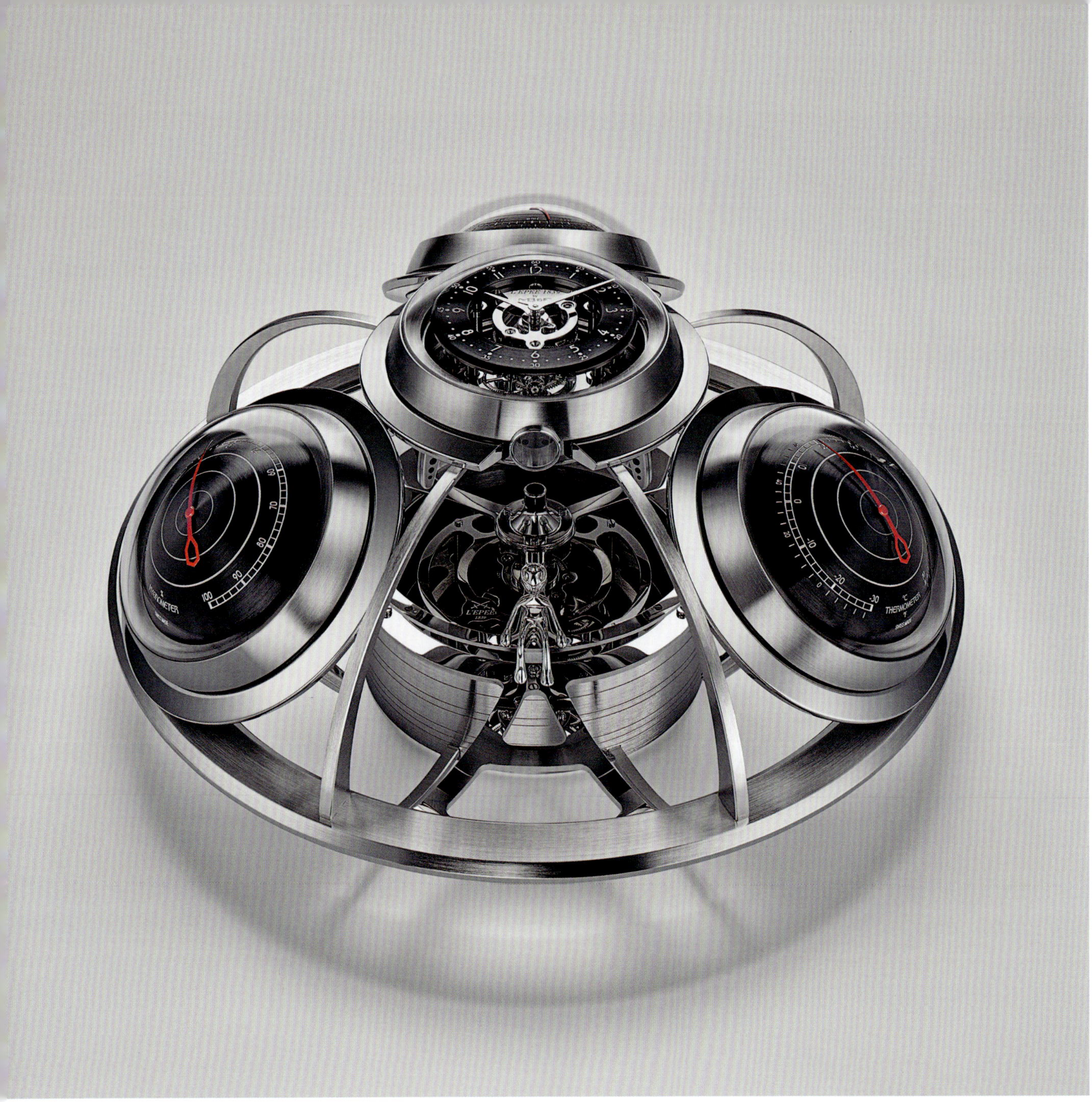
HYGROMETER
THERMOMETER

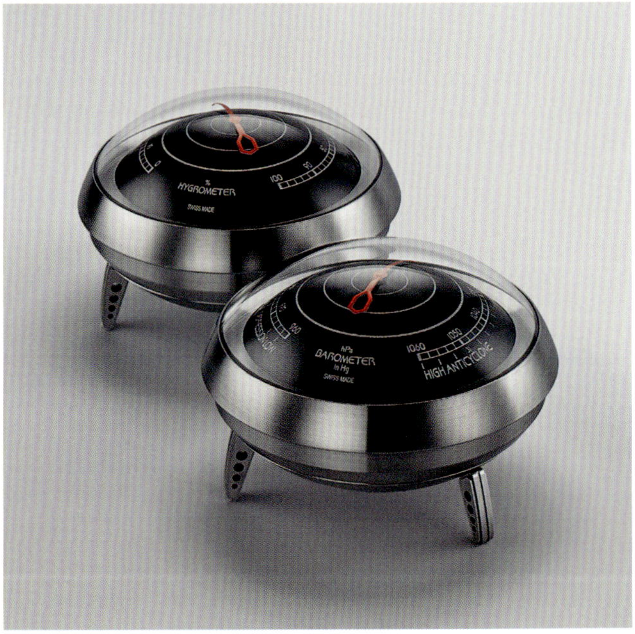
HYGROMETER
SWISS MADE
BAROMETER
HIGH ANTICYCLONE

L'ÉPÉE 1839 × MB&F

THE FIFTH ELEMENT

The Fifth Element combines four removable and interchangeable instruments (a clock element, a barometer, a hygrometer and a thermometer) to create a machine much greater than the sum of its parts. The frame, dubbed the "mothership", is manned by Ross, the alien pilot: a small, metallic figurine that rotates around the UFO cockpit powered by his own hand-wound, air-regulated movement. The Fifth Element was produced in three limited editions of 18 pieces each, in black, silver and blue. The Fifth Element is an intergalactic horological weather station enabling accurate weather forecasting even when the power goes down.

Collaboration
L'Epée 1839 × MB&F
Collection
Co-Creations
Series
The Fifth Element
Editions
The Fifth Element "Silver" (19.6000/114), The Fifth Element "Black" (19.6000/124), The Fifth Element "Blue" (19.6000/144), limited to 18 pieces each
Movement
L'Epée 1839 in-house designed and manufactured
Materials
Stainless steel, brass, mineral glass with alien in bronze
Power reserve
8 days/manual winding with double-depth square socket key
Dimensions/Weight
12.4 × 9.2 cm for each independent element; 37.6 × 20.9 cm for complete Fifth Element
Functions
Clock (hours and minutes), barometer (air pressure), thermometer (air temperature), hygrometer (air humidity)
Details
The Fifth Element's manually-wound, air-regulated movement rotates the alien Ross around the mothership's cockpit
Introduction
2018

L'ÉPÉE 1839 × MB&F

GRANT

Grant is a transformer-robot inspired table clock, with a time display on the shield and three working rubber tracks allowing it to take three different positions: lying flat horizontally, crouching at 45 degrees, or sitting upright at 90 degrees. Grant's right arm holds a detachable grenade launcher that doubles as a winding key. Grant was produced in three limited editions of 50 pieces each, in nickel, black and blue. The isochronal oscillations of the regulator keeping time in Grant's glass-domed "brain"are evidence of the clockwork's high precision.

Collaboration
L'Epée 1839 × MB&F
Collection
Co-Creations
Series
Grant
Editions
Grant "Nickel" (76.6005/104) Grant "Black" (76.6005/201) Grant "Blue" (76.6005/401), limited to 50 pieces each
Movement
L'Epée 1839 in-line eight-day movement, designed and manufactured in-house; Incablock shock protection system
Materials
Stainless steel, nickel-plated brass, palladium-plated brass, mineral glass, rubber tracks
Power reserve
8 days/manual winding with double-depth square socket key
Dimensions
Flat position: 11.5 × 21.2 × 23.1 cm; vertical position: 16.6 × 21.2 × 23.8 cm
Functions
Hours and minutes
Details
Transformer body with three operational tracks and three positions for clock/body
Introduction
2018

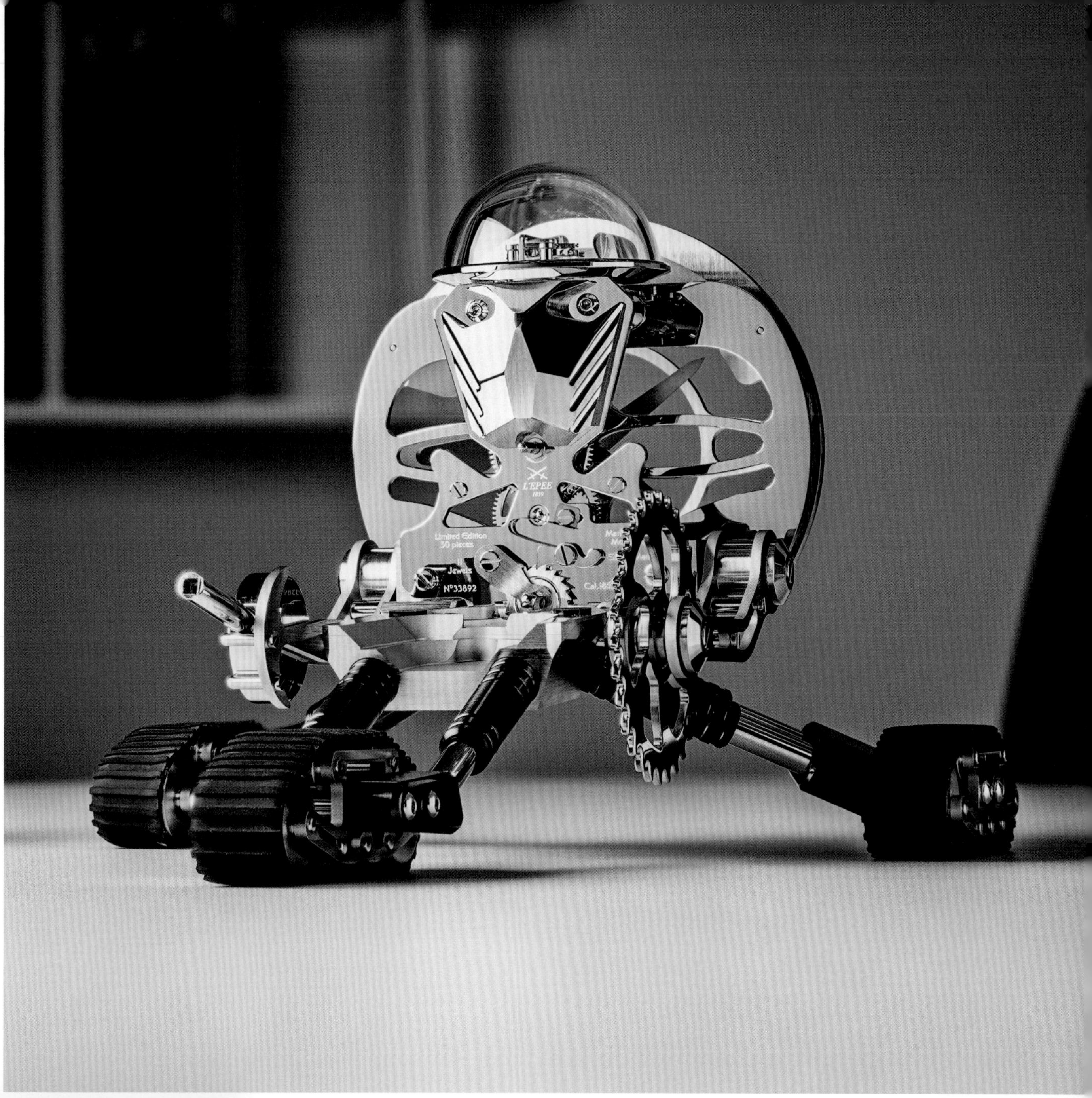
L'EPEE
Limited Edition
50 pieces
Jewels
N°33892

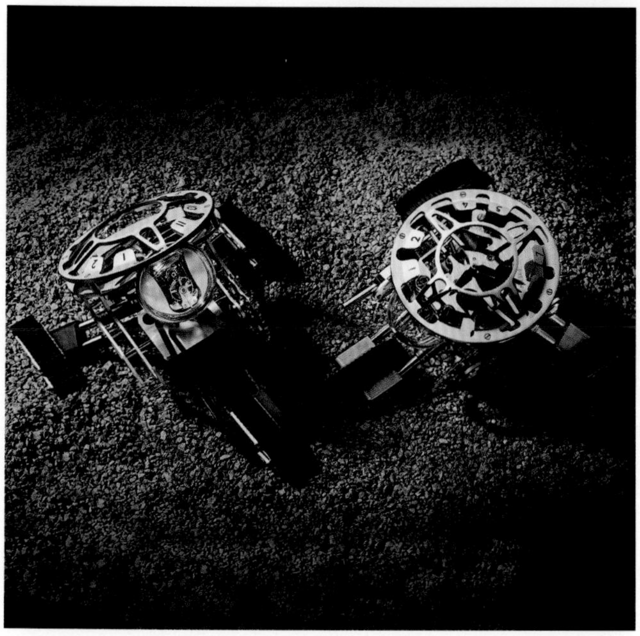

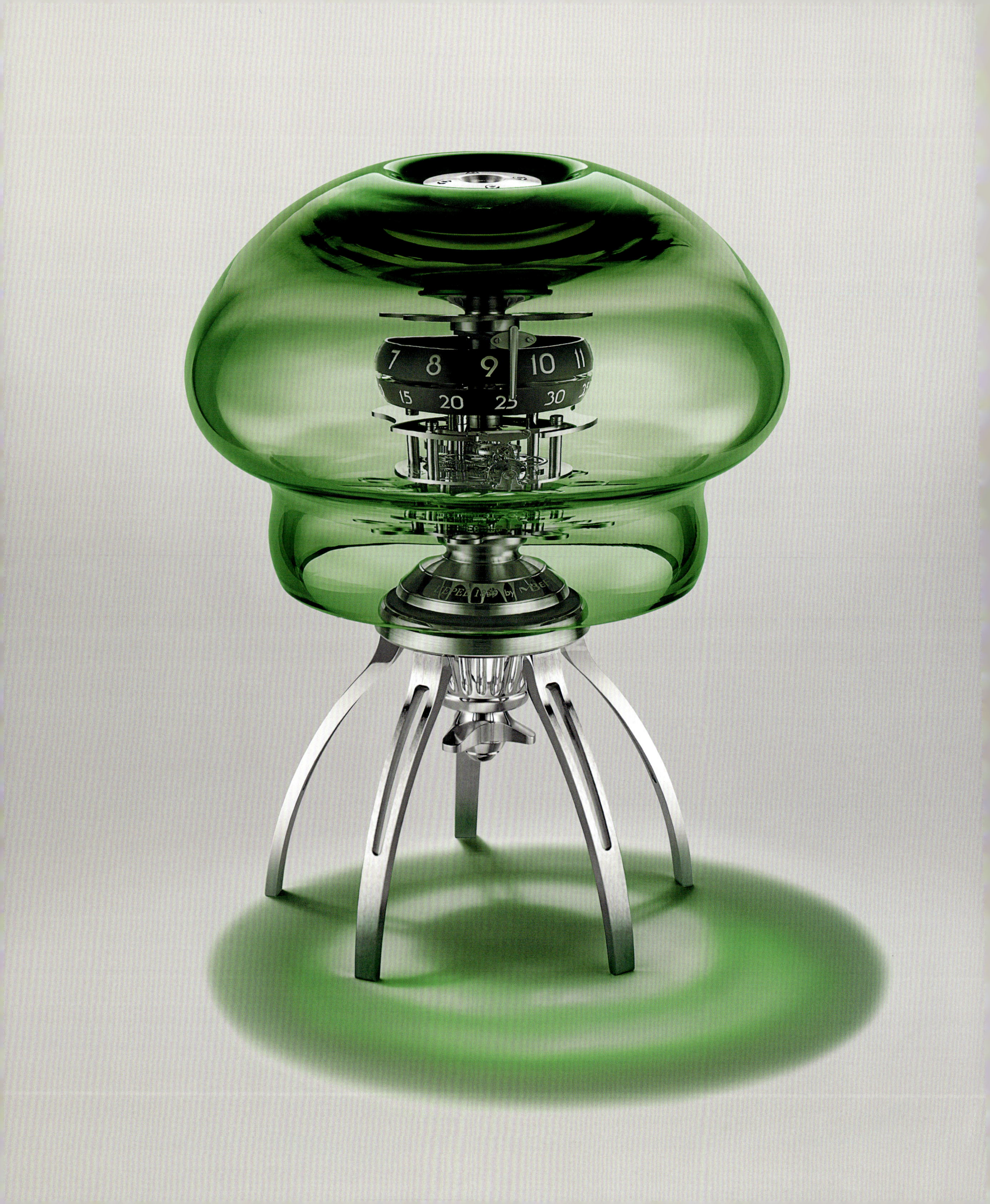
7 8 9 10 11
15 20 25 30

L'ÉPÉE 1839 × MB&F

MEDUSA

Medusa is a jellyfish-inspired dual-configuration clock, housed in hand-blown Murano glass, for ceiling mounting or placing on a desk. Time is displayed on two rotating rings, visible through the glass, with numerals painted in glowing Super-LumiNova, mimicking the natural bioluminescence of a jellyfish in the wild. Medusa was produced in three limited editions of 50 pieces each, in blue, green and pink. In its tenth collaboration with Switzerland's premier clockmaker, L'Epée 1839, MB&F plunges into warm ocean waters where the beautiful and ancient jellyfish proliferate.

Collaboration
L'Epée 1839 × MB&F
Collection
Co-Creations
Series
Medusa
Editions
Medusa "Green" (73.6000/134)
Medusa "Blue" (73.6000/144)
Medusa "Pink" (73.6000/164), limited to 50 pieces each
Movement
L'Epée 1839 in-house designed and manufactured; Incablock shock protection system
Materials
Hand-blown Murano glass, stainless steel and brass, indexed with Super-LumiNova
Power Reserve
7 days/manual winding with integrated key
Dimensions
Hanging: 28.6 × 25 cm standing 32.3 × 25 cm
Functions
Hours and minutes
Details
Medusa is a dual-configuration clock: it can be hung from the ceiling by its glass tentacles thanks to the special mounting system, or stood on a desk thanks to a metal frame with curved legs
Introduction
2019

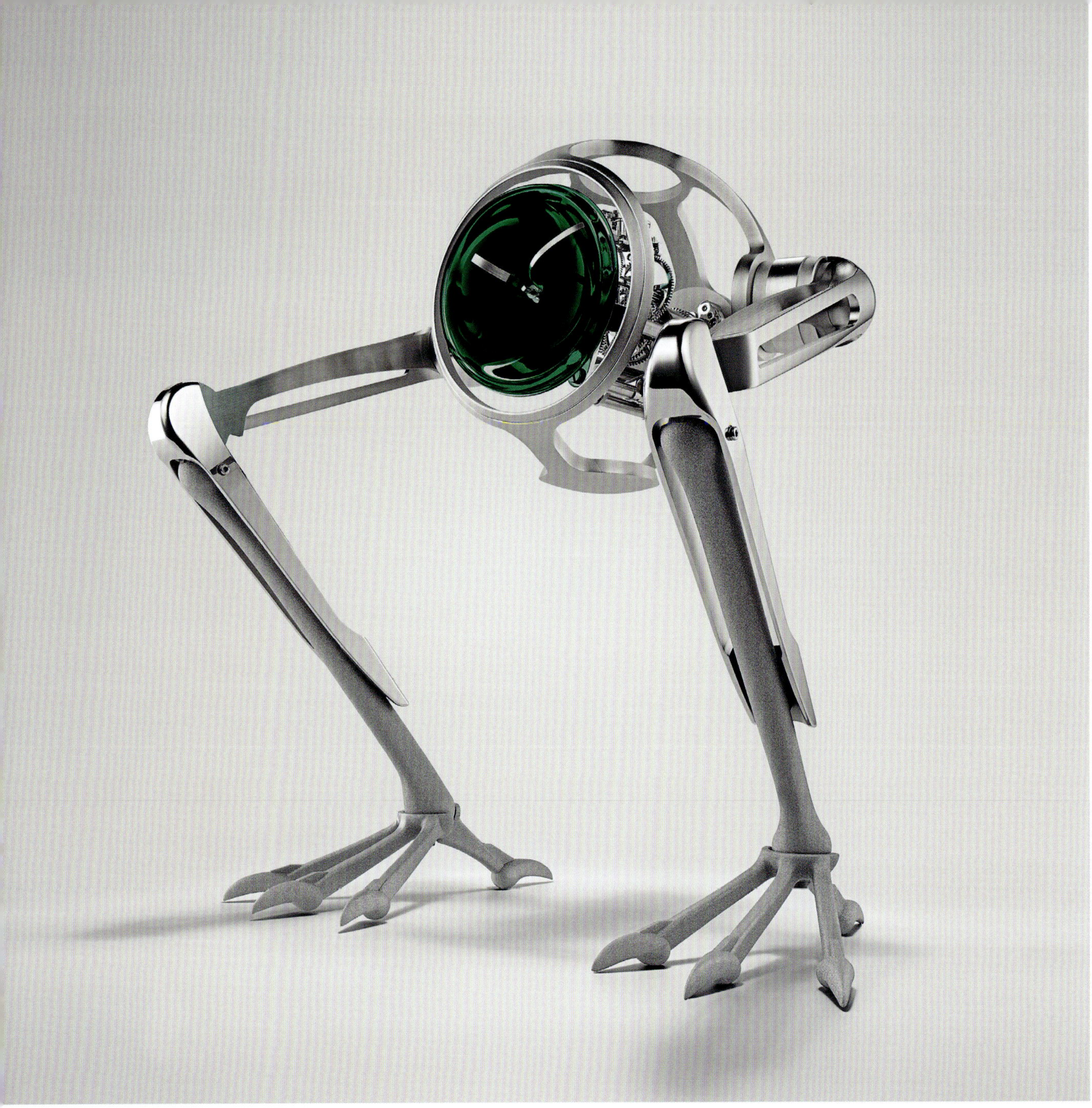
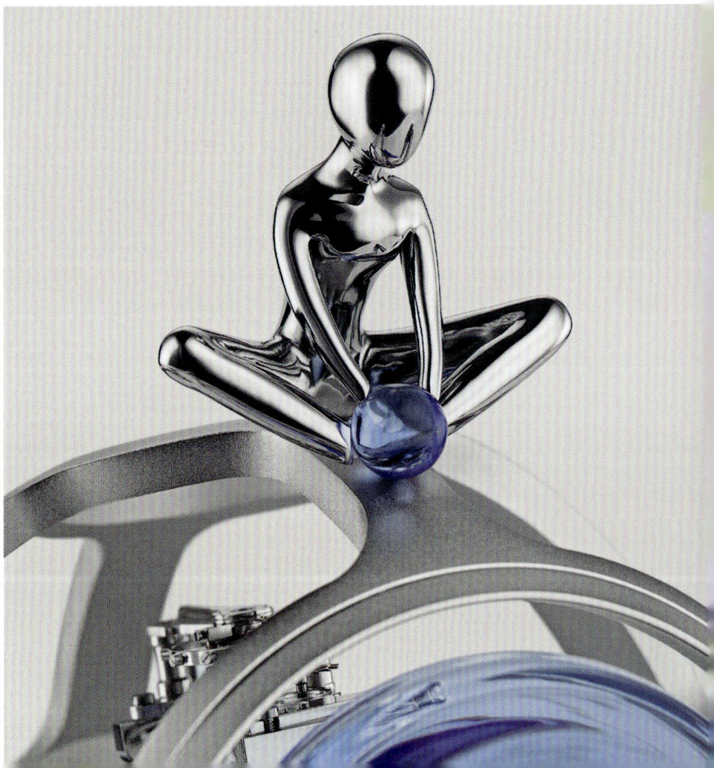

L'ÉPÉE 1839 × MB&F

T-REX

T-Rex is a table clock based on a design first conceived as a unique piece for the Only Watch charity auction: Tom & T-Rex. The legs are modeled on the real thing, using 3D scans of fossilized Tyrannosaurus Rex bones; the single, large, penetrating eye is made from Murano glass. T-Rex was released in three limited editions of 100 pieces each, with Murano glass dials in green, deep blue and red. There was also a limited edition of 15 pieces in bronze "reimagined" by Massena LAB. T-Rex is the 11th collaboration between MB&F and Switzerland's premier clockmaker L'Épée 1839.

Collaboration
L'Epée × 1839 MB&F
Collection
Co-Creations
Series
T-Rex
Editions
T-Rex "Red" (76.6008/160), T-Rex "Blue" (76.6008/140), T-Rex "Green" (76.6008/130), limited to 100 pieces each; Tom & T-Rex, unique piece for Only Watch 2019; T-Rex Massena Lab (76.6008/730), limited to 15 pieces
Movement
L'Epée 1839 in-house designed and manufactured; Incablock shock protection system
Materials
Hand-blown Murano glass, stainless steel, palladium-plated brass and bronze
Power Reserve
8 days/manual winding with double-depth square socket key
Dimensions
26.5 × 25.8 × 17.8 cm (30.8 × 25.8 × 17.8 cm for Tom & T-Rex)
Functions
Hours and minutes
Introduction
2019

STARFLEET EXPLORER

The Starfleet Explorer is a reprise of the Starfleet Machine, with important revisions. The size is more compact, enhanced with bright colours, and accompanied by a fleet of three small spacecraft, the machine's "explorers". Hours and minutes are indicated by means of two discs. The minute disc rotates, while the hour disc remains static, with an hour hand making a full rotation every hour. The Starfleet Explorer was produced in three limited editions of 99 pieces each, in blue, green and red. The original Starfleet Machine, launched in 2014, was the first clock ever co-created by MB&F and L'Epée 1839; it marked the 175th anniversary of L'Epée and the beginning of a long series of collaborations with MB&F.

Collaboration
L'Epée 1839 × MB&F
Collection
Co-Creations
Series
Starfleet Explorer
Editions
Starfleet Explorer "Green" (71.6000/131) Starfleet Explorer "Blue" (71.600/141) Starfleet Explorer "Red" (71.600/151) limited to 99 pieces each

Movement
L'Epée 1839 in-house designed and manufactured; Incablock shock protection system
Materials
Palladium-treated brass and stainless steel
Power Reserve
8 days/manual winding with double-depth square socket key

Dimensions
16.5 × 11 cm
Functions
Minutes indicated by a fixed curved aperture on the mobile upper dome, performing a complete rotation every 60 minutes; hours indicated by a mobile hand, performing a complete rotation every 12 hours on a fixed disc

Details
Starfleet Explorer features an animation in which three brightly coloured spacecraft perform a five-minute orbit of the station
Introduction date
2020

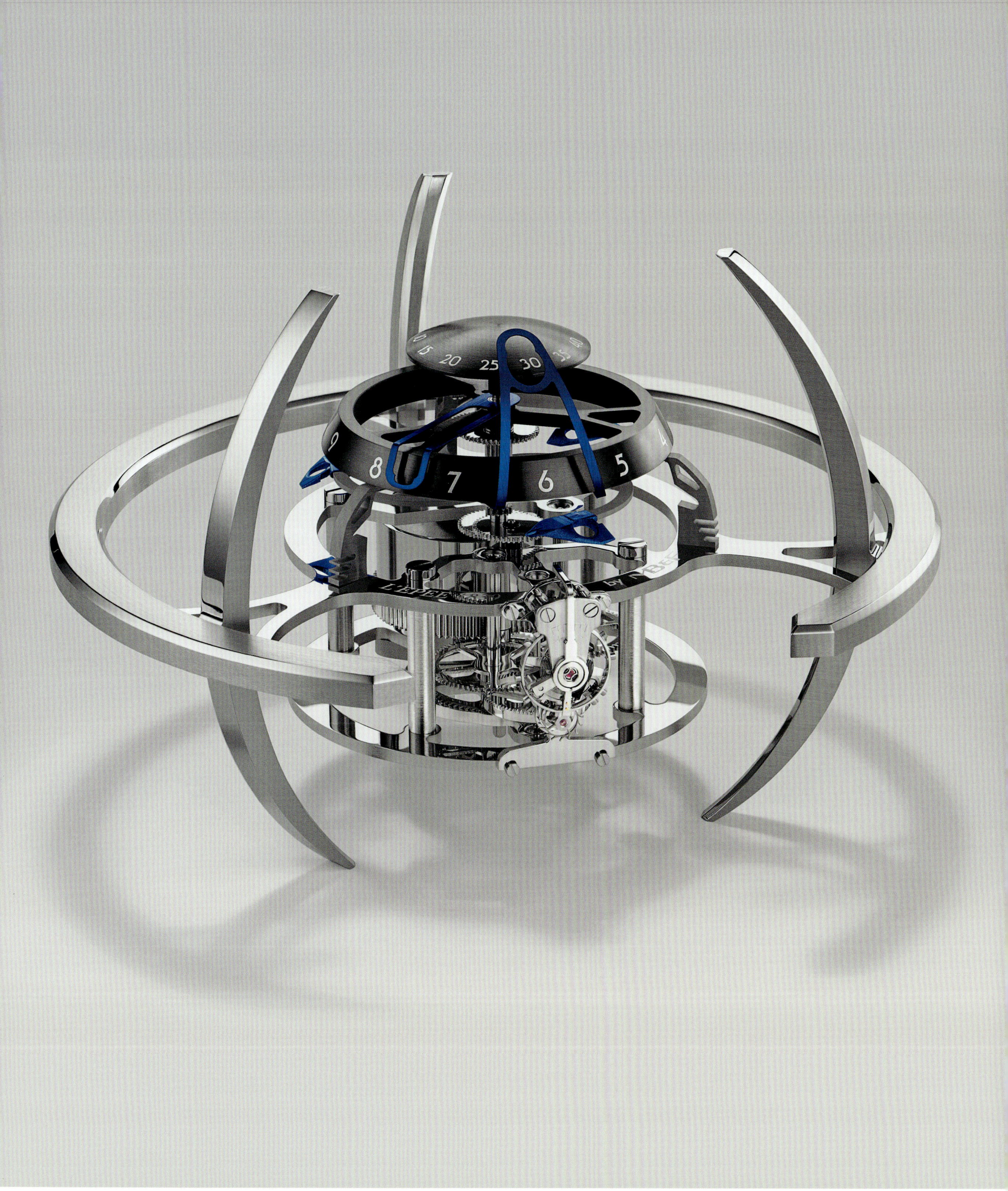
15 20 25 30
8 7 6 5

TRIPOD

The insect-like TriPod table clock has three spindle-like legs supporting a bulbous glass body and three glass eyes. It joins the MB&F family of what Maximilian Büsser calls "Robocreatures". Time is indicated on black discs on top of the TriPod's body, and then magnified and reflected upward into the TriPod's eyes, so making it possible to read the time from multiple angles. TriPod was released in three limited editions of 50 pieces each, in neon blue, neon green and neon red.

Collaboration
L'Epée 1839 × MB&F
Collection
Co-Creations
Series
TriPod
Editions
TriPod "Green" (74.6009/134), TriPod "Blue" (74.6009/144), TriPod "Red" (74.6009/164), limited to 50 pieces each
Movement
L'Epée 1839 in-house designed and manufactured; Incablock shock protection system
Materials
Palladium-plated brass, optical mineral glass and fluorescent acrylic
Power reserve
8 days/manual winding with double-depth square socket key
Dimensions
30 × 26 cm
Functions
Hours and minutes are indicated on two concentric dials visible from each of the three optical mineral glass spheres; the dials make one full rotation in 36 hours
Introduction
2020

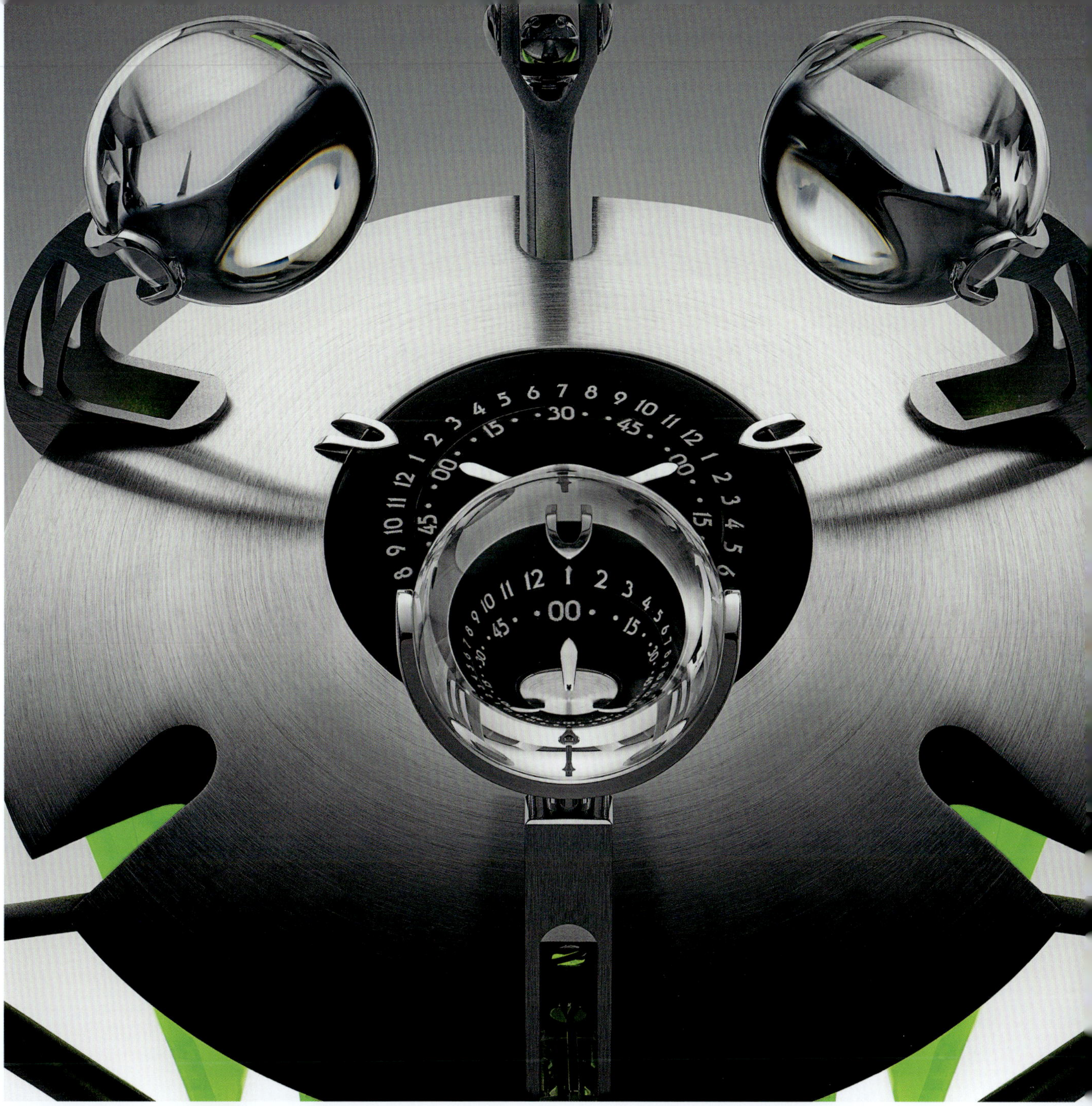

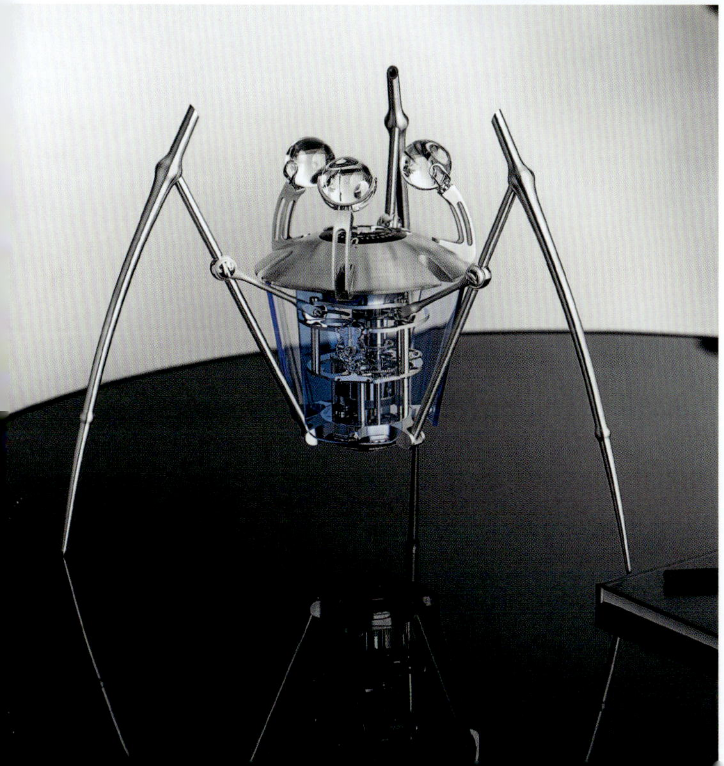

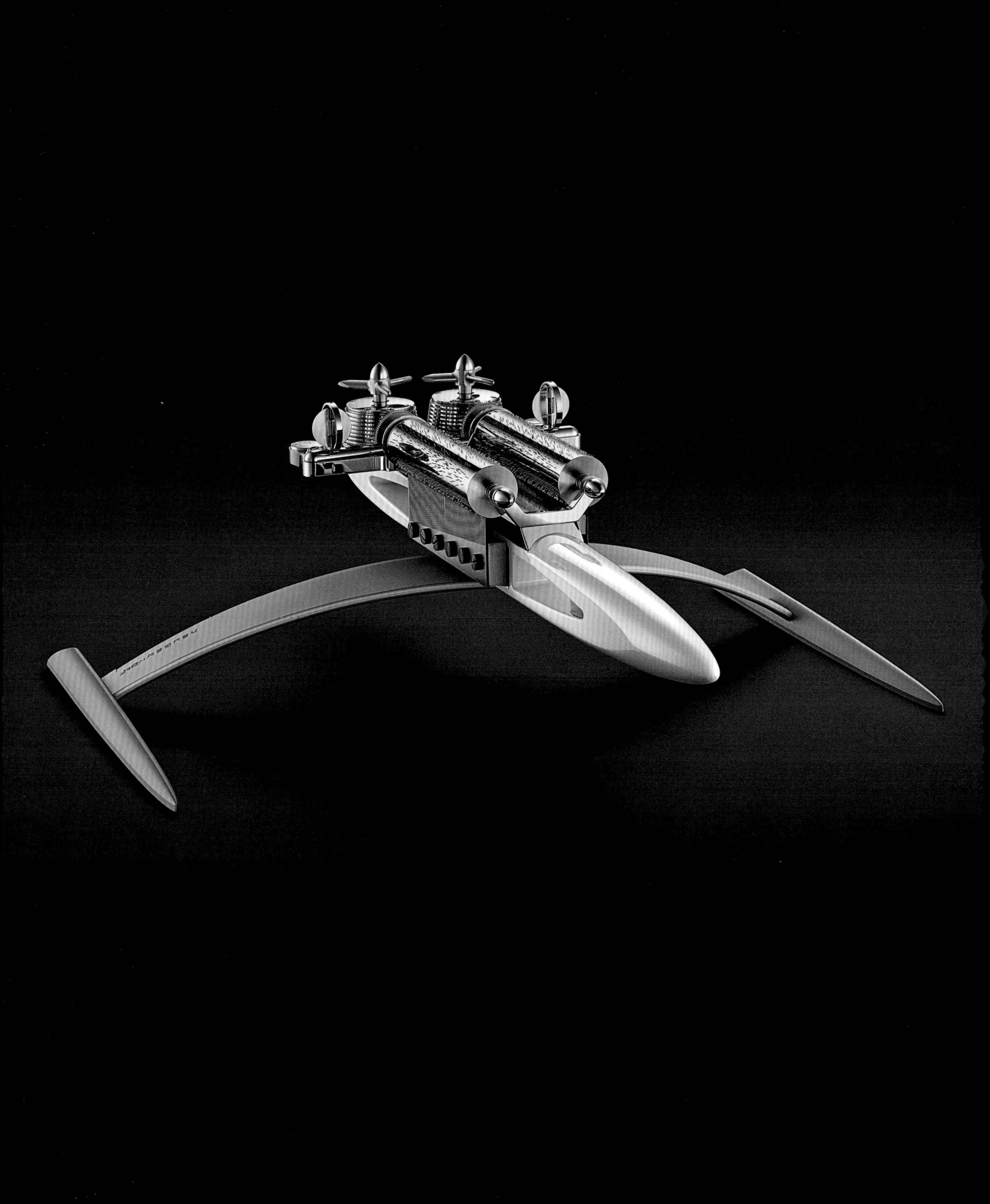

REUGE × MB&F

MUSIC-MACHINE 1

MusicMachine 1 is a music box, designed to look and sound like something from another world. Twin cylinders with dual propellers are mounted on a fuselage with landing gear, suggesting the fictional spacecraft in the science fiction fantasy classics much loved by MB&F founder Maximilian Büsser. Each cylinder plays three melodies – one cylinder, representing the Earth side, plays Pink Floyd, Deep Purple, and John Lennon – while its opposing cylinder plays themes from Star Wars and Star Trek. MusicMachine 1 was released in a limited edition of 66: 33 in white and 33 in black. Launched in 2013, it was the first of the MB&F co-creations. It contains all the traditional, time-honoured elements of a superlative high-end music box, but designed and configured in a totally unconventional way.

Collaboration
Reuge × MB&F
Collection
Co-Creations
Series
MusicMachine 1
Editions
MusicMachine 1 "Black" (AXA.72.5865.000), MusicMachine 1 "White" (AXA.72.5865.001), limited to 33 pieces each
Movement
Two 3.72 movements (3 refers to number of melodies on each cylinder; 72 refers to number of notes on each comb); one movement is "right" configured; one movement is "left" configured (they rotate in opposite directions)
Materials
Black lacquered walnut with black matte-anodized aluminium or white lacquered walnut with bead-blasted and anodized aluminium. Nickel-plated brass main plate
Power reserve
15 minutes per cylinder
Dimensions/Weight
39.5 × 47.5 × 16.5 cm
Functions
Plays 6 melodies (3 per cylinder): on the left, "May The Force Be With You" (the "Star Wars" theme), the "Imperial March" from "The Empire Strikes Back", and the theme from "Star Trek"; on the right, Pink Floyd's "Another Brick In The Wall", Deep Purple's "Smoke On The Water" and John Lennon's "Imagine"; start/stop and repeat/continue functions
Introduction
2013

REUGE × MB&F

MUSIC-MACHINE 2

MusicMachine 2 has a case design that conjures up the spacecraft in the Star Trek franchise. It features two cylinders, each powered by an independent movement in the ship's tail section. Each cylinder plays three melodies, including themes from Star Trek and Star Wars along with hits by The Rolling Stones, The Clash and Led Zeppelin. Its sound-emitting case is an innovation by Jean Michel Capt of JMC Lutherie: located in the hull of the spacecraft, it is kept separate from the musical cylinders. MusicMachine 2 was released in a limited edition of 33 in white and 66 in black; it is the second collaboration between MB&F and high-end mechanical music box specialists Reuge. Underneath its futuristic guise, MusicMachine 2 features all the traditional elements of a beautifully-crafted, high-end music box.

Collaboration
Reuge × MB&F
Collection
Co-Creations
Series
MusicMachine 2
Editions
MusicMachine 2 "Black" (AXA.72.5875.000) limited to 66 pieces; MusicMachine 2 "White" (AXA.72.5875.001) limited to 33 pieces
Movement
Two 3.72 movements (3 refers to number of melodies on each cylinder; 72 refers to number of notes on each comb); one movement is "right" configured; one movement is "left" configured (they rotate in opposite directions). Each movement includes a mainspring, cylinder, comb and regulator
Materials
Black or white lacquered aluminium dome, nickel-plated brass
Power Reserve
15 minutes per cylinder
Dimensions
30 × 51.1 × 16.8 cm
Functions
Plays 6 melodies (6 per cylinder): themes from "Star Wars", "The Empire Strikes Back" and "Star Trek" on one "channel"; Led Zeppelin's "Stairway To Heaven", The Rolling Stones' "Angie" and The Clash's "Should I Stay Or Should I Go?" on the other; start/stop and repeat/continue functions
Introduction
2014

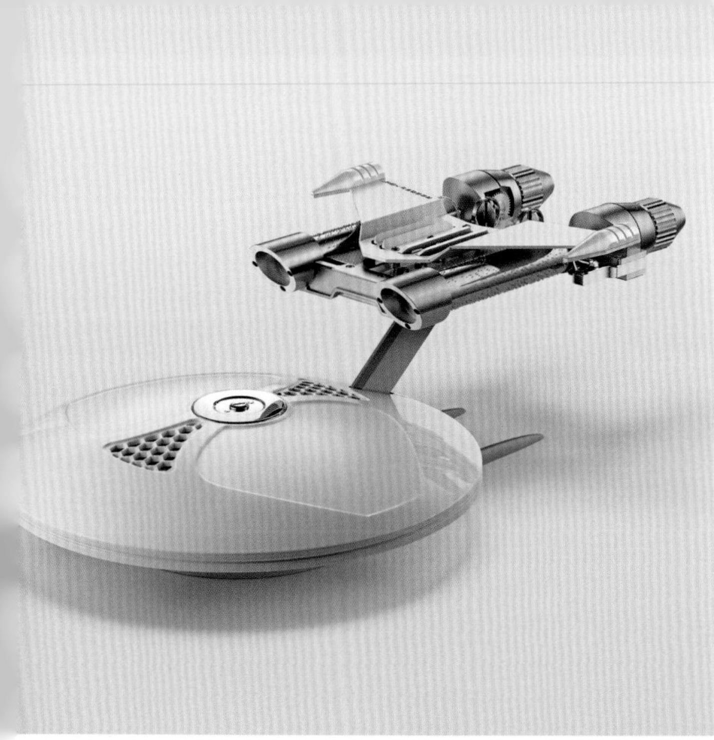

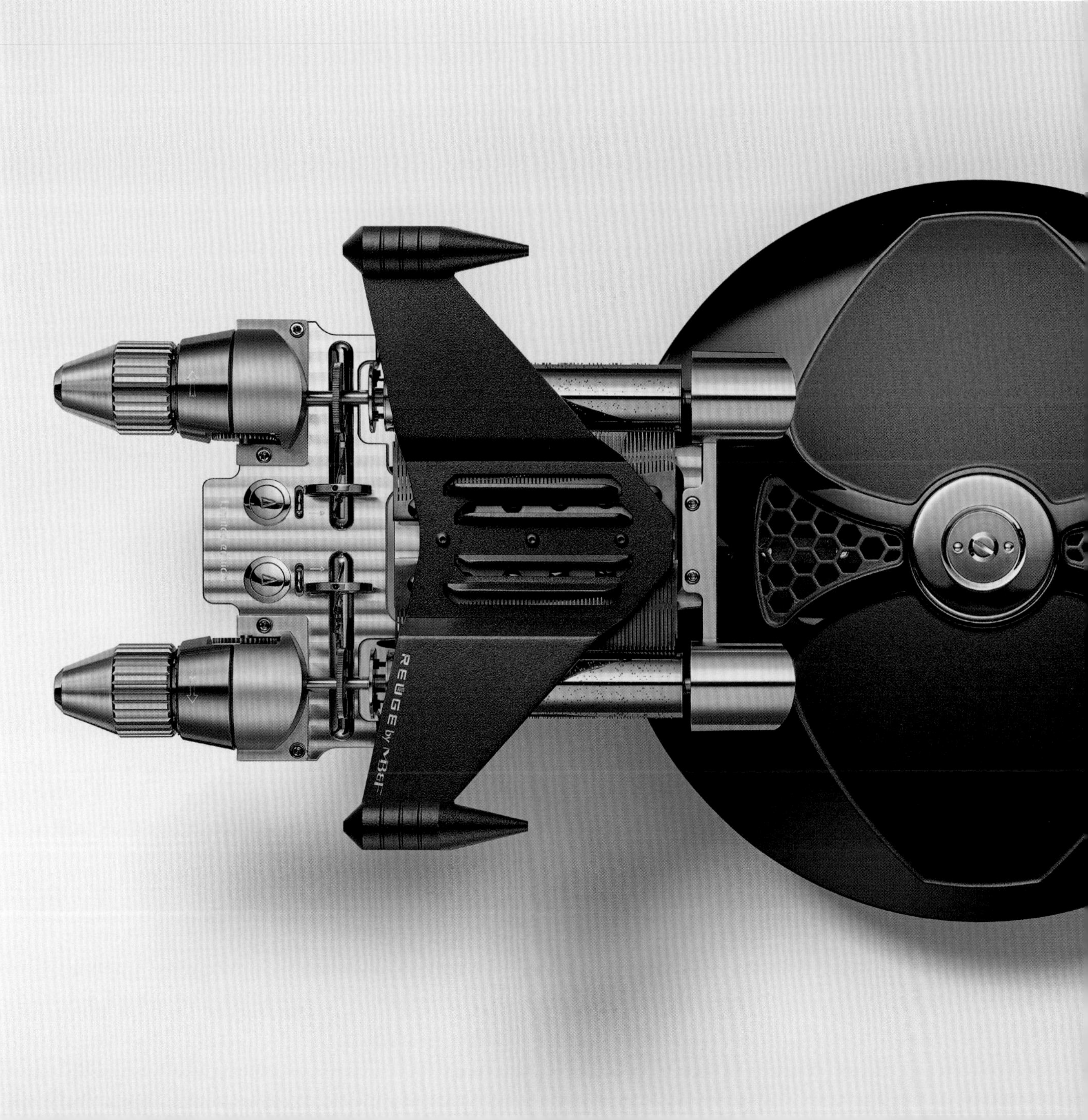
REUGE by MB&F

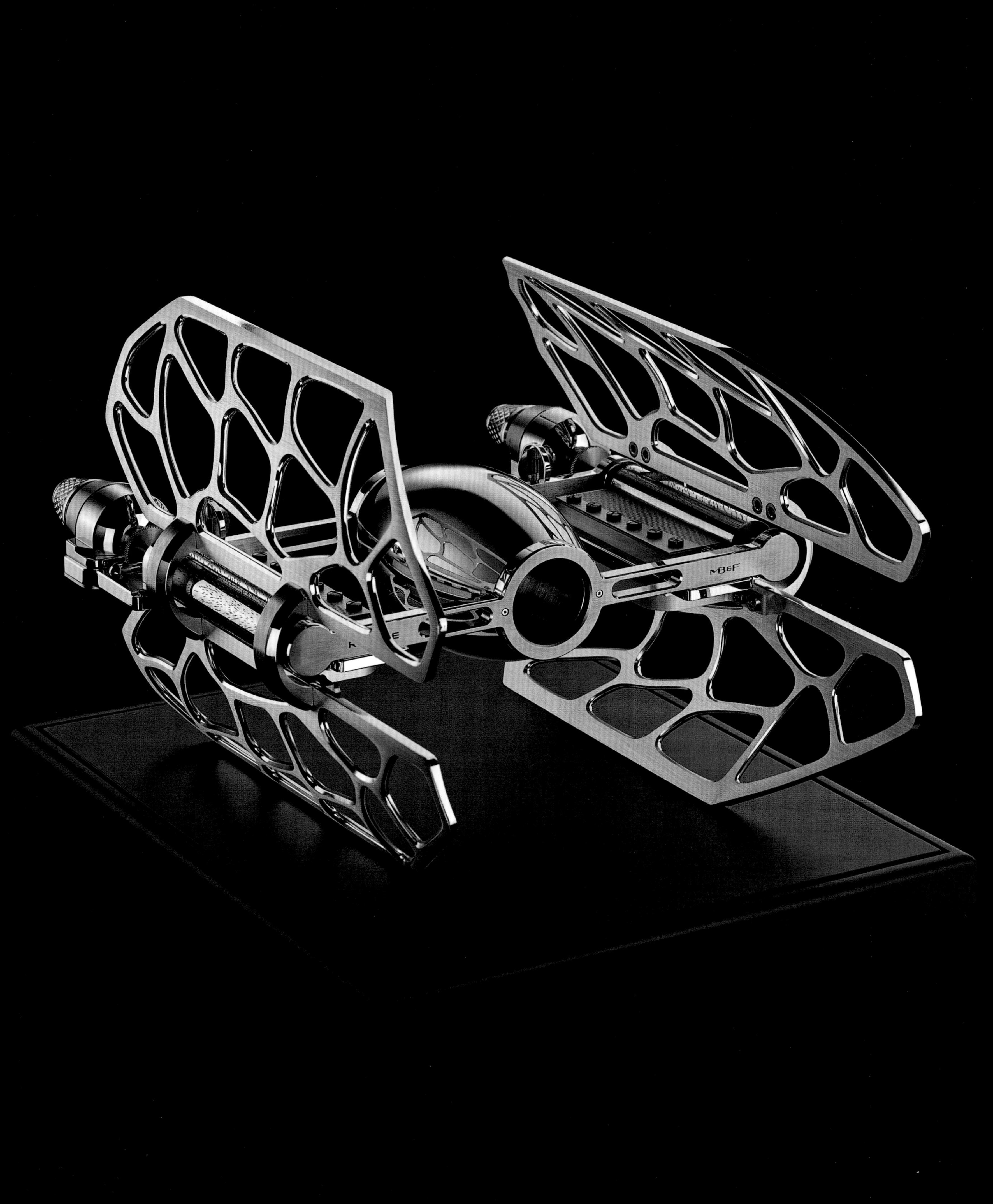
MB&F

REUGE × MB&F

MUSIC-MACHINE 3

The design of MusicMachine 3 is based on the TIE fighter aircraft of the Star Wars franchise, with two prominent vertical wings on either side supporting and protecting the machine's dual music cylinders, each spreading to amplify the machine's vibrations. The twin cylinders play melodies from Star Wars, Mission Impossible, and James Bond on the right, with tunes from The Godfather, Merry Christmas Mr. Lawrence and The Persuaders on the left. MusicMachine 3 was released in a limited edition of 99 pieces: 33 in white, 33 in black, and 33 in chrome.

Collaboration
Reuge × MB&F
Collection
Co-Creations
Series
MusicMachine 3
Editions
MusicMachine 3 "Black" (AXA.72.5885.000), MusicMachine 3 "White" (AXA.72.5885.001), MusicMachine 3 "Chrome" (AXA.72.5885.002) limited to 33 pieces each
Movement
Two 3.72 movements (3 refers to number of melodies on each cylinder; 72 refers to number of notes on each comb); one movement is "right" configured; one movement is "left" configured (they rotate in opposite directions)
Materials
Aluminium with lacquer finish and polished brass. Resonant amplifying base by JMC Lutherie: 350-year-old resonance spruce with 21st century composite materials like Nomex™ honeycomb Kevlar
Power Reserve
15 minutes per cylinder
Dimensions
40 × 34 × 28 cm (with soundboard base)
Functions
Plays 6 melodies (3 per cylinder): the theme tunes from "Star Wars", "Mission Impossible" and "James Bond" on the right; and "The Godfather", "Merry Christmas Mr. Lawrence", and "The Persuaders" on the left; start/stop and repeat/continue functions
Introduction
2015

KELYS & CHIRP

Two unlikely companions, a tortoise, Kelys, and a songbird, Chirp, are paired in this automaton developed by friends Reuge and Nicolas Court. Kelys, from the Greek chelys for tortoise, moves with a tortoise-like gait, its head turning slowly side-to-side. Chirp springs from a panel in the tortoise shell to sing her mechanical birdsong, the while flapping her wings and opening and closing her beak. Thankfully for both their sakes, a mechanical sensor is integrated within the automaton to prevent Kelys from walking off a tabletop, desk or any other elevated surface. Kelys & Chirp was released in four limited editions of 18 each: blue, green, yellow and ochre.

Collaboration
MB&F presents Reuge & Nicolas Court
Collection
Co-Creations
Series
Kelys & Chirp
Editions
Kelys & Chirp "Ochre" (AXT.92.6990.001), Kelys & Chirp "Green" (AXT.92.6990.002), Kelys & Chirp "Blue" (AXT.92.6990.003) Kelys & Chirp "yellow" (AXT.92.6990.004), limited to 18 pieces each
Movement
480 components Singing Bird: double bellows system (bi-directional air pressure). Security mechanism: if the bird or cover are pushed down while bird is singing, the bird automatically retracts Automaton tortoise: the mechanism is driven by the Singing Bird movement. Elliptical gearing enables the tortoise to advance with a realistically irregular gait. Security friction clutch
Materials
Chirp: 18K white polished gold, eyes in sapphire
Kelys: grained and polished finishing, rhodium-plated brass, black onyx gems for the eyes, and leather shell – all hand assembled
Power reserve
3 cycles of tortoise walking and bird singing. Circular winding key located on tortoise's belly
Dimensions
24 × 16 × 8 cm
Functions
Animation: the tortoise walks and moves its head; the bird emerges from back, moves and sings, flaps its wings, moves its tail, opens/closes its beak in time with the bird song, then vanishes
Introduction
2017

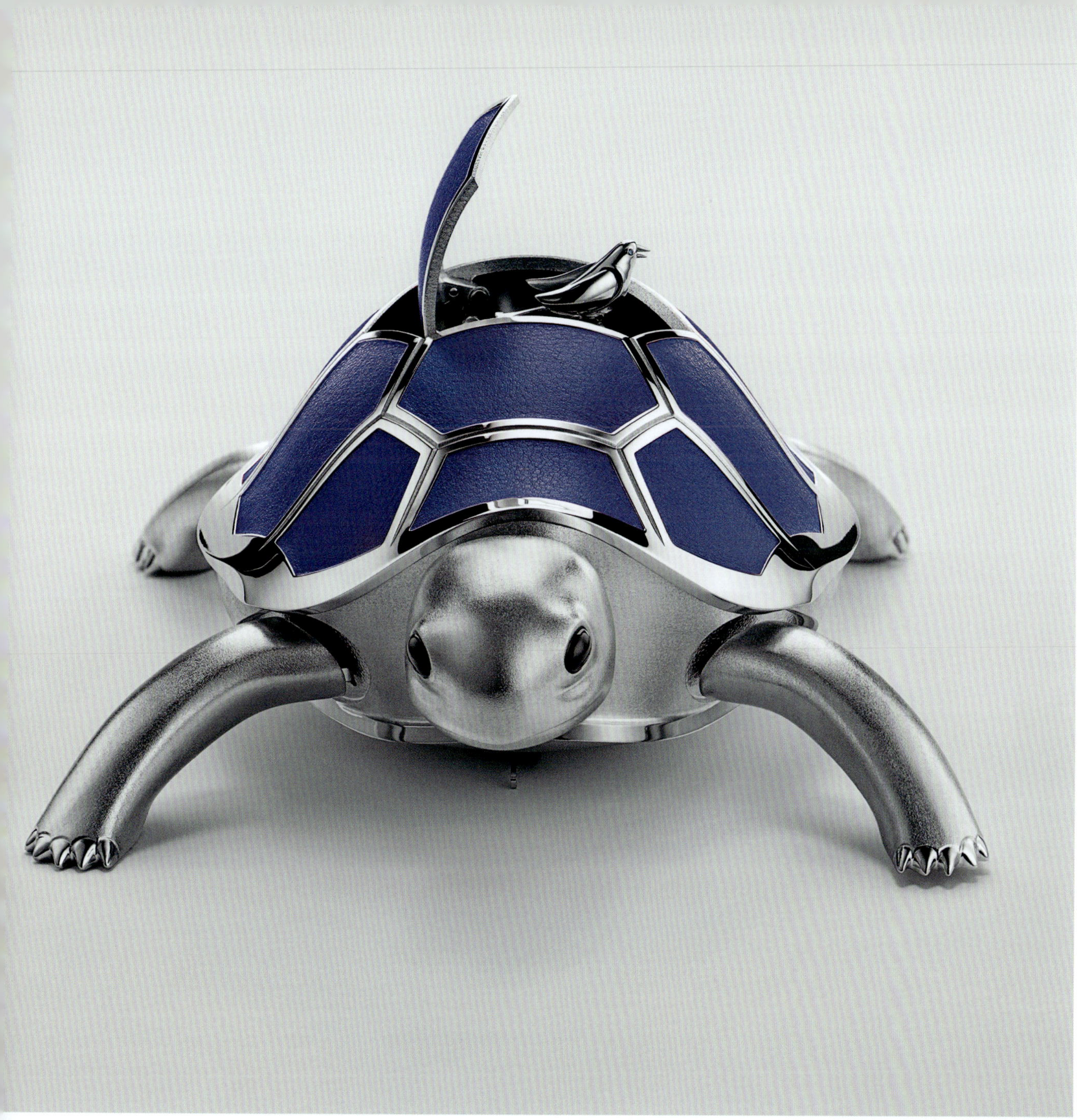

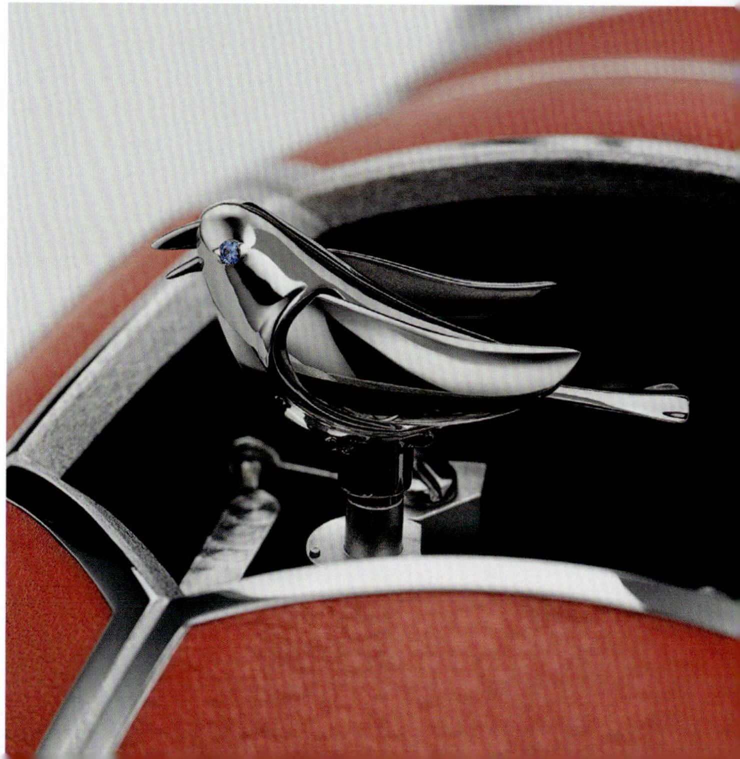

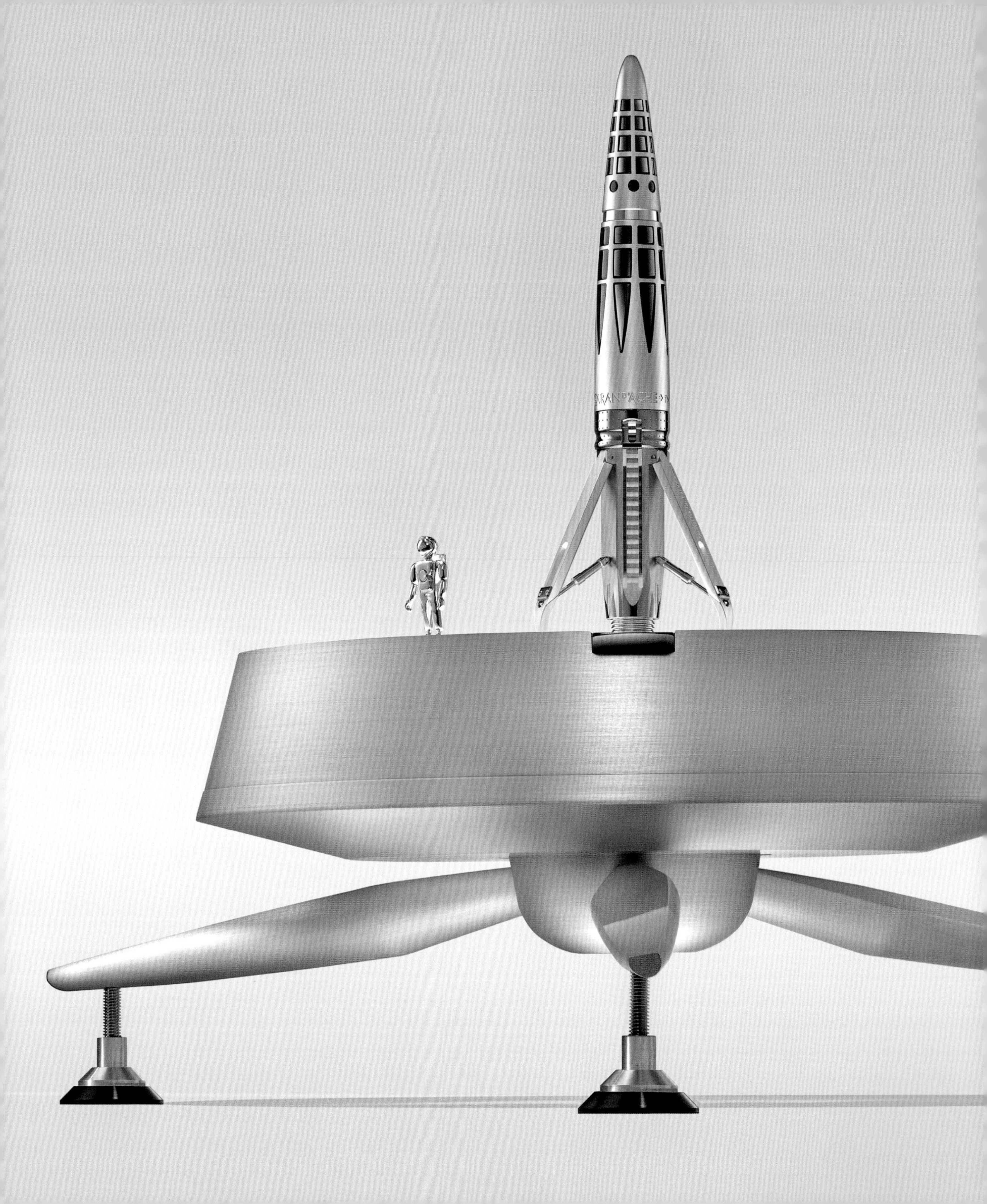

CARAN D'ACHE × MB&F

ASTROGRAPH

Projecting his dreams of conquering space onto a writing instrument was the original idea of MB&F founder Maximilian Büsser. The creative and technical forces within Caran d'Ache enthusiastically embraced this new adventure, both thrilling and complex. The base structure is articulated by an ingenious mechanism inspired by penknives: a miniature lever – the entry door to the rocket, concealed in the ring of the pen – activates the simultaneous lowering of three stabiliser legs. Once these have been deployed, it is possible to stand Astrograph up vertically, ready for take-off. A magnetic miniature astronaut figurine can be attached at will to the body of the writing instrument, ready to "climb on board". Despite its complex embellishments, the instrument is perfectly balanced and rests comfortably in the hand. The Astrograph was released in three limited editions: high-gloss rhodium, sandblasted matte rhodium and anthracite ruthenium.

Collaboration
Caran D'Ache × MB&F
Collection
Co-Creations
Series
Astrograph
Editions
Astrograph High Gloss Rhodium Fountain Pen (1654.481); Astrograph Matt Rhodium Fountain Pen (1655.481); Astrograph Matt Rhodium Roller Pen (1665.481); Astrograph Ruthenium Fountain Pen (1656.481), limited to 99 pieces each
Materials
The rocket-shaped pen body is ruthenium-plated and rhodium-plated with rhodium-plated 18-carat gold pen nib; silver and rhodium-plated magnetic astronaut
Details
Can be used with cartridges or ink pump. The structure is articulated by an ingenious mechanism, inspired by penknives: a miniature lever (the entry door to the rocket, concealed in the ring of the pen) activates the simultaneous lowering of three stabiliser legs.
Introduction
2016

LOUPE SYSTEM× MB&F

PROJET LPX

MB&F teamed up with Loupe System to create Project LpX: an intergalactic magnification station enabling accurate viewing of watch movement details, while simultaneously reminding you of the immensity of our galaxy. Project LpX is made up of five modular components packaged in a sturdy travel case. Together they form the body of a rocket ship that can accommodate up to three Loupe System loupes and protect them from dust. Project LpX is delivered with two loupes (one with 3x magnification and one with 6x magnification). It was released in four limited editions of 99 pieces each, in black, blue, red and green.

Collaboration
Loupe System × MB&F
Collection
Co-Creations
Series
Project LpX
Editions
Project LpX "Red" (AC-RKT-01-RED), Project LpX "Blue" (AC-RKT-01-BLUE), Project LpX "Green" (AC-RKT-01-GREEN) Project LpX "Black" (AC-RKT-01-BLACK) limited to 99 pieces each
Materials
Aluminium and stainless steel; tritium capsule under the rocket base; comes with 3x and 6x magnifying loupes, loupe system included.
Dimensions
28 × 33 cm/0.85 kg (with the loupes)
Details
The rocket body is composed of 5 elements with bayonet assembly: the base, three modular body parts, and the nose cone. Each rocket can therefore be configured to accommodate one, two or three magnifying loupes, hidden inside the body parts; loupe system universal clip included in the box
Introduction
2019

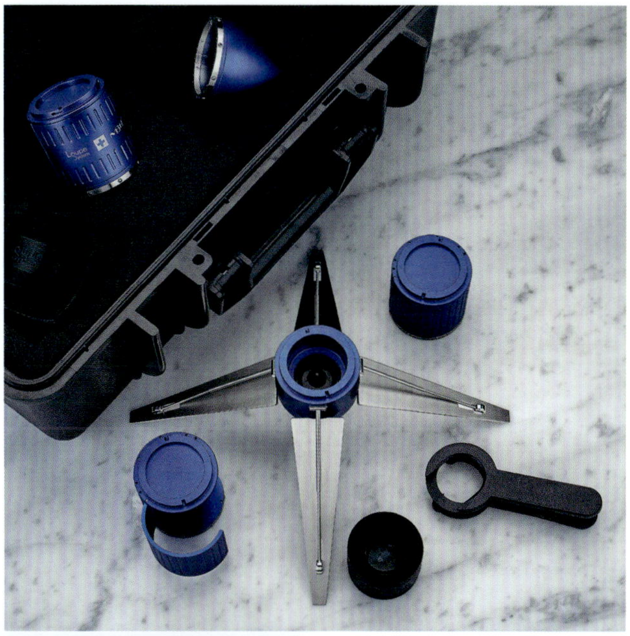

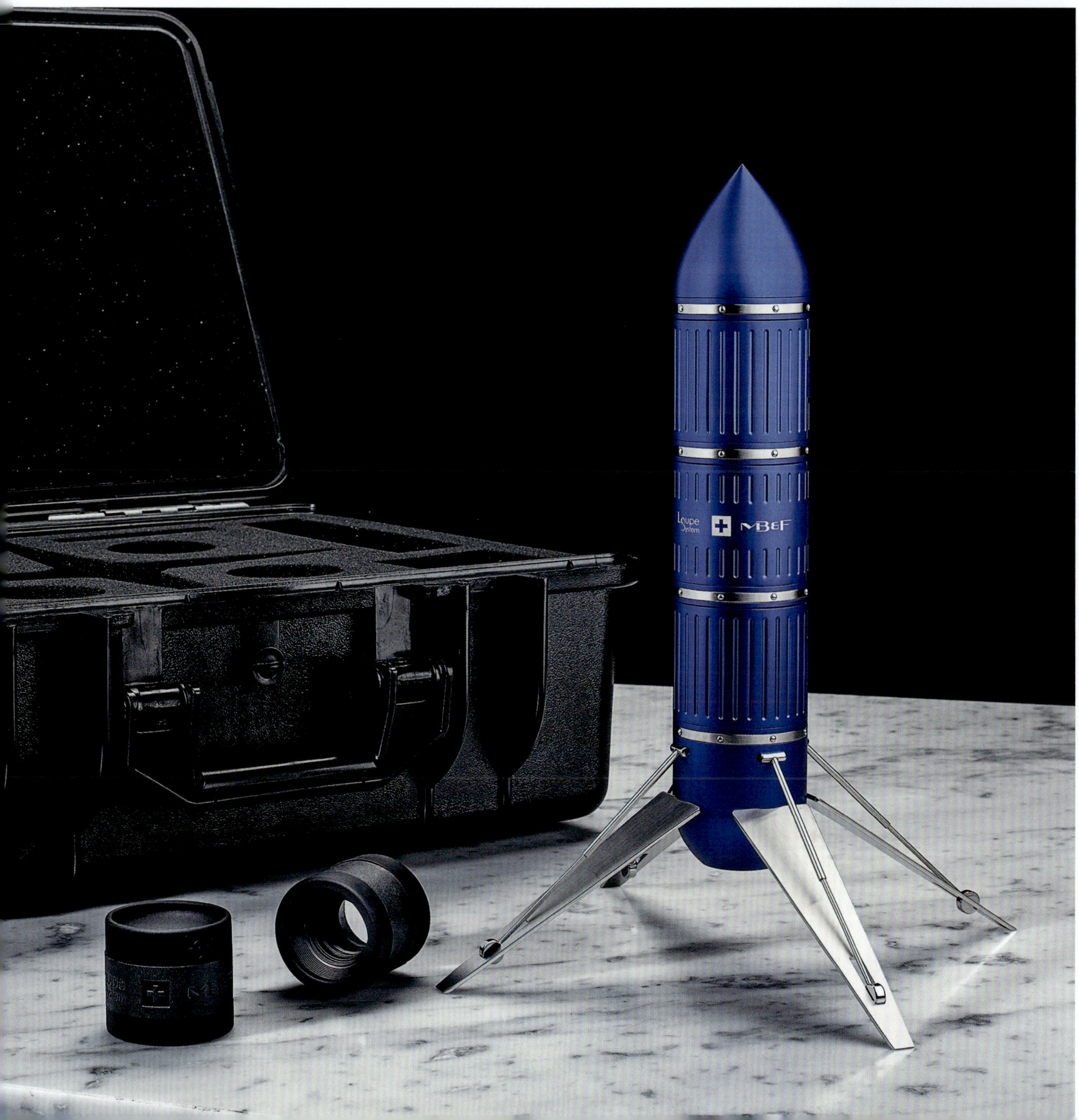
Loupe System
MB&F

HM

LM

INDEX

CREDITS

For all product pictures in this book, except those mentioned below, please note that they have been photographed by Maarten van Der Ende, Laurent-Xavier Moulin or Denis Hayoun. © MB&F

© Maximilian Büsser Archives; p.10 (up); p.13 (down left); p.14

© Alex Teuscher: p.46; p.48; p.49; p.180; p.183; p.283; p.291; p.308; p.310; p.311

© Laurent-Xavier Moulin: p.47; p.181; p.182; p 265 (down); p.303

© Régis Golay, Federal Studio: p.25; p.75; p.99 (up); p.161; p.195 (up); p.239 (up); p.307 (right)

© Marc Ninghetto: p.85

© Philippe Garcia: p.6; p.34 (down left)

© Collection Christophel © Dynamic Planning/Toei Animation: p.5 (up and down); p.10 (down right); p.13 (up)

© Collection Christophel/World History Archive: p.10 (down left)

© Kharbine Tapabor: p.13 (down right)

© David Mayenfisch: p.19

© RichVintage via Getty Images: p.22

© Jake Reeder: p.34 (up left); p.34 (center)

© Watch Anish @watchanish: p.26 (up)

© Vincent Calmel: p.26 (down)

© M.A.D.Gallery Geneva, Samuel Zeller: p.31 (down)

© Lee Yuen-Rapati @OneHourWatch: p.36 (up)

© Guy Allen for Phillips, in Association with Bacs & Russo: p.36 (down)

© Michael Pischetelli : p.71

© All Rights Reserved: p.117 (up); p.191 (up); p.213 (up)

© @Davidon: p.141 (up)

© Raúl Montes: p.159

© Venzin+Bühler Fotografen: p.212; p.309

© Christian Chaize: p.261; p.265 (up); p.300

© A Collected Man: p.307 (left)

Despite our best efforts, it has not been possible to locate all the copyright holders featured in this book. We apologize in advance to those concerned.

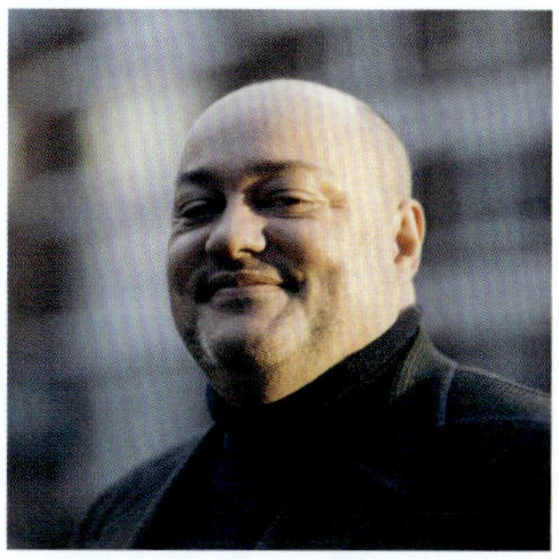

William Massena
Founder and owner of Massena LAB

William Massena is the founder of Massena LAB, a company which both designs and produces its own watches, and collaborates with independent watchmakers. William is also a partner at Digital Luxury Group, which advises on digital strategies for luxury brands.
Previously, William was the Chief Operating Officer and Managing Director of Antiquorum, a leading horological auction house. He was also the Managing Director of TimeZone.com, the world's largest online watch discussion forum. William was with TimeZone.com at its inception in 1995 and was an early advocate of the internet's ability to open up the exclusive world of watches to a broader base of passionate collectors.
William holds a B.S. and M.S. from New York University Stern School of Business. For six years, he was a jury member of the Grand Prix d'Horlogerie de Genève (GPHG); currently he is a member of the cultural council of the Fondation de la Haute Horlogerie, a Freeman of the Worshipful Company of Clockmakers, and a trustee of the Horological Society of New York.

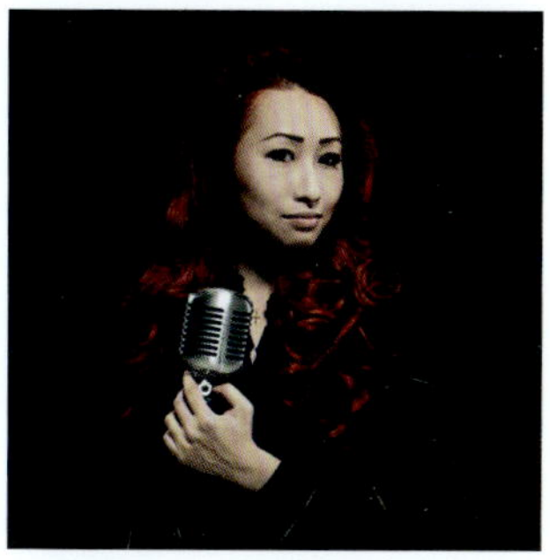

Suzanne Wong
Editor-in-chief, WorldTempus

Suzanne Wong is a veteran watch journalist, born and raised in Singapore, one of the world's most influential and mature watch markets. Her experience — which includes four years as editor-in-chief of Revolution magazine's original flagship Asia edition and her current position as editor-in-chief of online media WorldTempus — spans both print and digital publishing and ranges from directing editorial strategy to creating compelling watch-related content.
Few journalists are as equally proficient in both the technical aspects of fine watchmaking, and in the intricacies of traditional horological decorative arts; as familiar with 18th century mechanical innovations as they are with modern advances in material technology. The ability to transmit these various aspects of horological knowledge to contemporary audiences has allowed Suzanne to become one of the leading voices of opinion in the global community of watch enthusiasts today.
At the heart of Suzanne's work is the passion for furthering appreciation of fine mechanical watchmaking on a worldwide level. Her commitment to building relationships within the Swiss watch industry, allowing her to create stories that better reflect its core values, led her to relocate to Geneva in 2017, where she is now based.
Suzanne has been a member of the Cultural Council of the Fondation de la Haute Horlogerie since 2017, served three years on the jury of the prestigious Grand Prix d'Horlogerie de Genève (GPHG), is part of the pioneer cohort of the GPHG Academy, and is the co-founder of Watch Femme, a non-profit association dedicated to advancing women's voices in the watch world.

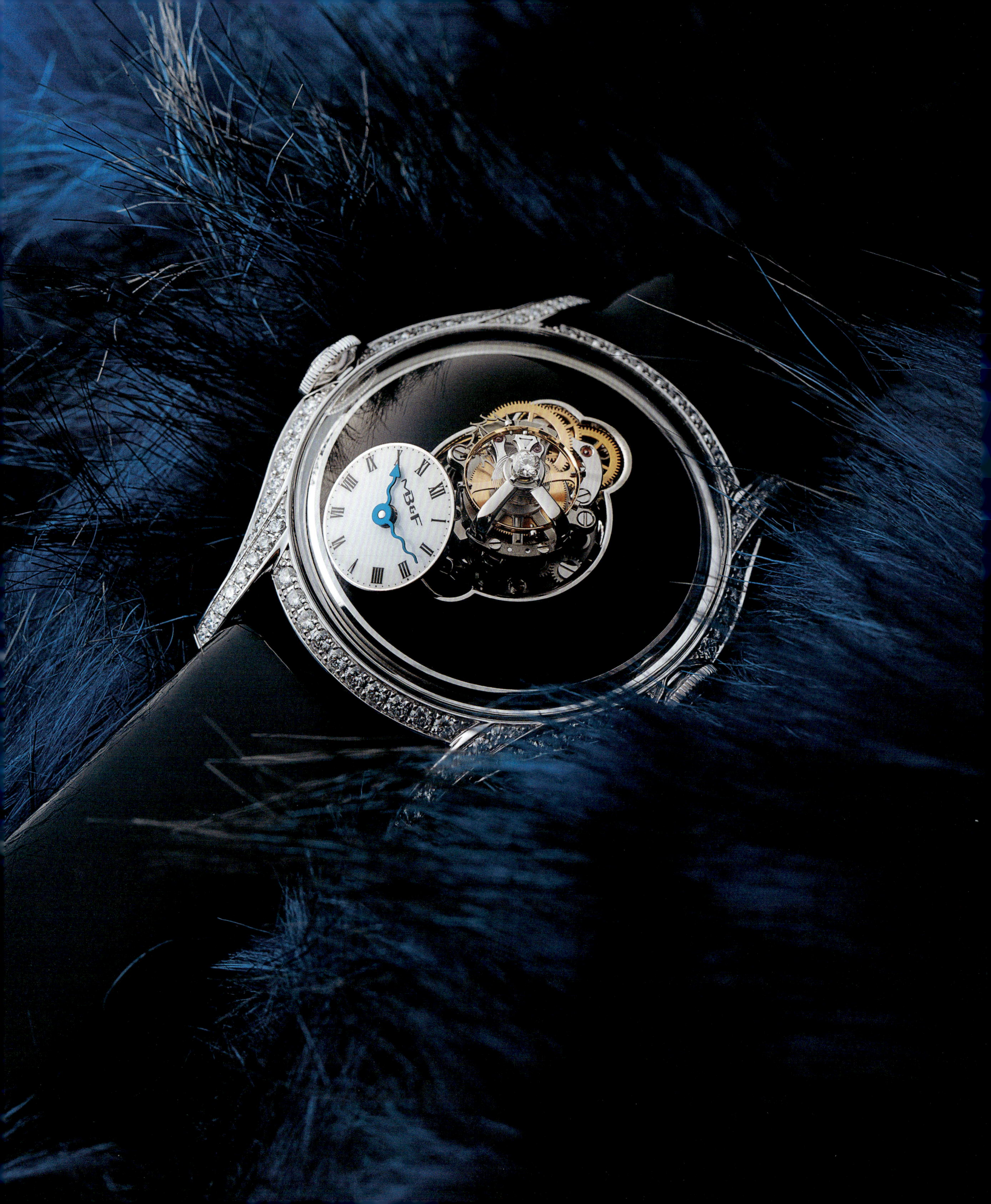
MB&F

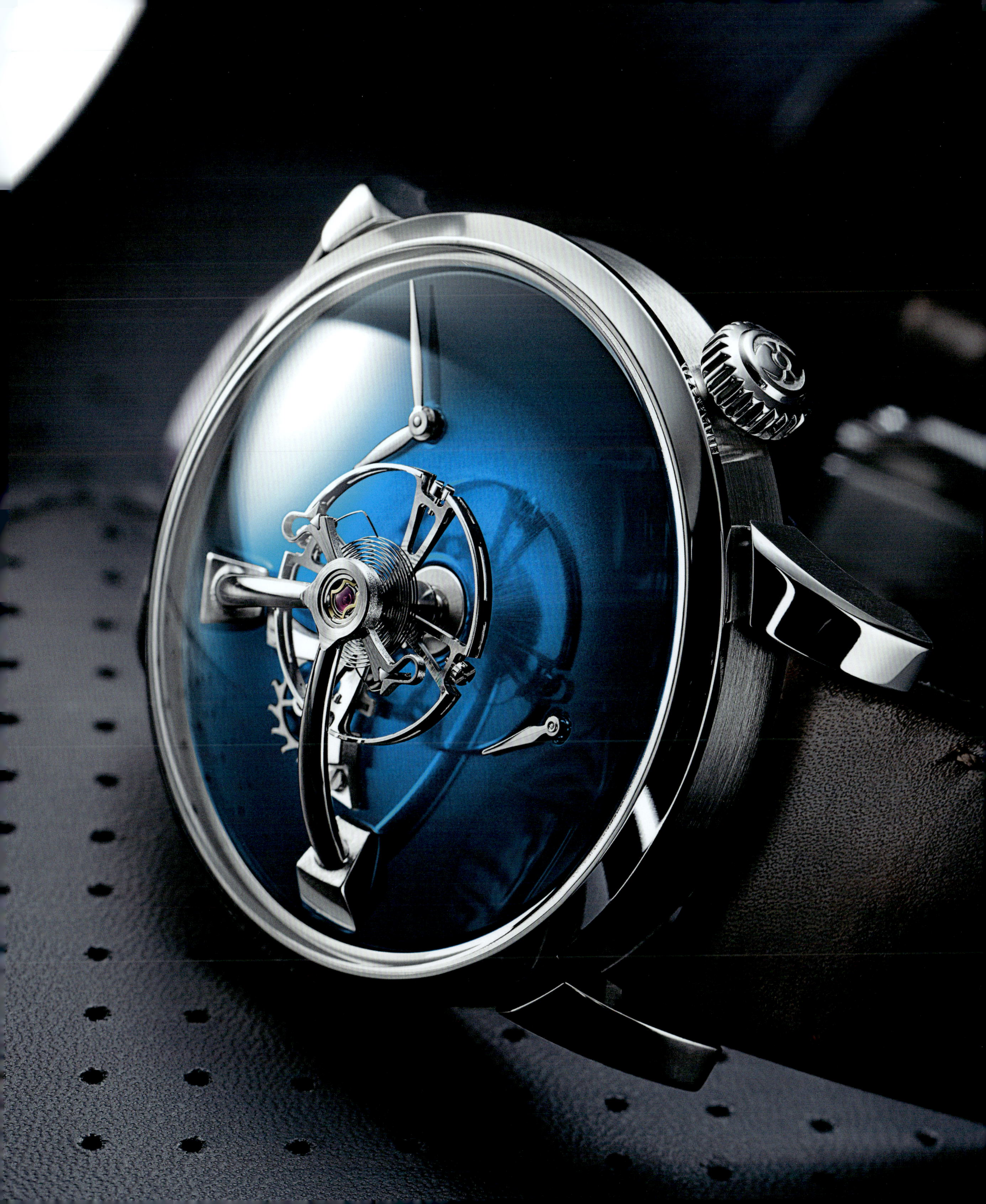

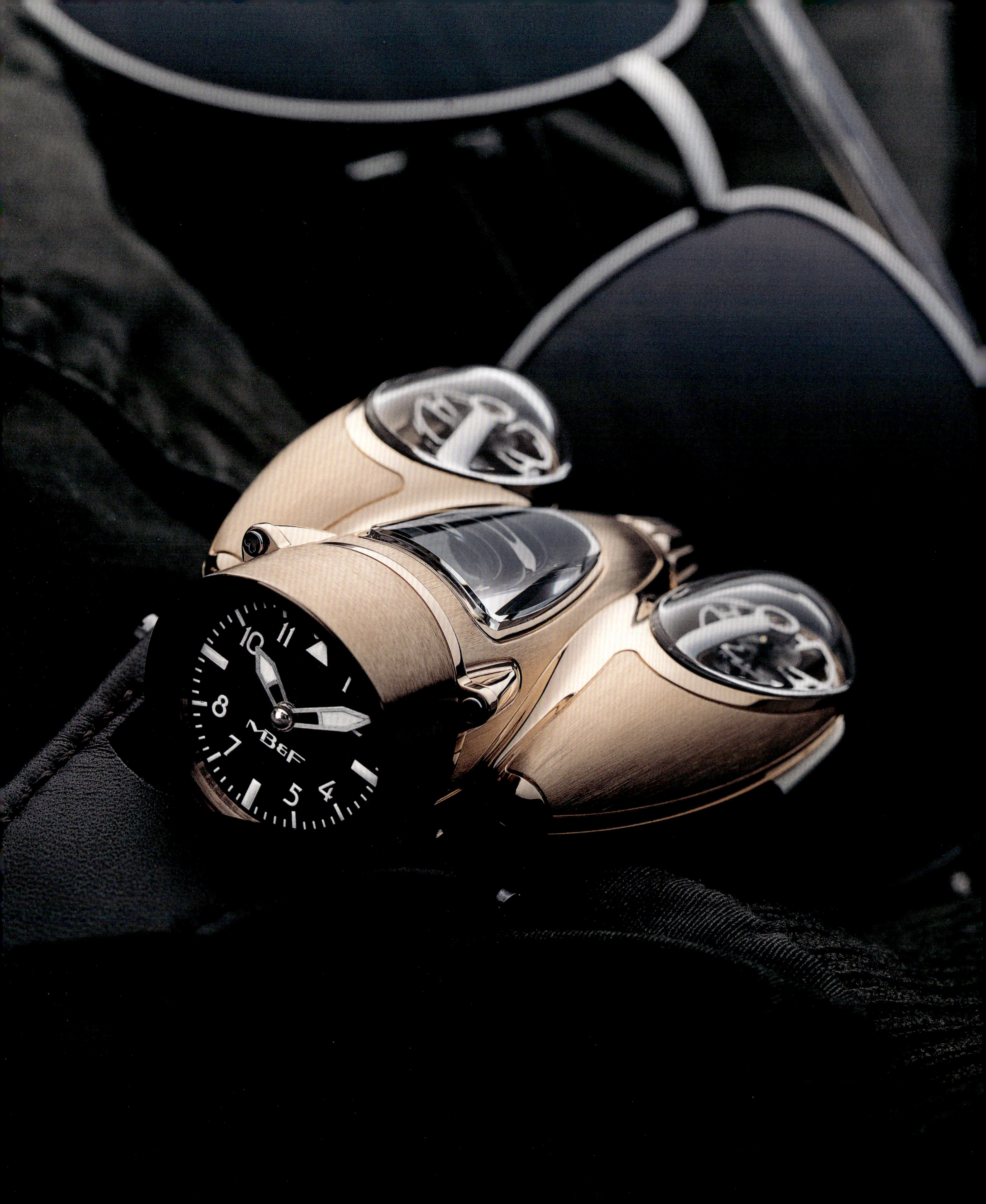
MB&F

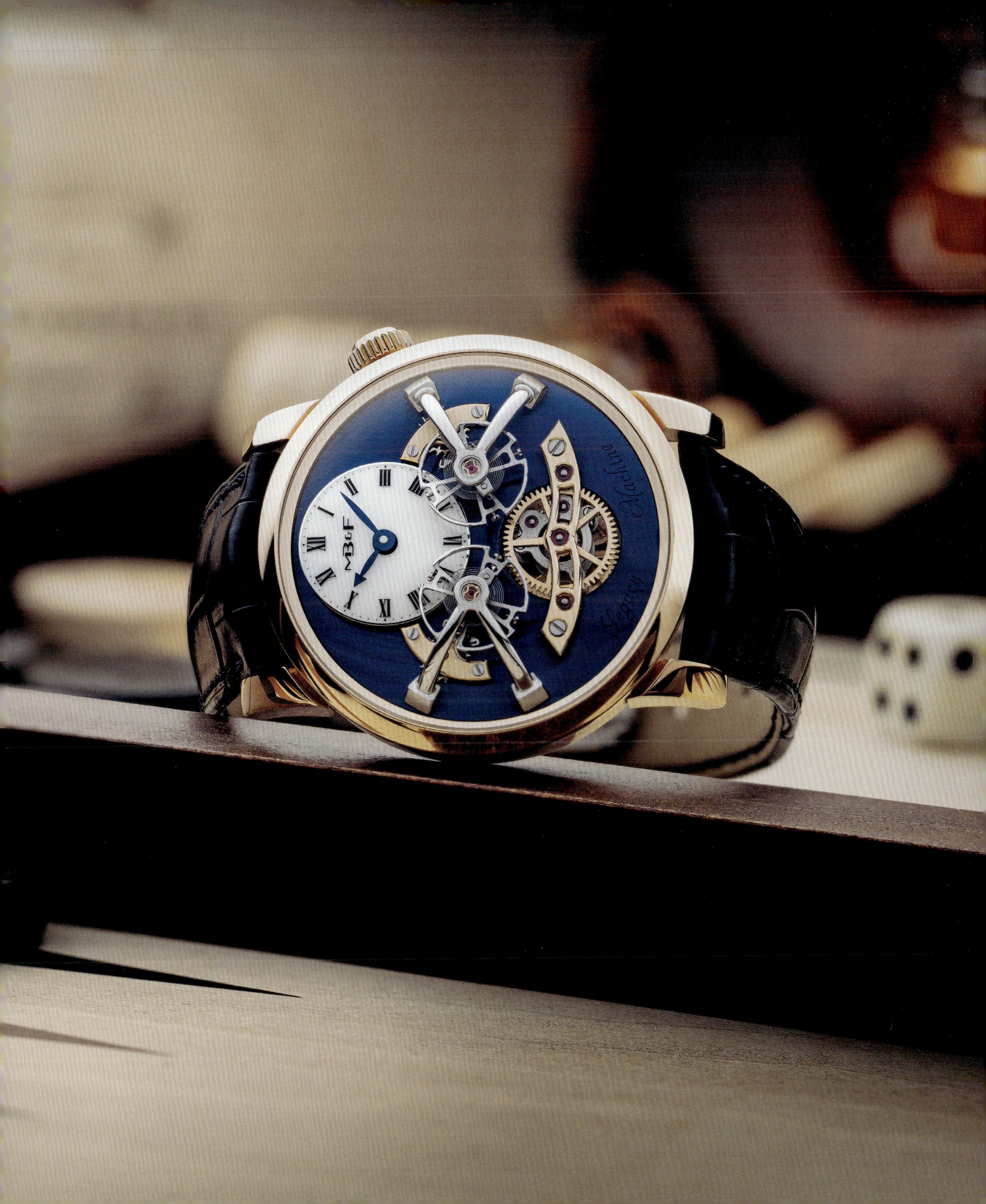
MB&F
Legacy Machine

Writers
Suzanne Wong
William Massena

Editorial coordination
Emmanuelle Halkin
Charlotte Court

Graphic design
Sidonie Bays
Lucas Haussener

MB&F project team
Maximilian Büsser
Charris Yadigaroglou
Vanessa André
Virginie Toral
Camille Reix
Cédric Roussel

Photography
Maarten van der Ende
Laurent Xavier Moulin
Alex Teuscher
Denis Hayoun
Philippe Garcia
Samuel Zeller
Jake Reeder

Proofreader
Florence Brutton

Photoengraving
Les Caméléons, Paris

Distributed in 2022 by Abrams, an imprint of ABRAMS
1098765432
Copyright © 2022 Éditions de La Martinière
ISBN : 9781419761072
All rights reserved. No portion of this book
may be reproduced, stored in a retrieval
system, or transmitted in any form or
by any means, mechanical, photocopying,
recording, or otherwise, without written
permission from the publisher.

This reprint of the book, gave us the opportunity to
make corrections regarding the quantity produced of
certain models (pp.92-104-174-208-218).

Printed and bound in February 2023 by Rotolito spa, Italy
Abrams® is a registered trademark
of Harry N. Abrams, Inc.

Recommended prices:
US$170 CAN$210 UK£130 €150

ISBN 978-1-4197-6107-2

ABRAMS The Art of Books
195 Broadway, New York, NY 10007
abramsbooks.com

PRINTED IN EUROPE